Advances in Science, Technology & Innovation

IEREK Interdisciplinary Series for Sustainable Development

Advances in Science, Technology & Innovation (ASTI) is a series of peer-reviewed books based on important emerging research that redefines the current disciplinary boundaries in science, technology and innovation (STI) in order to develop integrated concepts for sustainable development. It not only discusses the progress made towards securing more resources, allocating smarter solutions, and rebalancing the relationship between nature and people, but also provides in-depth insights from comprehensive research that addresses the **17 sustainable development goals (SDGs)** as set out by the UN for 2030.

The series draws on the best research papers from various IEREK and other international conferences to promote the creation and development of viable solutions for a **sustainable future and a positive societal** transformation with the help of integrated and innovative science-based approaches. Including interdisciplinary contributions, it presents innovative approaches and highlights how they can best support both economic and sustainable development, through better use of data, more effective institutions, and global, local and individual action, for the welfare of all societies.

The series particularly features conceptual and empirical contributions from various interrelated fields of science, technology and innovation, with an emphasis on digital transformation, that focus on providing practical solutions to **ensure food, water and energy security to achieve the SDGs.** It also presents new case studies offering concrete examples of how to resolve sustainable urbanization and environmental issues in different regions of the world.

The series is intended for professionals in research and teaching, consultancies and industry, and government and international organizations. Published in collaboration with IEREK, the Springer ASTI series will acquaint readers with essential new studies in STI for sustainable development.

ASTI series has now been accepted for Scopus (September 2020). All content published in this series will start appearing on the Scopus site in early 2021.

Ashootosh Mandpe · Sonam Paliya ·
Maulin P. Shah
Editors

Sustainable Waste Management

Challenges and Visions

Springer

Editors
Ashootosh Mandpe
Department of Civil Engineering
Indian Institute of Technology Indore
Indore, Madhya Pradesh, India

Sonam Paliya
Institute of Organic Biochemistry in Geo-Systems
RWTH Aachen University
Aachen, North Rhine-Westphalia, Germany

Maulin P. Shah
Department of Research Impact
and Outcome
Research and Development Cell
Lovely Professional University
Phagwara, Punjab, India

ISSN 2522-8714 ISSN 2522-8722 (electronic)
Advances in Science, Technology & Innovation
ISBN 978-3-031-85684-6 ISBN 978-3-031-85685-3 (eBook)
https://doi.org/10.1007/978-3-031-85685-3

This Springer imprint is published by the registered company Springer Nature Switzerland AG
The registered company address is: Gewerbestrasse 11, 6330 Cham, Switzerland

Acknowledgements

Authors are thankful to the contributors of these books for their cooperation and patience in the compilation of this task. Authors are also thankful to Springer team, for their constant support in the publication of this work.

Contents

Optimizing Sanitation Sustainability: Harnessing Biochar Co-digestion of Faecal Sludge for Resource Reclamation and Energy Generation

G. S. Salini Devi, C. S. Reshma, M. S. Chandana, B. Krishnakumar, and Saurabh Sakhre

Abstract

Efficient management of faeces and sludge is essential for sustainable sanitation practices. In sanitation systems, proper waste dispersal and management are critical to prevent environmental pollution, particularly surface and groundwater contamination, which can arise from inadequate or poorly maintained systems. Looking to global demand for energy and dwindling fossil fuel reserves, alternative energy sources are becoming increasingly important. Poor faecal waste management directly impacts human health, necessitating resource recovery methods over primary treatment alone. Anaerobic digestion of faecal sludge is highlighted as an environmentally friendly and promising method to produce biogas, addressing the world's growing energy needs. The fresh faecal matter in this study exhibited significant characteristics, including a chemical oxygen demand of 1820 mg/L, total phosphate of 26.5 mg/L, total nitrogen of 67.6 mg/L, and total carbon content of 23.68 mg/L. The study emphasizes the importance of the carbon-rich co-digestion substrate for enhancing biogas production, with lignocellulose agricultural residues like sorghum identified as an ideal source due to their disposability, renewability, and biodegradability. Through a pyrolytic process, sorghum-derived biochar was produced and found to interact synergistically with faecal sludge, enhancing waste degradation and biogas production due to its favourable C:N ratio. Water displacement measurements over a 10-day period during co-digestion experiments showed a consistent increase in biogas production, underscoring the positive correlation between biochar addition, biogas generation, and organic waste breakdown. This study proposes sustainable strategies for managing faecal sludge and suggests future directions for decentralized waste-to-energy systems. It also advocates for further research into biochar co-digestion applications to support international efforts in resource recovery and sustainable sanitation practices.

Keywords

Faecal sludge · Sanitation · Anaerobic digestion · Biochar · Inoculum · Pyrolysis · Biogas

1 Introduction

In many countries, faecal sludge (FS) is often disposed into surroundings without any authorization, posing serious threat to environment, public health, as well as economic welfare for neighbouring city dwellers (Rose et al. 2015). This eventually paves the way of health risk to human and livestock, lowers the oxygen level than normal to be required in aquatic ecosystems, creates eutrophication, shrinking opportunities for recreational use of water and increased cost on treatment plants (Nicholas et al. 2022). The recycling of FS as a fertilizer without any proper treatment also generates pathogenicity since it possesses numerous helminth eggs which affects people's health as well as agricultural practices (Sudrajat and Luqman 2020).

G. S. Salini Devi · C. S. Reshma · M. S. Chandana · B. Krishnakumar · S. Sakhre (✉)
CSIR-National Institute for Interdisciplinary Science and Technology, Thiruvananthapuram, India
e-mail: saurabh@niist.res.in

G. S. Salini Devi
e-mail: salinidevisalu@gmail.com

C. S. Reshma
e-mail: ponnuuthradam1988@gmail.com

M. S. Chandana
e-mail: chandanamsv@gmail.com

B. Krishnakumar
e-mail: krishna@niist.res.in

A. Mandpe et al. (eds.), *Sustainable Waste Management*, Advances in Science, Technology & Innovation,
https://doi.org/10.1007/978-3-031-85685-3_1

In this current aspect, faecal sludge management (FSM) and sanitation safety are essential components of preventing water-borne diseases, especially given the issues associated with the untreated discharge of FS. It urges the emergency in regions lacking a centralized sanitation facility. Whenever the onsite sanitation system requires emptying, due to the lack of adequate infrastructure often leads to the improper disposal of FS poses significant environmental and public health challenges (Heinss et al. 1998). Thus, these types of public restrooms without a link to sewage system creating serious health issues as well as results in polluting water resources too (Okoh et al. 2023). Depending on the source, FS may have diverse characteristics such as household sludge is usually different from non-household sludge (Strande and Brdjanovic (eds) 2014). Thus in the context of aforementioned situations, a reliable estimation methods are necessary for designing suitable treatment types and management strategies (Strande and Brdjanovic (eds) 2014).

The current FSM hinders challenges such as operational costs, technical complexity, inefficiency in pathogen reduction as well as organic matter degradation since it is associated with traditional treatment processes like anaerobic digestion (AD). In order to rectify this issue, there are several treatment alternatives; co-digestion is the prevalent one. It is one of the methods that implies anaerobic breakdown of organic substances facilitating biogas (Anderson et al. 2015). In a situation where FS are converted into beneficial materials using bio-methanation involving co-digestion, it produces the valuable biochar that can be applied as a soil amendment (Okoh et al. 2023) and also efficiently eliminates pathogens. In certain emergency sanitation scenarios, other treatment techniques such as urea treatment, lime treatment, and lactic acid fermentation have also included (Anderson et al. 2015). Similarly, for the management of FS in pit latrines, the potential use of biochar underlines the need for integrated approaches that enhance access to water and sanitation while addressing contamination challenges (Mamera et al. 2021). The challenges posed by untreated FS discharge can be addressed through reliable estimation techniques, innovative treatment technologies like bio-methanation, and sustainable strategies that include biochar utilization; thus, enhancing sanitation safety and resource recovery.

In conventional bio-methanation process, to make the process reliable, it is essential to maintain a suitable carbon-to-nitrogen (C:N) ratio since biomass is utilized for biogas production. However, if the desired C:N ratio is not achieved for effective methane production and microbial decomposition, there may be complications. In such cases, co-digestion is the suitable approach, as it enhances microorganism-mineralization as well as maintains a balanced C:N ratio through simultaneous anaerobic digestion of several organic substances. In short, the process elevates methane content and biogas yield in a way that is advantageous to both sides (Al-Samet et al. 2021). This is typically achieved by means of mixing the substances that have different nutrient compositions ensuring that the microbial community relying will have the necessary components for generating maximum biogas (Mouftahi et al. 2020). Furthermore, co-digesting different kinds of waste streams such as leftover foodstuff, waste water silt and farm produce remnants also be utilized for bioenergy sources like bio-methane (Aberra and Fufa 2016).

Co-digestion utilizes distinct options such as food waste, kitchen waste and biochar, which presents an effective means of enhancing anaerobic digestion. Biochar is a bulky carbon-rich material created from the pyrolysis of any type of organic matter and this explains its present fame since it has a positive influence on anaerobic digestion systems (Fagbohungbe et al. 2017). It is employed though it maintains a balanced microbial population in anaerobic digesters, which could result in long-term conservation of nutrients and sequestration of carbon (Santhosh et al. 2020). In addition to the above, an increased methane production and biogas development have been shown to occur with the advent of biochar during co-digestion (Santhosh et al. 2020). This is achieved due to the porous structure of biochar, which eventually supports the growth of bacteria and methanogens in digesters ultimately enhancing methane yields.

Biochar can be produced from various types of biomass with the help of contemporary thermochemical processes such as pyrolysis, gasification and torrefaction. Potential feedstock sources for biochar are agricultural residues, municipal solid wastes, paper mill wastes, organic wastes, animal manure, compost, and industrial or landfill waste. Previous research indicated that biochar can also change soil properties and influence plant growth or crop production made from rice husks, sawdust, sorghum silage, and other parts from trees (Mohamedi et al. 2022). Heavy metals leaching from industrial sewage sludge-derived biochar was another area of study emphasizing the need to understand its environmental impacts (Mohamedi et al. 2022).

Sustainable resource recovery is also becoming more concentrated on using biochar in FS treatment. The use of biochar during anaerobic digestion of FS enables to improve nutrient retention, stabilize microbial communities, and increase overall efficiency of this process (Shen et al. 2016). According to earlier studies, it has been shown that biochar-treated digestate had valuable nutrients including potassium, nitrogen, and phosphorus appropriate for use on soil and for the promotion of healthy crops (Shen et al. 2016). Moreover, these materials may lead to significantly increased biogas production of anaerobic methane from sludges (Wei et al. 2020). As observed, the regular contaminants, e.g. lead can be efficiently eliminated from digested wastes by applying biochar onto their surface (Wei et al. 2020). Moreover, earlier research concerning anaerobic co-digestion of pig dung and sewage sludge along with addition of biochar have displayed enhancement in digestion performance and copper immobilization (Ma et al. 2023).

Optimization of anaerobic digestion processes can be done through biochar based on its characteristic features like nutrient-rich composition and sorption capability thus contributing to sustainable waste management practices. Therefore, this particular book chapter begins by providing detailed characteristics about FS on the basis of different physico-chemical parameters. Elaborate deliberations on primary materials utilized for biochar, its preparation, its physico-chemical characteristics, and its numerous applications are also provided. The conclusions of initial investigations conducted using agro-residues-based biochar and its co-digestion with waste matter are also discussed in this chapter. This research will improve understanding of co-digestion of faecal sludge (FS) with biochar and its various scientific modalities through details given in the present work. Improved comprehension of the interactions between biochar and faecal sludge would be established as well as optimization of treatment processes, production of biogas from wastes and general concept about sanitation systems via circular economy.

2 Physico-chemical and Microbiological Characteristics of Fresh Faecal Matter

Exploring the microbiological and physico-chemical characteristics of FS is vital for safeguarding public health by identifying and mitigating the transmission of diseases caused by the microorganisms residing in it. This way of practises makes an aware to choose the right treatment, disposal and make sure that the environment is safe as per regulations. Additionally, the characterization facilitates better resource recovery and allows a sustainable management of valuable nutrients and organic substances that are present in FS. This section evaluates the bioavailability of nutrients within FS based on their physical/chemical characteristics including looks into harness organics in FS for resource recovery.

2.1 Physico-chemical Properties of FS

Fresh FS samples for the current study were collected from the staff quarters of CSIR-NIIST, Thiruvananthapuram, Kerala. A separate system of 100-L tank was installed for its conveyance and collection. The sampling was strategically conducted during the peak usage hours, between 9 and 10 a.m., to ensure the collection of fresh and representative samples that reflect the typical characteristics of the FS generated. The sample was taken from a sampling port with proper attention to precautions.

Initial characterization of FS was required for optimizing the conditions in the anaerobic reactors. Parameters such pH, conductivity, salinity, total solids, volatile solids, total nitrogen, chemical oxygen demand (COD), biological oxygen demand (BOD), total carbon, and phosphorus determine the composition of FS as previously mentioned and are crucial for understanding the management and treatment of the same (Fanyin-Martin et al. 2017). This initial characterization was performed following the standard methods outlined in IS 3025 and APHA.

In this study, the pH of FS was measured as 7.12 at room temperature, indicating a slightly alkaline nature. pH being a critical parameter in anaerobic digestion processes, as it impacts the activity of the bacteria that produce biogas. A pH near neutral is optimal for methane production highlighted that digester failure due to ammonia inhibition possesses the capacity to decrease the production of methane and a decrease in the methane content of biogas when the pH falls below 6 (Sun et al. 2020). Thus, the pH range inherent to FS creates an environment optimal for the microbial community participating in anaerobic digestion, thereby amplifying biogas yield. The conductivity of FS was found to be 1.253 mS/cm, which can be linked to the presence of ions and salts in the sludge. Conductivity affects the overall ionic strength of the medium, impacting microbial activity during anaerobic digestion.

The salinity of FS was determined to be 0.622 PSU, which can affect the osmotic balance within the digester. High salinity levels can inhibit microbial activity and reduce biogas production efficiency. The presence of salinity, particularly in waste streams, can lead to process inhibition and reduced biogas production as well as it has been discovered that, in comparison to non-saline samples, co-digestion of saline sludge with other organic substrates reduces methane production. The amount of volatile solids was found to be 210 mg/L and the total dissolved solids (TDS) to be 626.2 mg/l. TDS characterization of FS is important as it plays a significant role in estimating sludge generation after treatment (Ingallinella et al. 2002). The volatile solids content usually serves as a key indicator of organic matter available for anaerobic digestion, directly impacting biogas yield.

Total carbon, total nitrogen content, and total kjeldahl nitrogen (TKN) of FS were measured as 87.5 mg/L, 69.76 mg/L, and 4 mg/L, respectively. Both TKN and TC are crucial for understanding the nutrient composition of FS and its impact on microbial activity during anaerobic digestion. The Total Nitrogen concentration also influences the nitrogen cycle in the digester, affecting biogas yield and quality. The chemical oxygen demand (COD) of FS was found to be 1615 mg/L, indicating the amount of oxygen required to oxidize organic matter (Junglen et al. 2020). High COD levels suggest a higher organic load, which can enhance biogas production but may also require longer retention times. The BOD of FS was found to be 280.17 mg/L, reflecting the amount of oxygen needed by microorganisms to decompose organic matter. BOD is a key parameter in assessing the biodegradability of FS and its potential for biogas generation.

Overall, the FS can be categorized as high strength waste loaded with pathogens. The high strength of FS in terms of organic matter also denotes its suitability for anaerobic digestion and potential resource recovery. One of the important characteristics of FS is its heterogeneity and it is highly influenced by metabolism of an individual and dietary patterns of the area. Intuitively, the characteristics of water used for drinking or cooking purposes also plays a role in concentration of trace metals found in FS. Hence, looking to the high degree of variations in FS properties, its comparison with existing studies is imperative.

Analysing the previous studies done on FS characterization, the pH values were ranging from 4.01 to 9.34 (Appiah-Effah et al. 2014). Similarly, total solids (TS) varied greatly, ranging from 0.3 to 84.55 g/l (Talla et al. 2017). Meanwhile, there has been a reported value of total volatile solids (TVS) as 31.0–90.7% (Kengne et al. 2011) and total suspended solids (TSS) values ranging from 20 to 28,214 mg/l, as indicated in studies by (Strande et al. 2018). Ammonium nitrogen (NH_{4+}) concentrations varied notably, from 157 mg/l to 71,0071 g/l (Schoebitz et al. 2014; Lama et al. 2022). Nitrate (NO_2–N) concentrations showed a range from 0.3 to 17.7 mg/l (Kengne et al. 2011). Total Kjeldahl nitrogen (TKN) concentrations were reported earlier from 408.84 to 82,688 mg/l (Atwijukye et al. 2016). According to recent studies, total phosphorous(TP) concentrations were measured ranging from 24.21 to 521.1 mg/l (Forbis-Stokes et al. 2021) and potassium (K) levels varied from 24.4 to 474.7 mg/l (Ward et al. 2019). These findings underscore significant diversity in physico-chemical parameters across different studies, highlighting the importance of comprehensive monitoring and management strategies to address varied environmental conditions and anthropogenic impacts. The reviews mentioned above also concurs to the findings of the present research work.

2.2 Microbial Characterization of FS

Recent studies have explored various aspects of FS, including rheological characteristics of old FS versus synthetic human FS. By analysing the aforementioned property, researchers eventually become experienced in handling solid management of non-sewered sanitation facilities (Mercer et al. 2021). This current research any how sheds light on the physical as well microbiological properties of FS, since microbial characterization of faecal sludge is crucial for understanding its composition and potential risks.

In the present study, microbial diversity of FS samples was assessed by exploring the standard protocol of American Society of Microbiology. The presence of seven pathogenic organisms were effectively isolated and enumerated by employing the surface culture techniques such as spread plating and streak plating. Escherichia *coli, Salmonella spp., Shigella spp., Klebsiella spp., Staphylococcus spp., Vibrio spp., and Coliforms* were the mentioned organism, exhibiting significant growth on their respective culture media. For the analysis, a 10-ml sample was obtained, and repeated dilution was performed. Six bacteria's growth media were prepared, transferred to petri plates, and diluted samples were spread on the media. The petri plates were incubated for 24 h at 37 °C following inoculation. The experiment was carried out in an aseptic environment, and colony growth was obtained the next day (Fig. 1). The findings of the study are concurrent with some earlier studies carried out on the microbiological analysis of FS (Changara et al. 2019). The study investigated the bacterial population dynamics and physico-chemical characteristics of pit latrine faecal sludge during its mesophilic anaerobic digestion. Thus evaluating the microbial populations present in FS is crucial for assessing treatment efficiency and potential environmental impacts highlighted the presence of faecal indicator bacteria such as *faecal coliform, Salmonella spp., and faecal Streptococcus* in high levels in faecal sludge, emphasizing the importance of microbial monitoring for public health considerations and urgency of decontamination strategies before disposal to open spaces. (Yin et al. 2018).

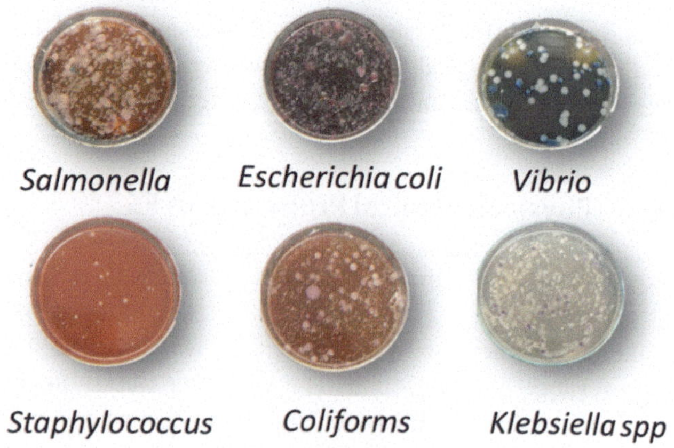

Fig. 1 Dominant bacteria identified in FS

3 Biochar: Properties, Mechanisms, Materials, and Applications

Biochar's versatility extends beyond its traditional role as a soil conditioner; it plays a pivotal role in co-digestion processes, enhancing biogas production and waste management efficiency. In this section, diverse base materials used for biochar production are enlisted, from agricultural residues to woody biomass, and also provides intricate physicochemical properties that underpin its functionality. Finally, the wide-ranging applications of biochar, from agricultural soil enhancement to environmental remediation, highlighting its potential to revolutionize sustainable practices across various sectors are described.

3.1 What is a Biochar?

Biochar is a carbon-rich product that possesses extraordinary environmental attributes (Lehmann 2007). Aside from its ability to sequester carbon thus reducing global warming, biochar has been studied as it has conflicting claims on its effectiveness in this regard. Furthermore, biochar can impact how organic compounds are absorbed into soils and how they are broken down by microorganisms. It can also be used for cleaning up contaminated land (Anyika et al. 2014). Moreover, adding biochar may alter microbial habitats, affecting microbial growth and community structures in the soil (Palansooriya et al. 2019). The characteristics of soil water are profoundly influenced by biochar properties such as porosity, shape, and size of particles (Liu et al. 2017). The carbon footprint of biochar production is another crucial consideration since the intended use of this material is mostly for soil improvement (Leppakoski et al. 2021). Variations in biomass chemistry and production conditions give rise to biochar's physico-chemical heterogeneity (Aziz et al. 2020). It has been found that use of biochar can increase retention of nutrients in the soil thus enhancing nutrient composition in these soils amended with it (Lee and Park 2020). An example on how to use these biochar-based products for environmental reclamation may involve their ability to remove inorganic contaminants from water (Lee and Park 2020). With functional groups being closely connected to the characteristics of biochar, the development of such characteristics under certain conditions are not yet clear (Li et al. 2020).

Researches have been conducted on biochar in relation to agriculture due to its ability to enhance nutrient availability and retention in soils; these effects differ according to feedstock's as well as production methods (Scott et al. 2014). Addition of biochar can influence microbial community composition, activity, and population size from soils (Hardy et al. 2019). Furthermore, studies have also investigated the influence of biochar on soil pore size distribution and compressibility (Petersen et al. 2016). In general terms, it is a multifarious material used for solving numerous environmental issues such as carbon storage in ground or removing contaminants out of water sources. This means that this study will cover a wide range of subjects including but not limited to agricultural science, environmental studies and material sciences among others.

3.2 Enhancement Mechanism of Co-digestion Through Biochar

Production of biogas can be improved by using the method of co-digestion where FS is combined with biochar that has been obtained from agro-residues. The intention of mixing biochar with FS aims to enhance biogas generation, stabilize heavy metals, and minimize the toxicity of biochar (Mohamed et al. 2022). Earlier studies reported that co-pyrolysis increases soil biota reproduction and improves overall efficiency of anaerobic digestion systems (Mohamed et al. 2022). In addition, during anaerobic digestion process, it has been established that synthesis of medium chain fatty acid was increased via electron transfer mediation by the presence of biochar (Wu et al. 2020). The charge distribution of biochar was found responsible for its ability to catalyse electron transfer reactions potentially leading to dehydrogenation of methane. Moreover, biochar has the ability to improve energy recovery since it can boost the production of methane from waste streams like primary sludge in anaerobic co-digestion systems. Additionally, it has been discovered that using biochar in anaerobic co-digestion procedures combining sewage sludge and swine manure can immobilize copper and enhance digestion effectiveness. Heavy metal stabilization in soil can be greatly aided by the use of biochar, as evidenced by research on the lead adsorption properties of biochar produced by co-pyrolysing rice straw and sludge (Wang et al. 2023). In summary, the incorporation of biochar agro-residues in co-digestion processes with faecal sludge offers a sustainable and effective strategy to enhance biogas production, stabilize heavy metals, and improve the overall performance of anaerobic digestion systems. (Fig. 2).

3.3 Methods of Biochar Production

Several literature describes a number of biochar production techniques such as torre faction, hydrothermal carbonization, pyrolysis, and flash carbonization, each with unique operating parameters and yields of finished products. Pyrolysis often performed within 300–700 °C with varying residence times ranging from less than 2 seconds–hours. The process yields biochar of approximately 35%, bio-oil

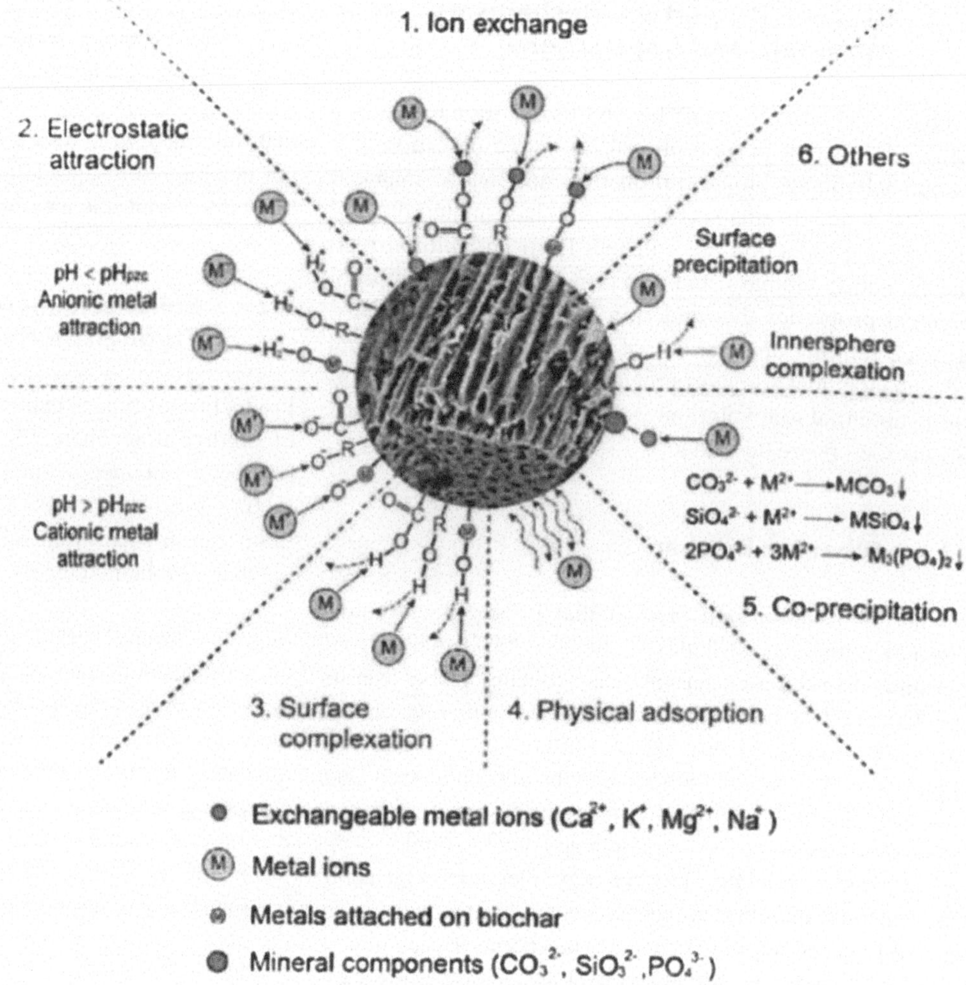

Fig. 2 Properties of biochar: enhances co-digestion. *Source* (Wang et al. 2023) Available via creative commons attribution 4.0

of about 30%, and syngas of about 35%, slight variations may occur above to this depending on the specific conditions utilized (Cantrell et al. 2012). Higher temperatures within the range of 500–1000 °C, with shorter residence times in the order of hours to days, result in decreased biochar yield (approximately 12%) but higher yields of bio-oil (approximately 75%) and syngas (approximately 13%) (Cantrell et al. 2012). Hydrothermal carbonization, operating at temperatures between 180 and 300 °C for residence times spanning from 1 to 16 h, typically yields biochar in the range of 50–80%, with lesser amounts of bio-oil (5–20%) and syngas (2–5%) (Funke and Ziegler 2010). Gasification, performed at temperatures of 750–900 °C and residence times of 10–20 s, primarily produces syngas (85%), with minor yields of biochar (10%) and bio-oil (5%) (Klinghoffer et al. 2015).

Torre faction, conducted at a temperature of 290 °C for residence times ranging from 10 to 60 min, yields predominantly biochar (80%) with no significant bio-oil production and a small amount of syngas (20%) (Klinghoffer et al. 2015). Flash carbonization, operated at temperatures between 300 and 600 °C for less than 30 min, yields approximately 37% biochar, though data on bio-oil and syngas production are not provided (Nunoura et al. 2006). Biochar can be produced through various pathways, each having its own advantages and disadvantages, this way allowing different approaches depending on the intended purpose in terms of product composition and process requirements.

3.4 Physico-chemical Properties of Biochar

Physical Properties of Biochar

Biochar's physical properties, includes high porosity, surface area, micropores, mesopores, and black colour, creates a conductive environment for microorganisms as they provide protection from predators, promote their colonization, serve as water store for their metabolism, and facilitate their growth and enzyme activity. Physical characteristics like size and porosity determine the effectiveness of biochar in numerous applications. They are crucial for catalytic processes, hydrogen gas storage, and adsorbent studies. The pore volume of the biochar is mainly related to the losses of volatile matter and dehydration during its production. When the surface area and pore volume of biochar are increased, this can improve its efficiency in many applications. In addition, biochar properties can be further enhanced by post-treatment methods, which can be either chemical or physical modifications. To increase the quality of biochar that will be obtained from raw biomass, before calcination raw biomass is usually blended with magnetic precursors, alkali compounds, and other functional agents. These alterations entail improving the porosity and surface area of biochar thus promoting the growth of microorganisms.

Chemical Properties of Biochar

The implementation of biochar has widely spread and sought because it has good chemical characteristics that contribute greatly to its usefulness. The chemical characteristics that influence the effectiveness of biochar include electrical conductivity, organic matter, total nitrogen, potassium, phosphorus, calcium, magnesium, sodium, and sulphur content. Biochar can be effective in improving soil quality and nutrient availability depending on its chemical composition that encompasses surface area, porosity, nutrient retention capacity, readily available nutrient contents, aromaticity as well as cation-exchange ability. Biochar is mainly comprised of carbon (C), hydrogen (H), nitrogen (N), sulphur (S), and oxygen (O) which dominate the chemical properties of biomass used to produce it. The character of the biomass and the process of carbonization are significant factors in determining these differences. Biomass comprises compositions of complex biomolecules such as cellulose, hemicellulose, and lignin which are present along with other chemical links and inorganic anions. The pyrolysis process breaks down atomic ratios like O/C, H/C and (O+N)/C in biomass resulting in the release of gases that can condense or not condense; these include methane, carbon monoxide, hydrogen, and carbon dioxide leaving behind some solid form rich in carbon called biochar. The ratios at atomically levels influence the properties of biochar, including its hydrophobic, hydrophilic, and aromatic qualities. By breaking down the lignocellulose structure and drying out oxygen-containing functional groups, high temperature carbonation methods lessen the amount of polar functional groups in biochar and increase its hydrophobicity. Nonetheless, biochar exhibits increased hydrophilicity at temperatures above 500 °C due to the elimination of *f* labile aliphatic functional groups, resulting in a high porosity and wettability within the material.

3.5 Base Materials Used for Biochar Production

This section focusses on various biomass or feedstocks that have been utilized to engineer biochar through synthesis methods such as pyrolysis. The pyrolysis temperature, a critical factor in the synthesis process, is carefully controlled to optimize the biochar's ability to remove pollutants. The performance and conditions under which the biochar effectively removes pollutants are meticulously documented, providing insights into its practical applications. The method of sorption examined in this study is based on the biochar's relationship with contaminants, providing an explanation for its mode of action on pollutants.

Cellulose was one among the most widely utilized forms of biomass or feedstock. By using a pyrolysis-cum-activation process, it was transferred into a CO_2 activated biochar using active charcoal. The process lasts for three hours at 500 °C. Chlorophenol and phenol were the targeted contaminants for elimination from which a removal performance of between 50 and 60% was attained. It was obtained under conditions that includes a dosage amount of 0.2 g/L; temperature held constant at 25 °C; pH range of 3–9; contact time was one hour. The proposed sorption mechanism for this performance was occurring under per sulphate-based advanced catalytic oxidation processes. This mechanism could aid degradation issues as well as increase the biochar ability in absorbing pollutants from surrounding media (Sun et al. 2020).

Wheat straw is another significant agricultural waste, which was utilized as a biomass feedstock with the intention of shaping Fe, N co-doped biochar through hydrothermal carbonization process. The method involved heating the material to 800 °C for one hour. Under the specific conditions of 0.4 g/L dosage, 25 °C temperature, pH level of 3-and 20-min contact time, this engineered biochar exhibited a stunning pollutant removal efficiency of 93.4%. The pollutants targeted included acid orange, methyl orange, phenol, bisphenol A as well as tetracycline chloride and for the removal of these contaminants via radical and non-radical mechanisms has been adopted. This implies that the adsorption and subsequent removal of pollutants are greatly influenced by chemical properties associated with Fe, N co-doped biochar (Li et al. 2013).

The hydrothermal carbonization of textile effluent treatment plant sludge at 400 °C for four hours produces a biochar commonly known as plant sludge-derived biochar that is used with iron oxide nanoparticles in order to achieve improved efficiency in the removal of pollutants from textile effluent. Specifically, this biochar acts as a highly efficient adsorbent for ofloxacin with an efficiency rate of 96% at a dosage 5 g/L treated for 5 h at 25 °C and pH 6. The proposed sorption mechanism includes π–π electron donor acceptor interactions and hydrogen bonding. This designed material's ability to remediate textile effluents gets demonstrated by showing how different contaminants like ofloxacin gets removed due to the fine balances between electron interactions and hydrogen bonds within its structure (Singh et al. 2019).

Woodchips and sludge are often utilized to produce a type of magnetic biochar, which has gained significant attention in recent years. Co-precipitation was utilized for the preparation of this biochar, where the temperature was maintained at 100 °C for a period of 2 h. The contaminants which magnetic biochar has removed included Cr (III) at 80.96 mg/g and Acid Orange at 110.27 mg/g. The test conditions were set at 1 g/L dosage, 25 °C temperature, pH 2, with a 3-h duration time. It is suggested that adsorption mechanism is the one that involves in removing both Cr (III) and Acid Orange pollutants (Santhosh et al. 2020).

In addition, biomass is formed from sludge found in a drying bed of an effluent treatment plant. Carbonization converts this sludge into biochar in four hours at 500 °C using a pyrolytic method. The suggested biochar was examined for the removal efficiency of tetracycline, which is a common contaminant and was conducted under the following conditions: 1 g/l of the chemical, a temperature of 25 °C, pH of 6 and time period of 5 h. Under these conditions, it was elucidated that 20 mg/l tetracycline could be absorbed by the said biochar. According to the proposed mechanism for this sportive process it involves electrostatic interactions during adsorption. This shows that tetracycline has high affinity towards the adsorbent formed from such biomass or that this pollutant can be effectively removed by this kind of biochar (Singh et al. 2019).

Fresh corn stalk is converted into a biochar loaded with iron through the process known as pyrolysis at 550 °C for a duration of 30 min. The biochar was tested on artificial wastewater containing Nitrogen (N) and Phosphorus (P). The test conditions were a dosage of 1 g/L, a temperature of 25 °C, and a duration of 12 h. Under this condition, the biochar was found to remove 98 mg/g of Phosphorus and 90 mg/g of Nitrogen. The mechanism involved in this process is nutrient absorption mechanism. Fresh corn stalk biochar has a high affinity for Nitrogen and Phosphorus making it effective in removal of such nutrients from the wastewater (Min et al. 2020).

In the production of biochar, the utilization of rice husk is similarly inevitable, as it is converted into biochar colloids. The biochar colloids are then utilized to produce mycelial pellets. Pyrolysis takes place for a period of two hours at a temperature of 600 °C. An examination of the ability of these biochar colloids to remove Cadmium (Cd (II) was done using 2g/L dosage, 25 °C temperature and 216 h' retention time. It was found that under these conditions' cadmium was removed by the biochar that had a mean removal capacity of 102.04 mg/g. The suggested mechanisms for such processes include: surface complexation, physical adsorption, electrostatic interaction, and hydrogen bonding (Bai et al. 2020).

The creation of a modified 3D biochar MnO_2 from rice husk includes foaming-polymerization process and pyrolysis at 1200 °C for 2 h, and tested the ability to remove Cadmium (Cd (II)) and lead ability (Pb(II)). It was dosed at 0.3 g/L for 72 h and temperatures of 25 °C. Therefore, this biochar has ability for removing 84.76 mg/g cadmium and 70.90 mg/g lead ions, respectively, from the solution above stated. In line with this, surface adsorption is supposed to involve some kind of electrostatic attraction or ion exchange or perhaps a combination of them all together during such a time frame. This form of biochar shows high affinity towards Cadmium and Lead which enhances their removal from the environment (Wu et al. 2020).

Wood of *Populus adenopoda* is a common feedstock for biomass and also can be converted to a Fe^- modified biochar through pyrolysis. This happens at 900 °C for 2 h. This biochar has the ability to remove Arsenic in both its forms As(V) and As(III), Chromium in its trivalent state Cr(III) as well as Mercury Hg(II) under conditions of 1 g/L dosage, 25 °C temperature and time duration of 48 h. Under these conditions, it was found that biochar could remove 42.0% of As(V), 62.5% of As(III), 19.6% of Cr(III), and 97.0% of Hg(II) (Feng et al. 2020).

Rice straw, which is one of the most common biomass sources, can also be targeted to produce an engineered biochar composite, specifically, a Fe_3O_4-graphene-biochar composite. This composite was synthesized using a thermal decomposition method where rice straw was subjected to a temperature of 600 °C for 1 h. The efficiency of the biochar composite towards removing contaminants under certain conditions was tested. The contaminant used during the tests is crystal violet and its removal rate has 436.68 mg/g corresponding to the test conditions such as dosage 0.3 g/L, temperature 40 °C, pH 6, time duration of 3 h. Crystal violet pollutant is removed through interactions occurring between functional groups in biochar developments; while Fe_3O_4-graphene-biochar composite's functional groups play vital roles in adsorbing and depleting crystal violet pollutants (Du et al. 2020).

Ulva prolifera a common seaweed was also utilized in making magnetic seaweed biochar/ sulphidated Fe (0) composite. This composite was synthesized by hydrothermal carbonization method where the seaweed underwent a temperature condition of 180 °C for four hours. The pollutant that was used for the test was tetrabromo bisphenol A, it had an adsorption capacity rate of 1.47 mg/g. The experiment conditions had a dosage of 0.3 g/L, temperature set at 30 °C, pH range 4–7 and duration of 24 h. The proposed sorption mechanism for the removal of tetrobromobisphenol A (TBBPA) is through chemical adsorption followed by reduction and enhancement of electron transfer. This indicates that the chemical properties of the magnetic seaweed biochar/sulphidated Fe (0) composite hold considerable importance in adsorption and removal of tetrabromo bisphenol A compound. (Zhang et al. 2020)

A magnetic biochar was created from densely nitrogenous biomass, in particular from the hydrophyte. The preparation of the biochar was carried out by means of a fast pyrolysis method followed by co-deposition at 700 °C for 1 h. The performance of the biochar in removing pollutants was tested under certain circumstances using U(VI) and has obtained a removal efficiency of 54.35 mg/g. The testing conditions were: 0.3 g/L dosage, 25 °C temperature, pH of 3, and time period of 2 h. Inner sphere coordination and reductive precipitation have been proposed as possible mechanisms for sorption with regards to U(VI). This implies that there is a strong chemical property influence by this magnetic biochar on adsorption onto and elimination of U(VI) pollutants. More research needs to be conducted to suggest this mechanism and enhance removal efficiencies of pollutants from the biochar (Hu et al. 2018).

The raw bamboo, bagasse, and hickory chips are also suggested as the base material to produce biochar. The ball-milled biochar was made from these materials. The synthesis of this biochar was done using a pyrolysis method followed by ball milling in which the biomass underwent a temperature of 450 °C for 2 h. Under specific conditions, the performance of the biochar in removing pollutants was tested. The pollutants tested were the Sulphamide antibiotics such as Sulfamethoxazole and Sulpha pyridine. The removal performance was measured to be 88% for Sulphathiazole and 89.6% for Sulpha pyridine. The test conditions included dosage of 0.3 g/L, temperature at 25 °C, pH at 3, and duration of 2 h. The proposed sorption mechanism for the removal of the Sulphide antibiotics includes hydrophobic interaction, π–π interaction, hydrogen bonding and electrostatic interaction respectively. It implies that chemical properties of ball-milled biochar have great impact on adsorption as well as elimination of Sulphide antibiotics (Huang et al. 2020).

On the other hand, a palm kernel shell was used to prepare a biochar from a microwave pyrolysis process. In this study, palm kernel shell was microwave pyrolysed at 700 W power and 8–15 kPa pressure for about 0.5 h. The contaminants used in the tests were total suspended solids (TSS) and ammonia. It was determined that 67% percent of ammonia removal was achieved together with 68%for TSS. A temperature of 25 °C and pH level of 6.8 were employed as the test parameters. The theory put forward for the removal of ammonia and TSS pollutants is having an aquaponics system. This implies that the properties present in microwave pyrolysis biochar have a great impact on these two pollutants' adsorptions and their removal process too. Future studies can investigate this process to develop efficient biochar for pollutant removal (Su et al. 2020).

Walnut is also a major feedstock for biochar production to produce Fe_2O_3. Biochar was prepared via a two-step process. Initially, a microwaves pyrolysis method was used at a power of 280 W for 15 min; then the material was heated at 600 °C for 2 h. The efficiency of the biochar in removing pollutants was tested under controlled conditions. The pollutants used during the experiment were nonsteroidal anti-inflammatory drugs such as salicylic acid, naproxen, or ketoprofen. Salicylic acid was removed at a rate of 683 mg/g while naproxen and ketoprofen were removed at rates of 533 mg/g and 444 mg/g, respectively. These tests were done with dosage of 0.5 g/L in temperature of 25 °C, pH of 3 and lasted for two hours (Anfar et al. 2020).

3.6 Applications of Biochar

A wide range of advantages are prevalent by using biochar but a careful evaluation of its environmental consequences must take to consideration to avoid unanticipated outcomes. Contaminated land remediation, biofuel production, energy storage, and waste management are among the many diverse uses for biochar. When transferred into soil, biochar serves to adsorb several kinds of organic pollutants such as volatile organic compounds, industrial and agricultural chemicals, antibiotics and dyes. As the quantity of biochar material increases in soils, it may accelerate the degradation and adsorption of those contaminants. By facilitating the process such as chemisorption, involving electrophilic contact and physisorption by electrostatic attraction/repulsion including pore diffusion, H-bonding, or even hydrophobic connections, pollutants are removed by biochar. The functional groups such as OH and R–C(O)OH eventually facilitate these processes. Furthermore, partitioning, biodegradation and chemical transformation also deserve mention. As a multi-functional catalyst, biochar finds application in agriculture, environment, and energy. Its high surface area provided with sufficient numbers of functional sites on its surface, making them a potential catalyst. Similarly, the contaminants like norfloxacin hg and ammonium are easily adsorbed by biochar due to the presence of functional groups such as –OH and C=O.

Biochar is applied as a catalyst for various applications which include production of biodiesel, energy generation, waste management, syngas production, as well as removal of pollutants from the earth's atmosphere. Transesterification of vegetable oil or esterification of free fatty acids with alcohols leads to biofuels that are perfect replacements for crude oil products since they possess biodegradable, nontoxic, and renewable nature. In that aspect biochar plays a key role in biofuel production primarily using solid alkali or solid acid type catalysts. Unlike other catalysts which can only undertake one process at a time; solid acid biochar catalysts such as gaseous SO_3 or liquid H_2SO_4-biochars can promote both esterification and transesterification simultaneously making them very versatile options. In addition, biochar-supported alkali catalysts such as CaO, K_2CO_3, or KOH functionalized biochar are also good option for biofuel production that eventually provide high yields and good reusability. Though calcium oxide is readily available and low-cost nature, it is not preferred as a catalyst due to its disadvantages such as loss of activity and smaller surface area, making them less efficient than biochar-based catalysts that have high efficiency, porosity, and cheaper.

Biochar, which is a solid with a high surface area and porosity, can effectively remove pollutants from wastewater and also act as an alternative for collecting nutrients during wastewater treatment. Due to porosity and high adsorption capacity in composition, it has the capability to absorb toxins, which results in nutrient-loaded biochar that can be applied as fertilizers while treating water effluents. In comparison with activated carbon, biochar is also better for the environment because it emits fewer greenhouse gases and uses much lower energy during production (6.1 MJ compared to 97 MJ/kg for activated carbon).

Remediation of Pollutants: When applied on the soil, biochar adsorbs a lot of organic contaminants, for instance, volatile organic compounds, agrochemicals, and chemicals from industries, antibiotics, and dyes. As biochar content increases in soil, these contaminants begin to degrade more rapidly as well as being adsorbed simultaneously. Biochar removes pollutants through chemisorption which includes electrophilic contacts, physisorption encompasses pore diffusion, hydrophobic interactions as well as electrostatic attraction/repulsion and H-bonding. Some functional groups that help in these processes include; OH and R–C(O)OH. Another way is through partitioning, biodegradation and chemical transformation. Pollutants become non-bioavailable and hence immobile with reduced environmental impacts due to interaction between pollutants and biochar. According to previous research findings, biochar could also influence remediation of contaminants through ion exchange, complexation, precipitation, and electrostatic interactions (Dai et al. 2016). In addition, these mechanisms promote better efficiency of remediation strategies by assisting in eliminating pollutants from both contaminated soils and aqueous solutions. Moreover, biochar can alter the pH nutrient content of soil, cation-exchange capacity that helps to eliminate contaminants from polluted land (Wang et al. 2023).

Catalyst: Biochar is a versatile catalyst with applications in fields like agriculture, environment, and energy. The large surface area of the biochar allows for many functional groups to be present at its surface, facilitating its role as a potential catalyst. Functional groups such as OH and C=O allow biochar to absorb pollutants such as norfloxacin and ammonium. In different circumstances, biochar acts as a catalyst such as the generation of biodiesel, energy, tar elimination, waste managements, production of syngas, synthesis of chemicals, microbial fuel cell electrodes, and removal of environmental contaminants. Since biochar is mainly obtained from biomass, the resulting catalysts are sustainable and renewable substances which positively influence the environment.

Biofuels: The renewable, nontoxic, and biodegradable property of biofuel makes it as an excellent alternative to petroleum products. They can be produced by transesterifying vegetable oils or esterifying free unsaturated fats (FFAs) with alcohols. Solid acid and solid alkali catalysts are the main ways, in which biochar catalysts are used in the generation of biofuel. Catalysing esterification and transesterification processes simultaneously is possible with solid acid biochar catalysts, which are created by sulfonating biochar with either gaseous SO_3 or liquid H_2SO_4. Biochar-supported alkali catalysts, such as CaO, K_2CO_3, or KOH functionalized biochar, are used for biofuel production and provide high yields and good reusability. During biofuel synthesis, biochar was produced as a by-product, which sheds light on the chemical processes involved in creating biofuel. Despite the availability and low cost of calcium oxide as a catalyst, its disadvantages such as loss of activity and less surface area make biochar-based catalysts a more preferable option due to their high efficiency, porosity, and low cost. Magnetic biochar-based catalysts have been explored for biodiesel production, offering benefits like recyclability and ease of recovery.

Wastewater Treatment: Being a solid with large surface area and porosity, biochar has the ability to extract pollutants, collect nutrients from wastewater and absorb toxins, producing nutrient-rich biochar which can be applied as a soil amendment and can also can be chosen as a best alternative for waste water treatment. Compared to activated carbon, biochar has a reduced environmental impact since it produces less greenhouse gases and requires a lot less energy to make (6.1 MJ/kg as opposed to 97 MJ/kg for activated carbon). Biochar-based sorbents have demonstrated effectiveness in reducing the presence of contaminants in water bodies, contributing to the remediation of polluted water sources (Dai et al. 2016).

Soil Amendentment: The application of biochar to soil enhances soil quality, microbial growth and helps in exchanging cations. It also helps in the retention of carbon and reduces greenhouse gases. Agricultural production is increased by the improvement of soil fertility and moisture retention. Moreover, biochar is useful in reclamation of contaminated lands by pesticides and metals. Microbial and soil health will be promoted by the presence of most nutritional elements in biochar. However, some types of biochar's, such as hydrochars may limit plant growth; hence, optimum application should be done to avoid negative effects. Addition of biochar reduces emissions from farmland and transforms carbon derived from biomass into a stable form that can be retain in the soil. Because of its distinct physico-chemical qualities, biochar serves as an excellent soil supplement which helps boost soil quality and fertility. Its ability to enhance soil ecology, nutrient content, and microbial community structure makes it a potent material for enhancing soils (Zhang et al. 2020).

Carbon Sequestration: The ring-shaped chemical makeup of biochar facilitates its microbial degradation, hence can potentially increase carbon sequestration. Nevertheless, there have been different results for example; some research shows elevated or lowered rates of carbon discharge. Generally, two kinds of carbon are found in biochar: liable (easily available) and recalcitrant (hard to decompose). Recalcitrant carbon remains in soil for a longer duration compared to easily dissolved ones that microbes can consume and initially cause higher mineralization rates of carbon. The general influence of biochar on the uptake of atmospheric CO_2 depends upon biomass used along with the operating conditions applied during pyrolysis cycle such as temperature and pressure. Ways to enhance biochar use for this purpose require a comprehension regarding how variables governing its production affect not only its yield but also what happens once it is applied back into the earth's surface. Biochar can boost soil fertility through increasing pH values, cation-exchange capacity together with organic matter content (Singh et al. 2019).

Energy Storage and Super Capacitors: While lithium-ion batteries have high energy density yet slow rate of charge and discharge; super capacitors are known for fast charging/discharging speed, high power density, and long cycle stability. The performance of these energy storage devices is primarily determined by the electrodes materials which should be porous in nature with a wide surface area. The prevalence of common carbon-based electrode materials like graphene, carbon nanotubes, and activated carbon may limit their utilization. Due to its high surface area, porosity and low cost compared to other materials; biochar is gaining interest as an electrode material for microbial fuel cells (MFCs) and super capacitors (SCs). Therefore, this makes it a hopeful alternative for energy storage uses. Biochar may also be manipulated to improve its porous structure thus suitable for different energy storage uses including electrodes for electrochemical energy storage devices and super capacitors.

Pollutant Control: Ammonium being a reducing agent are used alongside with biochar which acts as a low-temperature selective catalytic reduction catalyst. The researchers modified biochar produced from substrates such as sewage sludge and rice straw, either physically or chemically, to enhance their performance. Biochar particles can absorb contaminants from different environmental matrices because of their porous structures with large surface areas (Du et al. 2020). On the other hand, chemical activation is more effective than physical activation in removal processes due to the presence of certain characteristics such as adsorption sites and functional groups.

4 Co-digestion

The co-digestion of waste is a crucial strategy with multifaceted benefits. By mixing different organic waste streams, co-digestion improves biogas production rates and is especially beneficial for small-scale biogas plants that need constant feedstock sources. Also, this technique helps solve urgent waste management problems by keeping organic materials out of landfills and incinerators and hence diminishing greenhouse gas emissions and environmental pollution. Aside from being able to handle waste, co-digestion also plays a part in the recycling of nutrients, as the resultant digestate is an organic fertilizer that is rich in nutrients thus supporting sustainable farming practices and minimizing the quantity of nutrients lost into water bodies. Additionally, co-digestion supports the valuation of different organic substrates like industrial by-products and agricultural wastes which fall within the circle of circular economy since it aids in deriving value from waste streams while reducing dependence on limited resources. This means that co-digestion operates with a very broad perspective which addresses both sustainable waste management panning systems and energy generation/recovery thereby fitting into a circular economy model as well as an environmentally conscious society. Thus, this section discusses the anaerobic digestion stage, diverse types of substrates utilized also a section detailing the lignocellulosic waste used for the generation of biochar.

4.1 What is Anaerobic Co-digestion?

Anaerobic co-digestion denotes a way in which two or several substrates are digested together under anaerobic conditions. Concerning the role on improving methane production process than classical mono-digestion techniques this method has gained significant interest. For instance, combining some of the organic materials including activated sludge and other carbon-rich waste products can be used as diets for anaerobes, which will eventually lead to an increase in their rate of methane production (Borowski et al. 2018). In addition, this technique allows mixing of wastes with different characteristics leading to more synergies that not only enhance economic efficiency but also improve biogas yields during anaerobic digestion processes (Fernandez et al. 2005). Research has shown that anaerobic co-digestion may have synergistic impacts on the digestion medium resulting in enhanced biogas production (Neczaj et al. 2013). Similarly, the treatment of several wastes including sewage sludge, food waste, agricultural waste, and the organic part of urban solid wastes can improve methane yield and buffering ability (Chow et al. 2020). For instance, co-substrates such as microalgae, fallen teak leaves or liquid fractions from spent coffee bio-waste have been found to optimize methane formation when using anaerobic co-digestion processes (Ghaleb et al. 2020). Several different combinations have been tested under anaerobic co-digestion too; e.g. cow dung mixed with sewage sludge, fruit waste combined with cheese whey or animal manure mixed with sugar beet by-products confirm its versatility and efficiency in producing biogas (Aboudi et al. 2020). This procedure has also been investigated and evaluated in different thermophilic environments or within an operating system that is semi-continuous in order to optimize the balance of power and biogas production. Overall, the method of anaerobic co-digestion is commendable for waste treatment and bioenergy generation. The addition of various organic substrates may increase methane production, boost biogas production rates and improve the general performance of anaerobic digestion systems.

4.2 Co-digestion with Food Waste

According to researchers, co-digestion of food waste and dewatered sludge in anaerobic condition, methane was produced at an empirical value of 352 ml/g (VS) for the mixing ratio of 1:0.4 (Dai et al. 2016). But this approach eventually led to high-level accumulation of free ammonia nitrogen, hence it becomes necessary to monitor and control nitrogen content in the systems so that methanogenesis is not affected. Similarly, by providing a solids retention time (SRT) of 25 days, only 449 ml/g of methane was obtained through co-digesting food waste with sewage sludge (Sun et al. 2020). This recommends that the methane production rates in co-digestion systems are heavily influenced by difference in substrate compositions and operating conditions.

In earlier studies, it is found that a rate determining stage is prevalent in co-digesting fruit waste, food waste, and vegetable waste at an organic loading rate of 8 kg VS m^{-3} d^{-1} (Dai et al. 2016). For optimizing the performance of anaerobic digestion systems and maximizing biogas production, it is essential to have an understanding of the kinetics of microbial processes involved. As far as co-digestion ratios are concerned, there was an increase of 88% on methane generation from a 7:3 VS ratio of food waste to chicken manure when compared to mono-digestion (Chuenchart et al. 2020). However, a significant difference in methane output was not at all observed, when the ratio was adjusted to 2:1 VS emphasizing the importance of finding the optimal substrate ratio to achieve maximum biogas yields (Wang et al. 2023).

A nearly fourfold increase in methane output was earlier reported, when food waste when co-digested with dairy manure, a nearly fourfold increase in methane output was earlier reported compared to that of mono-digestion of food waste alone (Zhang et al. 2020). This marks the potential synergistic effects of combining different substrates in anaerobic digestion processes to enhance biogas production. Similarly, one of the earlier research at the pilot scale, demonstrated a successful co-digestion of food waste and sewage sludge, yielding approximately 284 ml/g of methane without any process inhibitions (Borowski et al. 2018). Thus, it provides a valuable insight into the scalability and feasibility of co-digestion processes for large-scale biogas production is often evaluated with these pilot-scale studies.

4.3 Co-digestion with Lignocellulosic Waste

Lignocellulosic substrates, gathering from various sources such as energy crops, forestry, and agricultural residues, offers immense potential for anaerobic digestion (AD) due to its abundant and continuous supply throughout the year (Sawatdeenarunat et al. 2016). When cassava pulp was co-digested with swine manure at a specific mixing ratio, notable enhancements in methane output were observed compared to individual digestion of cassava pulp alone (Panichnumsin et al. 2010). Specifically, at a 3:2 (VS) mixing ratio, methane production increased by 159%, indicating synergistic effects between the substrates. Even then managing co-digestion ratios is task, to avoid the adverse effects such as cyanide build-up. It was found that a mixing rate of 77:22(w/w), cyanide accumulation occurred was minimal (Glanpracha and Annachhatre 2016).

In addition, both trials-maintained carbon-to-nitrogen (C:N) ratios within optimal range for efficient anaerobic digestion, indicating how balanced nutrient compositions are important in biogas production. Approaches such as thermos alkaline pre-treatments combined with addition of microalgae biomass have shown potential to improve methane yields from lignocellulosic substrates like wheat straw (Solé-Bundó et al. 2017). This process improves substrate accessibility and biodegradability hence leading to more biogas production as opposed to untreated substrates.

On the other hand, co-digestion strategies involving complementary substrates have been effective in increasing methane yields. Earlier studies demonstrated that a notable increase of 174% methane production was achieved from co-digesting vinasse and sugar cane than mono-digestion of sugar cane pressed mud cake (López González et al. 2017). In order to retain microbial activity and maximization of biogas generation, a suitable pH levels including pH 7.4 is to be essentially maintained.

Additionally, advancements in pre-treatments methods such as acid-catalysed steam followed by enzymatic pretreatment have made a significant impact on improving the digestibility of difficult-to-digest materials such as wheat straw. The wheat straw (pre-treated) was co-digested with seaweed hydrolysate which led to an increase in methane production and a significant decrease in chemical oxygen demand (COD) at the best operational pH of 6.9.

4.4 Synergy Between Co-digestion of Biochar and Faecal Sludge

Co-digestion is a technique for enhancing the production of methane and it optimizes waste treatment systems through simultaneous anaerobic digestion of different organic substrates for biogas production. The method has several advantages, such as utilizing various feedstock's like kitchen waste, food waste, sewage sludge, and biochar which aid in increasing methane yield and improving biogas quality (Hastuti et al. 2021). Using co-digestion, it is able to balance C:N ratios, make nutrients available and promote separation between acidogenic and methanogenic phases during anaerobic digestion systems (Kalulu et al. 2021). For instance, co-digestion of cow dung and FS was given as an example of how different substrates with complementary characteristics could stabilize microbial communities, enhance biogas production rates, and decrease environmental effects (Dastpak et al. 2020). By acting as catalytic agent on anaerobic digestion, co-digestion of biochar and faecal sludge have been reported to increase biogas production (Shen et al. 2016). Microbial activity is enhanced by a carbon-rich substrate brought about by addition of biochar which accelerates decomposition of organic matter resulting into methane (Shen et al. 2016). In addition to increasing biogas efficiency, this synergism contributes towards sustainable waste management strategies through converting organic waste into an energy source (Shen et al. 2016).

4.5 Biochar Production of Agro-residues and Its Use in Co-digestion

Numerous investigations have recently focused on the generation of biochar from agricultural leftovers and its significance in co-digestion research. Experimental studies conducted on the conversion of quinoa and lupin agroresidues into biochar showcasing the possibility to produce biochar from residual biomass available locally (Salgado et al. 2022). Also, it was intended to produce and evaluate biochar for soil amendment and carbon sequestration using crop wastes like sorghum stems (Lompo et al. 2021). These kinds of investigations support the utilization of plant remnants in biochar production. Similarly, some earlier research pointed out how the soil properties and plant growth were impacted by biochar made from wood chips and animal bones in co-digestion experiments, showing that this can improve soils' physico-chemical characteristics and productivity (Gulyás et al. 2022). They have also examined whether applying biochar alone or with some organic/natural fertilizers would affect wheat yield and water utilization during droughts through an agro-ecosystem basis, demonstrating positive effects of biochar on rainwater retention as well as crop water productivity (Singh et al. 2019). Ultimately, more and more research are showing a growing interest in using agricultural leftovers for making biochar and demonstrating its advantages in enhancing soil properties, increasing crop yields as well as promoting agricultural system sustainability.

5 Insights from Experiments: Co-digestion of Faecal Matter with Sorghum-Based Biochar

This section provides the results from the experimental study carried out to find the impact of agro-residue based biochar on co-digestion with faecal matter. The section provides the brief background to the base material selection, biochar preparation, and discusses about the results.

5.1 Background

As discussed earlier, untreated disposal of FS in environment leads to increased pathogen load and water-borne diseases. Hence, the study attended to co-digest the faecal solids with agro-residues based biochar. The base material chosen for the study is Sorghum. Though, other agro-residues are also available in plenty in Kerala sorghum husk is used as it is one of the majorly cultivated crops among millets. Sorghum farming in Kerala has garnered significant research interest in recent years. Studies have explored various aspects such as genetic diversity, cultivation practices, environmental adaptability, and productivity of sorghum (Neczaj et al. 2013). Cultivating sweet sorghum due to its high biomass production and adaptability to environmental stressors like drought are some of its beneficial aspects.

5.2 Materials and Methods

Sorghum is one of the prominent agro-residue used as a choice for biomass utilization, on account to its readily available and widespread cultivation. In the current study, sorghum residues were meticulously selected as the primary biomass feedstock, admiring their inevitable suitability for bioenergy applications.

Initially, sorghum residues were harvested and dried to reduce moisture content, followed by crushing into smaller fragments using a crushing machine (Fig. 3a). Subsequently, the crushed sorghum residues were subjected to pyrolysis, a thermal decomposition process conducted at a controlled temperature of 300 °C in a muffle furnace which was already preheated to the desired temperature. To provide a pyrolysis environment free of oxygen, Argon gas was purged into the pyrolysis reactor. The reactor was maintained for half an hour, lead to the sorghum leftovers being converted into biochar (Fig. 3b). After pyrolysis, the reactor was cooled down, and the biochar was collected and weighed in desired quantities for further experimentation.

The obtained biochar of sorghum was co-digested with faecal sludge in a bioreactor. The biochar and faecal sludge were mixed in appropriate ratios, this symbiotic co-digestion aimed to leverage the synergistic interactions between the biochar and faecal sludge substrates, thereby enhancing the overall efficiency of biogas production. The bioreactor was sealed to maintain anaerobic conditions conducive to biogas production. Throughout the digestion period, the volume and composition of biogas generated were monitored using a gas detector. Additionally, the composition of the produced biogas, particularly the methane content, was analysed. Process parameters such as temperature, pH, and gas production rate were regularly monitored to ensure optimal conditions for biogas production (Fig. 3c).

Data collected from the experimentation were subjected to thorough analysis. The volume and biogas composition were recorded, and the methane content was measured to assess the quality of the produced gas.

5.3 Results

According to current investigations and conclusions, there is a promising possibility for co-digestion of faecal sludge with biochar. Consequently, two distinct combinations have

Fig. 3 **a** Sorghum residues crushed to smaller fragments; **b** Biochar obtained from sorghum residues; **c** Volume Displacement of combinations F_SB_CI and F_SB_CI

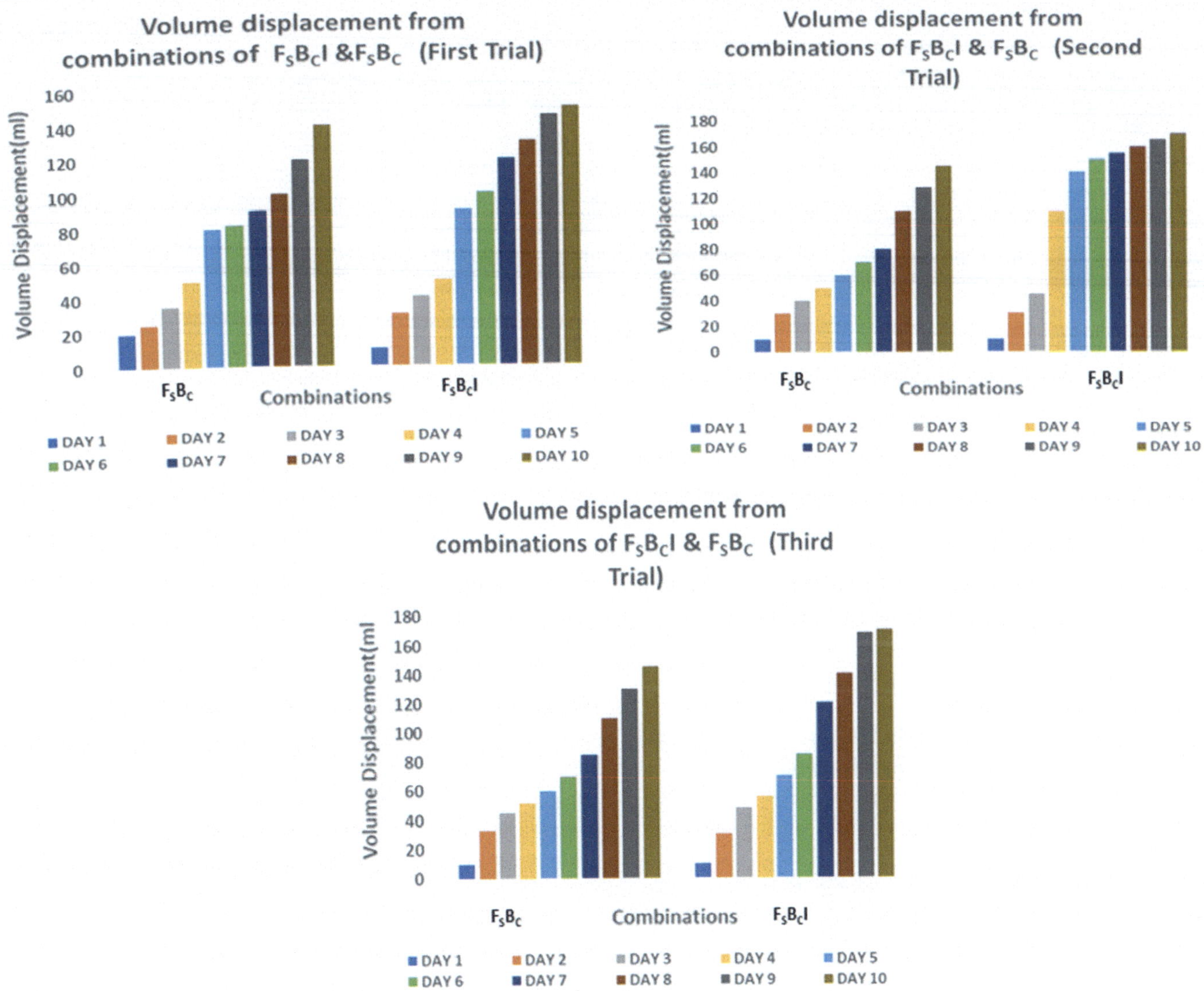

Fig. 4 Graph showing volume displacements from combination of F_SB_CI and F_SB_C

been studied using an experimental apparatus that has been developed. This configuration aims to increase the effectiveness of the co-digestion process and elucidate synergistic effects for enhanced waste management techniques.

In the experimental setup outlined above (Fig. 4), three independent trials were carried out to examine the volume displacement effectiveness of different combinations of biochar (BC), faecal sludge (FS), and an inoculum (I). The volume displacement serves as a proxy for assessing the efficacy of these combinations in enhancing biogas production. Two combinations were assessed in each trial, F_SB_CI (faecal sludge, biochar, and inoculum), and F_SB_C (faecal sludge and biochar). These combinations were monitored over a span of ten days, with daily measurements taken to capture any temporal trends or fluctuations in volume displacement. The volume displacement data showed recurring trends throughout all experiments, whereas the inclusion of an inoculum in the F_SB_C combination seemed to have a different effect than the F_SB_C combination alone. This implies that the addition of an inoculum might have a catalytic impact, improving the interaction between the constituents (FS and BC) and possibly stimulating certain chemical or biological reactions. This finding can be substantiated by the findings of previous studies, in which it is found that inoculum adaptation accelerated biogas production and enhanced sludge biodegradation rates in anaerobic digestion processes (Santhosh et al. 2020).

In Trial 1, both combinations show a lower volume displacement, with F_SB_C at 20 and F_SB_CI at 10. However, from Day 2, the volume displacement increases significantly for both combinations, indicating the commencement of anaerobic digestion processes. The volume displacement then steadily rises for both combinations, with F_SB_CI consistently exhibiting higher values compared to F_SB_C. This pattern indicates that a more persistent rise in volume displacement may have been the inclusion of inoculum enabling a slower and more effective breakdown or transformation of the element, leading to more efficient biogas production.

From Day 5 onwards, there is a substantial increase in volume displacement for both combinations, indicating a heightened rate of biogas production. This acceleration continues until the end of the observation period. Similarly, in Trials 2 and 3, there were discernible variations in volume displacement between the two combinations.

Similarly comparing the second and third trials with the first trial, several trends and differences emerge in terms of volume displacement of the combinations of F_SB_CI and F_SB_C.

In the second trial, both F_SB_C and F_SB_CI combinations start with similar volume displacements as in the first trial on Day 1. But throughout the subsequent days, there are notable variations. F_SB_C in the second trial shows slightly higher volume displacements compared to the first trial, F_SB_CI exhibits considerably higher values, indicating a more efficient biogas production process, particularly in the presence of inoculum.

Conversely, in the third trial, both F_SB_C and F_SB_CI combinations start with identical volume displacements as in the first trial on Day 1, as the trial progresses, there are differences in the volume displacements. F_SB_C in the third trial generally shows higher volume displacements compared to the first trial, suggesting a potentially more favourable anaerobic digestion process. On the other hand, F_SB_CI in the third trial exhibits comparable or slightly lower volume displacements compared to the first trial, indicating variations in the efficiency of biogas production when inoculum is introduced. Thus, the inclusion of inoculum appears to consistently enhance biogas production efficiency, as evidenced by higher volume displacements in F_SB_CI compared to F_SB_C across all trials.

6 Conclusions, Sustainability, and Way Forward

The study demonstrated the use of agro-residue based biochar for effective co-digestion of fresh faecal matter. Pyrolysis of sorghum husk was carried out and the biochar so produced was used as a co-digested with fresh faecal matter.

The research highlights the critical role that inoculum inclusion plays in enhancing the efficiency of biogas production when combined with biochar and FS. The volume displacement trends, that are constant across all experiments show that the introduction of inoculum leads to increased biogas production. This indicates the inoculum may facilitate the chemical and biological reactions thereby improving the interactions between FS and biochar components. Moreover, experiments have shown significant increase in biogas production volumes with F_SB_CI in terms of volume displacement than those of F_SB_C. The research demonstrates inoculum adaptation is relevant in enhancing substrate biodegradation rates thereby increasing biogas applied anaerobic digestion systems.

This study dwells on resource recovery and energy generation through co-digestion of faecal sludge with biochar with its broad implications. The dynamic interaction between biochar and FS has been shown in this study; for instance, how biochar's favourable C:N ratio stimulates microbial growth leading to waste degradation and increased biogas production. The results depicted a favourable association among the inclusion of biochar, biogas generation and organic refuse degradation in the co-digestion framework. Specifically, together with waste management processes, biochar significantly minimizes greenhouses gases emissions through immobilization of carbon. Enriching anaerobic digestion systems with biochar helps in preservation of sequestered carbon thus reducing their general carbon footprint. This double benefit emphasizes biochar as a relevant tool in mitigating climate change and encourage environmental stewardship within waste treatment processes. In the management of faecal sludge, renewable energy production and material recovery from sanitation systems; biochar co-digestion may be an alternative approach. The findings of this study indicate that decentralized waste-to-energy systems represent future directions within which improved waste disposal techniques that are not only more efficient but less harmful can also be developed.

The significance of employing biochar in co-digestion for optimization of sanitation sustainability is pegged on its multiple advantages in waste treatment, resource reclamation and energy generation. In this study, it is found that biogas production, nutrient recovery, and waste degradation may be enhanced by the use of biochar during anaerobic digestion. By using biochar as a bio-fertilizer for soil restoration and management purposes underscores the potential to promote soil health as well as maintain a sustainable agriculture practices. Going forward, these economic aspects should be incorporated into research carried out on larger scales so as to fill the knowledge gaps related to adsorption, process parameters, and cost estimation among others in biochar co-digestion processes. Biochar co-digestion is one of the alternatives to sustainable waste handling by utilizing it for energy generation and recovering resources. Consideration must be given as to how best to make bioreactors more sustainable and cheaper so that they can attract more users who may want them to protect the environment while conserving resources. A cost comparison on biochar-fuelled waste treatment systems will help determine their viability and promote other green initiatives. There is a possibility of making biochar co-digestion economically viable through resource recovery improvement and generation of energy from different wastes. To this end, it is necessary to significantly explore some of the areas highlighted in order to improve both understanding and use of biochar in waste-to-energy processes as research in this field progresses. The addition of biochar into waste-to-energy processes offers more than just renewable sources of energy; it also creates avenues for production of value-added products hence making waste management systems more economically viable and sustainable. The next step will involve addressing the existing gaps relating to how stable and effective biochar can be over a long period, whether various methods are optimal for its production, as well as looking at its cost-effectiveness if such objectives are ever going to be made possible in any effort to make sure that there are sustainable ways to manage garbage at household level while generating power.

Acknowledgements Authors are thankful to the contributors of these books for their cooperation and patience in the compilation of this task. Authors are also thankful to Springer team, for their constant support in the publication of this work.

References

Aberra D, Fufa F (2016) Bioenergy production from anaerobic co-digestion of sewage sludge and abattoir wastes. Adv Chem Eng Sci 06(03):281–287. https://doi.org/10.4236/aces.2016.63028

Aboudi K, Gómez-Quiroga X, Álvarez-Gallego CJ, Romero L (2020) Insights into anaerobic co-digestion of lignocellulosic biomass (sugar beet by-products) and animal manure in long-term semi-continuous assays. Appl Sci 10(15):5126. https://doi.org/10.3390/app1015512

Al-Samet MA, Goto M, Mubarak NM, Al-Muraisy SA (2021) Evaluating the biomethane potential from the anaerobic co-digestion of palm oil mill effluent, food waste, and sewage sludge in Malaysia. Environ Sci Pollut Res 28(47):67632–67645

Anderson C, Malambo DH, Gonzalez Perez ME, Nobela HN, De Pooter L, Spit J, Hooijmans CM, Van De Vossenberg J, Greya W, Thole B, Van Lier JB (2015) Lactic acid fermentation, urea and lime addition: promising faecal sludge sanitizing methods for emergency sanitation. Int J Environ Res Public Health 12(11):13871–13885

Anfar Z, Zbair M, Ahsiane HA, Jada A, El Alem N (2020) Microwave assisted green synthesis of Fe2O3/biochar for ultrasonic removal of nonsteroidal anti-inflammatory pharmaceuticals. RSC Adv 10(19):11371–11380

Anyika C, Majid ZA, Ibrahim Z, Zakaria MP, Yahya A (2014) The impact of biochars on sorption and biodegradation of polycyclic aromatic hydrocarbons in soils—a review. Environ Sci Pollut Res 22(5):3314–3341. https://doi.org/10.1007/s11356-014-3719-5

Appiah-Effah E, Nyarko KB, Gyasi SF, Awuah E (2014) Faecal sludge management in low income areas: a case study of three districts in the Ashanti region of Ghana. J Water Sanitation Hyg Dev 4(2):189–199

Atwijukye O, Nimanya C, Sugden S, Pillay S, Otaka G, Pietruschka B (2016) Development of low-cost decentralized faecal sludge treatment system for resource recovery

Aziz S, Yaseen L, Jamal ARA, Farooq U, Qureshi Z, Tauseef I, Haleem S, Ali MI (2020) Fabrication of biochar from organic wastes and its effect on wheat growth and soil microflora. Pol J Environ Stud 29(2):1069–1076. https://doi.org/10.15244/pjoes/99825

Bai S, Wang L, Ma F, Zhu S, Xiao T, Yu T, Wang Y (2020) Self-assembly biochar colloids mycelial pellet for heavy metal removal from aqueous solution. Chemosphere 1(242):125182

Borowski S, Boniecki P, Kubacki P, Czyżowska A (2018) Food waste co-digestion with slaughterhouse waste and sewage sludge: digestate conditioning and supernatant quality. Waste Manag 74:158–167

Cantrell KB, Hunt PG, Uchimiya M, Novak JM, Ro KS (2012) Impact of pyrolysis temperature and manure source on physicochemical characteristics of biochar. Bioresour Technol 107:419–428

Changara M, Sanyika W, Bangira C, Misi S (2019) Physico-chemical properties and bacterial community structure dynamics during the mesophilic anaerobic digestion of pit latrine faecal sludge. Water Sa 45(3 July)

Chow WL, Chong S, Lim JW, Chan YJ, Chong MF, Tiong TJ, Chin J, Pan GT (2020) Anaerobic co-digestion of wastewater sludge: a review of potential co-substrates and operating factors for improved methane yield. Processes 8(1):39. https://doi.org/10.3390/pr8010039

Chuenchart W, Logan M, Leelayouthayotin C, Visvanathan C (2020) Enhancement of food waste thermophilic anaerobic digestion through synergistic effect with chicken manure. Biomass Bioenergy 136

Dai X, Li X, Zhang D, Chen Y, Dai L (2016) Simultaneous enhancement of methane production and methane content in biogas from waste activated sludge and perennial ryegrass anaerobic co-digestion: the effects of pH. and c/n ratio. Biores Technol 216:323–330. https://doi.org/10.1016/j.biortech.2016.05.100

Dastpak H, Pasalari H, Jafari AJ, Gholami M, Farzadkia M (2020) Improvement of co-composting by a combined pretreatment ozonation/ultrasonic process in stabilization of raw activated sludge. Sci Rep 10(1):1070

Du C, Song Y, Shi S, Jiang B, Yang J, Xiao S (2020) Preparation and characterization of a novel Fe3O4-graphene-biochar composite for crystal violet adsorption. Sci Total Environ 711:134662

Fagbohungbe MO, Herbert BM, Hurst L, Ibeto CN, Li H, Usmani SQ, Semple KT (2017) The challenges of anaerobic digestion and the role of biochar in optimizing anaerobic digestion. Waste Manage 1(61):236–249

Fanyin-Martin A, Tamakloe W, Antwi E, Ami J, Awarikabey E, Apatti J, Mensah M, Chandran K (2017) Chemical characterization of faecal sludge in the Kumasi metropolis Ghana. Gates Open Res

Feng Y, Liu P, Wang Y, Finfrock YZ, Xie X, Su C, Liu N, Yang Y, Xu Y (2020) Distribution and speciation of iron in Fe-modified biochars and its application in removal of As (V), As (III), Cr (VI), and Hg (II): an X-ray absorption study. J Hazard Mater 384:12134IJ

Fernandez A, Sanchez A, Font X (2005) Anaerobic co-digestion of a simulated organic fraction of municipal solid wastes and fats of animal and vegetable origin. Biochem Eng J 26(1):22–28. https://doi.org/10.1016/j.bej.2005.02.018

Forbis-Stokes AA, Kalimuthu A, Ravindran J, Deshusses MA (2021) Technical evaluation and optimization of a mobile septage treatment unit. J Environ Manage 1(277):111361

Funke A, Ziegler F (2010) Hydrothermal carbonization of biomass: a summary and discussion of chemical mechanisms for process engineering. Biofuels Bioprod Biorefin 4:160–177

Ghaleb AAS, Kutty SRM, Ho Y, Jagaba AH, Noor A, Al-Sabaeei AM, Almahbashi NMY (2020) Response surface methodology to optimize methane production from mesophilic anaerobic co-digestion of oily-biological sludge and sugarcane bagasse. Sustainability 12(5):2116

Glanpracha N, Annachhatre AP (2016) Anaerobic co-digestion of cyanide containing cassava pulp with pig manure. Bioresour Technol 214:112–121

Gulyás M, Someus E, Klátyik S, Fuchs M, Varga Z, Dér S, Fekete G, Czinkota I, Székács A, Gyuricza C, Aleksza L (2022) Effects of combined application of solid pyrolysis products and digestate on selected soil properties of arenosol and plant growth and composition in laboratory experiments. Agronomy 12(6):1440

Hardy B, Sleutel S, Dufey J, Cornélis J (2019) The long-term effect of biochar on soil microbial abundance, activity and community structure is overwritten by land management. Front Environ Sci 7. https://doi.org/10.3389/fenvs.2019.00110

Hastuti E, Riyana R, Joy B, Supratman U, Pamekas R (2021) Integrated community onsite sanitation system for close loop faecal management. In: E3S Web of conferences, vol 249. EDP Sciences, France, p 01005

Heinss U, Larmie SA, Strauss M (1998) Solids separation and pond systems. For the treatment of faecal sludges in the tropics. In: Lessons learnt and recommendations for preliminary design

Hu Q, Zhu Y, Hu B, Lu S, Sheng G (2018) Mechanistic insights into sequestration of U (VI) toward magnetic biochar: batch, XPS and EXAFS techniques. J Environ Sci (China)

Huang J, Zimmerman AR, Chen H, Gao B (2020) Ball milled biochar effectively removes sulfamethoxazole and sulfapyridine antibiotics from water and wastewater. Environ Pollut 1(258):113809

Ingallinella AM, Sanguinetti G, Koottatep T, Montangero A, Strauss M (2002) The challenge of faecal sludge management in urban areas—strategies, regulations and treatment options. Water Sci Technol 46(10):285–294. https://doi.org/10.2166/wst.2002.0355

Junglen K, Rhodes-Dicker L, Ward BJ, Gitau E, Mwalugongo W, Stradley L, Thomas E (2020) Characterization and prediction of faecal sludge parameters and settling behaviour in informal settlements in Nairobi, Kenya. Sustainability 12(21):9040

Kalulu K, Thole B, Mkandawire T, Kululanga G (2021) Resource-efficient characterisation of pit latrine sludge for use in agriculture. Sustainability 13(9):4702

Kengne IM, Kengne ES, Akoa A, Bemmo N, Dodane PH, Koné D (2011) Vertical-flow constructed wetlands as an emerging solution for faecal sludge dewatering in developing countries. J Water Sanitation Hyg Dev 1(1):13–19

Klinghoffer NB, Castaldi MJ, Nzihou A (2015) Influence of char composition and inorganics on catalytic activity of char from biomass gasification. Fuel 157:37–47

Lama SL, Samal M, Luitel S (2022) Characterization of faecal sludge and design of faecal sludge treatment plant in Dhulikhel municipality. In: Sanitation for all by 2030 every year, the world water day highlights a specific aspect of freshwater

Lee JE, Park Y (2020) Applications of modified biochar-based materials for the removal of environment pollutants: a mini review. Sustainability 12(15):6112. https://doi.org/10.3390/su12156112

Lehmann J (2007) Bio-energy in the black. Front Ecol Environ 5(7):381–387. https://doi.org/10.1890/1540-9295(2007)5[381:bitb]2.0.co;2

Leppakoski L, Marttila M, Uusitalo V, Levanen J, Halonen V, Mikkila M (2021) Assessing the carbon footprint of biochar from willow grown on marginal lands in finland. Sustainability 13(18):10097. https://doi.org/10.3390/su131810097

Li X, Shen Q, Zhang D, Mei X, Ran W, Xu Y, Yu G (2013) Functional groups determine biochar properties (pH and EC) as studied by two-dimensional 13C NMR correlation spectroscopy. PLoS ONE 8(6):e65949

Li X, Jia Y, Zhou M, Su X, Sun J (2020) High-efficiency degradation of organic pollutants with Fe, N co-doped biochar catalysts via persulfate activation. J Hazard Mater 397:122764

Liu Z, Dugan B, Masiello CA, Gonnermann HM (2017) Biochar particle size, shape, and porosity act together to influence soil water properties. PLoS One12(6):e0179079. https://doi.org/10.1371/journal.pone.0179079

Lompo DJ, Yé L, Ouédraogo S, Sori SI, Nacro HB (2021) Production and characterization of crop residues derived biochars for soil amendment and carbon sequestration. J Agric Stud 9(1):130. https://doi.org/10.5296/jas.v9i1.1789

López González LM, Pereda Reyes I, Romero Romero O (2017). Anaerobic co-digestion of sugarcane press mud with vinasse on methane yield. Waste Manag 68:139–145

Strande L, Brdjanovic D (eds) (2014) Faecal sludge management: systems approach for implementation and operation. IWA publishing, UK

Ma W, Han R, Zhu L, Chen L (2023) Effects of biochar addition on anaerobic co-digestion of sewage sludge and swine manure: copper immobilization and digestion performance. J Chem Technol Biotechnol 98(9):2247–2256

Mamera M, van Tol JJ, Aghoghovwia MP, Nhantumbo AB, Chabala LM, Cambule A, Chalwe H, Mufume JC, Rafael RB (2021) Potential use of biochar in pit latrines as a faecal sludge management strategy to reduce water resource contaminations: a review. Appl Sci 11(24):11772

Mercer E, Usher S, McAdam E, Stoner B, Fernández Y (2021) Rheological characterisation of synthetic and fresh faeces to inform on solids management strategies for non-sewered sanitation systems. J Environ Manage 300:113730. https://doi.org/10.1016/j.jenvman.2021.113730

Min L, Zhongsheng Z, Zhe L, Haitao W (2020) Removal of nitrogen and phosphorus pollutants from water by FeCl3- impregnated biochar. Ecol Eng 149

Mohamed BA, Ruan R, Bilal M, Khan NA, Awasthi M, Amer MA, Leng L, Hamouda MA, Vo DVN, Li J (2022) Co-pyrolysis of sewage sludge and biomass for stabilizing heavy metals and reducing biochar toxicity: a review. Environ Chem Lett 21(2):1231–1250. https://doi.org/10.1007/s10311-022-01542-6

Mohamedi B, Ruan R, Bilal M, Khan N, Awasthi M, Amer M, Li J (2022) Co-pyrolysis of sewage sludge and biomass for stabilizing heavy metals and reducing biochar toxicity: a review. Environ Chem Lett 21(2):1231–1250. https://doi.org/10.1007/s10311-022-01542-6

Mouftahi M, Tlili N, Hidouri N, Bartocci P, Alrawashdeh K, Gul E, Liberti F, Fantozzi F (2020) Biomethanation potential (BMP) study of mesophilic anaerobic co-digestion of abundant bio-wastes in southern regions of tunisia. Processes 9(1):48. https://doi.org/10.3390/pr9010048

Neczaj E, Grosser A, Worwąg M (2013) Boosting production of methane from sewage sludge by addition of grease trap sludge. Environ Prot Eng 39(2). https://doi.org/10.37190/epe130212

Nicholas H, Mabbett I, Apsey H, Robertson I (2022) Physico-chemical properties of waste derived biochar from community scale faecal sludge treatment plants. Gates Open Res 6:96. https://doi.org/10.12688/gatesopenres.13727.2

Nunoura T, Wade SR, Bourke JP, Antal MJ (2006) Studies of the flash carbonization process. 1. Propagation of the flaming pyrolysis reaction and performance of a catalytic afterburner. Ind Eng Chem Res 45:585–599

Okoh E, Patrick I, Yelebe Z, Oruabena B (2023) Co-composted faecal sludge as organic fertilizer to restore crude oil- contaminated soil. https://doi.org/10.21203/rs.3.rs-2437560/v1

Palansooriya KN, Wong JTF, Hashimoto Y, Huang L, Rinklebe J, Chang SX, Bolan N, Wang H, Ok YS (2019) Response of microbial communities to biochar-amended soils: a critical review. Biochar 1(1):3–22. https://doi.org/10.1007/s42773-019-00009-2

Panichnumsin P, Nopharatana A, Ahring B, Chaiprasert P (2010) Production of methane by co-digestion of cassava pulp with various concentrations of pig manure. Biomass Bioenergy 34:1117–1124

Petersen CL, Hansen ES, Larsen HH, Hansen L, Ahrenfeldt J, Hauggaard-Nielsen H (2016) Pore-size distribution and compressibility of coarse sandy subsoil with added biochar. Eur J Soil Sci 67(6):726–736. https://doi.org/10.1111/ejss.12383

Rose C, Parker A, Jefferson B, Cartmell E (2015) The characterization of feces and urine: a review of the literature to inform advanced treatment technology. Crit Rev Environ Sci Technol 45(17):1827–1879. https://doi.org/10.1080/10643389.2014.1000761

Salgado MAH, S Coba JAC, Cianferoni A, Säumel I, Tarelho LAC (2022) Conversion of quinoa and lupin agro-residues into biochar in the Andes: an experimental study in a pilot-scale auger-type reactor. Front Bioeng Biotechnol 10. https://doi.org/10.3389/fbioe.2022.1087933

Santhosh C, Daneshvar E, Tripathi KM, Baltrėnas P, Kim T, Baltrėnaitė E, Bhatnagar A (2020) Synthesis and characterization of magnetic biochar adsorbents for the removal of Cr (VI) and acid orange 7 dye from aqueous solution. Environ Sci Pollut Res 27:32874–32887

Sawatdeenarunat C, Nguyen D, Surendra KC, Shrestha S, Rajendran K, Oechsner H, Xie L, Khanal SK (2016) Anaerobic biorefinery: current status, challenges, and opportunities. Bioresour Technol 215:304–313

Schoebitz L, Bassan M, Ferré A, Vu THA, Nguyen VA, Strande L (2014) Conference paper: FAQ: faecal sludge quantification and characterization-field trial of methodology in Hanoi, Vietnam

Scott HL, Ponsonby D, Atkinson C (2014) Biochar: an improver of nutrient and soil water availability—what is the evidence? CABI Rev, 1–19. https://doi.org/10.1079/pavsnnr20149019

Shen Y, Linville JL, Ignacio-de Leon PA, Schoene RP, Urgun-Demirtas M (2016) Towards a sustainable paradigm of waste-to-energy process: enhanced anaerobic digestion of sludge with woody biochar. J Clean Prod 1(135):1054–1064

Singh R, Singh P, Singh H, Raghubanshi AS (2019) Impact of sole and combined application of biochar, organic and chemical fertilizers on wheat crop yield and water productivity in a dry tropical agro-ecosystem. Biochar 1(1):229–235

Solé-Bundó M, Eskicioglu C, Garfí M, Carrère H, Ferrer I (2017) Anaerobic co-digestion of microalgal biomass and wheat straw with and without thermo-alkaline pretreatment. Bioresour Technol 237:89–98

Strande L, Schoebitz L, Bischoff F, Ddiba D, Okello F, Englund M, Ward BJ, Niwagaba CB (2018) Methods to reliably estimate faecal sludge quantities and qualities for the design of treatment technologies and management solutions. J Environ Manage 1(223):898–907

Su MH, Azwar E, Yang Y, Sonne C, Yek PN, Liew RK, Cheng CK, Show PL, Lam SS (2020) Simultaneous removal of toxic ammonia and lettuce cultivation in aquaponic system using microwave pyrolysis biochar. J Hazard Mater 5(396):122610

Sudrajat A, Luqman Y (2020) Scheduled faecal sludge service (SFSS) for sustainable faecal sludge management in Cilacap. In: E3S Web of conferences, vol 202. EDP Sciences, France, p 06031

Sun C, Chen T, Huang Q, Zhan M, Li X, Yan J (2020) Activation of persulfate by CO2-activated biochar for improved phenolic pollutant degradation: performance and mechanism. Chem Eng J 15(380):122519

Talla A, Sezawo R, Ngohe-Ekam PS (2017) Characterization of depotted faecal sludge into the environment and design of a suitable treatment system: case of Nomayos area in Yaounde city. British J Appl Sci Technol 21(2):1–2

Wang H, Ge S, Ye L, Zhou C (2023) Study on the Adsorption Characteristics of Lead in Soil by WoScience. IOP Conf Ser Earth Environ Sci 1171(1):012062. https://doi.org/10.1088/1755-1315/1171/1/012062

Ward BJ, Traber J, Gueye A, Diop B, Morgenroth E, Strande L (2019) Evaluation of conceptual model and predictors of faecal sludge dewatering performance in Senegal and Tanzania. Water Res 15(167):115101

Wei W, Guo W, Ngo HH, Mannina G, Wang D, Chen X, Liu Y, Peng L, Ni BJ (2020) Enhanced high-quality biomethane production from anaerobic digestion of primary sludge by corn stover biochar. Biores Technol 1(306):123159

Wu Z, Chen X, Yuan B, Fu ML (2020) A facile foaming-polymerization strategy to prepare 3D MnO2 modified biochar-based porous hydrogels for efficient removal of Cd (II) and Pb (II). Chemosphere 1(239):124745

Yin F, Dong H, Shang B, Zhang W (2018) Effect of time and mixing in thermal pretreatment on faecal indicator bacteria inactivation. Int J Environ Res Public Health 15(6):1225. https://doi.org/10.3390/ijerph15061225

Zhang C, Lu J, Wu J (2020) One-step green preparation of magnetic seaweed biochar/sulfidated Fe0 composite with strengthen adsorptive removal of tetrabromobisphenol A through in situ reduction. Biores Technol 1(307):123170

Analysis of E-Waste Management on Environmental Hazards and Population Health Impact

Sumanta Bhattacharya, Bhavneet Kaur Sachdev, and Arkadyuti Seth

Abstract

The management of electronic trash (e-waste) is a significant concern because of its harmful environmental effects and possible risks to public health. The increasing number of electronic devices worldwide leads to substantial environmental and health problems due to inappropriate disposal and insufficient management of e-waste. This research examines e-waste management strategies, with a particular focus on the environmental risks associated with e-waste build-up, recycling, and disposal, as well as how they affect public health. This paper investigates the origins and components of e-waste, such as lead, mercury, cadmium, and brominated flame retardants, through an in-depth analysis of literature, regulations, and case studies. These toxic substances have the potential to seep into soil and water, polluting ecosystems, and posing risks to human health via exposure routes. Furthermore, informal e-waste recycling methods, which are commonly carried out in underdeveloped countries without proper safety protocols, worsen health hazards for workers and local residents by exposing them to dangerous chemicals and contaminants. The article assesses the efficiency of existing e-waste management solutions, such as recycling, reuse, and safe disposal procedures, in reducing environmental and health consequences. Identified issues include insufficient infrastructure, low knowledge, and poor enforcement of rules, emphasising the necessity for complete integrated measures to tackle e-waste management.

S. Bhattacharya (✉)
Asian International University, Manipur, India
e-mail: sumanta.21394@gmail.com

B. K. Sachdev
Suresh Gyan Vihar University, Jaipur, India

A. Seth
IGNOU, Kolkata, India

Keywords

Electronic trash · E-waste management · Environmental risks · Public health · Recycling · Sustainable growth

1 Introduction

Society is moving towards extensive digitalisation, particularly after the Covid-19 pandemic. The rapid development of artificial intelligence and machine learning significantly revolutionises every human activity. Nowadays, human society is entirely dependent on electronic gadgets in every aspect of their daily lives. As the digital world advances, society is increasingly digitalising, as the advancement of the digital world is bringing sustainability to every sector by providing innovative eco-friendly solutions. Digitalisation causes innovations in data storage, logistics, agriculture, quality control of natural resources, waste management, prevention of environmental degradation, protection of biodiversity, financial transactions, etc. This growing digitalisation of society also boosts the demand for electronic gadgets. Furthermore, rapidly changing technology leads to frequent gadget updates, making the old gadgets outdated. Therefore, all the old and out-of-date gadgets end up in the garbage. All these factors lead to an enhancement in the number of electronic trashes (e-waste). The World Health Organisation conducted a study in 2019 that estimated the global production of 53.6 million metric tonnes of electronic trashes, with only 17.4% collected and recycled (Mor et al. 2021). These large volumes of e-waste contain hazardous inorganic contaminants, including toxic metals like lead, cadmium, mercury, copper, tin, lithium, cobalt, antimony, chromium, etc., which cause serious effects on ecosystems as well as human health. The contamination of toxic contaminants in the soil affects the population of the innocuous microbial community, which decreases the rate

A. Mandpe et al. (eds.), *Sustainable Waste Management*, Advances in Science, Technology & Innovation,
https://doi.org/10.1007/978-3-031-85685-3_2

of photosynthesis in plants. It leads to a decrease of soil fecundity and a reduction in agricultural crop yield. E-waste also contains a significant amount of non-biodegradable plastic components, which cause plastic pollution in soil and water bodies, greatly affecting the ecosystem and biodiversity (Mmereki et al. 2016).

The disposal of electronic trash in water bodies leads to the bioaccumulation of poisonous substances in the aquatic ecosystem, which further reaches the terrestrial ecosystem and human body through biomagnification. People typically landfill, incinerate, or dispose of e-waste in water bodies. The landfilling of e-waste leads to the release of toxicants into the groundwater reserve, which further enters the human body through drinking water. The toxic substances also accumulate on the soil's porous surface and affect soil microbe populations. The shredding and melting of electronic materials during their unscientific disposal produce dust particles containing toxins like dioxins, which contaminate the air and lead to respiratory issues. The incineration of e-waste at high temperatures causes the emission of poisonous gases, including greenhouse gases, which damage air quality and enhance the risk of climate change. The accumulation of metallic toxicants like mercury, lead, cadmium, and pollutants like barium, lithium, polybrominated flame retardants, and so on in e-waste in the human body causes acute health issues as well as genetic disorders. These toxicants exhibit carcinogenic effects. It also causes serious injury to the nervous, metabolic, cardiac, and skeletal systems of the human body (Verma and Prakash 2020). The paper highlights the need for sustainable e-waste management practices, such as extended producer responsibility, public education, and capacity building. It emphasises the importance of collective action by governments, industries, and civil society to reduce environmental hazards and protect public health from the negative impacts of e-waste pollution.

2 Conceptual Background

2.1 E-Waste Composition and Hazards

The classification of e-waste based on its source of origin will allow us to develop an effective management strategy. Today, every sector of the economy will be a potential source of electronic trash due to the extensive use of electronic appliances. Below, we describe the classification of e-waste based on its generation source. Figure 1 depicts the classification of different types of e-wastes.

Spare household appliances: This type of waste originates from the electrical and electronic appliances used in the household. It contains discarded electronic gadgets, viz., computers, laptops, mobile phones, TVs, fridges, microovens, fans, bulbs, tubelights, air conditioners, toys, leisure and sports equipment, etc. The household e-waste contains several hazardous items like infrared detectors, semiconductors, chip resistors, condensers, printed circuit boards, batteries, transformers, capacitors, cathode ray tubes, and different types of plastic items that consist of contaminants like cadmium, lead, barium, mercury, nickel, polyvinyl chloride, polychlorinated biphenyl, brominated flame retardants, etc. These acute toxicants cause serious health issues like neural damage, nervous system breakdown, kidney damage, heart and liver damage, muscle weakness, lung damage, bronchitis, endocrine system function disruption, type 2 diabetes, etc. (Attia et al. 2021).

Discarded office electronics: The electronics items obtained in different offices are of voluminous amount due to their growing demand in the daily office work. The household electronics items are also available here, but in large amounts as per the requirements. It contains photocopying machines, calculators, typewriters, telephones, fax machines, CPUs, screens, monitors, keyboards, networking equipment, audio amplifiers, USB cables, pen drives, memory card readers, hard discs, graphics drawing tablets, drones, cameras, etc. It also includes all of the other advanced gadgets that are coming to market. Apart from the toxicants found in spare household electronics, the discarded advanced office electronic gadgets also contain nanomaterials, which increase the risk of nanotoxicity when contaminating soil and water. Advanced flexible hybrid electronic gadgets contain plastic items like polyethylene terephthalate and a variety of inks, such as nanoparticle inks, silver inks, and carbon inks, among others. The toxic

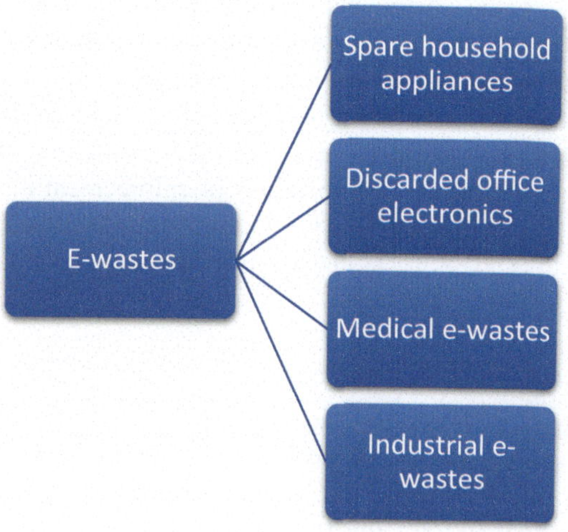

Fig. 1 Classification of e-wastes

nanoparticles affect the human body and ecosystem more effectively as these toxicants interact with the substrates with higher effectiveness than normal toxicants (Tong 2022).

Medical e-wastes: The discarded biomedical wastes also contain electronic trash, such as expired medical devices, cosmetic surgical devices, biomedical devices, CRT monitors, LCD screens, and all types of advanced medical devices. The advanced electronic gadgets used in the healthcare industry contain toxic metals like nickel, copper, lead, mercury, chlorinated plastics, brominated flame retardants, etc. Advanced biomedical equipment contains nanometals that lead to nanotoxocity. Moreover, to adhere to the provisions of the Health Insurance Portability and Accountability Acts, it is necessary to break and shred the memory devices used to store patient data after use, thereby ensuring patient security and privacy. The lack of proper disposal guidance during the destruction of these electronic devices resulted in the generation of voluminous e-waste (Shantaram et al. 2014).

Industrial e-waste: The automation and digitalisation of industrial processes increase the demand for advanced electronic equipment in the industry. The industrial sector generates more e-waste than the household sector, as the electronics machines used there are vulnerable to experiencing more damage due to their high usability. Furthermore, the manufacturing techniques of industry are upgrading at a higher rate due to the continuous upgradation of technology, which compels the industries to discard the old machines. Silicon, cobalt, antimony, bismuth, cobalt, talc, germanium, petroleum-based plastic components, ceramics, etc., are some important components obtained in microchips, transducers, semiconductors, capacitors, resistors, insulators, and other industrial electronic trash components. The electronics industry also generates waste (Mades 2021).

3 Methodology

The paper examines the impact of e-wastes on environment and public health and analyses the need for the development of sustainable e-waste management strategies. It also discusses about the implication of e-waste management on environment, economy, public health, and industry. The paper aims to develop an effective regulatory framework and strategy for sustainable e-waste management by analysing the nature of e-wastes and its impact on environment and public health.

4 Results

4.1 Environmental Impacts of E-Wastes

Typically, we incinerate, landfill, or dispose of electronic waste in garbage and water bodies. Otherwise, informal ragmen typically handle these wastes without adhering to any safety protocols, extracting the metals from them in a highly unhygienic manner. These unscientific ways of handling electronic trash lead to the emergence of serious environmental issues as depicted in Fig. 2. The advanced electronic items contain semiconductors, nanometals, specially engineered nanocomposites, insulators, magnetic items, etc. The advanced electronic items include LCD displays, cathode ray tubes, DC power supplies, batteries with innovative energy storage mechanisms, and LED screens, among others. When these things are taken apart, the right amount of contaminants are extracted, such as cadmium, arsenic, antimony, copper, mercury, and more. Organic toxicants, such as polybrominated biphenyl, tetrabromobisphenol A, polychlorinated biphenyls, and more, are also extracted (Ankit et al. 2021).

Soil in places where e-waste is dumped has harmful chemicals in it, such as polychlorinated and polybrominated dibenzo-p-dioxins, polybrominated diphenyl ethers, polychlorinated and polybrominated dibenzofurans, polycyclic aromatic hydrocarbons, dechlorane plus, hexabromocyclododecanes, and more (Zhang et al. 2012). The plant species absorb the contaminants through the roots and translocate them to the leaves and other plant tissues. The excessive contamination of soil pores with e-waste results in an increase in calcium concentration. However, calcium is an essential element of the cell wall, and its excessive presence causes a decrease in the concentration of other essential elements. The deficiency of nutrients in the soil also leads to the improper growth of plant species and a decrease in the rate of photosynthesis and transpiration. Toxic metal accumulation also affects the population of soil microbes, leading to the depletion of soil fertility (Dharini et al. 2017).

Electronic trash disposal in water bodies depletes water quality and threatens aquatic ecosystems. The chemicals present in the e-wastes, such as polybrominated diphenyl ethers, dichlorodiphenyltrichloroethane, polybrominated diphenyl ethers, brominated flame retardants, polychlorinated biphenyls, and per- and polyfluoroalkyl substances, pose a serious threat to the aquatic ecosystem. Heavy metals that stay in rivers for a long time change the way animals behave, mess up their metabolism and nervous systems, lower the death rates of organs, stop them from growing, and make it hard for them to reproduce. The liquid crystal monomers, mostly prevalent in LCD screens, cause

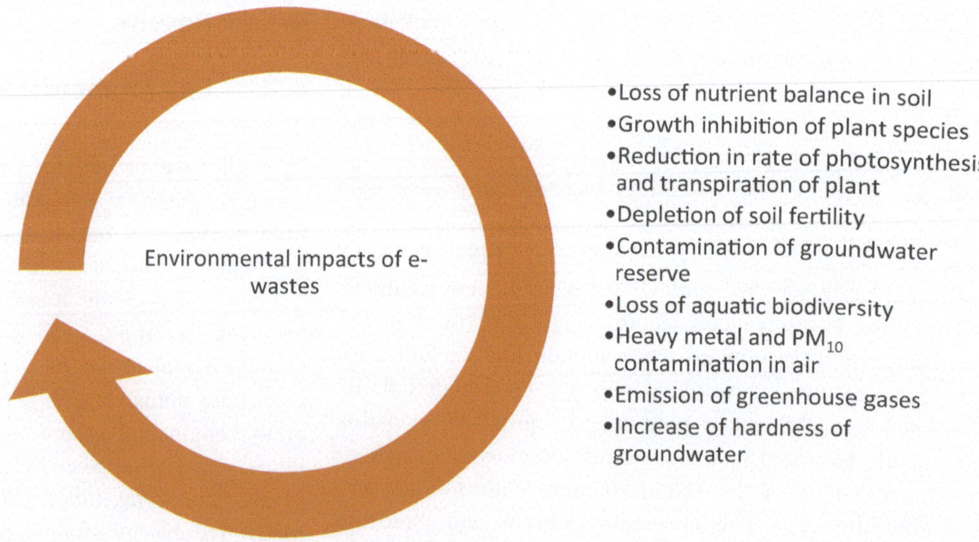

Fig. 2 Environmental impacts of e-wastes

detrimental effects on aquatic life (Shaaban et al. 2024). The percolation of metallic contaminants into the groundwater reserve through the soil pores leads to the depletion of the groundwater's quality. Metal contamination increases the groundwater's hardness, total dissolved solid contents, pH, and conductivity (Dharini et al. 2017).

The smelting of e-waste emits dioxins, increasing the risk of cancer and other genetic disorders. The people from the informal sector associated with the dismantling of e-waste are the most vulnerable to the air pollution associated with e-waste. Dismantled batteries and photographic equipment expose people to lithium vapour over an extended period. The burning of e-waste also emits greenhouse gases, increasing the risk of climate change (Jain et al. 2023). Some studies on the air quality of the areas near e-waste processing sites indicate the presence of heavy particulate matter (PM10) and heavy metallic toxins like lead, copper, zinc, nickel, and chromium (Gangwar et al. 2019).

4.2 Implications of E-Wastes on Public Health

The microwave-assisted acid digestion of human blood samples collected from the population near e-waste disposal sites revealed the presence of heavy metallic toxicants such as lead, cadmium, chromium, mercury, nickel, and cobalt. These toxicants lead to chronic hypoxia and hypertension. These metallic ions form a stable complex with haemoglobin because of their higher chelating capacity, which restricts the flow of oxygen in the blood. The population's physical health measurements indicate their susceptibility to hypertension and acute cardiovascular diseases. Apart from that, a higher rate of systolic and diastolic blood pressure fluctuation increases the risk of blood pressure-related health issues. Researchers also found that the area near the e-waste burning area displays more prominent effects (Gangwar et al. 2019). The International Labour Organisation and the World Health Organisation estimate that millions of women and children are involved in the management of e-waste. Urban slums process around 95% of the e-waste, and the adverse health effects associated with e-waste management primarily affect children and women workers (Annamalai 2015). The impacts of the toxicants present in e-wastes have been depicted in Table 1.

The plastic parts in electronic waste make harmful organic pollutants like polychlorinated dibenzo-p-dioxins and furan-based compounds like polychlorinated dibenzofurans. These pollutants don't break down naturally and are biodegradable. Exposure to these potential organic contaminants leads to severe skin diseases like chloracne, darkening of skin colour, and damage to liver function, the endocrine system, the nervous system, immune system, and the reproductive system. Long-term exposure to these contaminants also leads to cancer and other genetic disorders. Lead is one of the most effective metallic contaminants in electronic circuit board components. Short-term exposure to lead causes vomiting, diarrhoea, coma, convulsions, and death due to its accumulation in the brain, kidneys, and nervous system. Children exposed to lead cause improper intellectual development, growth inhibition, behavioural anomalies, hearing disabilities, and so on. Exposure to beryllium causes acute lung diseases, including lung cancer. The exposure to toxic mercury and cadmium also causes serious damage to the brain, spinal cord, kidneys, liver, etc. These metallic toxicants also affect the developing foetus, which leads to serious genetic issues for newborn children. Flame retardants, such as polybrominated diphenyl ethers, increase the risk of cancer and alter thyroid gland functions (Annamalai 2015). The advanced electronic items contain specially engineered

Table 1 Impacts of different toxicants extracted from e-wastes on human health

Pollutants in e-wastes	Effects on human health
Dioxins and furan-based contaminants	Acute skin infection; alteration of the function of liver, and kidney; damage to the immune, endocrine, and nervous system of the body; cancer and genetic disorder
Nonferrous metallic toxicants	Restrict oxygen flow in blood, cardiovascular disease, increase the risk of blood pressure
Lead	Damage nervous, and endocrine system, intellectual disability, behavioural disorder, hearing and vision impairment, issue in reproductive system
Mercury	Damage in brain, kidney, spinal cord, liver; Effect the growth of developing foetus; damage immune system; memory problems; neurological disturbances; skin rashes; insomnia; neuromuscular dysfunction
Cadmium	Damage in kidney, bone, and pulmonary tissues; increase the vulnerabilities of cancer
Beryllium	Lung cancer and other severe infection in lung, dermatitis, chronic pulmonary diseases, acute pneumonitis
Brominated flame retardants	Increase the risk of cancer and thyroid infection; hormonal dysfunction; neurological damage; immune dysfunction; damage in reproductive organ
Nanomaterials	Affect lymphatic, circulatory, and nervous system

nanomaterials that translocate into the human body through the circulatory, lymphatic, and nervous systems and affect the organs adversely (Viswanath and Kim 2016).

4.3 Economic Implications of E-Wastes

E-waste contains industrially important metals like lead, cadmium, chromium, lithium, cobalt, copper, silver, carbon, iron, aluminium, plastics, etc., which are valuable resources in different sectors. The economic implications of e-wastes have been depicted in Fig. 3. The successful extraction of these resources lowers the cost of raw materials in the industry, leading to economic growth. Moreover, the developing nations do not have an adequate reserve of transition metals and lanthanides, which are important components of modern electrical and electronic industries. Therefore, they have to import these metals in good quantities in order to fulfil the demand for electronics products in the nation. The extraction of these metals from the e-waste will reduce their tendency to import the metals, and they will become self-sufficient in the electronics industry (Liu et al. 2023).

The lower operational cost of collecting raw materials results in higher economic outcomes for various industries, particularly electrical and electronic industries. The reduction in operational costs will encourage industrialists and investors to increase their investment in labour welfare and

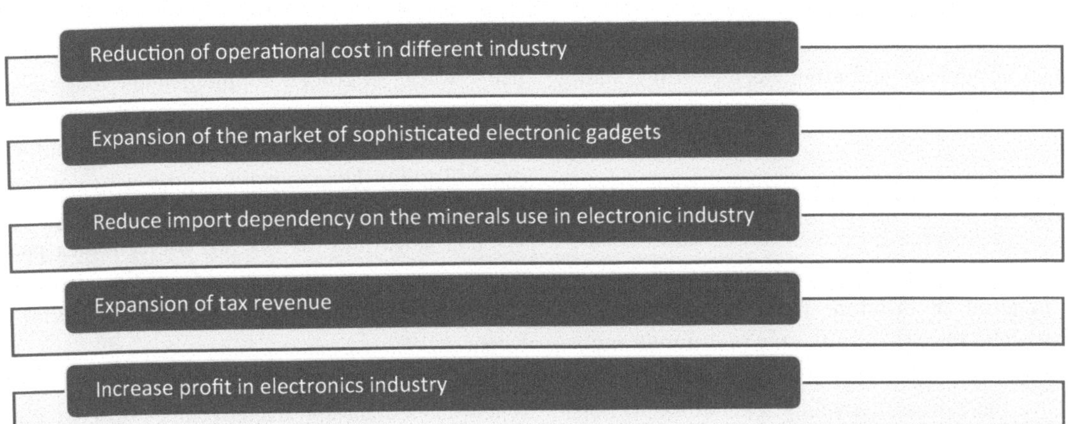

Fig. 3 Economic implication of E-wastes

corporate social responsibility, which will lead to economic growth as well as social empowerment (Seif et al. 2023).

E-waste has become a potential source of precious metals in the populous developing industries, as these are the sources of a voluminous number of E-waste. It lessens the reliance on imports for developing economies and fosters industrial growth within those nations. This accelerates the growth of developing nations' economies, particularly in the electrical and electronics industries.

The increase in industrial demand for e-waste will cause an increase in income for the informal sector associated with its handling. It will lead to an improvement in the financial status of the informal ragmen. Furthermore, the increase in income will formalise their business, which will also increase the country's tax revenue. The economic upgradation of the informal sector associated with it will also encourage the social growth of the urban poor and marginal sections and improve their quality of livelihood. It will subsequently lead to an increase in government revenue in the form of property tax, water tax, road tax, and so on. The formalisation of the informal e-waste handlers will encourage them to participate in different financial schemes, viz., the MSME loan scheme, the property insurance scheme, etc., which will enhance the income of the financial sector and promote the overall growth of the economy. Furthermore, the growth of the e-waste handling sector will also encourage them to register under the GST, increasing the government's indirect tax revenue (Thukral and Singh 2023).

The use of e-waste as ingredients will reduce the operational costs of the electrical and electronics industries, thereby enabling the affordability of advanced electronics gadgets for lower-income groups. It will increase the market size of sophisticated electronics products and stimulate economic growth. Moreover, the introduction of advanced gadgets among lower-income groups will accelerate technological advancement in traditionally backward sectors such as agriculture, small business, and small industrial sectors, thereby promoting the growth of these sectors and improving the economic status of these groups. The technological advancement of agriculture and allied sectors will enhance their contribution to the nation's GDP (Cardenas 2023).

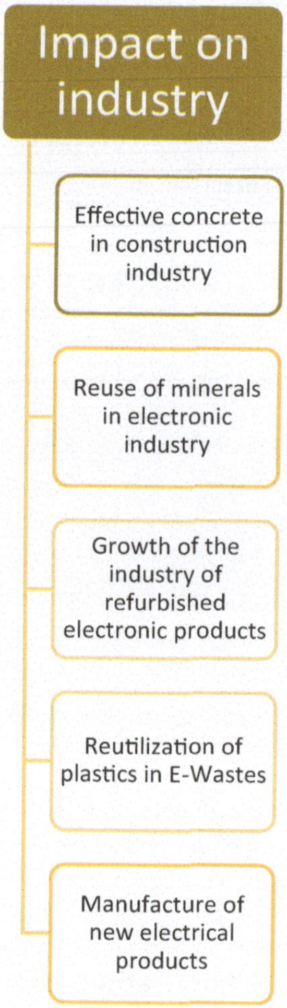

Fig. 4 Impact of e-wastes in industry

4.4 The Impacts of E-Wastes on Industry

The potential of e-waste as an important primary and secondary raw material in different industries increases its market demand. The impacts of e-wastes in different industrial sectors have been depicted in Fig. 4. However, we need to improve the e-waste recycling technology to boost its large-scale commercialisation and acceptance in the industrial sector. The electrical and electronics industries can reuse the rich mineral contents in e-waste to manufacture other electronic gadgets. It is possible to remove and repair the parts from the electronic scraps for reutilisation. Advanced technology enhances the use of e-waste in various industries, particularly in the construction and infrastructural sectors (Liu et al. 2023).

Construction and infrastructural sectors can largely utilise e-waste as crucial supplementary materials to produce green concrete. Effective physical, chemical, and metallurgical recycling improves concrete's thermal strength and durability. Further innovations in e-waste enable the development of sustainable and cost-effective green concretes. We can substitute the discarded electronics parts for natural fine or coarse aggregates. Concrete can incorporate the extracted plastics as a crucial coarse material (Luhar and Luhar 2019).

The electrical and electronics industries can effectively reuse the metals extracted from e-waste as important raw materials. The introduction of refurbished electronic items enables the expansion of the electronic gadget market due to the availability of advanced technology at a low price. Parts of the discarded electrical and electronic items contribute to the growth of the refurbished item manufacturing industry, which in turn stimulates the growth of the electronic industry in developing nations such as India. The promotion of refurbished electronics will facilitate digital literacy at the grass-roots level by enabling the dissemination of advanced gadgets to a small and marginal section of society (Cardenas 2023).

Due to their non-biodegradability, different industrial sectors can use the plastics extracted from e-waste after proper modification. Different plastic industries can convert these plastics into small plastic scraps and use them as raw materials to manufacture containers and other plastic products. After necessary technological modifications, we can reuse electrical scraps such as circuit boards and switches to manufacture new electrical products. Different industrial and commercial sectors can use the metals extracted from the discarded circuit boards through mechanical, hydrometallurgical, and pyrometallurgical methods as secondary materials (Priyan et al. 2023).

4.5 Ethical Consideration of the Application of E-Wastes

While the application of useful metals extracted from e-waste shows great promise, their potential influence on the environment and long-term sustainability raises serious ethical questions. The extraction of metals from e-waste and improper treatment methods, especially if inadequately handled, may result in negative environmental effects. The possible ecological impacts of releasing toxic chemicals into the environment raise ethical considerations. We raise concerns about resource exploitation, geopolitical conflicts, and the ethical implications of using these resources sustainably. Hazardous levels of materials, such as toxic metals, non-metals, and metalloids, may expose workers and consumers in the e-waste management sector. For ethical reasons, we must prioritise the health and safety of anyone working with or exposed to toxicants (Ozzie Mascarenhas et al. 2016).

Everyone should have access to the advantages of e-waste processing technologies. Restricting these technologies to a select few instead of making them available to all nations raises ethical questions and exacerbates existing global disparities. Advanced e-waste processing technologies should incorporate the informal methods typically used for e-waste management. To mitigate the severe health impacts, the policy should prioritise enhancing their expertise in the field. The e-waste processing industry's corporatisation may limit the commercial rights of small, informal ragmen. We need to frame policy to protect their legal and commercial rights. Research transparency, risk communication, and responsibility for adverse effects on the environment and human health are all crucial ethical considerations. Both academics and businesses need to be forthright about their methods and results. Such technologies must be carefully created and implemented in order to be ethically treated. Ethically, we have a responsibility to secure the long-term viability of nanomaterials for use in energy harvesting. We must consider the environmental impact of these technologies, including recycling and waste management. Resolving ethical issues with e-waste management requires teamwork and open communication. International collaboration and standardisation can aid in the reduction of risks. To ensure adherence to moral standards, research and uses involving nanomaterials require ethical oversight from governments, regulatory bodies, and independent organisations. Ethical issues like these require ethical research and development practices, strict rules, environmental impact assessments, and international cooperation (Hawari and Hassan 2008).

4.6 Current E-Waste Management Practices

Different e-waste management practices existing today are depicted in Fig. 5. Landfilling is one of the most common methods of disposing of e-waste. In this method, we dispose of the electronic waste in trenches made on a flat surface and bury it under a thick layer of soil. This method has the risk of contaminating soil pores with organic and inorganic toxicants extracted from waste materials. Non-biodegradable pollutants like metallic toxicants, brominated flame retardants, polychlorinated biphenyls, etc., mix with the groundwater reserve and soil pores, affecting the quality of the groundwater and the population of microbes. The chemical reaction of contaminants with soil particles affects the soil's nutrient balance, resulting in soil fertility degradation (Annamalai 2015). The landfills are also vulnerable to leakage, producing leachate containing all the toxic components present in the buried e-waste, like volatile contaminants, heavy metals, phenols, nitrogen, chloride, etc. It causes the soil's ionic balance to be lost and plant species to grow incorrectly (Dagwar and Dutta 2024).

The specially designed incinerators also burn the e-waste at temperatures between 900 and 1000 °C. Other sectors can also harness this high energy as a potential energy source. It also converts hazardous e-waste into less dangerous waste. The combustion of e-waste also produces toxic gases such as dioxin, as well as other greenhouse gases such as carbon monoxide and carbon dioxide. The emission of the oxides of mercury, lead, cadmium, etc., also pollutes the atmosphere. In addition to that, the slag, fly

Fig. 5 Current e-waste management practices

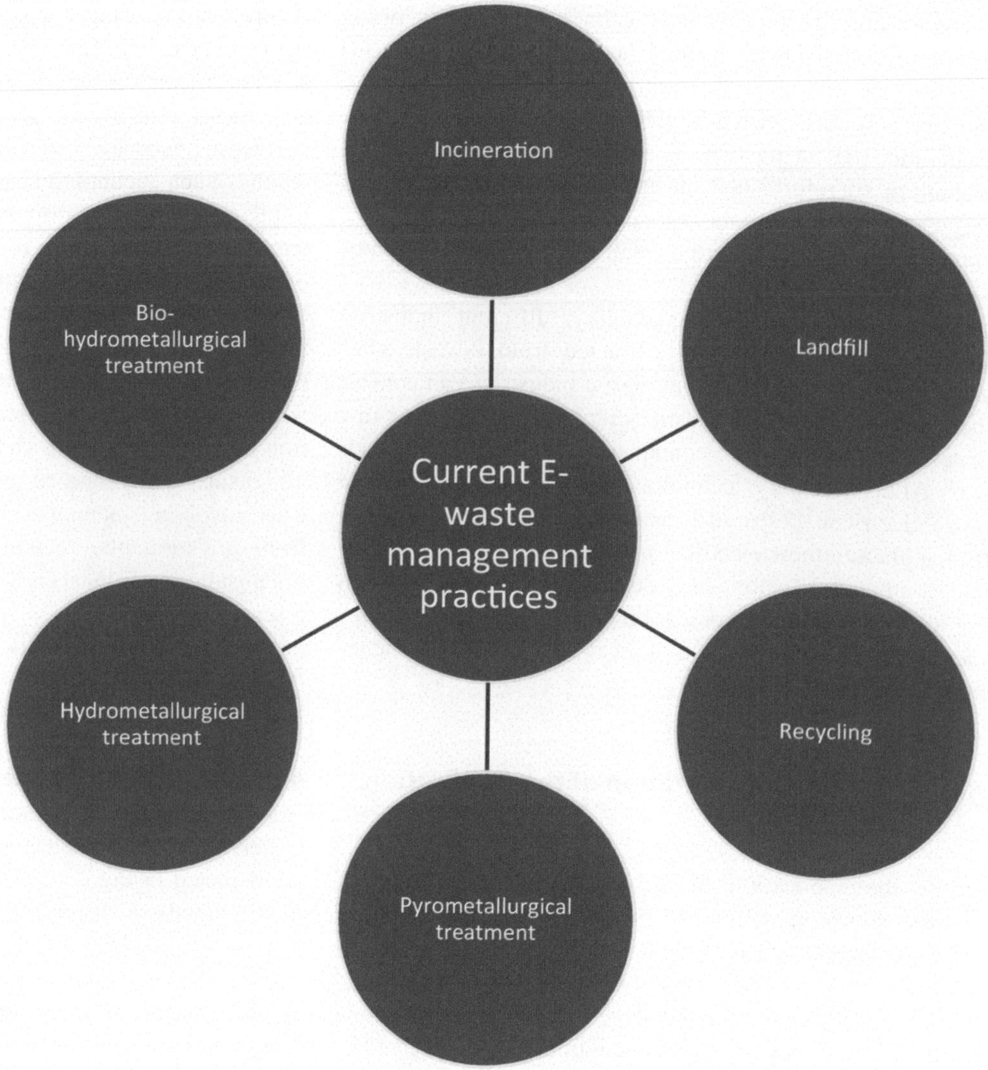

ashes, and flue ashes contaminate the soil and water bodies (Annamalai 2015).

The recycling process for e-wastes involves dismantling and shredding the waste, as well as extracting precious metallic substances such as gold, silver, PCBs, mercury, and copper from various components such as cathode ray tubes and circuit boards. After necessary modifications, you can reuse monitors, printers, floppy discs, hard discs, compact discs, mobile phones, fax machines, CPUs, memory chips, connecting wires, motherboards, modems, keyboards, and other components. Informal ragmen, lacking proper training and awareness of safety protocols, primarily handle the recycling of electronic waste and conduct the dismantling process. They dismantle the e-waste with hammers and other old-fashioned utensils, generating fly ash and voluminous dust particles. These not only contaminate the atmosphere and soil but also result in severe health issues. Inhalation of heavy metals like lead and mercury causes serious damage to the nervous system, circulatory system, kidney, and endocrine systems (Annamalai 2015).

Separating metals and noble elements from e-waste through pyrometallurgical, hydrometallurgical, and bio-hydrometallurgical processes is a technologically advanced method of extracting precious metallic components from e-waste. The pyrometallurgical process employs eddy current, air current, magnetic, and vacuum metallurgical separation techniques, while the hydrometallurgical process uses chemical processes to separate metals. An environmentally friendly version of the hydrometallurgical process, the bio-hydrometallurgical process uses microbes instead of chemicals to separate metallic and noble elements. We specifically use the electrometallurgical process to recover metals, then treat them with acid. Developed states in Europe and North America, along with Saudi Arabia and the UAE, typically employ these methods (Madkhali et al. 2023).

5 Regulatory Frameworks

The European Union published the Waste from Electrical and Electronic Equipment (WEEE) Directive in 2012 to facilitate the collection, segregation, and treatment of e-waste in a sustainable manner. According to this directive, the European Union's members must adopt a separate collection method for ensuring the collection of all e-waste generated in order to avoid unscientific disposal. It makes distributors accountable for collecting electronic waste from consumers. The directives also focus on the proper treatment of the collected e-waste to minimise its environmental impact. It also emphasises the reuse of electronic equipment by repairing the items (Mmereki et al. 2016).

In 2011, the Government of Australia introduced the Product Stewardship Act, which imposed a legal obligation on each industry in Australia, including the electronics industry, to formulate and implement sustainable product management strategies for the products they produce. This act is based on the idea that each product is different in nature, and as a result, its management strategies must be different. However, most of the provisions under this regulation are voluntary in nature (Cardenas 2023). The developed nations in Europe and North America generate e-waste, which the African nations then dump. These nations do not have specific rules and regulations for e-waste management due to a lack of infrastructure and expertise in the governing bodies. In Africa, the informal sectors in Ghana, Madagascar, Egypt, Rwanda, Nigeria, and South Africa are particularly involved in the handling of such wastes (Andeobu et al. 2023).

The E-waste Management and Handling Rules, 2011, introduced by the Government of India, state that producers are responsible for ensuring the collection of waste in a sustainable and eco-friendly manner. The producers must collect and recycle the e-waste generated during the manufacturing of electrical and electronic products. It also includes the provision of Extended Producer Responsibility, which extends the responsibility of the producers to collect the discarded electronic products after consumption and send them to the authorised e-waste processing facilities. The e-waste dismantler and recycler must obtain approval from the concerned State Pollution Control Bureau, and they must ensure the implementation of safe e-waste handling processes to reduce the negative effects on the environment (Bhadra and Mishra 2021). The E-waste (Management) Rules, 2016, make the roles of the Extended Producer's Responsibility stronger and more prominent. The E-waste (Management) Rules, 2016, impose a moral obligation on producers of electrical and electronic items to establish a dedicated department for e-waste handling, collection, disposal, and recycling. It also proposes conducting an exclusive training programme on the safe handling of e-waste for informal e-waste handlers to develop a sustainable and effective e-waste management mechanism at the grass-roots level (Porwal and Chatterjee 2018). The E-waste (Management) Rules, 2022, introduced the concept of a refurbishing certificate, which promotes the sale of refurbished electronic items by separating different parts from the discarded electrical and electronic items. The management of discarded solar photovoltaic cells, panels, and other e-waste from a solar energy installation is a key focus (Kumar et al. 2022).

6 Findings and Discussion

The current e-waste management strategies, i.e. landfilling, incineration, and disposal of e-waste in water bodies, are not environmentally sustainable in nature. E-waste landfilling contaminates soil and groundwater reserves with toxic metals and plastic-based organic pollutants. The percolation of toxic metals into the groundwater through soil pores contaminates the groundwater's quality. The contaminated groundwater increases the risk of water borne diseases to the human society as groundwater is an important source of drinking water. The loss of nutrient balance in the soil and contamination in groundwater also restrict the growth and other physiological activities of plant species due to the translocation of contaminants to the plant cells. It decreases the rate of photosynthesis and transpiration due to the insufficient growth of the plant leaves and the reduction in the number of photosynthetic pigments. The incineration of e-waste causes toxic substances to pollute the air and emit greenhouse gases. The emission of dioxin exhibits carcinogenic activities when it enters human cells through inhalation. The disposal of e-waste in waterbodies also affects the aquatic ecosystem and aquatic biodiversity by affecting the physiological function of aquatic species (Gopalakrishnan et al. 2023).

Electronic products nowadays use specially engineered nanomaterials and nanocomposites in their manufacture. The disposal of these electronic products as e-wastes leads to the contamination of toxic nanometals with the soil, groundwater, and terrestrial water bodies. The nanometals interact with the substrates with more effectiveness than the normal toxicants, which exhibit more toxicity to the environment. Contaminated nanometals may enter plant species through soil and human bodies through contaminated groundwater. The nanomaterials may also contaminate surface water resources while disposing of the electronic equipment in the surface water. Through biomagnification in the food web, it affects aquatic ecosystems as well as terrestrial ecosystems (Rautela et al. 2021).

In 2019, the world generated 53.6 metric tonnes of e-waste globally, which is growing at a rate of around 7.3 kg per year (Ray et al. 2009). The emergence of the Covid-19 pandemic expands the scope of digitalisation in the economy, which increases the demand for electronic equipment like mobile phones, laptops, modems, Wi-Fi routers, etc., at a higher rate. It increases the risk of e-waste contamination in the near future (De et al. 2020). The digitalisation of every sector of our daily lives leads to the innovation of new gadgets like electronic toothbrushes, smart watches, augmented reality glasses, fall-detecting AI smart lamps, self-cleaning upright vacuums and mops, AI haptic headsets for visually impaired individuals, smart indoor and outdoor wall lights, noiseless karaoke microphones, transparent TV screens, nail watches, smart electronic jewellery, robot suitcases, 3D pens, smart drones, etc. (UN Environment Programme 2021). The application of AI and ML in every sector will also cause an increase in the generation of e-waste in the near future due to the increase in demand. The governments must adopt a sustainable e-waste management policy to deal with these increasing amounts of e-waste.

Several studies indicate that a large number of people from the informal sector are involved in e-waste management facilities. We also observe a significant presence of women and children in e-waste management facilities, contributing to the deterioration of public health. The vulnerability of poor men to chronic health issues also increases due to their 8–10 h of daily exposure to electronic trash. The incineration of e-waste makes them more vulnerable to serious cardiovascular infections, neurological disorders, behavioural issues, cancer, and acute genetic problems. The African nations are more vulnerable to e-waste contamination issues because they import electronic trash from Europe and North America for dismantling. The majority of Africa's population lacks technological literacy and lacks adequate training in managing e-waste. Awareness of the severe negative impacts of e-waste also needs to increase in order to encourage more poor people to take training and empower themselves in the workplace (Hassan and Dhusia 2022).

E-wastes are found to contain precious metals like gold, silver, cadmium, chromium, carbon tubes, different types of semiconductors, lead, etc. Effective recycling allows for its use in other sectors. We use metallurgical processes to extract metals from electronic trash for potential use in other sectors. Dismantling e-waste yields plastics for use in various sectors. Refurbishment allows for the use of various components from a discarded electronic item to manufacture another. We can reuse the extracted lead from e-waste to manufacture another new electronic item (Rajesh et al. 2022).

New innovative ways are emerging for using the products obtained from electronic trash recycling. If attached to trees, discarded mobile phones can naturally generate electricity by absorbing sunlight. We can also use the metals extracted from the e-waste to synthesise effective nano-filters, aiding in water purification and waste water treatment. The jewellery sector can utilise the precious metals extracted from e-waste, such as gold, silver, and platinum. You can also use old, discarded electronic items for effective interior design in a house. For instance, after modification, you can use an old TV as an aquarium. You can also use the plastics extracted from the electronic trash for 3D printing (Bhattacharya et al. 2021).

The market for refurbished electronic items is growing rapidly at a CAGR of 4.53%. In 2022, the market size for refurbished items was US$ 29,913.17 million. Big corporate manufacturers like Samsung, Asus, Dell, Lenovo, Acer, Microsoft, Nintendo, Sony, Apple, Hewlett-Packard, etc., are manufacturing and selling refurbished electronic gadgets. Online e-commerce platforms like Amazon, eBay, Flipkart, etc., are promoting refurbished electronic items to customers. Manufacturers prepare these refurbished electronic items by using parts from old electronic items, thereby reducing production costs and enabling more consumers to access the latest technology. Promoting the market for refurbished electronic items is an effective e-waste recycling strategy. The scope of market growth is high in the near future due to an increase in demand for electronic gadgets. It is also necessary to develop technology to convert old, discarded electronic gadgets into new ones (Resmi and Fasila 2017). The new E-Waste (Management) Rules, 2022, enforced by the Government of India, give special emphasis to refurbished items. In developing nations like India, refurbished electronic items are a good option to deal with the increasing demand for electronic gadgets. The Indian economy's rapid digitalisation will broaden the scope of refurbished electronic items.

Extended Producer Responsibility is a new provision that enables effective e-waste management. It extends the producer's responsibility to collect old-fashioned electronic gadgets from consumers and send them to proper e-waste recycling facilities. Companies offer discounts on new products in exchange for old gadgets. Apart from that, platforms like Citify receive old electronic gadgets from consumers and send them to e-waste recycling facilities. The improved e-waste recycling infrastructure will broaden the scope of extended producer responsibility. In the new E-Waste (Management) Rules, 2022, the government of India places special emphasis on the effective enforcement of the extended producer responsibility policy. The government's tax exemption and other promotional strategies will encourage the corporate sector to effectively enforce the extended producer responsibility provisions (Esenduran et al. 2019).

7 Case Studies

Case Study 1: Producer Responsibility Organisation

The Producer Responsibility Organisation (PRO) is an enterprise that aims to implement the concept of Extended Producer's Responsibility (EPR) in India. The Central Pollution Control Board registers PROs to collect and deliver e-waste to producers for recycling. The Central Pollution Control Board estimates that by the end of May 2020, around 34 PROs will have registered in India (Enter Climate 2022). Under PRO, many individuals and organisations registered and received proper training regarding the handling of e-waste. A large number of women who belong to marginal economic groups became financially empowered after registering themselves. The idea will significantly transform the rural economy and foster economic growth in small and marginal areas. By implementing the concept of Extended Producer Responsibility (EPR), PROs make manufacturers accountable for the environmental impacts of their products throughout their entire lifecycle, including disposal and recycling. Registered by the Central Pollution Control Board, PROs collect and deliver e-waste to producers for recycling, ensuring that e-waste is handled and disposed of in an environmentally responsible manner. The impact of PROs is multifaceted and far-reaching. Many individuals, including women from marginal economic groups, have registered with PROs and received training on e-waste handling, leading to financial empowerment and new economic opportunities. This initiative has the potential to transform the rural economy by generating employment and income opportunities in e-waste management, contributing to economic growth in small and marginal areas. Moreover, PROs promote responsible e-waste management, reducing the environmental impacts associated with e-waste disposal, such as pollution and health risks. By facilitating the recycling of e-waste, PROs also help conserve natural resources, reduce the need for new raw materials, and mitigate the environmental impacts of mining and processing. Overall, the PRO initiative is a significant step towards promoting sustainable e-waste management in India, with benefits extending beyond environmental protection to social and economic empowerment. As the number of registered PROs continues to grow, their positive impact is likely to amplify, contributing to a cleaner, greener, and more inclusive future for India (Allegra Lab 2020).

Case Study 2: Green E-Waste Recyclers Pvt. Ltd., an informal e-waste ragman, turns into a formal e-waste recycling agency

Mohammad Sabir, a Delhi-based e-waste ragman, established Green E-Waste Recyclers Pvt. Ltd. in 2011 in technological collaboration with a German-based company, Gesellschaft für Internationale Zusammenarbeit (GIZ). He received training on how to handle e-waste by following safety protocols and has now developed an effective e-waste management ecosystem. This organisation dismantles e-waste by following safety protocols and sells it to large corporate agencies. The organisation's commitment to responsible e-waste management has earned them a reputation as a trusted partner for large corporate agencies, who purchase the dismantled e-waste from them. This not only supports the circular economy but also reduces the environmental and health risks associated with e-waste disposal. Under Sabir's leadership, Green E-Waste Recyclers Pvt. Ltd. has become a model for sustainable e-waste management in India, demonstrating that responsible e-waste disposal can be both environmentally beneficial and economically viable. By creating a safe and efficient e-waste management ecosystem, Sabir has made a significant contribution to reducing India's e-waste burden and promoting a cleaner environment for future generations (Vasudeva and Sohail 2013).

Case Study 3: PARO-CI: an innovative e-waste recycling approach in West Africa

Africa is the place to dump e-waste from Europe and North America, which makes the African people vulnerable to e-waste contamination. Evariste Aohoui, a young entrepreneur from Cote d'Ivoire, established PARO-CI (Programme Assainissement Recyclage Ordures, Côte d'Ivoire), an organisation that transforms the electronic wastes of West Africa into potential resources and inspires African youth to transform these wastes into resources and sources of economic growth. Evariste Aohoui, a visionary young entrepreneur from Cote d'Ivoire, has taken a bold step towards addressing the pressing issue of e-waste management in West Africa. Recognising the region's vulnerability to e-waste contamination, Aohoui established PARO-CI, an organisation that transforms electronic waste into valuable resources. PARO-CI collects and sorts e-waste, dismantles and refurbishes electronic devices, recycles hazardous materials, and creates employment opportunities for young people in the e-waste management sector. Through his initiative, Aohoui aims to promote awareness and education on responsible e-waste management practices, inspiring African youth to transform e-waste into resources and sources of economic growth. PARO-CI has already made a significant impact, creating a network of e-waste collection centres across Cote d'Ivoire, training over 100 young people in e-waste management and recycling techniques, and partnering with local authorities, schools, and businesses to raise awareness about the importance of e-waste management. Aohoui's work has inspired other African entrepreneurs to explore e-waste management opportunities, fostering a growing movement towards responsible e-waste disposal across the continent. By transforming e-waste into resources, PARO-CI is mitigating the environmental and health risks associated with e-waste contamination, unlocking economic potential, and empowering African youth to drive sustainable growth. Evariste Aohoui's pioneering work serves as a beacon of hope for a cleaner, greener, and more prosperous Africa, demonstrating that innovative solutions can emerge from the continent's most pressing challenges (ZeWaLS Academy 2023).

Case Study 4: Emory University generates $31,840 in revenue from selling e-waste

Emory University, an educational institution in Georgia, used to lose around $64,000 per year to clear the excess electronic items on campus. They used to give these items to local vendors, who collected and handled them in an environmentally unsustainable manner. Since 2017, the authority has partnered with Premier Surplus, an organisation that specialises in e-waste recycling. Now the authority sells the excess electronic items to the company, earning around $31,840 per year. Emory University's collaboration with Premier Surplus serves as a model for other educational institutions and organisations to adopt sustainable e-waste management practices, demonstrating that environmentally responsible actions can also generate financial benefits. The programme has helped divert over 100 tons of electronic waste from landfills since its inception. Premier Surplus uses state-of-the-art technology to refurbish and recycle electronic devices, ensuring data security and compliance with environmental regulations. Emory University has integrated e-waste recycling education into its sustainability initiatives, raising awareness among students, faculty, and staff about the importance of responsible electronic waste management (Leonard 2020).

8 Future Outlook and Recommendations

A sustainable strategy for e-waste management is required as society moves towards extensive digitalisation as shown in Fig. 6. To develop a sustainable e-waste management ecosystem, we must undertake extensive technological improvements to the existing e-waste recycling method, implement eco-friendly e-waste handling strategies to minimise environmental impact, foster multilateral and multisectoral collaboration among government and corporate bodies, and establish effective e-waste recycling facilities.

The electronic item manufacturing industry must enforce environmental restrictions and waste management policies to ensure minimal waste generation and environmental contamination. It will encourage the industry to adopt an environmentally sustainable method for producing electrical and electronic items. The use of bio-based products will result in fewer waste products being produced. The fridge with biopolymer gel for cooling is a sustainable alternative that will reduce the negative impact on the environment and generate less waste. The strict environmental restrictions also encourage electronic item manufacturers to produce items that consume less energy and have a minimal environmental impact.

Effective campaigning is necessary to enforce the e-waste management rules at the grass-roots level. We need to launch awareness programmes on the adverse effects of e-waste. The government must actively engage with citizen's associations, local groups, non-governmental organisations, corporate agencies, electronic gadget manufacturing agencies, electronics and telecommunication research organisations, among others, to improve the facilities of e-waste collection, segregation, disposal, and recycling.

The current e-waste handling processes are not eco-friendly in nature. It leads to the contamination of soil, water, and other natural resources. Technological advancement is necessary to develop the environment and e-waste handling processes effectively. Bio-based metallurgical processes, such as bio-hydrometallurgy, are environmentally sustainable methods for extracting metals from e-waste. We need to widely adopt and implement these environmentally sustainable processes. Environmental authorities and the government must strictly implement the safety protocols, rules, and regulations on e-waste management in the e-waste handling units.

Most of the people associated with e-waste handling belong to the informal sector and are technologically unempowered, making them vulnerable to serious health issues. We need to provide them with proper training through skill development programmes, encouraging them to handle e-waste by adhering to safety protocols. We can encourage local-level entrepreneurship to collect and process e-waste in an environmentally friendly manner, then sell the waste to corporate electronic item producers. The establishment of start-ups in this field as micro, small, and medium enterprises will enable economic growth at the grass-roots level and also facilitate the eco-friendly management of e-waste in every locality. The financial institutions need to actively participate by providing effective financial schemes like loan disbursement at low interest and providing life and non-life insurance. Microfinancial groups such as self-help groups, cooperative associations, citizen's clubs, and so on will encourage inclusive participation in the area. Moreover, the availability of local-level e-waste management facilities will ensure the collection of e-waste at the grass-roots level. This inclusive participation will also increase awareness

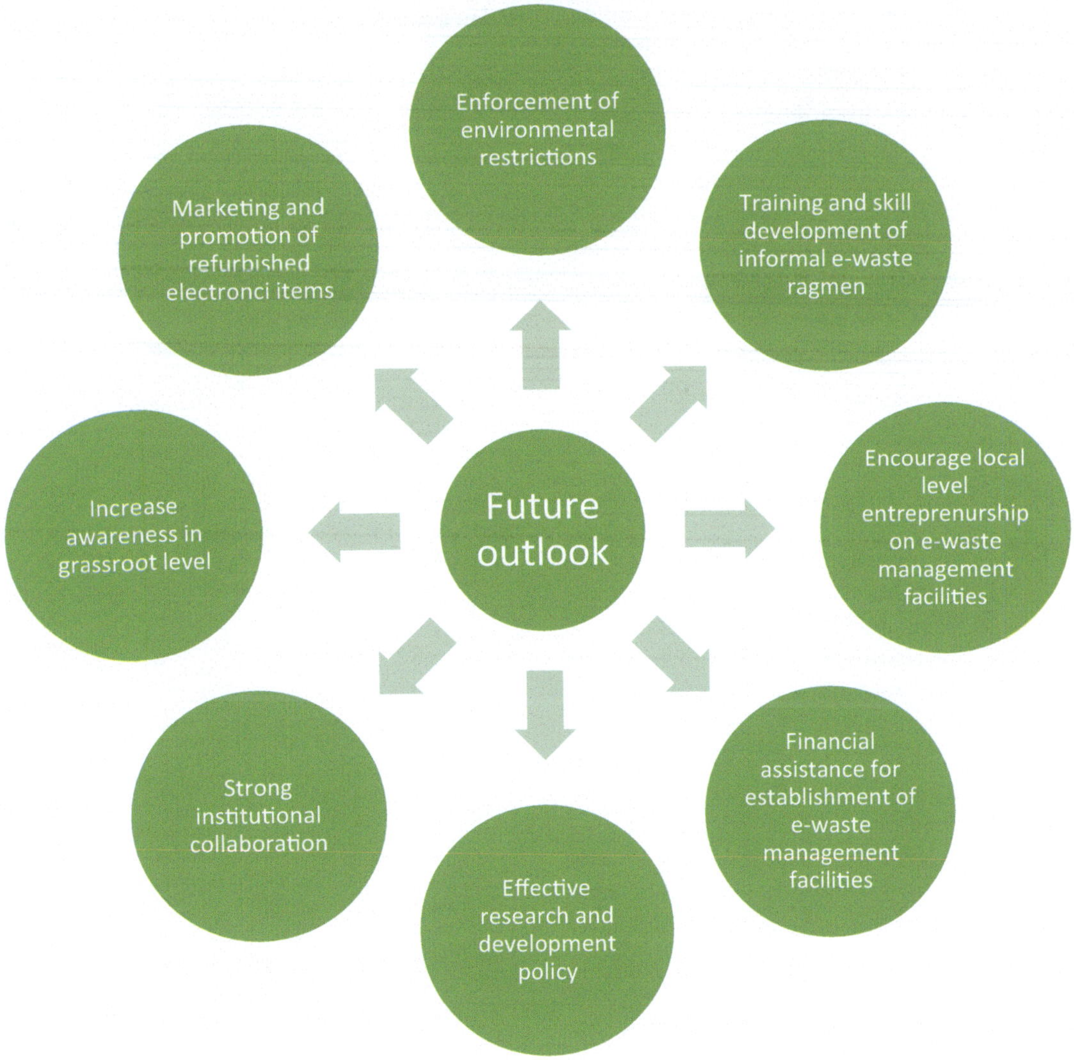

Fig. 6 Future outlook

of the adverse effects of e-waste on the environment and human health.

The government and corporate agencies need to encourage research and development to advance e-waste recycling. The establishment of e-waste recycling research facilities in every electronic waste manufacturing agency will facilitate the technological advancement in the manufacturing of environmentally sustainable electronic items and also the recycling of e-waste. The enforcement of extended producer's responsibility provisions makes the electronic item manufacturers responsible for the collection and processing of e-waste. The establishment of e-waste handling laboratories by the government will empower the informal sector associated with e-waste handling technologically. We need to establish policy research groups to formulate a sustainable ecosystem for the collection of discarded electrical and electronic items. The introduction of academic programmes to train human resources in this field will enable the formulation of sustainable policies at the national, state, and local vels. It will also help to formulate a real-time policy for the implementation of safety protocols in informal e-waste handling units.

Refurbished electronic items are one potential idea for effective e-waste recycling. We need to promote the market for refurbished electronic items through active physical and digital marketing strategies. Corporate agencies should provide price exemptions, MRP concessions, and other facilities to promote refurbished items. People's attitude towards refurbished items is still defensive due to their misconceptions about their quality. Technological advancement is necessary to improve the quality and efficiency of refurbished electronic items. We need to adopt a dedicated research and development policy in this sector. In this field, corporate agencies and the government can introduce dedicated scholarship schemes for research.

Institutional collaboration and cooperation facilitate the establishment of a strong infrastructure at all levels of society. The international cooperation will enable the transfer of advanced technologies from developed countries to developing countries. The integration among governments of all levels will encourage the establishment of e-waste management facilities at the local-level. It will also facilitate the development of the world's overall infrastructure. Addressing these issues on a global scale will impose a moral obligation on all member states to adopt policies for the establishment of environmentally sustainable e-waste management facilities.

9 Conclusions

Research on the sustainable management of e-waste will be required in parallel with the technological advancement of electronic gadgets. The public awareness of the adverse effects of e-waste also needs to be increased to transform the attitude of the public towards the disposal of damaged electrical and electronic items. The training and skill development of the people associated with the handling of e-waste also needs to be increased to restrict the contamination of e-waste and improve public health. The corporate agencies associated with the manufacture of e-waste need to work collaboratively with the informal sector associated with e-waste handling to formulate an effective e-waste management ecosystem. Provisions such as extension producer responsibility and promotion of refurbished electronic items are some key steps that encourage the manufacturer, consumer, ragmen, and all other stakeholders to handle e-waste responsibly. The technological advancement in the manufacture of new electronic items needs to be focused on the minimisation of environmental impact and sustainability. The concept of refurbished electronic items emerges as a potential idea for reducing and recycling e-waste. The future of electronic gadgets is focused on the use of bio-based materials, which will improve the biodegradability of e-waste. The development of bioplastics and other biodegradable plastics will also bring sustainability to the production of electronic items. Sustainability and its impact on the environment and public health need to be an important component of research and development in the field of electronics and telecommunication. Separate research infrastructure needs to be developed for technological innovation in e-waste recycling and to formulate effective policies for sustainable e-waste management.

Acknowledgements Authors are thankful to the contributors of these books for their cooperation and patience in the compilation of this task. Authors are also thankful to Springer team, for their constant support in the publication of this work.

References

Allegra Lab (2020) Narratives of dysfunction and success stories in e-waste recycling in India. https://allegralaboratory.net/narratives-of-dysfunction-and-success-stories-in-e-waste-recycling-in-india/#:~:text=The%20woman%2C%20Vivek%20said%2C%20is,to%20support%20her%20six%20children

Andeobu L, Wibowo S, Grandhi S (2023) Informal e-waste recycling practices and environmental pollution in Africa: what is the way forward? Int J Hyg Environ Health 252:114192. https://doi.org/10.1016/j.ijheh.2023.114192

Ankit, Saha L, Kumar V, Tiwari J, Sweta, Rawat S, Singh J, Bauddh K (2021) Electronic waste and their leachates impact on human health and environment: global ecological threat and management. Environ Technol Innov 21:102049. https://doi.org/10.1016/j.eti.2021.102049

Annamalai J (2015) Occupational health hazards related to informal recycling of E-waste in India an overview. Indian J Occup Environ Med 19(1):61–65. https://doi.org/10.4103/0019-5278.157013

Attia Y, Soori PK, Ghaith F (2021) Analysis of households' e-waste awareness, disposal behavior, and estimation of potential waste mobile phones towards an effective e-waste management system in Dubai. Toxics 9(10):236. https://doi.org/10.3390/toxics9100236

Bhadra U, Mishra PP (2021) Extended producer responsibility in India: evidence from Recykal, Hyderabad. J Urban Manag 10(4):430–439. https://doi.org/10.1016/j.jum.2021.07.003

Bhattacharya S, Chatterjee SM, Sachdev BK, Seth A, Mondal A (2021) An analysis on the e waste management system in the world in particular reference to India. Int J Res Appl Sci Eng Technol 9(XI). https://doi.org/10.22214/ijraset.2021.38962

Cardenas J (2023) Role of refurbished electronics in reducing e-waste: circular economy perspective. Forest Nation. https://forestnation.com/blog/role-of-refurbished-electronics-in-reducing-e-waste-circular-economy-perspective/

Dagwar PP, Dutta D (2024) Landfill leachate a potential challenge towards sustainable environmental management. Sci Total Environ 926:171668. https://doi.org/10.1016/j.scitotenv.2024.171668

De R, Pandey N, Pal A (2020) Impact of digital surge during Covid-19 pandemic: a viewpoint on research and practice. Int J Inf Manag 55:102171. https://doi.org/10.1016/j.ijinfomgt.2020.102171

Dharini K, Cynthia JB, Kamalambikai B, Celestina JPAS, Muthu D (2017) Hazardous e-waste and its impact on soil structure. IOP Conf Ser Earth Environ Sci 80:012057. https://doi.org/10.1088/1755-1315/80/1/012057

Enter Climate (2022) Functions and role of producer responsibility organisation (PRO. https://enterclimate.com/blog/functions-and-role-of-producer-responsibility-organisation-pro/#:~:text=PRO%20or%20Producer%20Responsibility%20Organisation,of%20E%2DWaste%20Management%20Rules

Esenduran G, Atasu A, Van Wassenhove LN (2019) Valuable e-waste: implications for extended producer responsibility. IISE Trans 51(4). https://doi.org/10.1080/24725854.2018.1515515

Gangwar C, Choudhari R, Chauhan A, Kumar A, Singh A, Tripathi A (2019) Assessment of air pollution caused by illegal e-waste burning to evaluate the human health risk. Environ Int 125:191–199. https://doi.org/10.1016/j.envint.2018.11.051

Gopalakrishnan R, Deb S, Vishwanathan V, Pattanayak A (2023) India: the e-waste (management) rules 2022. https://www.mondaq.com/india/waste-management/1374872/the-e-waste-management-rules-2022#:~:text=The%202022%20Rules%20also%20introduce,of%20electrical%20and%20electronic%20items.&text=Each%20refurbishing%20certificate%20will%20correspond%20to%20a%20certain%20quantity%20of%20waste

Hassan Z, Dhusia DK (2022) Hazardous e-waste recycling practices affecting informal recycler health in India: a case study. Int J Circ Econ Waste Manag 2(1). https://doi.org/10.4018/IJCEWM.302205

Hawari M, Hassan MH (2008) E-waste: ethical implications for education and research. IIUM Eng J 9(2). https://doi.org/10.31436/iiumej.v9i2.97

Jain M, Kumar D, Chaudhary J, Kumar S, Sharma S, Verma AS (2023) Review on e-waste management and its impact on the environment and society. Waste Manag Bull 1(3):34–44. https://doi.org/10.1016/j.wmb.2023.06.004

Kumar A, Gaur D, Liu Y, Sharma D (2022) Sustainable waste electrical and electronic equipment management guide in emerging economies context: a structural model approach. J Clean Prod 336:130391. https://doi.org/10.1016/j.jclepro.2022.130391

Leonard JV (2020) Electronic waste: a LITS success story in recycling, sustainability, and security. https://scholarblogs.emory.edu/lits/2020/05/22/electronic-waste-a-lits-success-story-in-recycling-sustainability-and-security/

Liu K, Tan Q, Yu J, Wang M (2023) A global perspective on e-waste recycling. Circ Econ 2(1):100028. https://doi.org/10.1016/j.cec.2023.100028

Luhar S, Luhar I (2019) Potential application of E-wastes in construction industry: a review. Constr Build Mater 203:222–240. https://doi.org/10.1016/j.conbuildmat.2019.01.080

Mades (2021) Top advanced materials for electronics components manufacturing. https://mades.es/blog/top-advanced-materials-for-electronics-components-manufacturing/#:~:text=Silicon%20is%20perhaps%20the%20most,bismuth%2C%20cobalt%2C%20and%20talc

Madkhali H, Duraib S, Nguyen L, Prasad M, Sharma M, Joshi S (2023) A comprehensive review on e-waste management strategies and prediction methods: a Saudi Arabia perspective. Knowledge 3(2):163–179. https://doi.org/10.3390/knowledge3020012

Mmereki D, Li B, Baldwin A, Hong L (2016) The generation, composition, collection, treatment and disposal system, and impact of e-waste. In: Mihai FC (ed) E-waste in transition—from pollution to resource. https://doi.org/10.5772/61332

Mmereki D, Li B, Baldwin A, Hong L (2016) The generation, composition, collection, treatment and disposal system, and impact of e-waste; e-waste in transition—from pollution to resource edited by Florin-Constantin Mihai. ISBN: 978-953-51-2500-6, https://doi.org/10.5772/61332

Mor RS, Sangwan KS, Singh S, Singh A, Kharub M (2021) E-waste Management for environmental sustainability: an exploratory study. Proc CIRP 98:193–198. https://doi.org/10.1016/j.procir.2021.01.029

Ozzie Mascarenhas SJ, Doris D'Souza AC, S George SJ (2016) Ethics of e-waste management: an input–process–output analytic approach. Manag Labour Stud 41(1):Sage J. https://doi.org/10.1177/0258042X16649467

Porwal P, Chatterjee S (2018) Extended producer responsibility on e-waste management in India: challenges and prospects. Int J Sci Res 8(5):1026–1034

Priyan MV, Annadurai R, Onyelowe KC, Alaneme GU, Giri NC (2023) Recycling and sustainable applications of waste printed circuit board in concrete application and validation using response surface methodology. Sci Rep 13

Rajesh R, Kanakadhurga D, Prabaharan N (2022) Electronic waste: a critical assessment on the unimaginable growing pollutant, legislations and environmental impacts. Environ Challenges 7:100507. https://doi.org/10.1016/j.envc.2022.100507

Rautela R, Arya S, Vishwakarma S, Lee J, Kim KH, Kumar S (2021) E-waste management and its effects on the environment and human health. Sci Total Environ 773:145623. https://doi.org/10.1016/j.scitotenv.2021.145623

Ray PC, Yu H, Fu PR (2009) Toxicity and environmental risks of nanomaterials: challenges and future needs. J Environ Sci Health Part C 27(1):1–35. https://doi.org/10.1080/10590500802708267

Resmi NG, Fasila KA (2017) E-waste management and refurbishment prediction (EMARP) model for refurbishment industries. J Environ Manage 301:303–308. https://doi.org/10.1016/j.jenvman.2017.06.065

Seif R, Salem FZ, Allam NK (2023) E-waste recycled materials as efficient catalysts for renewable energy technologies and better environmental sustainability. Environ Dev Sustain 26:5473–5508

Shaaban M, Wang XL, Song P, Hou X, Wei Z (2024) Microplastic pollution and e-waste: unraveling sources, mechanisms, and impacts across environments. Curr Opin Green Sustain Chem 46:100891. https://doi.org/10.1016/j.cogsc.2024.100891

Shantaram M, Shafy S, Fatimah R (2014) A concern of E-waste in the hospital setting and its ways of disposal. Int J Appl Biol Pharm Technol 5(2):5–7

Thukral S, Singh M (2023) An exploratory study on producer's perspective towards E-waste management: a case of emerging markets. Clean Waste Syst 5:100090. https://doi.org/10.1016/j.clwas.2023.100090

Tong C (2022) Advanced materials for printed flexible electronics. In: Springer series in materials science. https://doi.org/10.1007/978-3-030-79804-8

UN Environment Programme (2021) New report: the impact of the COVID-19 pandemic on e-waste in the first three quarters of 2020. https://www.unep.org/ietc/news/story/new-report-impact-covid-19-pandemic-e-waste-first-three-quarters-2020#:~:text=On%20the%20other%20hand%2C%20consumption,caused%20by%20high%2Dincome%20countries

Vasudeva C, Sohail S (2013) Life of a licenced e-waste collector. Down Earth. https://www.downtoearth.org.in/news/life-of-a-licenced-ewaste-collector-42033

Verma AK, Prakash S (2020) E-wastes and their impact on environment and public health. Int J Appl Res Part C 6(9)

Viswanath B, Kim S (2016) Influence of nanotoxicity on human health and environment: the alternative strategies. Rev Environ Contam Toxicol 242:61–104. https://doi.org/10.1007/398_2016_12

ZeWaLS Academy (2023) Making money from e-waste: a success story of Evariste Aohoui. Linkedin: https://www.linkedin.com/pulse/making-money-from-e-waste-success-story-evariste-aohoui?trk=article-ssr-frontend-pulse_more-articles_related-content-card

Zhang WH, Wu YX, Simonnot MO (2012) Soil contamination due to e-waste disposal and recycling activities: a review with special focus on China. Pedosphere 22(4):434–455. https://doi.org/10.1016/S1002-0160(12)60030-7

Streamlining E-waste Management for a Greener Tomorrow

Syed Suffia Iqbal, Pranav Prashant Dagwar, and Deblina Dutta

Abstract

The advancement in technology and growing population is responsible for the rise in the volume of electronic waste (E-waste) that has led to significant environmental issues. It has been evaluated that the production of E-waste will rise 74.7 billion kgs by 2030. This chapter aims to clarify the governance of E-waste, its impacts on health and the environment, and the associated challenges. It also discusses various recycling technologies and processes while exploring their relevance across diverse geographical regions, promoting the implementation of sustainability. It also focuses on informal and formal sectors. Enhancements to global waste electrical and electronic equipment (WEEE) management are proposed, emphasizing the integration of circular economy principles into the manufacturing and production of electrical and electronic equipment (EEE) as well as the overall management of WEEE. This includes embracing concepts such as urban mining. E-waste can provide a good business opportunity which are available through metal recovery and the better improvement in the waste management sector. Developing an environmentally friendly regulatory framework for recycling is crucial in establishing an inventory of end-of-life electronic products. The environmental agencies need to innovate and develop environmentally sound strategies for resource recovery from E-waste which can lead to the business advantage converting the concept of linear economy to circular economy. The chapter concludes with recommendations for policymakers and researchers to promote sustainable E-waste management practices.

Keywords

Electronic waste · Sustainability · Metal recovery · Management strategies · Circular economy · Metal recycling · Environmental impact · Informal sector

1 Introduction

Electronic waste, commonly referred to as E-waste, comprises obsolete electrical and electronic apparatus that are deemed non-operational. This category of waste is also known as WEEE. E-waste is recognized as the most significant and rapidly escalating challenge within the solid waste management domain, and its proliferation poses severe threats to environmental integrity and public health (Shahabuddin et al. 2023). The world generated 62 billion kgs of electronic waste in 2022, which amounts to an average of 7.8 kgs per person. These projections also suggest that by 2030, this figure could escalate to 74.7 billion kgs (Baldé et al. 2024; Forti et al. 2018). However, only 22.3% of this E-waste, equivalent to 13.8 billion kgs, has been officially recorded, including its collection and proper recycling. In 2022, the global E-waste generated consisted of 17 billion kgs of plastics, 31 billion kgs of metals, and 14 billion kgs of other materials, such as minerals, glass, and composite materials, have a monetary value of USD 91 billion. Copper, gold, and iron represent some of the most economically significant secondary raw materials, with respective valuations of USD 19 billion, USD 15 billion, and USD 16 billion. The efficient recovery of these metals is achievable through established electronic waste management technologies, which indicates that improving collection rates could markedly enhance the existing value recovery rates. In general, developed nations experience annual growth rates ranging from 1 to 5% based on weight,

S. S. Iqbal · P. P. Dagwar · D. Dutta (✉)
Department of Environmental Science and Engineering, SRM University-AP, Amaravati, Andhra Pradesh, India
e-mail: deblina.d@srmap.edu.in

A. Mandpe et al. (eds.), *Sustainable Waste Management*, Advances in Science, Technology & Innovation,
https://doi.org/10.1007/978-3-031-85685-3_3

while developing countries typically exhibit growth rates between 10 and 25% (Baldé et al. 2017).

Recycling of E-waste presents a challenge due to its diverse physicochemical properties, due to the metals it contains, which are hazardous halogen metals. Improper treatment of substances containing hazardous wastes may threaten human health and contribute to environmental degradation. Proper waste management and safe disposal has become a global concern as it mitigates the hazardous wastes that impact human and environmental health. Traditional recycling methods like acid stripping of metals, incineration, and acid baths generate substances such as dioxins, furans, and heavy metals, which potentially cause environmental degradation. Although resources recovery from E-waste offers job and commercial opportunities, its effective management can be more difficult by poor infrastructure and a shortage of efficient technologies. The EEE sector has experienced a surge as these devices are increasingly recognized as essential components of daily life (Forti et al. 2018). The contemporary world is equipped with a diverse array of basic and sophisticated tools, appliances, and equipment which are being utilized not only in homes and businesses but also in various sectors like transportation, energy production, defence, manufacturing, and others. This category of equipment encompasses a variety of items, including household and kitchen appliances, information, and communication technology (ICT) devices, laptops, mobile phones, toys, music players, and more. Additionally, traditional items such as clothing and furniture are found to have links to EEE (Kumar and Nayar 2021). The increased dependence on electronic devices for remote work, education, and communication amid COVID-19 has resulted in a potential rise in the production of E-waste. Amid pandemic many individuals have been observed purchasing electronic devices such as laptops, smartphones, and personal computers to support their work from home setup. However, the disposal of these gadgets post utilization could pose long-term risks to both people and the environment if not handled appropriately. The COVID-19 pandemic has prompted an increased reliance on equipment that enables remote activities, leading to regional variations in the consumption of EEE (Ribeiro et al. 2022). Electronic devices have become popular due to their ability to meet the rapidly changing needs of the global population, including both time periods of normal and pandemic, offering features and capabilities necessary for remote operations.

This investigation delineates the challenges inherent in E-waste management, including insufficient infrastructure, a lack of public awareness, and issues related to regulatory enforcement. It also investigates the repercussions on environmental and human health, encompassing soil and water contamination, air pollution, and exposure to hazardous materials, which present considerable risks to labourers and adjacent communities (Kumar and Nayar 2021). Moreover, the study underscores the critical importance of sustainable E-waste management methodologies which are designed to mitigate environmental degradation, enhance resource recovery, and safeguard the health and safety of individuals engaged in E-waste processing. Inadequate management of E-waste results in ecological harm, encompassing soil deterioration, aquatic contamination, and the release of hazardous gaseous emissions, which harm living organisms. A substantial quantity of metals is essential for manufacturing electronic items, which includes copper, iron, gold, manganese, aluminium, silver, zinc, chromium, and various rare earth elements, regardless of the current requirements, sustainability can only be realized by restricting the inconsiderable way of using (Suárez-Eiroa et al. 2019). Only $10.0 billion in the $57.0 billion worth of E-waste produced globally is recovered and recycled sustainably, according to estimations (Shahabuddin et al. 2023). Within the Asian region, India is the world's second-largest generator of E-waste, generating an annual volume of 400,000 tons. Approximately 60% of these contribution stems from just 65 cities in different states which includes Maharashtra holding the top position as the leading state in E-waste production, followed by Tamil Nadu, West Bengal, Andhra Pradesh, and Uttar Pradesh. Furthermore, the aggregate E-waste production from Delhi, Gujarat, Karnataka, Madhya Pradesh, and Punjab accounts for 70% of the total national output (Vats and Singh 2014). As per United Nations Environment Program (UNEP), In Indian subcontinent there has been a significant rise in mobile phones disposal, reaching eighteen times the amount recorded in 2007. Asia generates 6.4 kg of E-waste per capita, with 0.76 kg formally collected and a recycling rate of 11.8% (Baldé et al. 2024). In India, only 17% of E-waste is formally collected and recycled, while the remaining 83% is either disposed of in landfills or handled by informal recyclers, commonly referred to as "raddiwalas or kabadivalas". This uptrend has anticipated to accelerate the overall rate of E-waste generation, and significant quantity of E-waste is shipped to nations like China, India, Ghana, and Kenya. Where processing is more profitable due to better labour conditions and laxer environmental regulations. There are plenty of metals in E-waste, including gold, zinc, copper, nickel, aluminium, palladium, and rare earth metals, which can be reused and can provide new economic prospects (Arya et al. 2021). The metallic components existing in

E-waste can be used as resources by using the correct scientific techniques, and it will assist in addressing the global resource scarcity of metals (Zeng et al. 2017). E-waste has certain parts that can be repurposed, such as various capacitors and circuit boards, while the plastics present in the bodies of electronic devices can be effectively utilized as a valuable resource for pyrolysis, enabling efficient recycling for both energy and materials (Debnath et al. 2016). The component of E-waste contains 28% non-ferrous materials, 38% ferrous materials, 4% glass, 19% plastics, and 11% rubber, wood, and other materials (Shittu et al. 2021; Manomaivibool 2009). About 800,000 tons of E-waste are produced per annum in India, and it is predicted to grow about 10–15% (Meenakshi and Harini 2012). Considering the potential for recycling E-waste, it offers a revenue stream for industries and creates opportunities for new employment. Metals employing scientific methods for their recycling has the potential to address global shortages of metal resources (Zeng et al. 2017).

This book chapter explores the significant commercial importance of E-waste and investigates the difficulties, barriers, business prospects, and advantages associated with it. Consequently, the creation of an efficient E-waste management facility necessitates the formulation of country-specific standards and legislation. This chapter also offers a brief overview of E-waste recycling methods covering aspects such as collection, pre-treatment, challenges, and prospects. This study aimed to assess the status of E-waste management across various states in the country. It focused on evaluating E-waste collection, dismantling, and recycling practices within both the formal and informal sectors in India. Data on these practices were gathered from existing literature, providing insights into the operations of both sectors. Moreover, the study emphasizes the critical importance of sustainable E-waste management methodologies, which are designed to mitigate environmental degradation, enhance resource recovery, and safeguard the health and safety of individuals engaged in E-waste processing.

2 Bibliometric Analysis

It is a method used for the qualitative analysis of literature in a specific field. This tool is employed as a standard for the methodological examination evaluated across diverse domains of inquiry within the field of science and engineering (Suffia and Dutta 2023). This is employed as a standard for the methodological examination evaluated across diverse domains. The data was obtained from the Scopus database on 19 November 2024. The data was retrieved from the Scopus database by using different keywords; those keywords were more relevant for this study "E-waste, Metal recovery, Circular economy", and "E-waste, Management strategies, Environmental impact". Using the keywords E-waste, Metal Recovery, and Circular Economy, 159 documents were identified between 2006 and 2024 found over 18-year period shown in Fig. 1a. Similarly, using E-waste, Management Strategies, and Environmental Impact as search terms, 175 documents were retrieved spanning the past 18 years shown in Fig. 1b.

Based on the bibliometric analysis, this study provides a comprehensive overview of publications, including their authors, and keywords related to E-waste. This bibliometric analysis provides valuable insights for identifying appropriate methodologies and guiding future research directions.

3 Human Health Impacts and Environmental Hazardous Due to E-waste Disposal

The electric and electronic equipment sector is experiencing growth and expansion worldwide due to the advancement in technology, industrialization, and urbanization from rural areas. Technological advancements have reduced the life span of electronic equipment that leads to the production of waste. E-waste sites are swelling at a concerning rate in developing nations, presenting serious risks to the surrounding environment and the well-being of their inhabitants. The contaminants have the potential to readily penetrate and infiltrate groundwater and surrounding ecosystems, leading to detrimental environmental consequences (Brindhadevi et al. 2023). Generally, there are three primary forms of E-waste that originate from EEE are as follows: home appliances, consumer equipment's machinery and information technology (IT), and telecom (Ádám et al. 2021). Although E-waste contributes over 70% of the overall toxicity despite making up just 2–5% of the total solid waste volume. Substances like chromium (Cr), lead (Pb), beryllium (Be), palladium (Pd), gallium (Ga), among others, pollute the water, air, and soil, thus contaminating the food chain (Ankit et al. 2021).

Inadequate handling of E-waste results in significant environmental challenges, including soil deterioration, contamination, and the release of hazardous gases from the burning of electronic waste in open environments. These pollutants directly affect the health of living organisms as shown in Fig. 2. E-waste leachates have been shown to affect soil microorganisms, such as fungi and bacteria, as well as other organisms living in soil, aquatic environments, plants, and humans (Ankit et al. 2021).

For instance, mercury is a vital component in batteries, displays, and switches of electronic equipment. In soil, microbes have an ability to convert mercury into methyl mercury, a neurotoxic substance that can build up in the fatty tissues of humans, posing risks especially to young

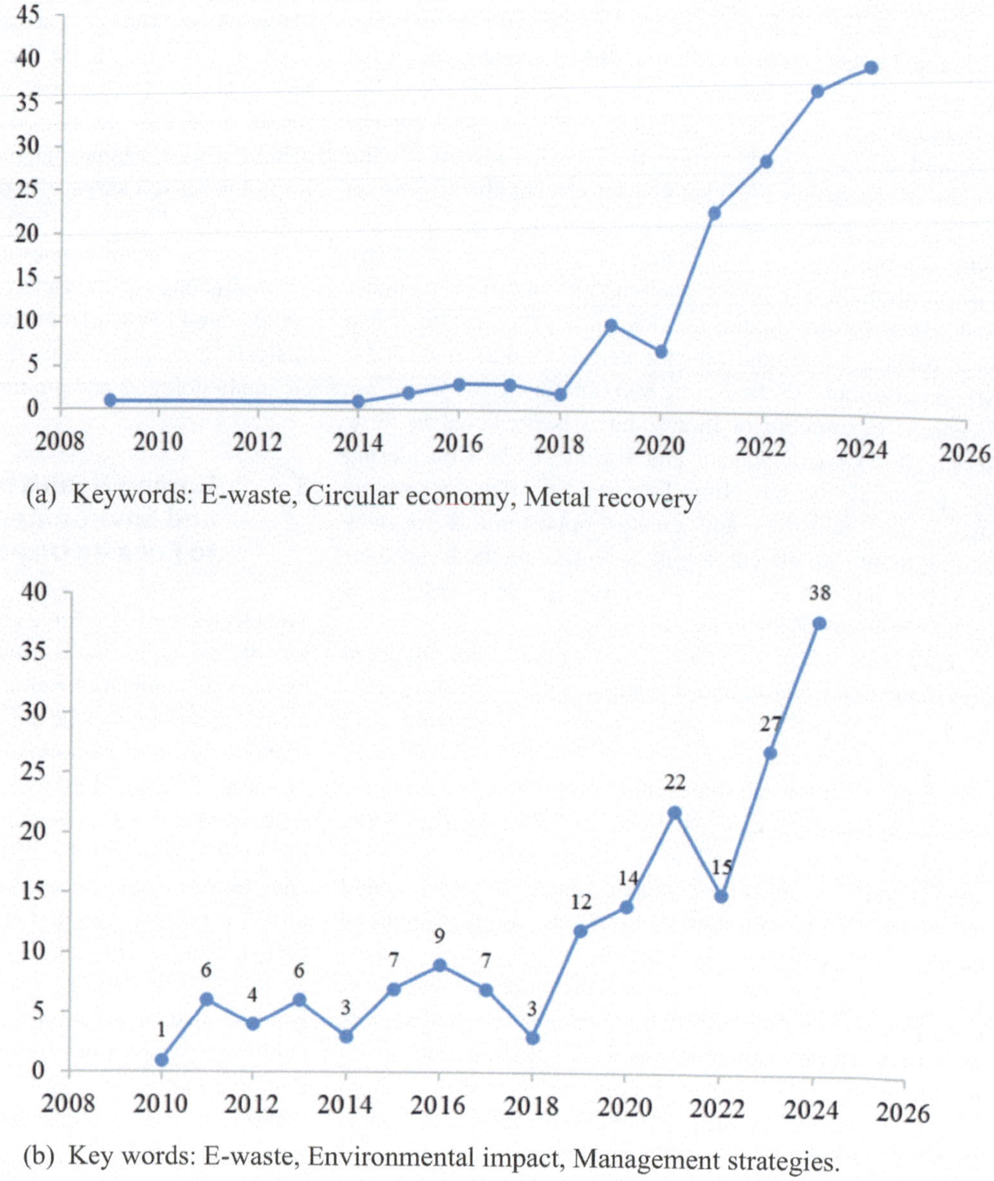

Fig. 1 Graphical representation using different key words **a** Keywords: E-waste, Circular economy, Metal recovery **b** Key words: E-waste, Environmental impact, Management strategies

children and foetuses. This compound is associated with various issues related to health including skin irritation, neurological disorders, memory loss, and cognitive impairments. Similarly, other heavy metals such as chromium (Cr), arsenic (As), copper (Cu), and nickel (Ni) are equally detrimental to humans and other organisms, either through direct or indirect effects (Sharma et al. 2020). Currently, various studies are examining different aspects of E-waste across numerous provinces worldwide, covering its entire life cycle from production to disposal (Debnath et al. 2016). Potential contaminants found in E-waste can have several negative impacts on both human health and the environment. Some of these are shown in Table 1.

4 Current Challenges in E-waste Management

The realm of E-waste and its management offers a variety of challenges and opportunities. A substantial hurdle is the absence of E-waste regulations, especially common in developing nations. As of 2014, only 61 countries, accounting for 44% of the global population, had implemented E-waste legislation, with developed nations leading the way in this regard. Managing large quantities of E-waste presents a major challenge. In Indian cities and towns due to factors such as the substantial volume of E-waste produced, illegal disposal practices, inadequate infrastructure,

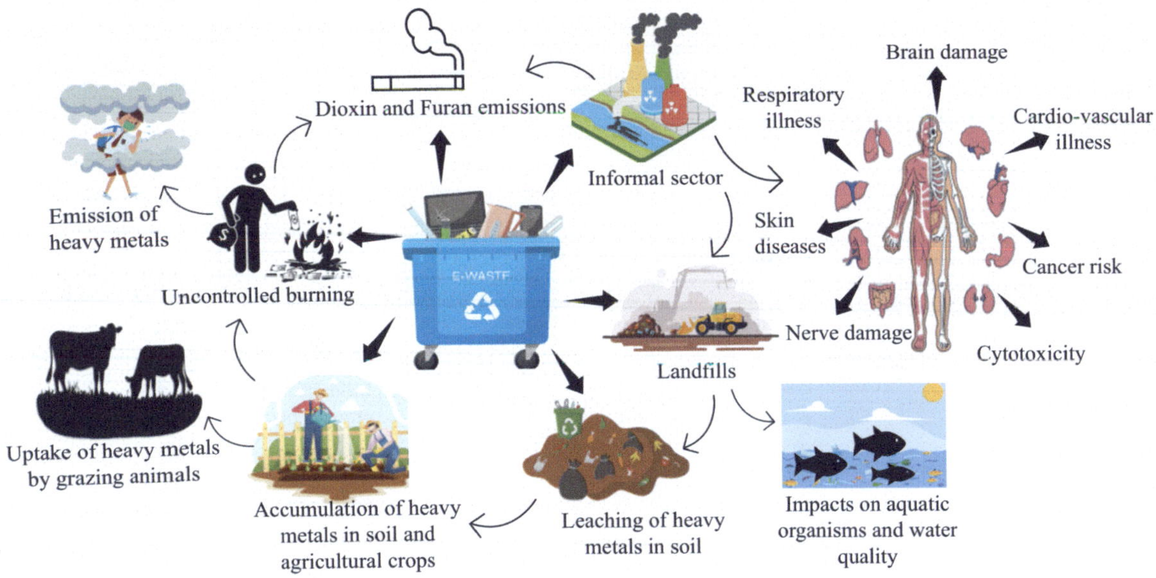

Fig. 2 Routes and pathways of E-waste that affect the environment and human health

ineffective legislation, administrative hesitance in enforcement, and the resulting implications (Dutta et al. 2023). The approach to E-waste collection varies across regions and is mainly influenced by factors such as local laws and existing E-waste management systems, public awareness, and economic circumstances. There are frequently no official records available on E-waste figures in several developing nations like North Africa and Southeast Asia, where E-waste regulation is either non-existent or severely constrained. Governments in these provinces depend on independent contractors and self-employed individuals to gather and process E-waste. These people engage in door-to-door collection efforts, personally sorting the collected items, and ultimately selling them to urban vendors for refurbishment or recycling purposes. By 2019, E-waste legislation was enacted in 78 countries, extending its coverage to 71% of the global population. In the Asian subcontinent, the populated countries like India and China have implemented E-waste laws, while others with large populations exceeding 160 million, such as Pakistan and Bangladesh, currently lack E-waste regulations (Forti et al. 2018).

Recently, the Government of Pakistan has instituted a ban on the import of E-waste from other nations. Nevertheless, uncertainties remain regarding the efficacy of this regulatory measure. Concurrently, Bangladesh, which holds the 8th position globally in terms of population, has initiated considerations for analogous legislation (Shahabuddin et al. 2023). The Department of Environment (DoE) in Bangladesh has established regulatory frameworks, as articulated in the Hazardous Waste Management Rules of 2019, aimed at restricting the utilization of fifteen hazardous materials within designated classifications of EEE. The predominant challenge in the administration of E-waste resides in the effective extraction of valuable resources, encompassing precious metals and rare earth elements. While ensuring that hazardous materials are processed in a safe manner. Hazardous materials in E-waste management incorporate the handling of dangerous chemicals such as polychlorinated biphenyls (PCBs), CFC fluids, and mercury. Additionally, there are concerns regarding the manual handling of significant objects, compliance with electrical safety regulations, and the inherent risks associated with fire and explosive occurrences. Additional hurdles include the limited availability of eco-friendly chemical solutions for E-waste management, insufficient infrastructure facilities. Moreover, the thermodynamic limitations associated with separating complex material mixtures, leading to cost-ineffective recovery processes. There is also a need for financial support and policy assistance, particularly in developing nations, and challenges arising from varying legislative frameworks across different countries. There are various challenges in the management of WEEE; some are highlighted in Fig. 3. Addressing these challenges requires a thorough approach involving cooperation among governments, manufacturers, recyclers, and consumers. Together they need to develop and put sustainable E-waste management strategies into action. This may include policy interventions, investment in infrastructure, public education campaigns, and technological innovations to expand recycling processes and lessen the generation of E-waste.

Developing effective approaches for the handling of WEEE on a global scale is proving highly challenging.

Table 1 Negative impacts of potential contaminants on human health and environment found in electronic waste (Ghulam and Abushammala 2023)

Toxic substances	Existence in E-waste	Health effects and environment
Poly vinyl chloride (PVC)	Cable insulation, monitors, and keyboards	Birth defects, endometriosis, diabetes, and immune system abnormalities
Poly chlorinated biphenyl's (PCBs)	Transformer, capacitors in plastics, and adhesives	Cancer, dysfunction of the immune system, and impairment of the endocrine system
Arsenic (As)	Solar cells, semiconductors, and LED	Causes harm to the digestive system, skin issues, and increases the risk of lung cancer
Antimony (Sb)	Plastics and fire retardants	Inflammatory conditions of the lungs, persistent bronchitis, and long-term emphysema
Beryllium (Be)	Circuit boards, motherboard, and connectors	Lung cancer, Carcinogenic, and dust causes chronic beryllium disease
Chromium (Cr)	Protective coating against corrosion, data storage tapes, floppy disks, colouring agents, and pigments	Induces irritation to the eyes, skin, and mucous membranes, leading to bronchitis, as well as potential damage to the kidneys and liver
Silver (Ag)	Switches, batteries, and capacitors	Digestive discomfort, renal impairment, and brain damage
Nickel (Ni)	Rechargeable Ni–Cd batteries and electron guns used in cathode ray tubes (CRTs)	Skin allergies, respiratory infections, behavioural abnormalities, and cancer
Iron (Fe)	Tvs, laptops, mobile phones, computers, and refrigerators	Damages liver
Lithium (Li)	Lithium ion batteries	Impact neurologic system and gastrointestinal
Zinc (Zn)	CRT screens	Cytotoxicity, trauma, and ischemia
Selenium (Se)	Older models of photocopy machines	High concentration causes selenosis
Lead (Pb)	CRTs, cables and wires, and batteries	Asthma and weakened immune function, diminished cognitive abilities in children, and impacts on learning capabilities and brain damage
Mercury (Hg)	Batteries, lamps, switches, and thermostat	Damages the kidneys, brain, and foetuses
Chlorofluorocarbons	Cooling units and insulation foam	These compounds can damage the ozone layer, potentially resulting in an increased risk of skin cancer
Brominated flame retardants (BFRs)	BFRs are utilized to minimize the risk of fire in plastic casings, printed circuit boards, and insulation for cables	It causes hormonal disorders when burned in open atmosphere
Dioxins, including polychlorinated dibenzodioxins (PCDDs) and polychlorinated dibenzofurans (PCDFs)	Released as a by-product of combustion	Inhalation, ingestion, and dermal contact
Hexavalent chromium (Cr) VI, barium (Ba)	Soldering materials, coatings, and paints used in circuit boards, as well as in cathode ray tubes (CRTs)	DNA damage, respiratory inflammation, and muscular fatigue

There are numerous approaches and scenarios for WEEE management worldwide, each varying from region to region. Nonetheless, certain standard management scenarios have been recognized and categorized.

Scenario 1: Involves the formal documentation and collection of WEEE, adhering to the legal requirements outlined in existing WEEE legislation. Typically, WEEE collection in this context is conducted through municipal collection centres, EEE producers and vendors, or scheduled pick-up services.

Scenario 2: Contains the combined WEEE disposal along with mixed household waste. Consumers discard WEEE alongside unsorted household waste. The mixed waste is then potentially sent to landfill or incineration, dependent on the prevailing discarding practices.

Scenario 3: Is predominantly observed in developing nations and entails the informal gathering of WEEE from consumers through untrained recyclers. These practices operate without regulation due to the lack of or inadequate enforcement of WEEE management laws. As a result,

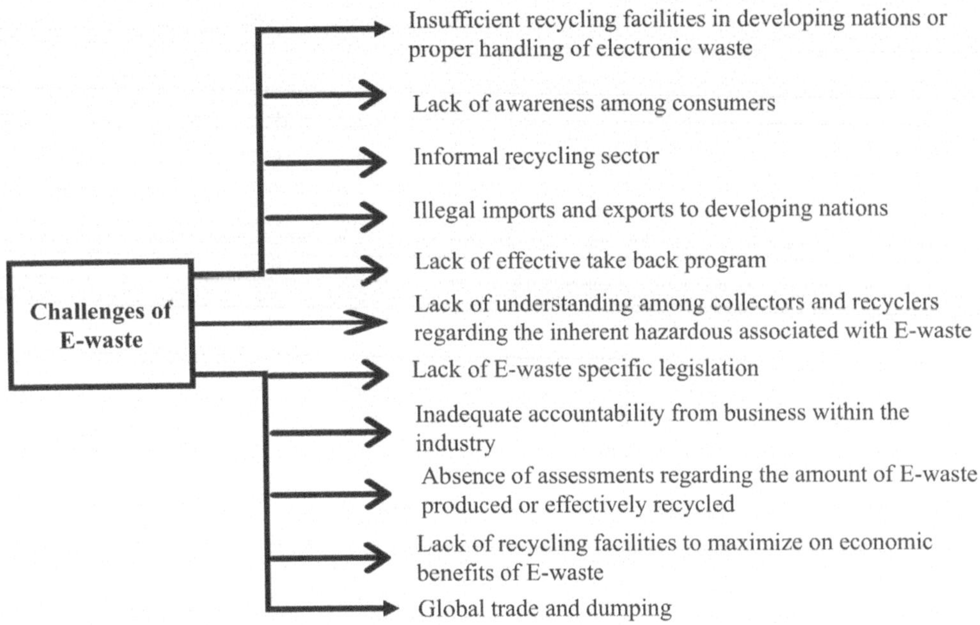

Fig. 3 Global challenges in the management of E-waste

treatment techniques are generally primitive and simplistic. Collectors primarily focus on retrieving metal components from WEEE and commonly employ practices to methods like acid leaching and open burning for metal extraction. This approach, which includes reusing, repairing, and recovering parts from WEEE, is also found in Europe.

Scenario 4: This includes the informal collection of electronic waste (WEEE), often involving independent recyclers and dealers. The collected WEEE might be processed at specialized facilities, refurbished, or sent to developing countries. This informal collection is not officially recorded, making it difficult to track or monitor the amount gathered. The absence of proper documentation may result from a lack of legal mandates or an established WEEE management system. Consequently, the treatment of collected WEEE may not prioritize environmental safety, and there is a risk of illegal exportation.

Managing electronic waste involves recognizing the obstacles to its proper and safe disposal. Many countries are currently updating rules and regulations to tackle local issues caused by negative environmental and health impacts. Measures are being implemented to protect people and biodiversity across various regions. E-waste management is an expanding sector offering significant opportunities for innovation and improving current conditions. This advancement promotes sustainability and encourages a shift towards a circular economy.

5 E-waste Recycling Technologies and Processes

E-waste presents significant challenges due to its complex structure and the wide variety of materials it contains. In 2022, the E-waste produced contained plastics, metals, and other materials such as minerals, glass, and composite materials (Baldé et al. 2024). E-waste generation has significantly increased for the last few years on a worldwide level. When compared to earlier predictions of a 3–4% annual growth rate, the rate of global E-waste generation has grown to about 6% till 2019 (Forti et al. 2018; Alam et al. 2022). The recovery techniques not only reduce waste but also enable the extraction of valuable metals. E-waste, which constitutes a significant portion of mined metallic resources, contains approximately 95% metals. These metals are commonly used in electrical and electronic devices. Furthermore, the metal recovery rates from electronic wastes surpass those from traditional mining operations. These recovered metals from E-waste not only conserve energy but also reduce the need for substantial investments in mining. This approach is particularly advantageous for the retrieval of precious and other valuable metals from E-waste (Arya and Kumar 2023). Due to cost-effective techniques and lack of infrastructure, the informal sector handles E-waste primarily utilizing unsafe and ineffective resource recovery practices (Ikhlayel 2018). Recycling of E-waste is the major problem due to its hazardous nature as harmful concentrated acids are used for the leaching processes (Mmereki et al. 2016). Hence, safety measures are desperately needed for safe, cost-effective, and environmentally friendly recycling and recovery methods to handle electronic waste. The formal recycling of electronic trash is still lacking in many Asian countries, despite different E-waste management legislation has seen only marginal growth, failing to keep pace with the remarkable surge in total electronic waste production (Ghulam and Abushammala 2023; Wei and Liu 2012). The composition of toxic elements found in electronic waste varies depending on the specific EEE. The release of these toxic components and their byproducts depend on the type of method used for processing. E-waste consists of a variety of metals and non-metals, and the typical composition is 30% ceramic, 40% inorganic, and 30% organic. The electronic waste category of inorganic materials consists of base metals (zinc, chromium, cadmium, mercury, nickel, beryllium, and lead), rare earth metals (tantalum, platinum, and gallium) and valuable metals (Au, Ag, Pd, and Pt). Many E-waste recycling techniques target gold due to its economic value even though it is there in trace amounts. Resource recovery from E-waste can create jobs and business opportunities, but management is hindered by inadequate infrastructure and inefficient scientific processes (Lakshmi and Raj 2017). The primary stages in the formal electronic waste handling sector encompass collection, preprocessing, subsequent processing, and final disposal (Ashiq et al. 2019). A brief comparison of informal and formal recycling processing of E-waste recycling sector is shown in Fig. 4.

In 2022, the total production of E-waste globally reached an astonishing 62 billion kgs, which corresponds to an annual per capita rate of approximately 7.8 kgs. However, only 22.3% of this E-waste was properly collected and processed through eco-friendly recycling strategies. Since the year 2010, there has been a notable rise in the amount of E-waste produced worldwide, which commenced at 34 billion kgs and has experienced an average annual increment of 2.3 billion kgs. Concurrently, during this timeframe, the formal collection and recycling of electronic waste have also exhibited a positive trajectory, initiating at 8 billion kgs in 2010 and progressively increasing by an average of 0.5 billion kgs each year to reach 13.8 billion kgs in 2022 (Wang et al. 2022).

5.1 Informal Sector

E-waste disposal pathways vary widely, encompassing informal and formal recycling, dumping, and storage practices across both developed and developing nations. In developed countries, the formal collection sector is typically well-managed, utilizing advanced manual and semi-automated dismantling and separation systems. In contrast, developing countries often rely on informal sectors to manage their E-waste. Unfortunately, in some cases, proper precautions are not taken during E-waste collection in developing countries. In India, the informal E-waste sector operates extensively across major urban centres and various locations within them. Individuals, frequently from disadvantaged backgrounds and without formal education,

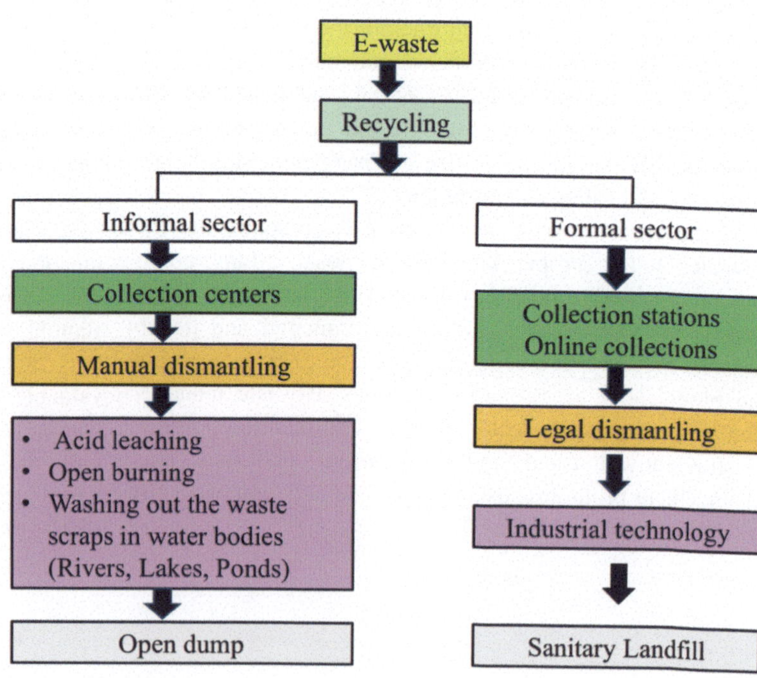

Fig. 4 Comparison of informal and formal recycling processing of E-wastes

including women and kids, participate in recycling and recovery activities, which pose health risks to themselves and others. In every state, numerous sites gather E-waste, where it is disassembled and then transported to larger informal sector hubs for further processing (Dutta and Goel 2021).

Moradabad

Moradabad city, situated alongside the Ramganga River in Uttar Pradesh, serves as a significant hub for E-waste recycling in India. It is projected that around half of the country's generated waste printed circuit boards (PCBs) are processed in Moradabad. With over 50,000 individuals, including children, involved in these activities, the area faces numerous occupational health risks, and the environment is adversely affected by pollution.

Dismantling and recycling activities predominantly occur within the basements or rooftops of residential homes owned by individuals involved in this trade. The labour involves the manual extraction of copper, silver, and gold from dismantled gadgets and motherboards through hammering. Motherboards containing heavy metals are often burned in open piles, releasing highly toxic fumes. Circuit boards are heated in shallow pans or over open flames to retrieve copper, exposing workers to harmful lead fumes. Additionally, gold extraction from circuit board chips involves the use of acid baths, emitting further toxic gases into the atmosphere. There are serious risks associated with these procedures for public health and the environment.

Sangrampur

Sangrampur, a small town in West Bengal, India's South 24 Parganas area is renowned for its substantial informal E-waste treatment facilities. In this town, a significant number of individuals, including men, women, and children, are involved in E-waste recycling activities without the use of protective gear or attire. There will not be easy access to information about their work, and taking pictures is absolutely forbidden. A limited number of soil samples were obtained, revealing the presence of various metals such as chromium, cadmium, lead, manganese, copper, cobalt, nickel, and iron at the sampling sites (Alam et al. 2022). Furthermore, the soil samples contained substantial levels of polyaromatic hydrocarbons such as naphthalene, phenanthrene, anthracene, and fluoranthene.

Seelampur

The largest E-waste dismantling market in India is at Seelampur, which is on the outskirts of New Delhi, where approximately 50,000 people earn a livelihood by extracting metals. Seelampur, situated in East Delhi, holds the distinction of being India's largest hub for dismantling of E-waste. Over the years, it has served as the workplace for numerous informal labourers who dedicate approximately 10–12 h daily to extract valuable from discarded electronic devices (Dey et al. 2023).

In Seelampur, all activities related to E-waste are entirely illegal. The operations involve the employment of children, burning of electronic waste, and disposal in open areas. The workforce lacks skills and proper infrastructure, where both children and adults are engaged in the dismantling process, metal extraction, and utilization of methods such as acid burning and open incineration. These practices not only endanger the health of those involved but also pose significant environmental risks. It is common to observe the exploitation of very young children in these activities. For instance, tasks requiring delicate handling, like extracting small elements from laptop motherboards, often enlist children due to their smaller fingers. Moreover, extracting metals from items like mortars, compressors, and wires is done without access to technology or machinery, leading to the burning of wires. This burning process releases toxic gases due to the PVC coating found in electric wires, posing severe health risks to those inhaling them in the vicinity (Hussain et al. 2020). Where both children and adults are engaged in the dismantling process, metal extraction, and utilization of methods such as acid burning and open incineration. These practices not only endanger the health of those involved but also pose significant environmental risks.

This can result in dangerous backyard practices such as open-air incineration, cyanide leaching, and basic smelting being employed to recover the valuable metals having high economic value and disposing the leftover waste alongside regular municipal solid waste in open landfill sites. Heavy metals like lead (Pb) and cadmium (Cd) have harmful effects on human health, leading to permanent damage to the nervous system and resulting in brain injury. Additionally, Cd is identified as a potential carcinogen due to its ability to alter human cell DNA. Improper landfilling of E-waste through informal treatment processes further disrupts the fertility of soil and groundwater.

The extraction of metals by unscientific and unethical means often lacking adequate safety measures or without taking environment and human health into consideration is the informal method for the recovery of metals. Such recovery approach of metals leads to environmental degradation and human health hazard. For the recovery of copper by informal way open combustion is preferred over all other methods that leads to the generation of hazardous emissions like furans and dioxins (Song and Li 2015). The presence of mercury in E-waste affects the immune system and significantly harms the liver and nervous system. Inappropriate disposal of E-waste has a significant detrimental effect on soil fertility, because landfilling the heavy metals might leach soil and affect the soil quality and degrade the groundwater quality. After preprocessing valuable metals present are recovered from E-wastes by hydrometallurgy, pyrometallurgy, and biohydrometallurgy. The unregulated recycling activities for electronic waste within developing nations such as Bangladesh, India, Pakistan, Nigeria, Brazil, various areas of China, and other similar regions play a significant role in their domestic economy (Julander et al. 2014).

The informal sectors have expanded because of the immediate availability of resources and expanded opportunities for profiting from E-waste. The informal E-waste recovery process results in degradation of the surrounding environment and natural resources, posing adverse effects on animal health. Typically, valuable metals like copper (Cu), mercury (Hg), aluminium (Al), lead (Pb), cadmium (Cd), and platinum (Pt) are extracted through open burning of E-waste, leading to the release of significant amounts of hydrocarbons and other toxic compounds (Song and Li 2015). Similarly, precious metals are reclaimed using highly hazardous acids like hydrochloric acid (HCl), nitric acid (HNO_3), sulphuric acid (H_2SO_4), and cyanide. The toxic gases and fumes are emitted into the environment without adequate treatment, resulting in harm to the atmosphere and posing risks to human health. The informal recyclers, known as waste pickers, scavengers, or scrap dealers, are active across the globe in various locations, especially in developing countries where formal E-waste management infrastructures are deficient or insufficient.

5.2 Formal Sector

Every year, the world generates around 20–50 million tons of E-waste, and this amount is projected to grow by 3–5% annually. It is expected that only around 10% of this global E-waste is processed in facilities specifically designed to minimize the exposure to hazardous substances. These facilities incorporate both technical measures and worker safety (Fujimori et al. 2012). The level of exposure is expected to differ based on the location where the work is carried out. In Europe and North America, workers typically engage in recycling activities within facilities specially constructed for this purpose, ensuring adequate ventilation, and safeguarding the health of workers. This is commonly referred to as formal recycling (Sengupta et al. 2022). The prevailing notion is that formal recycling, in contrast to informal recycling, is expected to be more beneficial for both the workers and the environment. The subsequent stage within the formal recycling procedure involves the "comprehensive retrieval" of the components/elements from LIBs. This step may consist of multiple sub-steps, and, crucially, it must be resilient enough to handle variations in feedstock composition, source, type, and other factors. Formal recycling becomes highly reliant on fixed expenses such as setting up facilities, building infrastructure, and acquiring machinery. However, this substantial investment results in significantly improved efficiency and more attractive profit margins, especially considering the prevailing market conditions (Sengupta et al. 2022).

In 22 states of India, there are 569 authorized dismantling and recycling sites as shown in Table 2. Among them, Maharashtra has the highest number of dismantling sites, with 140 sites having a combined capacity of approximately 118,031.5 metric tons per annum. Uttar Pradesh follows with 121 dismantling sites, boasting a capacity of 624,219.47 metric tons. Karnataka is home to 72 dismantling sites, with a capacity of about 126,015.48 metric tons. The total capacity across all 22 states stands at 1,790,348.27 metric tons per annum, distributed among the 569 sites (Baldé et al. 2024).

On the other hand, the informal sector's function is akin to small-scale cottage industries. For the formal sector, the beginning of the production process is based on sound, scientifically validated procedures that are the result of careful research and frequently leasing patents. The recovering of metals from E-waste holds substantial ecological and economic significance. Pyrometallurgy, hydrometallurgy, and biohydrometallurgy are important metallurgical processes applied for metal recovery (Zeng et al. 2022). Pyrometallurgy and hydrometallurgy have traditionally been the two methods used by both the formal and informal sectors to extract the metals. Later, biohydrometallurgy emerged. The ability of the countries to pay for these methods determines their acceptability; typically, industrialized countries use hydrometallurgy or pyrometallurgy depending on their industrial scalability (Faraji et al. 2022).

Pyrometallurgy involves heating waste samples in an atmosphere of inert gas. During this procedure, organic materials such as rubber, wood, and plastics are broken down at elevated temperatures, producing volatile substances. These substances can be utilized as high-value chemical products like oil or as sources of energy generation. Most widely used pyrometallurgical methods are incineration, smelting, and roasting. In the smelting process, WEEE containing a high concentration of copper is subjected to melting at elevated temperatures, approximately 1250 °C, and subsequently undergoes a conversion step to yield copper blisters of 99.1% purity. After this process, a few valuable metals remain, which can be reclaimed using various techniques like electrowinning and electrorefining However, the emission of secondary pollutants such as furan and dioxin, along with the disposal of crude and unwanted materials in open fields and near water banks, is deemed unsustainable (Pant et al. 2014).

Table 2 List of E-waste formal dismantlers and recyclers and their installed capacities on a state-wise basis in India (CPCB 2023; Thakur and Kumar 2020)

S. no.	State	Number of authorized recycler/dismantler	State-wise capacity metric tons per annum (MTA)	Percentage state-wise capacity (MTA)
1	Andhra Pradesh	10	44,002.8	2.45
2	Assam	1	120	0.0067
3	Chhattisgarh	2	6750	0.37
4	Delhi	6	1989	0.11
5	Gujarat	41	158,604.92	8.85
6	Goa	2	153	0.008
7	Haryana	43	157,187.87	8.77
8	Himachal Pradesh	2	1500	0.08
9	Jammu & Kashmir	3	705	0.039
10	Jharkhand	2	660	0.03
11	Karnataka	72	126,015.48	7.03
12	Kerala	1	1200	0.067
13	Maharashtra	140	118,031.5	6.59
14	Madhya Pradesh	3	13,600	0.759
15	Orissa	7	9050	0.50
16	Punjab	8	10,092	0.56
17	Rajasthan	27	82,007.67	4.58
18	Tamil Nadu	42	130,636	7.29
19	Telangana	23	148,115	8.272
20	Uttar Pradesh	121	624,219.47	34.86
21	Uttarakhand	8	153,068	8.5
22	West Bengal	5	2640	0.14
Total		569	1,790,348.27	100

Hydrometallurgy is a highly effective method for recovering metals from electronic waste, particularly from used lithium-ion batteries. This process utilizes chemical agents such as mineral acids, bases, and aqueous and organic solvents to dissolve the metals present in E-waste. Processes entail the targeted dissolution of a specific component into a liquid phase using a suitable solvent, followed by a separation step to isolate solids from the liquid solution. Hydrometallurgy encompasses chemical leaching techniques, including ligands, acid leaching, and etching methodologies (Chauhan et al. 2018). Hydrometallurgy, established in the industrial sector for many years, has gained broad acceptance among recyclers because of its cost-effective performance. Hydrometallurgical treatments generate substantial wastewater pollution that must be treated before being released into natural water bodies. Due to the constraints of a pyrometallurgical procedure, the adoption of hydrometallurgical techniques becomes necessary. The benefits of hydrometallurgical processes include reduced emissions of harmful gases, lower energy requirements, straightforward implementation in laboratory settings, decreased operational expenses, and improved efficiency, making them an appealing choice (Zhu et al. 2012). Some researchers recommend an initial treatment with acids like nitric and sulfuric acid, combined with an oxidizing agent, to improve the extraction of precious metals while limiting the simultaneous dissolution of copper. Although there is abundant knowledge regarding the toxicity and safe management of cyanide, environmental concerns persist and restrict its application in industrial processes. Hydrometallurgical processes are easily manageable and cost-effective than pyrometallurgical ones, and they also have less impact on the environment (Desmarais et al. 2020). The hydrometallurgical process involves the use of specific lixiviants such as thiourea and sulfuric acid to extract metals. This method creates a stable metal complex, facilitating the simultaneous leaching of metals.

Biohydrometallurgy is a technique by which microorganisms help in dissolving the metals, changing solid materials into a liquid state precedes the separation-based recovery of the desired metal. The transformation of metals from solid forms into a liquid state precedes the extraction of the target metal during the separation process. The reduced energy consumption and limited reliance on chemicals have transformed this procedure into an emerging and hopeful domain (Parajuly et al. 2020). Biohydrometallurgy presents abundant opportunities for both research and business. Moreover, this technology is often described as eco-friendly and environmentally friendly. Even when utilized at low concentrations, this microbial-driven approach proves to be highly effective, environmentally sustainable, economically viable, and energy-efficient. Moreover, it demands minimal industrial infrastructure for the execution of the metal recovery process.

6 E-waste Scenario in Different Regions

Electronic waste across the globe has seen a significant rise in the recent years, as per Shittu et al. (Shittu et al. 2021), the global generation of WEEE or electronic waste in 2019 was found to be approx., 54 million tons, but with constant rise of population across the globe the E-waste generation in 2022 is recorded to be 62 billion kgs on a global scale (Global E-waste Monitor 2024). This has been projected that E-waste production will increase by 82 billion kgs until 2030. This ever-increasing E-waste is proving to be a significant challenge, due to the dire need of different strategies to combat the rising global issue (Shittu et al. 2021). There are different management scenarios across the world for mitigating WEEE generation, and there is a variation in management strategies region by region. According to Jain et al. (Jain et al. 2023) the constant technological advancements and uncertain designs of electronic devices has increased the disposal rate across the globe, despite the availability of different rules and regulations for internal and external administration of E-waste, the improper implementation of these policies is a grave area of concern. With region-wise the production and collection of electronic waste varies significantly where Europe has been observed to be the highest producer of electronic waste (17.6 per capita), pursued by Oceania (16.1), and America (14.1), while region like Asia (6.4) is one of the leading producers followed by Africa (2.5). With the availability of advances systems to facilitate the target of collections, the regions have observed a good proportion of recycling and documentation of the per capita production (Javed and Singh 2024). Figure 5 provides the distribution of region-wise E-waste generation and recycling.

Recycling the electronic waste plays a very critical role as it acts as a source of metals for the different global economies, reducing reliance on the mining. It has been estimated that the patent applications in this field of E-waste management have seen an inflexed rise 148 per million in 2010 to 787 million in 2022. Majority of nations across the world are focusing on recycling the various materials that are present inside E-waste to recover the secondary raw materials (Islam et al. 2024). According to Ilyas et al. (2021), various developed nations across the globe have formed practices to reclaim metals from the electronic waste, as it is an economical tool to uncover monetary benefits. As per the Gollakota et al. (2020) approximately 80% of E-waste concludes down in the open environment, including landfills, incineration facilities leading towards a concern for a need of efficient management system. As E-waste holds an important value, but due to the inappropriate treatment methods or managing methods, a heavy financial loss has been bearded by the different countries as in many developing and under-developed nations the practices like incineration, landfilling, or substandard treatment lead to the loss of materials which can be in benefit for circular economy and economic sustainability (Islam et al. 2024). The substandard methods when applied for management cause an unknowing loss to environment due to release of heavy metals and other contaminants (Anuardo et al. 2023). The global E-waste problem has reached critical levels, with both developed and developing countries struggling to manage the large amounts of electronic waste produced each day in an effective and sustainable manner. Several legislative bodies have attempted to regulate the international transportation of hazardous substances, such as electronic waste, due to their potential threats to both global ecosystems and human well-being. International agreements and treaties aimed at monitoring waste transportation included the Basel Convention (1989) pertains to the regulation of transboundary transfers of hazardous waste and the management of such waste disposal, and the Stockholm Convention (2001) on persistent organic pollutants. From 2019 to 2023, there was a slight uptick in the number of countries with legislation regarding electronic waste, rising from 78 to 81. Among these 81 countries, 67 had legal frameworks for waste management that included provisions advocating the environmental principle of extended producer responsibility (EPR). As of June 2023, 81 out of the 193 countries examined had implemented some form of policy, legislation, or regulation concerning electronic waste (Thakur and Kumar 2020). Asian countries produce nearly half of the production of electronic waste, amounting to (30 billion kg) when compared to the world, yet they have made minimal progress in managing this waste effectively. Furthermore, only a small number of these countries have implemented laws or set clear targets for collecting E-waste. In 2022, Europe, Oceania, and the Americas were the places with the highest per capita electronic waste generation, with figures standing at 17.6 kg, 16.1 kg, and 14.1 kg, respectively.

It has been observed that there has been a rise in implementing legislation in different countries, according to

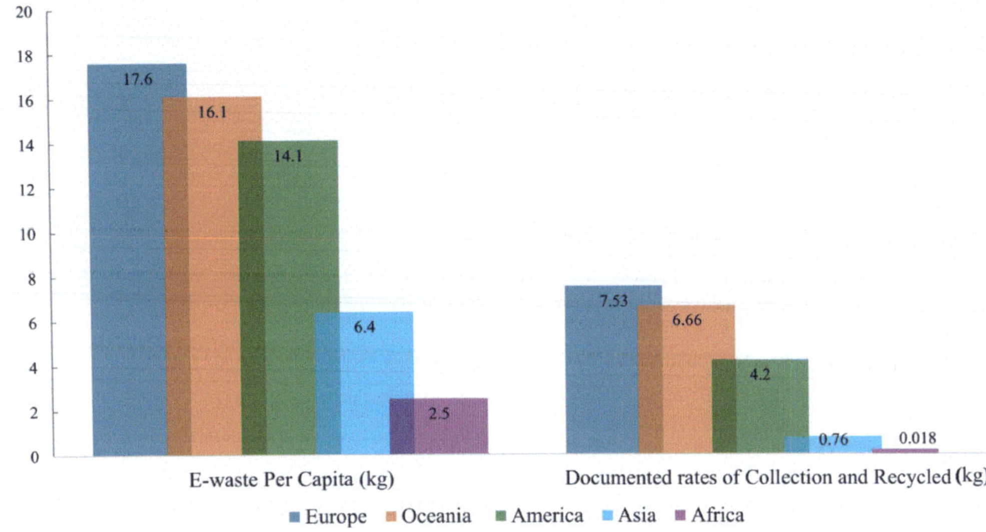

Fig. 5 Global distribution of E-waste generation and recycling

Global E-waste Monitor (2024), Baldé et al. (2024) there has been a significant growth in the countries that are adopting laws and regulations for successful mitigation of E-waste. Between years 2019 and 2023, the number has witnessed an increase from 78 to 81, all these countries make the 72% of the global population (Baldé et al. 2024). Figure 6 provides increasing knowledge about the E-waste management legislation implemented in different countries in different years.

Currently global economies have started implementing different legislations including extended producer responsibility (EPR), which aims to ensure that producer should take responsibility of end of life and post life management of a product (Islam et al. 2024). It has been noted that across the world only 67 countries out of 81 have EPR policy, while there are different national and international legislations, i.e. environmental, health and safety standard being followed by 62 countries, 42 countries out of the 81 have national E-waste collection targets, and only 36 countries have the E-waste recycling targets at national levels.

Different countries have various policies implemented as to manage the problem of E-waste, depending upon every continent, the policies and legislation have been modified and implemented. As per the Global E-waste Monitor, 2024 report Europe has the maximum emission of E-waste, followed by Oceania and America. According to Gollakota et al. (2020) in these continents the popular laws have been enacted which includes European union Directive, EPR, Basel convention, Organization for Economic Cooperation and Development (OECD). All these directives or amendments aim towards directing the electronic waste in a scientific and nature responsive manner. Table 3 provides the

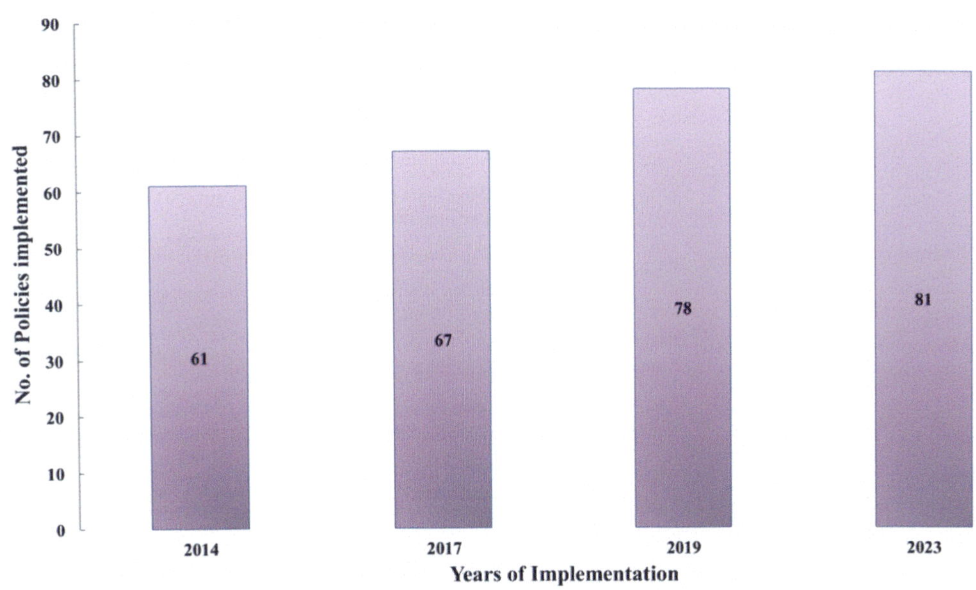

Fig. 6 Number of countries with legislations and rules for E-waste management

Table 3 Information about the E-waste laws of different countries according to the continents (Baldé et al. 2024)

Region of continent		E-waste law implemented
Europe	European Union (EU) and Non-EU member countries, Western Balkans	WEEE Directive—The main aim is to provide a set criterion for the overall management which includes collection to recovery of WEEE
		RoHS Directive—It aims towards minimizing and preventing the hazards associated to human and environmental health caused from E-waste
		Critical Raw Minerals Act—The goal is to strengthen the stream of critical raw metals
		Extended Producer Responsibility (EPR)—It aims towards to ensuring that producer should take responsibility of end of life of a product
	United Kingdom	Waste Electrical and Electronics Waste Regulation—It aims towards collecting single-use devices
	Belarus	E-waste management and regulations standard—Electronic waste is reported as a legal framework under the general waste management
	Russia	E-waste management using By-laws
	Moldova	Law of Waste No. 209—It aims towards the waste management and electronic refuse management
	Ukraine	Law no. 187/98-BP (66)—It covers the legal frameworks for waste management along with E-waste
	Eurasian Economic Union (Russia, Belarus, and Member states)	TR EAEU 041/2017—Technical Regulation—It regulates the substances to manufacture electronic products
Oceania	Australia	National Waste Policy 2009—It aims to enhance waste and resource management and facilitate resource recovery
		Product Stewardship Act 2011—Its aims towards producer responsibility (television and computers)
		Go for Zero 2023—It aims towards to preventing disposal of broken mobiles to landfills
		Australian Mobile Recycling Programme "MobileMuster"—Aimed towards recycling the mobile phones and its parts
	New Zealand	Waste Minimization Act 2008—It aims towards minimizing waste and reducing disposal
		Product Stewardship Framework (2014)—It aimed towards the overall management of E-waste data from collection to treatment
	Micronesia	The 2016–2025 Pacific Regional Waste Pollution Management Strategy (Cleaner Pacific 2025)—It aims to manage all waste streams including E-waste
	Samoa	PacWaste (Pacific Hazardous Waste)—It aims towards collecting on E-waste management strategies
		Samoa PacWastePlus—It aims towards establishing an E-waste product stewardship program for E-waste management
	Pacific Islands	Implementing Sustainable Low and Non-chemical Development in Small Island Developing States (ISLANDS)—It assists in control of importing hazardous materials and disposal in eco-friendly manner
North America	United States	Extended Producer Responsibility (EPR)—It aims towards to ensuring that producer should take responsibility of end of life of a product
		Recycling Standards and Certification and E-Steward—It aims towards minimizing the inappropriate handling and unlawful export of E-waste
		Right to Repair—It aims as the producers are must to provide repair manuals
	Canada	Product Stewardship for targeted EEEs—To control the toxic substances and manage E-waste
		Resource Recovery and Circular Economy Act, 2016—It states that producer is responsible finically for collecting and recycling the products
		Provisional EPR Programme—It aims to take forward community education initiatives for E-waste management
Central America	Costa Rica	The Integrated Waste Management General Law No. 8839
	Mexico	National Standard for End-of-Life Management (NOM-161)—It provides a clause of special handling for EEE products
		NADF-019-AMBT-2018 Environmental Standard for the Federal District—It aims to focus on E-waste management
		USA-Mexico Environmental Program: Border 2025—It aims to develop strategies for promoting material recovery and reducing illegal dumping
	Central America—UNIDO-GEF	Peral Project—It aims to improve the electronic waste initiatives nationally and encourage cooperation for management of POPs from E-waste

(continued)

Table 3 (continued)

Region of continent		E-waste law implemented
Caribbean	Jamaica	National Solid Waste Management Act—It aims to mitigate waste, including E-waste
	Dominican Republic	The Integrated Waste Management Law No. 225-20
South America	Columbia	Ministry of Commerce, Industry and Tourism resolution—It aims to modify the registration process for sellers and producers
	Ecuador	Legal framework (2002)—It aims to classify E-waste into categories
	Peru	National E-waste Policy—It aims to revise the collection targets and E-waste management
	Bolivia	National Legislation—It aims to bestow responsibility primarily on producer
Eastern South America	Venezuela	Hazardous Waste Law—It aims to regulate E-waste management
	Guyana	Environmental Protection Act (1996)—E-waste management is covered under this act
		National Solid Waste Management—It contains E-waste management provision; it relates to legal framework
	Paraguay	Integrated Waste Management Law—It aims to manage E-waste; it outlines the collection to disposal principle
	Brazil	Policy for National Solid Waste (Law 12.305/2010)—Its goal is to offer a framework for managing trash, particularly electronic waste
		Decree No. 10 240—It established a mandatory reverse logistics system (SisResiduos) which traces E-waste through its lifecycle
		ABNT NBR (16156:2014)—It aims to provide sound management of E-waste
Southern Cone	Chile	Law 20.920—It aims to manage the electronic waste by the lifecycle of the manufactured products
	Uruguay	Law No. 19.829—It aims to serve as compressive legal framework for waste management including E-waste
	Argentina	Resolution No. 92/2019 and Law No. 25.916—It provides guidelines for hazardous waste management
Eastern Asia	China	EPR Principle (2008)—It aims towards to ensuring that producer should take responsibility of end of life of a product
		E-waste Recycling Pilot Programme—It aimed to determine a standard and measured system for E-waste management
		National Hazardous Waste List—It aims to specify the hazardous substances present in EEE
	Taiwan	Waste Disposal Act—Its goal is to control the laws governing the management of E-waste
		Recycling Fund Management Act—Its goal is to control the laws governing the management of E-waste
	Japan	Resource Circulation Act—It aims towards providing a comprehensive framework for E-waste management from collection to recycling
		Eco-labelling Programme—It aims towards identifying the products that fulfil the environmental criteria. (EcoMark)
		Home Appliance Recycling Law—It aims towards recycling the household appliances
	Republic of Korea	Act on the circulation of resources for electrical and electronic devices and automobiles
		Green Card Program—It aims to connect the producer and consumers for effective management of electronic waste
	Mongolia	National Programme on Environmental Protection—It aims to provide provisions on E-waste management
		The program for managing and recycling electrical and electronic equipment is designed to reduce electronic waste through a holistic strategy
		Eco Town Project—It is an approach to promote E-waste recycling

(continued)

Table 3 (continued)

Region of continent		E-waste law Implemented
Southeastern Asia	Brunei	Waste management strategy—It plans to progress a sustainable waste management practice
		Management of E-waste—It provides a framework for E-waste management. It also gives priority to the waste hierarchy approach
	Cambodia	National Environmental Strategy and Action Plan—it recognizes the E-waste concern and focuses on its management and disposal
		Law of Environmental protection and Natural Resource Management—It provides requirements for E-waste management and works to addressing the challenge of hazardous materials
	Indonesia	Hazardous and Toxic Waste Management Regulation—It aims to manage E-waste so that it does not harm human and environment
		National Action Plan on E-waste Management—It seeks to manage E-waste in a sustainable manner
		E-waste Management System—It aims to provide the comprehensive database of EEE producer
	Lao People's Democratic Republic	Promoting environmentally responsible waste management and the recycling of E-waste are the two main objectives of the National Solid Waste Management Strategy and Action Plan for 2019–2028
	Malaysia	Environmental Quality Regulation and National Solid Waste Management Act and Communications and Multimedia Act—These are different measures taken over time for ensuing the compliance, for managing and disposal of different waste
		The goal of the National Strategy Plan for Solid Waste Management (2018–2030) is to advance environmentally friendly waste management techniques
	Myanmar	EPR Schemes for EEE—This E-waste management scheme is ensured to manage proper disposal
	Philippines	The Ecological Solid Waste Management Act requires the proper management of solid waste, encompassing electronic waste management
		DENR—Administrative order No. 2013–22—It aims to outline recommendations for the ecologically responsible handling of E-waste
		National Waste Management Strategy—It aims towards providing provisions for E-waste management
	Singapore	National EPR Scheme—It aims towards reusing of EEE via repairs and refurbishments programmes
		Network of E-waste Recycling Points and Programme—It aims to boost the individual disposal and recycling of electronics products
		E-waste Management System—Recycling is the national environmental agency's 2030 goal for Singapore
	Thailand	National E-waste Management Plan—It aims to determine a sustainable and efficient system for E-waste management
		Thailand's Strategic Plan for Integrated Waste Management Its goal is to provide a long-term structure for managing e-waste
		WEEE act—A draft by pollution control department to develop the concept of EPR
	Vietnam	National Action Plan for Electronic Product Waste Management
		Decree No. 31/2020/ND-CP—It aims to control the trade of E-waste and regulates its components
Central Asia	Central Asian Countries and Eurasian Economic Union	The Basel Convention intends to establish legal guidelines for the transboundary transfer of trash, including hazardous waste
		Technical Standard on the Prohibition of Dangerous Materials in Electronic and Electrical Products
		Consent to Collaborate on Electronic and Electrical Equipment Management—It aims to organizes the establishment of regional system for E-waste management
		UNITAR—It aims to manage and monitor E-waste collection and recycling
Western Asia	Gulf and Arab Countries	Israel—Electrical and Electronic Equipment and Batteries Law
		Turkish Regulation for E-waste Management
		Qatar's Law—Treatment and disposal of hazardous waste
		Georgia—Waste Management Code
		Agreement between the European Union and Armenia on Comprehensive and Enhanced Partnership
		Azerbaijan—National Strategy for Improving Solid Waste Management
		Palestine—1999 Environmental Act covered a component of hazardous waste

(continued)

Table 3 (continued)

Region of continent		E-waste law implemented
Southern Asia	India	E-waste Management and Handling Rules 2011—It included 21 IT devices and introduced EPR concept
		The 2016 E-waste Management and Handling Regulations Its objective was to reinforce the producer responsibility principle and the EPR framework
		Right to Repair Framework—It aims to promote circular economy
	Afghanistan	No specific legislation on E-waste
	Bangladesh	E-waste Management Rules 2021—It aims for E-waste management and mitigation of environmental impact
		E-waste Management Plan
		Guidelines on E-waste Management and Recycling
	Bhutan	Waste Prevention and Management Act
		National Waste Management Strategy
	Maldives	E-waste Management Guidelines
	Nepal	Waste Management Act—Aims towards mitigating the E-waste, along with the waste generated
	Pakistan	National Hazardous Waste Management Policy
	Sri Lanka	National E-waste Programme—It aims to mitigate the E-waste
Africa	Northern Africa	E-waste management Law no 202, 2020
	Egypt	
	Western Africa	National Environmental Regulations, 2022
		Hazardous and Electronic Waste Control Act, 2016
	Central Africa	No such legislation
	East Africa	Extended producer responsibility (EPR) framework
	Southern Africa	National Environmental Management Waste Act, 2021
	South Africa	
	Zambia	EPR Framework 2018
	Malawi	National E-waste Management Policy, 2022

information about the E-waste laws of different countries according to the continents.

7 Sustainable Development Goals and E-waste Management

The sustainable development goals (SDGs) provide a representation of worldwide initiatives aimed at achieving a future that is more sustainable, fair, and inclusive. The SDGs address a wide array of interrelated issues such as health care, access to clean water, sanitation, and renewable energy (Suffia and Dutta 2023). Among the 17 goals, SDG 3 is centred on advancing health and wellness, encompassing targets related to maternal health, reducing child mortality, and combating non-communicable diseases. In the year 2015, United Nations, along with its Delegate States, get on an ambitious plan for sustainable development. This agenda aims to accomplish 17 significant goals by the year 2030. The SDGs encompass various aims, including eradicating poverty and promoting sustainability. To achieve all these goals, efficient waste management is essential. E-waste and its management are closely linked to at least six of the SDGs, namely SDG 3 (good health and well-being), 6 (clean water and sanitation), 8 (decent work and economic growth), 11 (sustainable cities and communities), 12 (responsible consumption and production), 14 (life below water), and 17 (partnerships for the goals). As prosperity increases, the production of waste electrical and electronic equipment also rises. The UN introduced the 2030 Agenda to propel sustainability forward, utilizing its 17 (SDGs) to advocate for responsible practices across the spectrum of production, consumption, and waste management. These SDGs advocate for the effective handling of E-waste through measures such as promoting recycling, advancing technological innovations, fostering partnerships among diverse societal stakeholders, and more. By mitigating the adverse effects of E-waste on the environment and society, these actions support and advance the goals outlined in the 2030 Agenda. Target 3.9 aims to significantly reduce fatalities and diseases resulting from exposure to harmful chemicals and pollution of air, water, and soil (Anuardo et al. 2023). E-waste contains various hazardous components which, if not properly dismantled and processed, pose risks

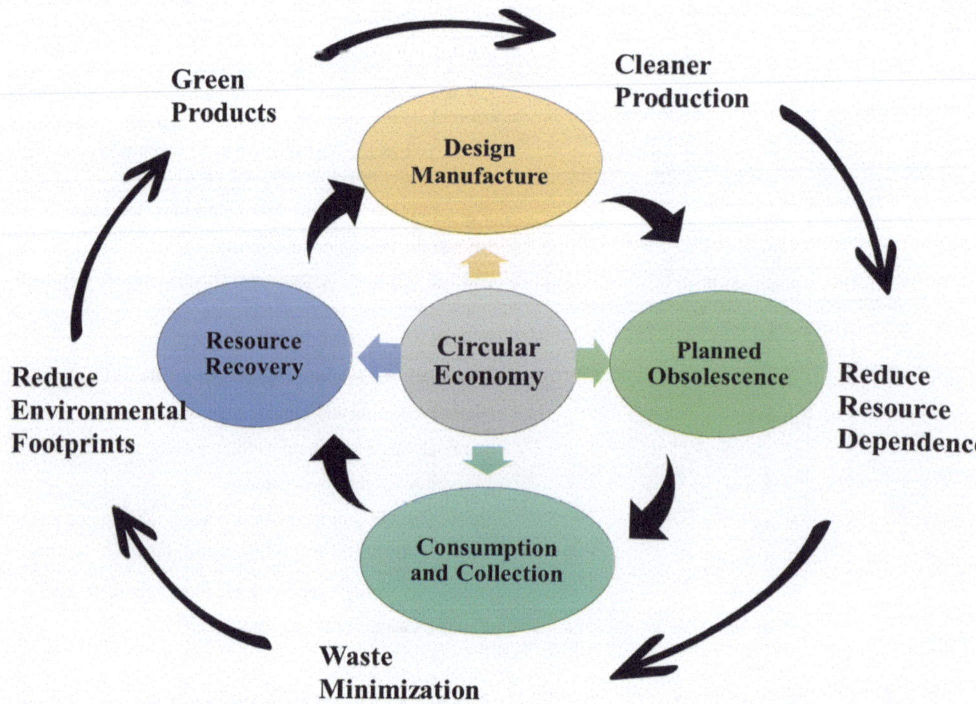

Fig. 7 E-waste management and sustainable development goals

to human health by contaminating air, water, and land. It is essential to adopt environmentally responsible practices during the design, production, dismantling, and processing of EEE to remove hazardous compounds. Unfortunately, in some areas, demolition activities are conducted using primitive methods that pose health risks. For instance, the burning of cables in open fires to extract copper releases furans and dioxins into the environment. Among the 8 targets in the goal 6, Target 6.1 aims to confirm that everyone has equal access to safe and affordable drinking water. Meanwhile, Target 6.3 undertakes to enhance quality of water through the reduction of pollution, eliminating dumping, and minimization of the release of dangerous chemicals and substances (Kundu et al. 2022). Target 8.3 aims to advocate for the establishment of policies that bolster productive endeavours, generate quality employment opportunities, stimulate entrepreneurial ventures, and facilitate the expansion of small- and medium-sized enterprises. Target 8.8 endeavours to safeguard labour rights and ensure the implementation of safe working environments for all employees, including working environments for all employees, including migrant labourers, with a particular emphasis on female migrants and those engaged in precarious employment circumstances. Central to achieving this goal is the formalization of their status, which involves state recognition and their integration into waste management systems. Figure 7 shows the relation between SDGS and E-waste management.

Target 4.6 aims to decrease the harmful environmental impact per person in cities, specifically aiming to enhance the quality of the air and implementing effective waste management practices. Given that about half of the global population resides in urban areas, these cities consume 75% of the Earth's natural resources, and addressing these issues is critical for sustainable urban development. Goal 14: Life Below Water and Goal 15: Life on Land: Preventing E-waste from entering water bodies and landfills helps protect marine and terrestrial ecosystems from pollution, preserving biodiversity and supporting ecosystem health. Goal 17 emphasizes the importance of cooperation between governments, industries, civil society, and communities to create efficient E-waste management policies, infrastructure, and awareness initiatives. Collaboration among these stakeholders is necessary for developing successful strategies to attempt the E-waste challenge. Through comprehensive E-waste management, countries can significantly advance progress towards various sustainable development goals concurrently, promoting sustainable development and environmental responsibility. India faces a dual challenge in E-waste management, characterized by two main dimensions. Firstly, there is the significant volume of E-waste generated within the nation itself. Secondly, there are issues surrounding imported E-waste, both through legal channels and illegal means (Kumar et al. 2021). It is widely recognized that crude methods like uncontrolled burning and recycling of WEEE present health hazards for individuals directly involved and the nearby environment. Pollution of air, water, and soil from such practices hinders efforts towards reaching SDGs 3 and 6, which focus on health and sanitation. Furthermore, adopting a closed-loop approach to managing WEEE has the potential to support Goal 12 by ensuring the recovery of secondary resources and reducing waste. Effectively recovering and reusing or recycling WEEE can also make a notable contribution in achieving a net positive impact on climate change, as outlined in SDG 13.

8 Discussion

This chapter highlights that E-waste contains both hazardous and valuable materials. If not properly managed, valuable metals may be lost forever, and harmful impacts on human health and the environment can arise. Improper disposal of E-waste releases toxic substances such as lead, mercury, and cadmium, which can contaminate soil and water, posing serious health risks to humans and wildlife. Additionally, the rapid pace of technological advancements has led to shorter lifespans for electronic devices, further contributing to the growing E-waste problem. Developing durable and easily recyclable products is essential to address this critical issue effectively. Governments and private companies around the world are exploring ways to protect the environment in response to the growing threats of climate change and rising energy costs. There is a growing global effort to adopt eco-friendly computing practices to help address these challenges (Manna et al. 2024). Developing and low-income countries are often the main targets for illegal waste dumping. This practice creates serious problems, particularly E-waste is handled without following proper safety measures. In addition, proper recycling standards are often missing. Therefore, it is crucial to ensure safe and efficient recycling and management of used consumer products to prevent harm to both people and the environment (Nandy et al. 2022).

The formal sector should be encouraged to manage hazardous waste, ensuring environmental health is prioritized. This approach will promote sustainability and ultimately contribute to building a circular economy. Formalizing the E-waste sector improves monitoring and ensures adherence to health standards. This significantly reduces the risk of chronic diseases and occupational hazards linked to E-waste processing. Transitioning to a circular economy for E-waste requires collaborative efforts to advance recycling technologies, implement effective policies, and enhance consumer awareness. These actions collectively drive environmental sustainability.

8.1 Future Implications and Recommendations

The management of electronic waste has emerged as a formidable challenge, especially within the context of developing nations. Future research should improve recycling, collection, and disposal technologies for a more sustainable environment (Murthy and Ramakrishna 2022). Proper documentation of policies and practices related to E-waste recycling is essential, along with adherence to government regulations to establish integrated E-waste management facilities. These facilities should include ongoing monitoring, control, and data management systems to track process variables, ensure safety compliance, and regularly document the outflow and inflow of materials and waste. Additionally, recycling systems that optimize economic value while promoting safety awareness should be explored to achieve environmentally friendly outcomes. Public involvement will be crucial in implementing household E-waste guidelines and regulations, particularly through the support of take-back and extended producer responsibility (EPR) programs (Kumar et al. 2022). UN reports emphasize reducing electronics in the waste stream and enhancing end-of-life management to support a circular economy.

Here are some recommendations for effective management of E-waste to achieve the sustainable development goals and support a circular economy.

1. Increase and promote E-waste management infrastructure by primarily shifting towards a circular economy.
2. Utilize available technology to implement strategies that extend the lifespan of products. The focus should be on postponing the "end-of-life" stage for products, appliances, and equipment.
3. Collaborating with E-commerce platforms to facilitate the "pick-up" of electronic equipment that needs to be disposed of, directing them straight to formal recyclers.
4. Achieving the global SDGs by 2030 may depend on integrating E-waste management with a sustainable development framework. It can help to reduce health issues and deaths caused by exposure to the hazardous components of E-waste and mitigate the environmental risks from air pollution, chemical and acid leaching in soil, and toxic effluent contamination in water. Additionally, it can provide secure and safe working environments (Arya and Kumar 2020).
5. A sustainable roadmap can be implemented to develop an accomplished E-waste management system by creating sustainable and resilient environments, bridging the gap between informal and formal recycling sectors,

expanding job opportunities, and enhancing occupational health and safety.

6. Addressing E-waste challenges necessitates international collaboration, as E-waste is a global issue. Working together can result in standardized practices, shared expertise, and coordinated strategies for managing E-waste around the world (Gaur et al. 2024).
7. The effective recycling and reclamation of materials from E-waste can safeguard precious resources including valuable metals, and rare earth elements. This practice diminishes the necessity for the extraction of new raw materials and mitigates the ecological consequences associated with mining activities.

9 Conclusion

The advancement of appropriate and scientific methods for managing electronic waste is crucial to implement the circular economy's guiding principles, aiming to safeguard the environment. The formal sector management of electronic waste in India encounters numerous obstacles. Presently, many states lack systems for regularly tracking the production of E-waste. Before COVID-19, inadequate metal recovery from essential technology was a concern. The pandemic worsened this globally, with more outdated electronics circulating and disruptions in material markets. India contributes major amount to global E-waste, but effective management of E-waste has been lacking attention. The hazardous substances present in E-waste pose significant risks to the environment. Consequently, there is an urgent need for the government to establish a thorough regulatory framework for E-waste management, combat illegal E-waste imports, and protect people from potential hazards. Furthermore, there is a pressing need to raise awareness about the harmful effects of E-waste and educate the public on proper disposal methods instead of simply discarding devices in landfills without proper handling. The implementation of regulations faces inefficiencies at the grassroot level due to various issues and challenges, such as insufficient skills, negligence by producers, recyclers, and electronic waste dealers, as well as improper recycling practices. The environmental agencies need to innovate and develop environmentally sound strategies for extraction of resources from E-waste which can lead to the business advantage converting the concept of linear economy to circular economy.

Acknowledgements Authors are thankful to the contributors of these books for their cooperation and patience in the compilation of this task. Authors are also thankful to Springer team, for their constant support in the publication of this work.

References

Ádám B, Göen T, Scheepers PT, Adliene D, Batinic B, Budnik LT, Duca RC, Ghosh M, Giurgiu DI, Godderis L, Goksel O, Hansen KK, Kassomenos P, Milic N, Orru H, Pashalidou A, Petrovic M, Puiso J, Radonic J, Sekulic MT, Au WW (2021) From inequitable to sustainable e-waste processing for reduction of impact on human health and the environment. Environ Res 194:110728

Alam T, Golmohammadzadeh R, Faraji F, Shahabuddin M (2022). E-waste recycling technologies: an overview, challenges and future perspectives. Paradigm Shift E-waste Manag, 143–176

Ankit, Saha L, Kumar V, Tiwari J, Rawat S, Singh J, Bauddh K (2021) Electronic waste and their leachates impact on human health and environment: global ecological threat and management. Environ Technol Innov 24:102049

Anuardo RG, Espuny M, Costa ACF, Espuny ALG, Kazançoğlu Y, Kandsamy J, de Oliveira OJ (2023) Transforming E-Waste into opportunities: driving organizational actions to achieve sustainable development goals. Sustainability 15(19):14150

Arya S, Kumar S (2020) E-waste in India at a glance: current trends, regulations, challenges and management strategies. J Clean Prod 271:122707

Arya S, Rautela R, Chavan D, Kumar S (2021) Evaluation of soil contamination due to crude E-waste recycling activities in the capital city of India. Process Saf Environ Prot 152:641–653

Arya S, Kumar S (eds) (2023) Global E-waste management strategies and future implications. Elsevier, Netherlands

Ashiq A, Kulkarni J, Vithanage M (2019) Hydrometallurgical recovery of metals from e-waste. In: Electronic waste management and treatment technology. Butterworth-Heinemann, pp 225–246

Baldé CP, Forti V, Gray V, Kuehr R, Stegmann P (2017) The global E-waste monitor. United Nations University (UNU), International Telecommunication Union (ITU) and International Solid Waste Association (ISWA), Bonn/Geneva/Vienna, pp 1–109

Baldé CP, Kuehr R, Ymamoto T, Mcdobald R, Angelo E, Althaf S, Bel G, Deubzer O, Fernandez C, Forti V, Gray V, Herat S, Honda S, Lattoni G, Khetriwal DS, Cortemiglia VLD, Lobuntsova Y, Nnorom I, Pralat N, Wager M (2024) The global E-waste monitor 2024. United Nations Institute for Training and Research (UNITAR)—Sustainable Cycles (SCYCLE) Programme

Baldé CP, Kuehr R, Ymamoto T, Mcdobald R, Angelo E, Althaf S, Bel G, Deubzer O, Fernandez C, Forti V, Gray V (2024) The global E-waste monitor 2024. International Telecommunication Union(ITU); United Nations Institute for Training and Research (UNITAR), Geneva, 2024

Brindhadevi K, Barceló D, Chi NTL, Rene ER (2023) E-waste management, treatment options and the impact of heavy metal extraction from e-waste on human health: scenario in Vietnam and other countries. Environ Res 217:114926

Central Pollution Control Board (2023) Annual report for implementation of E-Waste (Management) Rules, 2016(for the year 2021–2022). Ministry of Environment, Forest and Climate Change, Government of India

Chauhan G, Jadhao PR, Pant KK, Nigam KDP (2018) Novel technologies and conventional processes for recovery of metals from waste electrical and electronic equipment: challenges and opportunities–a review. J Environ Chem Eng 6(1):1288–1304

Debnath B, Roychoudhuri R, Ghosh SK (2016) E-waste management–a potential route to green computing. Procedia Environ Sci 35:669–675

Desmarais M, Pirade F, Zhang J, Rene ER (2020) Biohydrometallurgical processes for the recovery of precious and base metals from waste electrical and electronic equipments:

Current trends and perspectives. Bioresource Technology Reports 11:100526

Dey S, Veerendra GTN, Padavala SSAB, Manoj AP (2023) Recycling of e-waste materials for controlling the environmental and human heath degradation in India. Green Analytical Chemistry 7:100085

Dutta D, Goel S (2021) Understanding the gap between formal and informal e-waste recycling facilities in India. Waste Manage 125:163–171

Dutta D, Rautela R, Gujjala LKS, Kundu D, Sharma P, Tembhare M, Kumar S (2023) A review on recovery processes of metals from E-waste: a green perspective. Sci Total Environ 859:160391

Faraji F, Golmohammadzadeh R, Pickles CA (2022) Potential and current practices of recycling waste printed circuit boards: a review of the recent progress in pyrometallurgy. J Environ Manage 316:115242

Forti V, Baldé K, Kuehr R (2018) E-waste statistics: guidelines on classifications, reporting and indicators

Fujimori T, Takigami H, Agusa T, Eguchi A, Bekki K, Yoshida A, Terazono A, Ballesteros FC Jr (2012) Impact of metals in surface matrices from formal and informal electronic-waste recycling around Metro Manila, the Philippines, and intra-Asian comparison. J Hazard Mater 221:139–146

Gaur TS, Yadav V, Mittal S, Sharma MK (2024) A systematic review on sustainable E-waste management: challenges, circular economy practices, and a conceptual framework. Manag Environ Qual Int J 35(4):858–884

Ghulam ST, Abushammala H (2023) Challenges and opportunities in the management of electronic waste and its impact on human health and environment. Sustainability 15(3):1837

Gollakota AR, Gautam S, Shu CM (2020) Inconsistencies of e-waste management in developing nations—facts and plausible solutions. J Environ Manage 261:110234

Hussain A, Koli SK, Tripathi R, Pandey S (2020) India's lethal informal e-waste recycling: a case study of Delhi and NCR region. In: Smart cities—opportunities and challenges: select proceedings of ICSC 2019. Springer, Singapore, pp 363–375

Ikhlayel M (2018) An integrated approach to establish e-waste management systems for developing countries. J Clean Prod 170:119–130

Ilyas S, Kim H, Srivastava RR (2021) Sustainable urban mining of precious metals: an introduction. In: Sustainable urban mining of precious metals. CRC Press, USA, pp 1–33

Islam MK, Khatun MS, Mourshed M (2024) An in-depth analysis and review of management strategies for E-waste in the south Asian region: a way forward towards waste to energy conversion and sustainability. Heliyon

Jain M, Kumar D, Chaudhary J, Kumar S, Sharma S, Verma AS (2023) Review on E-waste management and its impact on the environment and society. Waste Manag Bull

Javed A, Singh J (2024) Process intensification for sustainable extraction of metals from e-waste: challenges and opportunities. Environ Sci Pollut Res 31(7):9886–9919

Julander A, Lundgren L, Skare L, Grandér M, Palm B, Vahter M, Lidén C (2014) Formal recycling of e-waste leads to increased exposure to toxic metals: an occupational exposure study from Sweden. Environ Int 73:243–251

Kumar A, Nayar KR (2021) COVID 19 and its mental health consequences. J Ment Health 30(1):1–2

Kumar S, Agarwal N, Anand SK, Rajak BK (2022) E-waste management in India: a strategy for the attainment of SDGs 2030. Mater Today Proc 60:811–814

Kumar P, Singh RK, Kumar V (2021) Managing supply chains for sustainable operations in the era of industry 4.0 and circular economy: analysis of barriers. Resourc Conserv Recycl 164:105215

Kundu D, Dutta D, Samanta P, Dey S, Sherpa KC, Kumar S, Dubey BK (2022) Valorization of wastewater: a paradigm shift towards circular bioeconomy and sustainability. Sci Total Environ 848:157709

Lakshmi S, Raj A (2017) A review study of e-waste management in India. Asian J Appl Sci Technol (AJAST) 1:33–36

Manna S, Mukherjee S, Das D, Saha A (2024) Influence of E-waste management in green-computing. J Inf Technol Policy

Manomaivibool P (2009) Extended producer responsibility in a non-OECD context: the management of waste electrical and electronic equipment in India. Resour Conserv Recycl 53(3):136–144

Meenakshi DT, Harini V (2012) Entrepreneurship opportunities in managing e-waste. Asia Pac J Manag Entrepreneurship Res 1(3):47

Mmereki D, Li B, Baldwin A, Hong L (2016) The generation, composition, collection, treatment and disposal system, and impact of E-waste. E-Waste Trans Pollut Resour 10:61332

Murthy V, Ramakrishna S (2022) A review on global E-waste management: urban mining towards a sustainable future and circular economy. Sustainability 14(2):647

Nandy S, Fortunato E, Martins R (2022) Green economy and waste management: an inevitable plan for materials science. Prog Nat Sci Mater Int 32(1):1–9

Pant D, Singh P, Upreti MK (2014) Metal leaching from cathode ray tube waste using combination of Serratia plymuthica and EDTA. Hydrometallurgy 146:89–95

Parajuly K, Fitzpatrick C, Muldoon O, Kuehr R (2020) Behavioral change for the circular economy: a review with focus on electronic waste management in the EU. Resour Conserv Recycl X 6:100035

Ribeiro JN, Barbosa AFM, Ribeiro AVFN, de GoPereira M, de Oliveira JP, Zordan AB, da Silva AR (2022) E-waste and its consequence for environment and public health: perspectives in covid-19 pandemic times. Glob J Health Sci 14(3):54–76

Sengupta D, Ilankoon IMSK, Kang KD, Chong MN (2022) Circular economy and household e-waste management in India: integration of formal and informal sectors. Miner Eng 184:107661

Shahabuddin M, Uddin MN, Chowdhury JI, Ahmed SF, Uddin MN, Mofijur M, Uddin MA (2023) A review of the recent development, challenges, and opportunities of electronic waste (e-waste). Int J Environ Sci Technol 20(4):4513–4520

Sharma S, Wakode S, Sharma A, Nair N, Dhobi M, Wani MA, Pottoo FH (2020) Effect of environmental toxicants on neuronal functions. Environ Sci Pollut Res 27:44906–44921

Shittu OS, Williams ID, Shaw PJ (2021) Global E-waste management: can WEEE make a difference? A review of e-waste trends, legislation, contemporary issues and future challenges. Waste Manage 120:549–563

Song Q, Li J (2015) A review on human health consequences of metals exposure to e-waste in China. Environ Pollut 196:450–461

Suárez-Eiroa B, Fernández E, Méndez-Martínez G, Soto-Oñate D (2019) Operational principles of circular economy for sustainable development: linking theory and practice. J Clean Prod 214:952–961

Suffia S, Dutta D (2023) Applications of deep eutectic solvents in metal recovery from E-wastes in a sustainable way. J Mol Liq, 123738

Thakur P, Kumar S (2020) Metallurgical processes unveil the unexplored "sleeping mines" e-waste: a review. Environ Sci Pollut Res 27(26):32359–32370

Vats MC, Singh SK (2014) E-Waste characteristic and its disposal. Int J Ecol Sci Environ Eng 1(2):49–61

Wang K, Qian J, He S (2022) Global destruction networks and hybrid e-waste economies: practices and embeddedness in Guiyu, China. Environ Plann Econ Space 54(3):533–553

Wei L, Liu Y (2012) Present status of e-waste disposal and recycling in China. Proc Environ Sci 16:506–514
Zeng X, Wang F, Li J, Gong R (2017) A simplified method to evaluate the recycling potential of e-waste. J Clean Prod 168:1518–1524
Zeng Y, Zou J, Liao C, Liu F, Zhou X (2022) Selective separation and recovery of selenium from copper anode slime by compound leaching followed by sulfate roasting. Miner Eng 186:107749
Zhu Z, Han Z, Bi X, Yang W (2012) The relationship between magnetic parameters and heavy metal contents of indoor dust in e-waste recycling impacted area, Southeast China. Sci Total Environ 433:302–308

Utility of Plastic Waste in the Construction Industry as Carbon–Neutral Initiatives

Manish Chaudhary, Rajat Singh Rana, and Surindra Suthar

Abstract

Due to wide applications and durability, plastic has become an essential part of human life. In recent years, plastic production has reached up to 353 million metric tonnes globally; only 9% of the total is recycled. Most plastic waste ends up at landfill sites, which pose severe environmental contamination and toxicity issues in terrestrial and aquatic environments. The thermal treatment of waste plastic (pyrolysis, gasification, and incineration) has been widely recommended for its disposal, but it is also not cost-effective and environmentally safe. Recent studies indicate that waste plastic could be used effectively in the construction industry for concrete blocks, composite blocks, and plastic tiles. However, developing new construction materials using plastic waste is crucial for the construction and recycling industries. Research has shown that recycling plastic waste for construction purposes could greatly enhance climate sustainability and provide dependable source material for construction supplies. Meanwhile, the manufacturing cost of such building materials is lower than that of traditional construction materials. This chapter summarized the potential of waste plastic substances in eco-friendly and cost-effective building material development, contributing to a carbon–neutral economy under the flagship of the circular economy process. This chapter also examines the advantages of newly developed construction materials over conventional ones regarding economic viability, environmental sustainability, and construction quality.

M. Chaudhary · R. S. Rana · S. Suthar (✉)
School of Environment & Natural Resources, Doon University, Dehradun, Uttarakhand, India
e-mail: suthariitd@gmail.com

Keywords

Sustainability · Green economy · Plastic pollution · Concrete block · Sustainable development goals

1 Introduction

In recent years, the building sector has been looking for ways to turn a waste problem into part of the solution in sustainable development efforts. Hence, scientists are addressing plastic pollution issues through sustainable plastic waste recycling and working for a better and environmental friendly world. Several studies have investigated the advantages and roles of plastic waste in construction materials to help sustainability and reduce environmental impact. Not surprisingly, recycling and reusing plastic waste in Aville building materials is a significant area of interest, which could be an enabler for modest growth (Lamba et al. 2022). The sector can horizon building waste and pollution by recycling plastic trash into the built form, redefining greenfield building techniques, and influencing the economy (Rahman et al. 2023). Plastic waste can effectively be used for various construction applications like paver blocks and coarse aggregate replacement. This new method exemplifies how construction methods can help the environment and industry (Ahmed 2023).

Plastic waste in construction is not limited to conventional materials. Researchers have hypothesized that plastic waste could be used innovatively in cement binders, road-building aggregates, insulating materials, and eco-bricks (Tahir et al. 2024). These applications of plastic waste in buildings show the multi-faceted contribution that concrete as a material can make a sustainable issues. Consequently, studies have investigated how post-consumer plastics can be brought into waste plastic composites in building elements, and hence the limits and possibilities for the use of

A. Mandpe et al. (eds.), *Sustainable Waste Management*, Advances in Science, Technology & Innovation,
https://doi.org/10.1007/978-3-031-85685-3_4

plastics recycled composites in the improvement of energy efficiency and sustainability in construction (Ribeiro et al. 2023). Indeed, incorporating plastic waste into building materials has been promoted as a critical strategy to combat the planet-wide plastic problem. The building industry offers copious opportunities for these long-term efforts (Tang 2022). In the long run, it is a solution to manage plastic waste. Also, this will encourage green practices in construction management systems, and a less green country such as Malaysia will eventually benefit the construction industry of such developing countries (Chong et al. 2023). Plastic waste can change the face of construction industries for a sustainable environment and economic development. One of the potential technologies for plastic waste utilization is using plastic waste as part of the ingredients of building materials. It was proved that using plastic waste in building materials can enhance the sustainability of building materials supply (Rahat et al. 2022). A promising future in the use of recycled plastic in concrete was discovered through extensive studies (Kamal et al. 2021). Waste plastic, a potential substitute for sustainable construction material, could address plastic pollution by proposing a green building strategy (Adiyanto et al. 2022). Studies have reported that different types of plastic waste can be utilized to enhance plastic recycling techniques for various industries, including the construction sector (Lahtela et al. 2019). Polypropylene (PP)-based waste aggregate composite material has been identified as a potential sustainable alternative for various construction materials. Záleská et al. (2018) explored recycled plastic materials in con struction to reduce waste by providing thermal insulation, enabling energy recovery.

Additionally, experimental attempts were made to recycle plastic wastes and to use them as coarse aggregates in concrete involving PET plastic bottles, where fly ash was also included to enhance the properties of concrete and to reduce the use of natural concrete materials (Ahmed et al. 2024; Sau et al. 2024). Efforts were made to manufacture recycled plastic coarse aggregates for concrete, and various plastic wastes were used as construction material. As per the latest study, it is proposed to use recycled plastic in subsequent bricks to construct buildings to eradicate environmental issues of high-density polyethylene (HDPE) plastics. Many studies are being conducted on using plastic waste in the construction sector, and these are very advantageous in developing methods to create an eco-friendly environment (Belay Wendimu et al. 2021). To mitigate plastic waste sustainably, engineers also studied and developed policies and resource assignments for sustainable practice (Kataki et al. 2022). This chapter discusses the environmental issue, supports environmentally sustainable development, and analyses the plastic trash for sustainable development projects in building materials. The construction industry can contribute to reducing waste production, pollution control, and energy conservation by recycling plastic into building materials. The sustainable development goal and green built environment could benefit the construction industry if they opt for creativity and recycled plastic material for their construction purpose.

2 Application of Different Plastic Waste in the Construction Industry

The building sector is the country's foundation and a significant force behind economic expansion. Making use of waste materials can improve sustainable construction methods and procedures. Utilizing plastic waste in construction offers both financial benefits and promotes environmental sustainability. The amount of plastic garbage that enters the maritime environment can be significantly reduced by creatively integrating it into construction applications. This approach also offers substitute materials to satisfy industry demands.

However, waste plastic must meet the mechanical and durability requirements of the intended application to be used in construction. Plastic trash should function well as a building material, reasonably priced and environmental friendly to promote its use instead of alternative materials. Table 1 illustrates the mechanical properties of construction materials made from different types of plastics in various proportions. Plastic trash was used as a fine aggregate in self-consolidating concrete at a weight percentage of 12.5% of the fine aggregate, which improved the material's fresh qualities, particularly its filling and passing capacities (Hama and Hilal 2017). Non-biodegradable waste plastic bags have been repurposed to create wall and floor tiles with improved tensile strength and reduced flammability (Dhawan et al. 2019). A range of plastic wastes, including high-density polyethylene (HDPE), low-density polyethylene (LDPE), polystyrene (PS), polypropylene (PP), polyethylene terephthalate (PET), and others, are produced based on their original application (Fig. 1). Recent studies reveal that LDPE makes up the majority of plastic trash.

There is also a noticeable resemblance between LDPE and PP. The effectiveness of plastic waste in fibre-reinforced concrete beams was studied by (Khalid et al. 2018, 2022). In their study, the concrete was loaded with ring-shaped PET to increase its splitting tensile, fracture energy, compressive strength, and beam flexural strength. Additionally, it demonstrated how adding plastic fibres to the concrete enhanced the beams' mechanical performance regarding load and initial crack strength while having little to no impact on the failure mode. Plastic trash has its uses beyond the building, such as indoor use and engineering.

Table 1 Mechanical properties of plastic waste used as a substitute in construction material

S. No.	Application of plastic waste	Used plastic waste type	Plastic waste (%)	Tests conducted	Compressive strength (MPa)	References
1	Used as aggregate	PET	20, 30, 40	Tensile strength	38.14	Aneke and Shabangu (2021)
2	Used as aggregate	PET	20, 30, 40	Tensile strength	42.01	Ikechukwu and Shabangu (2021)
3	Used as aggregate	PET	0, 1, 3, 7	Durability	244.4% increment	Akinwumi et al. (2019)
4	Used as additives	HDPE and PET	0, 1, 3, 7, 15, 20	Bulk density	5–28%	Limami et al. (2020)
5	Used as aggregate	LDPE	–	Flexural strength, density, and thermal conductivity test	27	Paihte et al. (2019)
6	Used as aggregate	PP	20, 50, 30	Hardness, water absorption, efflorescence, and fire resistance	22.85	Velmurugan (2019)
7	Used as aggregate	PET: PU	60:40	Flexural test and impact test	84.54%	Alaloul et al. (2020)
8	Additive	HDPE and PET	–	Specific heat capacity and thermal conductivity	–	Limami et al. (2020)
9	Replacement of sand	PET	1, 3, 5, 7, 10	Compressive strength, flexural strength, density, and tensile strength	20.72	Hameed and Ahmed (2019)
10	Binder	Waste polyethene stability value, flow value, and air voids	5–11%	Flow value, stability value, and air voids	–	Seghiri et al. (2017)
11	Substitute	Waste plastic and E-waste	Waste plastic 4.5–6, shredded E-waste 7.5, 10, 12.5, 15	Crushing value, loss abrasion value test, Marshall stability test, and affect value	–	Shrikant et al. (2020)
12	Modifier	Thermoplastics and elastomers	–	Bending strength	320	Belmokaddem et al. (2020)
13	Modifier	Shredded plastic waste	6, 8, 10, 12, 14	VFB and flow value	18.55–10.72 (Decreased)	Rajput and Yadav (2016)
14	Aggregate coating	Plastic waste	–	Impact, abrasion, soundness, and water absorption	–	Dawale (2016)

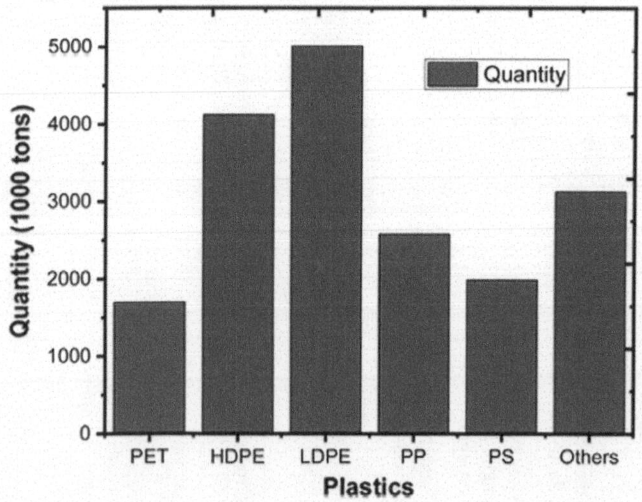

Fig. 1 Types of plastic wastes and estimated quantities (Hama and Hilal 2017)

Use of plastic waste instead of aggregate in base and subbase construction improves pavements' shear, stiffness, and bearing capacity. A study by Henriksen et al. (2022) showed that incorporating plastic waste lowers the blends' stiffness, bearing capacity and resilient modulus; however, acceptable performance was still achieved. Bricks were produced utilizing waste thermoplastic materials such as polycarbonates, polystyrene, mixed plastics, and varying proportions of sand, ash, and ordinary Portland cement. Thermoplastic constituted 10–15% of the total weight; sand accounted for 70%, and the remaining 15% combined fly ash and Portland cement (Mondal et al. 2019). Similar to how plastic waste is used as an aggregate in cementitious composites, plastic waste can also be used in asphalt mixtures (Heydari et al. 2021).

According to available data, 1.7 trillion metric tons of asphalt are produced annually worldwide, primarily for use in asphalt pavements as a binder, waterproofing materials, roofing, or sealing (Heydari et al. 2021; Tejaswini et al. 2022) because of their outstanding durability and load-bearing capabilities, bitumen and aggregates, and the most common materials used to pave roadways, typically makeup 5–8% of asphalt pavement. Adding polymers, including waste plastics and rubber, to an asphalt pavement is a typical performance improvement method.

There are two primary methods for incorporating waste plastics or rubber into asphalt: the wet and dry processes. Waste materials serve as modifiers in the wet process, combining with bitumen to produce polymer-modified bitumen. Plastic trash is used as aggregates or fillers in the asphalt mixture during the dry process (Soni et al. 2022). We can significantly enhance asphalt mixtures' performance at high temperatures by using plastic waste, which will benefit the economy and the environment. The kind of plastic debris utilized and its incorporation into the asphalt mixture are essential factors that affect how well the combination performs.

For instance, plastic waste increases stability and reduces noise when driving on a road constructed with asphalt and plastic waste mixed (Heydari et al. 2021). Moulding processes increase the plastic waste viable substitute for materials like wood, blocks, and bricks (Bae et al. 2019).

3 Benefits of Plastic Waste in the Construction Industry

Integrating plastic waste in the construction industry has a deeper meaning that has become a significant issue. However, substituting concrete with plastic waste is a viable method of recycling. Construction material made of plastic waste causes least harms to the environment and reduces natural resource exploitation, depending on Tayebani et al. (2023). It is a substitute for concrete, which enables the control of plastic waste and provides rigid and elastic concrete required for construction purposes. As for the type of material, this very rationale explains how plastic waste is highly favoured for the production of materials such as plastic timber for decking, outdoor furniture and similar products due to the high durability, weather resistance and low maintenance (dos Santos et al. 2020).

Additionally, a successful integration of plastic waste with asphalt mixture in road construction has been observed (Heydari et al. 2021). The road becomes more flexible and durable, indicating a more extended service with low maintenance. More sustainable infrastructure can be developed using plastic waste to redirect waste from landfills. However, various challenges may appear while using plastic waste in construction projects. One main obstacle involves maintaining the regular quality of the products. Plastic waste variability and diversity make it difficult to produce a valuable building construction material, and efficient sorting and processing technologies are essential to address this issue. Chemical compatibility between plastic waste and construction material is another challenge. Cementitious substances and plastic components interaction can alter the mechanical properties of the final product (Babafemi et al. 2018; Mahmood and Kockal 2020).

Another primary concern is the possibility of harmful substances being released from plastic garbage while the building is underway. Research and development initiatives are essential to guarantee that waste plastic satisfies industrial standards and maximizes material performance (Tahir et al. 2024). Environmental and human health hazards may arise from contaminants and additives found in waste plastic (Maddela et al. 2023). Consequently, these problems must be addressed when developing long-lasting and safe building materials. Despite these challenges, using plastic waste in the building sector provides a workable technological answer to the shortage of resources and pollution in the environment. More research and technology developments are required to realize this novel method's potentially.

4 Analysis of Some Case Studies

The analysis of multiple case studies concerning the application of plastic trash in the construction sector offers a thorough comprehension of the possible advantages and obstacles linked to this inventive methodology. Several countries and groups have shown that reusing plastic trash into construction materials has a potential to minimize environmental pollution, save natural resources, and build more durable and cost-effective infrastructure.

A review of various case studies demonstrates the merits and drawbacks of this strategy, providing significant insights into its practicality and future potential. Through the effective integration of plastic waste into materials like concrete, asphalt, and bricks, these case studies demonstrate the flexibility and adaptability of plastic in diverse building scenarios. The projects also highlight the financial benefits of this strategy by providing affordable options that can lower building costs overall while concurrently addressing urgent environmental issues. The knowledge obtained from these case studies presents a strong justification for the broader acceptance of plastic waste as a crucial element in the future of sustainable infrastructure as the global construction industry continues to look for sustainable alternatives. Some case studies have been discussed here.

4.1 Bottle House in Nigeria

The Development Association for Renewable Energies (DARE), a Nigerian non-governmental organization, has pioneered an innovative method of sustainable construction by building a two-bedroom home out of plastic bottles filled with sand. In this approach, used plastic bottles are repurposed and filled with sand to form a robust and lasting building block. These sand-filled bottles are then compacted and organized in a structural framework strengthened by a mud-cement combination. The bottles are linked by a network of strings knotted around their necks to improve structural integrity, ensuring the whole assembly remains cohesive and robust (see Fig. 2).

According to documented studies, this construction method can yield building materials up to 20 times stronger and far more durable than traditional bricks. The resulting constructions are more resistant to environmental stressors and potential damage because of the sand-filled plastic bottles' remarkable resistance to external pressures. Furthermore, the bottle caps that projected from the wall surfaces enhances the building's visual attractiveness by giving the facade a distinctive texture and a splash of colour.

Building homes with plastic bottles is a cost-effective and environmentally beneficial alternative to typical

Fig. 2 Bottle house in Nigeria. *Source* Rosenfield (2015)

building materials. According to (Rosenfield 2015), the cost of building a home with plastic bottles is roughly one-third that of traditional concrete and bricks. This considerable cost reduction is primarily due to the availability of plastic trash and the reduced requirement for costly materials such as cement and bricks. Numerous studies highlight the benefits of using sand or mud-filled plastic bottles instead of conventional bricks. These advantages include higher thermal resistance, reduced cement and water usage, faster construction times, ease of relocation, minimal need for skilled labour, zero construction waste, and enhanced seismic bearing capacity (Dehghani et al. 2019; Kadam 2008; Mansour and Ali 2015; Safinia and Alkalbani 2016).

4.2 Beach House in Canada

According to Canadian Plastics, a Canadian construction company uses 6,12,000 recycled plastic bottles to construct a 2000-square-foot beach house in Nova Scotia (Othalo—The Future of Housing), showcasing an inventive use of PET foam technology in residential construction. Using this technique, lightweight composite panels are made by layering a foam core between fibreglass skins after plastic bottles have been melted, shredded, and treated with gases. This procedure efficiently changes recycled plastic into a building material that is both structurally sound and acceptable. The building process, which usually takes a while, was highly effective since the panels were put up in just two days following three months of planning.

Compared to traditional building materials, the resultant panels, mainly composed of PET (polyethene terephthalate) foam, have several advantages. With a continuous R-30 rating, indicating excellent thermal resistance, the PET foam utilized in the beach home offers exceptional thermal insulation. Compared to conventional insulation techniques, this feature helps to manage interior temperatures more successfully, potentially lowering energy use for heating and cooling. The panels' aluminium and vinyl siding, featuring a cedar print, are applied to improve their visual appeal and longevity. The 200 prefabricated PET wall panels are built primarily to match the roof's borders, demonstrating the material's adaptability and versatility in many building scenarios (see Fig. 3). Because of its structural strength and resistant qualities, PET foam technology, which was employed in this project, is highly respected in the insulation and aerospace industries. However, its use in house construction is a new application for the material.

Additionally, the PET foam panels offer enhanced overall energy efficiency, resistance to mould and decay, and the possibility of fire resistance when certain resins are employed. The material's capacity to reduce carbon emissions is awe-inspiring because it uses recycled plastics, which supports sustainability initiatives (Othalo—The Future of Housing). The panels' lack of thermal breaks guarantees that the house's thermal performance is maintained, which improves energy efficiency and adds to the building envelope's overall effectiveness.

In conclusion, the use of PET foam technology in the construction of this beach house shows how it can

Fig. 3 Beach house in Canada. *Source* Othalo—The Future of Housing

transform building methods by providing an energy-efficient, ecologically friendly, and structurally sound substitute for conventional materials.

4.3 Road Construction Using Plastic Waste in India

The viability of using waste plastics in the construction of roads while at the same time contributing to the environment's benefits has been the area of focus in several research works. Plastic waste can be added as an additive to binders or as a replacement for aggregates. The studies demonstrate that it increases the strength and fatigue life, deformation, and stability and enhances the pavement life, durability, and performance (Aldagari et al. 2022; Appiah et al. 2017; Manju and Sheema 2017).

According to Plastic Waste Management and Handling Regulation, 2015, issued by the Ministry of Environment, Forest and Climate Change (MoEF&CC), India, a noteworthy step was taken to lessen plastic waste. These regulations detail the production, usage, and disposal of products to complete the display of this management system.

One of the most important clauses is extended producer responsibility (EPR), which mandates that importers, manufacturers, and brand owners handle the plastic trash their products produce. To promote recycling, the regulation also requires plastic manufacturers to register with State Pollution Control Boards (SPCBs), raise the minimum thickness of plastic carry bags to 50 microns, and hold urban local bodies responsible for the collection, segregation, and disposal of plastic waste. The utilization of plastic waste in road construction is a remarkable suggestion. To make roadways more durable, mixing shredded plastic waste and bitumen is advisable, which could also effectively reduce plastic waste. The plastic waste management and handling considerations promote the use of this technology as pilot road projects. Training, capacity building, public–private partnership, and community collaboration are also encouraged.

India has excellent expertise in building roads with waste plastic, especially its national highways and rural roads. It has even approved the use of plastic trash in bituminous mixes, setting it as the standard procedure for hot mix road reconstruction every 50 kms near cities with a population of more than 500,000. Over 2500 km of roads have been built with waste plastic since 2002, and ten years later, there are still no potholes, ravelling, or rutting on these roads (Vasudevan et al. 2019). Compared to traditional methods, a ton of scrap plastic was used for every kilometre of road construction, reducing carbon dioxide emissions by three tons per kilometre. $670/kilometre of plastic waste in road construction was calculated in 2012 (Vasudevan et al. 2019).

5 Expected Income Potential from Using Plastic Waste

Plastic waste has a significant and diverse potential as a source of money. Here are a few ways to generate income:

a. **Sale of plastic waste resources**: Manufacturers can purchase plastic waste resources to make new construction products like composite blocks, paving blocks, wall bricks, and flooring tiles. These construction materials have a market value, bringing the manufacturers money. The selling of plastic waste resources is governed by several factors, such as transportation and gate fees, administrative expenses, tariff and non-tariff barriers, variations in environmental taxes and policy strictness, variations in treatment capacities, variations in incentives for recycling and energy recovery, variations in legislation and classification, the necessity of particular technologies, and regional or national geographic features. On the other hand, in nations with lax environmental laws governing recycling and disposal practices, the trade in plastic trash can result in significant environmental problems (Kellenberg 2012; Zhao et al. 2021).
b. **Government Grants**: The government may provide grants to encourage plastic waste collection centres; such funds could help set up start-ups and run recycling facilities, allowing driving money. In India, the four cities, Surat, Varanasi, Bengaluru, and Patna, would receive Rs. 6 crore each from the central government to construct plastic waste management centres. Research on the recycling and disposal of plastic will also be conducted at this centre (The Indian Express 2019).
c. **Contractual endorsement**: Companies can work with plastic waste management centres to effectively manage their waste organization. Usually, these collaborations entail official contracts under which the management centre is responsible for gathering, sorting, and recycling the business's plastic waste. The financial parameters of the agreements, including the expenses related to each step of the waste management process, should be spelt out. The amount and kind of plastic garbage collected, the frequency of pickups, and the intricacy of the recycling procedure can all influence how payments are made. Clauses for any value-added services, such as on-site waste audits or returning recycled goods to the business, may also be included in the agreement. The plastic trash centre and the collaborating business may guarantee a win–win deal that promotes sustainable waste practices by precisely outlining the operational and financial components of the waste management services.
d. **Product sales**: Plastic waste can be processed at specific recycling centres into consumer products sold for a profit, such as outdoor furniture, lumber, or packaging

materials. Recycle India, Hyderabad firm constructed homes, covered bus stops, and water tanks using plastic waste bottles in 2015. Additionally, dining tables were made with thousands of chip packets woven into ropes and fastened to metal frames as part of this project. With the CSIR-National Physical Laboratory in Delhi, Shayna Ecounified Ltd. in Delhi transformed 340 tonnes of waste plastics (HDPE, LDPE, and PP) into 11 lakh tiles. The tiles were then sold to firms like Tata Motors, L'Oréal International and other cities. Moreover, not many recyclers turn plastic waste like shower curtains, oil containers, milk pouches, and household plastics into poly-fuel, a blend of gasoline and diesel. A small number of them gather PET waste and recycle it into carpets, cans, battery cases, clothing, and vehicle parts, among other things (Shanker et al. 2023).

e. **Corporate and public sponsorship/awareness**: Recycling programs can draw funding from government agencies, private companies, or nonprofit groups to promote a sustainable and eco-friendly project. Such sponsorships may contribute to cash assistance and brand awareness. World Environment Day showcases various community-led programs that aim to reduce plastic pollution worldwide. Examples include beach clean-ups in Mumbai, workshops in Ghana that teach people how to knit cloth bags, Atlanta concerts with zero plastic waste, and more. Furthermore, the hashtags #WorldEnvironmentDay and #BeatPlasticPollution are trending on Twitter on World Environment Day, indicating a high level of participation. These programs highlight how people and communities promote environmental action, motivating governments, banks, and businesses to fund comprehensive solutions to plastic pollution (Nayanathara Thathsarani Pilapitiya and Ratnayake 2024).

6 Conclusion

Utilizing waste plastic in the construction and building industry represents a promising opportunity to achieve carbon neutrality while addressing the environmental challenges of plastic pollution. The construction sector minimizes plastic waste development, lowers its carbon footprint, and promotes a more sustainable built environment by incorporating plastic waste into products and activities. Overcoming technological, regulation, and perception hurdles will need to be considered for coordinated efforts and cross-sector cooperation. As research and innovation continue to advance, the potential for plastic waste to transform the construction industry into a more sustainable and carbon–neutral sector is increasingly within reach.

Additionally, specific problems are associated with the tendency to use plastic waste in construction industry. There are specific technological challenges that one must overcome to realize the principle of plastic waste use in building construction, including the need to employ complex processing techniques to ensure that the building materials made from plastic waste are strong and cannot easily be compromised. The regulations must open and encourage using recycled plastics in infrastructures without compromising building and safety standards. In addition, it is critical to respond to such issues, including those that hit the public perception, because some stakeholders may have certain suspicions concerning the stability and environmental effect of the products developed from plastic. In this case, it becomes all-important for people in the different sectors to work together to ensure that plastic waste is effectively incorporated into the building sector. An example is the interaction between waste management corporations, material scientists, building companies or firms, and the regulating agencies. Such a partnership can also help to accelerate the development of such things as the best of breed, documented processes, and the funding of innovative research and development.

7 Future Perspective

Utilizing plastic waste in the construction sector has a great prospect because more effective solutions are being discussed due to constant research. As the technologies advance, the capability to turn waste plastic into valuable building materials improves, making the strategy workable and common. In addition, this will likely be more lavish spending on organic and sustainable construction materials, such as plastic waste, in the construction industry, depending on the emphasis on attaining carbon neutrality worldwide.

Research on the durability of plastic waste building materials in different climates and conditions may be performed in the following years, enhancing the utilization. Other research areas include using the waste plastics and incorporating them with other wastes like slag/fly ash to produce improved quality hybrid building materials. In addition, the precise and efficient application of plastic waste in construction works can also be made, owing to new trends in digital construction technologies like 3-D printing, which would also minimize the waste and energy consumption rate. Altogether, the escalation of sustainability in the construction industry might help assimilate the plastic waste into making the business environmentally friendly and carbon neutral.

References

Adiyanto O, Mohamad E, Razak JA (2022) Systematic review of plastic waste as eco-friendly aggregate for sustainable construction. Int J Sustain Constr Eng Technol 13(2):243–257. https://doi.org/10.30880/ijscet.2022.13.02.022

Ahmed N (2023) Utilizing plastic waste in the building and construction industry: a pathway towards the circular economy. Constr Build Mater 383:131311. https://doi.org/10.1016/J.CONBUILDMAT.2023.131311

Ahmed A, Ahmad S, Musa AES et al. (2024) Properties of concrete incorporating recycled coarse aggregates and recycled plastic fine aggregates. Innov Infrastruct Solut 9:165. https://doi.org/10.1007/s41062-024-01470-0

Akinwumi II, Domo-Spiff AH, Salami A (2019) Marine plastic pollution and affordable housing challenge: Shredded waste plastic stabilized soil for producing compressed earth bricks. Case Stud Constr Mater 11:e00241. https://doi.org/10.1016/J.CSCM.2019.E00241

Alaloul WS, John VO, Musarat MA (2020) Mechanical and thermal properties of interlocking bricks utilizing wasted polyethylene terephthalate. Int J Concr Struct Mater 14(1). https://doi.org/10.1186/s40069-020-00399-9

Aldagari S, Kabir SF, Fini EH (2022) A comparative study on efficacy of waste plastic and waste rubber in bitumen. Constr Build Mater 325:126724. https://doi.org/10.1016/J.CONBUILDMAT.2022.126724

Aneke FI, Shabangu C (2021) Green-efficient masonry bricks produced from scrap plastic waste and foundry sand. Case Stud Constr Mater 14:e00515. https://doi.org/10.1016/J.CSCM.2021.E00515

Appiah JK, Berko-Boateng VN, Tagbor TA (2017) Use of waste plastic materials for road construction in Ghana. Case Stud Constr Mater 6:1–7. https://doi.org/10.1016/J.CSCM.2016.11.001

Babafemi AJ, Šavija B, Paul SC, Anggraini V (2018) Engineering properties of concrete with waste recycled plastic: a review. In: Sustainability (Switzerland) 10(11). https://doi.org/10.3390/su10113875

Bae JS, Oh SK, Pedrycz W, Fu Z (2019) Design of fuzzy radial basis function neural network classifier based on information data preprocessing for recycling black plastic wastes: comparative studies of ATR FT-IR and Raman spectroscopy. Appl Intell 49(3):929–949. https://doi.org/10.1007/s10489-018-1300-5

Belay Wendimu T, Neguse Furgasa B, Mohammed Hajji B (2021) Suitability and utilization study on waste plastic brick as alternative construction material. J Civ Constr Environ Eng 6(1):9. https://doi.org/10.11648/j.jccee.20210601.12

Belmokaddem M, Mahi A, Senhadji Y, Pekmezci BY (2020) Mechanical and physical properties and morphology of concrete containing plastic waste as aggregate. Constr Build Mater 257:119559. https://doi.org/10.1016/J.CONBUILDMAT.2020.119559

Chong JH, Liu MS, Hernandes E, Albescu M (2023) Implementation of green materials in construction management system in Malaysia. Civ Sustain Urban Eng 3(1):51–69. https://doi.org/10.53623/csue.v3i1.212

Dawale SA (2016) Use of waste plastic coated aggregates in bituminous road construction

Dehghani M, Montazeri Z, Malik OP, Dhiman G, Kumar V (2019) BOSA: binary orientation search algorithm. Int J Innov Technol Exploring Eng 9(1):5306–5310. https://doi.org/10.35940/ijitee.A4215.119119

Dhawan R, Bisht BMS, Kumar R, Kumari S, Dhawan SK (2019) Recycling of plastic waste into tiles with reduced flammability and improved tensile strength. Process Saf Environ Prot 124:299–307. https://doi.org/10.1016/J.PSEP.2019.02.018

Dombe S, Tapase AB, Ghugal YM, Konnur BA, Patil A (2020) Investigation on the use of E-waste and waste plastic in road construction. In: Badawy Sherif D-H, Chen (ed) Recent developments in pavement engineering. Springer International Publishing, pp 85–99

dos Santos FA, Canto LB, Lúcia N. da Silva A, Lea Yuan Visconte L, Vasques Pacheco E BA (2020) Processing and properties of plastic lumber. In: Thermosoftening plastics. IntechOpen. https://doi.org/10.5772/intechopen.82819

Hama SM, Hilal NN (2017) Fresh properties of self-compacting concrete with plastic waste as partial replacement of sand. Int J Sustain Built Environ 6(2):299–308. https://doi.org/10.1016/J.IJSBE.2017.01.001

Hameed AM, Ahmed BAF (2019) Employment the plastic waste to produce the light weight concrete. Energy Procedia 157:30–38. https://doi.org/10.1016/J.EGYPRO.2018.11.160

Henriksen ML, Karlsen CB, Klarskov P, Hinge M (2022) Plastic classification via in-line hyperspectral camera analysis and unsupervised machine learning. Vib Spectrosc 118:103329. https://doi.org/10.1016/J.VIBSPEC.2021.103329

Heydari S, Hajimohammadi A, Haji Seyed Javadi N, Khalili N (2021) The use of plastic waste in asphalt: a critical review on asphalt mix design and Marshall properties. Constr Build Mater 309:125185. https://doi.org/10.1016/J.CONBUILDMAT.2021.125185

Ikechukwu AF, Shabangu C (2021) Strength and durability performance of masonry bricks produced with crushed glass and melted PET plastics. Case Stud Constr Mater 14:e00542. https://doi.org/10.1016/J.CSCM.2021.E00542

Kadam SS (2008) Application of plastic bottle as a wall structure for green house 1 Nikhil Datir. Int Res J Eng Technol 4162. www.irjet.net

Kamal MA, Moussa RR, Guirguis MN (2021) Recycled plastic as an aggregate in concrete. Civ Eng Architect 9(5):1289–1294. https://doi.org/10.13189/cea.2021.090502

Kataki S, Nityanand K, Chatterjee S et al. (2022) Plastic waste management practices pertaining to India with particular focus on emerging technologies. Environ Sci Pollut Res 29:24478–24503. https://doi.org/10.1007/s11356-021-17974-6

Kellenberg D (2012) Trading wastes. J Environ Econ Manag 64(1):68–87. https://doi.org/10.1016/J.JEEM.2012.02.003

Khalid FS, Irwan JM, Ibrahim MHW, Othman N, Shahidan S (2018) Performance of plastic wastes in fiber-reinforced concrete beams. Constr Build Mater 183:451–464. https://doi.org/10.1016/J.CONBUILDMAT.2018.06.122

Khalid MY, Arif ZU, Ahmed W, Arshad H (2022) Recent trends in recycling and reusing techniques of different plastic polymers and their composite materials. Sustain Mater Technol 31:e00382. https://doi.org/10.1016/J.SUSMAT.2021.E00382

Lahtela V, Hyvärinen M, Kärki T (2019) Composition of plastic fractions in waste streams: toward more efficient recycling and utilization. Polymers 11(1). https://doi.org/10.3390/polym11010069

Lamba P, Kaur DP, Raj S, Sorout J (2022) Recycling/Reuse of plastic waste as construction material for sustainable development: a review. In: Environmental science and pollution research, vol 29, no 57. Springer Science and Business Media Deutschland GmbH, pp 86156–86179. https://doi.org/10.1007/s11356-021-16980-y

Limami H, Manssouri I, Cherkaoui K, Saadaoui M, Khaldoun A (2020) Thermal performance of unfired lightweight clay bricks with HDPE & PET waste plastics additives. J Build Eng 30:101251. https://doi.org/10.1016/J.JOBE.2020.101251

Maddela NR, Kakarla D, Venkateswarlu K, Megharaj M (2023) Additives of plastics: entry into the environment and potential risks to human and ecological health. J Environ Manag 348:119364. https://doi.org/10.1016/J.JENVMAN.2023.119364

Mahmood RA, Kockal NU (2020) Cementitious materials incorporating waste plastics: a review. In: SN Applied Sciences, vol 2, no 12. Springer Nature, Berlin. https://doi.org/10.1007/s42452-020-03905-6

Manju SS, Sheema K (2017) Use of plastic waste in bituminous pavement. https://www.researchgate.net/publication/320243162

Mansour AMH, Ali SA (2015) Reusing waste plastic bottles as an alternative sustainable building material. Energy Sustain Dev 24:79–85. https://doi.org/10.1016/J.ESD.2014.11.001

Mondal MK, Bose BP, Bansal P (2019) Recycling waste thermoplastic for energy efficient construction materials: an experimental investigation. J Environ Manag 240:119–125. https://doi.org/10.1016/J.JENVMAN.2019.03.016

Nayanathara Thathsarani Pilapitiya PGC, Ratnayake AS (2024) The world of plastic waste: a review. Cleaner Mater 11:100220. https://doi.org/10.1016/J.CLEMA.2024.100220

Othalo—The Future of Housing. (n.d.). Retrieved 8 June 2024. From https://othalo.com/

Paihte PL, Lalngaihawma AC, Saini G (2019) Recycled aggregate filled waste plastic bottles as a replacement of bricks. Mater Today Proc 15:663–668. https://doi.org/10.1016/J.MATPR.2019.04.135

Rahat MHH, Massarra C, Wang GC (2022) Using plastic wastes in construction: opportunities and challenges. In: ASC2022. 58th annual associated schools of construction international conference, vol 3, pp 785–774. https://doi.org/10.29007/6369

Rahman MH, Bhoi PR, Menezes PL (2023) Pyrolysis of waste plastics into fuels and chemicals: a review. Renew Sustain Energy Rev 188:113799. https://doi.org/10.1016/J.RSER.2023.113799

Rajput PS, Yadav RK (2016) Use of plastic waste in bituminous road construction. Int J Sci Technol Eng 2:509–513. https://api.semanticscholar.org/CorpusID:54939994

Ribeiro LS, Stolz CM, Amario M, da Silva ALN, Haddad AN (2023) Use of post-consumer plastics in the production of wood-plastic composites for building components: a systematic review. Energies 16(18). https://doi.org/10.3390/en16186549

Rosenfield K (2015) This house was built in 5 days using recycled plastic bricks. ArchDaily. https://www.archdaily.com/606704/this-plastic-bottle-house-turns-trash-into-affordable-housing-in-nigeria

Safinia S, Alkalbani A (2016) Use of recycled plastic water bottles in concrete blocks. Procedia Eng 164:214–221. https://doi.org/10.1016/J.PROENG.2016.11.612

Sau D, Shiuly A, Hazra T (2024) Utilization of plastic waste as replacement of natural aggregates in sustainable concrete: effects on mechanical and durability properties. Int J Environ Sci Technol 21:2085–2120. https://doi.org/10.1007/s13762-023-04946-1

Seghiri M, Boutoutaou D, Kriker A, Hachani MI (2017) The possibility of making a composite material from waste plastic. Energy Procedia 119:163–169. https://doi.org/10.1016/J.EGYPRO.2017.07.065

Shanker R, Khan D, Hossain R, Islam MT, Locock K, Ghose A, Sahajwalla V, Schandl H, Dhodapkar R (2023) Plastic waste recycling: existing Indian scenario and future opportunities. Int J Environ Sci Technol 20(5):5895–5912. https://doi.org/10.1007/s13762-022-04079-x

Soni A, Das PK, Hashmi AW, Yusuf M, Kamyab H, Chelliapan S (2022) Challenges and opportunities of utilizing municipal solid waste as alternative building materials for sustainable development goals: a review. Sustain Chem Pharm 27:100706. https://doi.org/10.1016/J.SCP.2022.100706

Tahir F, Alzahrani S, Noori Y, Al-Ghamdi SG (2024) Environmental impacts and the future prospects of waste utilization in the concrete production. Mater Today Proc. https://doi.org/10.1016/J.MATPR.2024.05.150

Tang KHD (2022) Valorization of plastic waste through incorporation into construction materials. Civ Sustain Urban Eng 2(2):96–109. https://doi.org/10.53623/csue.v2i2.141

Tayebani B, Said A, Memari A (2023) Less carbon producing sustainable concrete from environmental and performance perspectives: a review. Constr Build Mater 404:133234. https://doi.org/10.1016/J.CONBUILDMAT.2023.133234

Tejaswini MSSR, Pathak P, Ramkrishna S, Ganesh PS (2022) A comprehensive review on integrative approach for sustainable management of plastic waste and its associated externalities. Sci Total Environ 825:153973. https://doi.org/10.1016/J.SCITOTENV.2022.153973

The Indian Express (2019) Plastic waste management: Surat among four cities chosen for central project. https://indianexpress.com/article/india/plastic-waste-management-surat-among-four-cities-chosen-for-central-project-6127752/

Vasudevan V (2019) Guidelines for the use of plastic waste in rural roads construction. 1.0 Background specifications for waste plastic

Velmurugan V (2019) Rebuilding of plastic waste to pavement bricks. Int J Res Appl Sci Eng Technol 7(4):927–931. https://doi.org/10.22214/ijraset.2019.4165

Záleská M, Pavlíková M, Jankovský O, Lojka M, Pivák, A, Pavlík Z (2018) Experimental analysis of MOC composite with a waste-expanded polypropylene-based aggregate. Materials 11(6):931. https://doi.org/10.3390/ma11060931

Zhao C, Liu M, Du H, Gong Y (2021) The evolutionary trend and impact of global plastic waste trade network. Sustainability 13(7):3662. https://doi.org/10.3390/SU13073662

Navigating the Microplastic Challenge: Unraveling the Dynamics in Anaerobic Digestion and Treatment Processes of Sewage Sludge

Sandhya Kumari Gupta and Sankar Ganesh Palani

Abstract

Microplastic (MP) pollutants, defined as polymers smaller than 5 mm, are increasingly recognized as a widespread environmental concern. While their impacts are becoming clearer, the full extent of their consequences remains uncertain. MPs vary significantly in size and are prevalent in domestic sewage sludge (DSS), largely due to synthetic fibers from textiles and personal care products like toiletries and cosmetics. DSS is rich in organic matter, biomass, and inorganic chemicals, which complicates the extraction and identification of MPs. Proper characterization is essential for effective removal. Anaerobic digestion (AD), a leading treatment method for organic-rich sewage sludge, offers multiple benefits: biogas production, pathogen elimination, odor reduction, and sludge volume minimization. However, studies reveal mixed findings on how MPs influence AD, highlighting both potential benefits and challenges that remain poorly understood. Therefore, the present study aims to characterize microplastics using various methodologies and outlining sampling and extraction techniques for microplastics from sewage sludge. It also examines MPs' degradation during AD and their behavior within sewage treatment systems. The study highlights that MPs are found universally in sewage sludge, varying in size, composition, and concentration, with polyethylene, nylon type and fibers/fragments being the most common. Removal efficiencies at wastewater treatment plants range from 93.8 to 99.8%, yet the regional and study-specific variations are significant. Traditional sewage sludge management methods, such as composting, landfilling, and agricultural applications, inadvertently reintroduce MPs into the environment. In agriculture, MPs alter soil properties, affecting microbial populations, plant health, and soil biota. Hence, it is essential to prevent MPs from impacting the environment and to exclude them from sewage sludge.

S. K. Gupta · S. G. Palani (✉)
Environmental Biotechnology Laboratory, Department of Biological Sciences, Birla Institute of Technology and Science, Pilani, Hyderabad Campus, Hyderabad, Telangana, India
e-mail: p20210100@hyderabad.bits-pilani.ac.in

A. Mandpe et al. (eds.), *Sustainable Waste Management*, Advances in Science, Technology & Innovation,
https://doi.org/10.1007/978-3-031-85685-3_5

Graphical Abstract

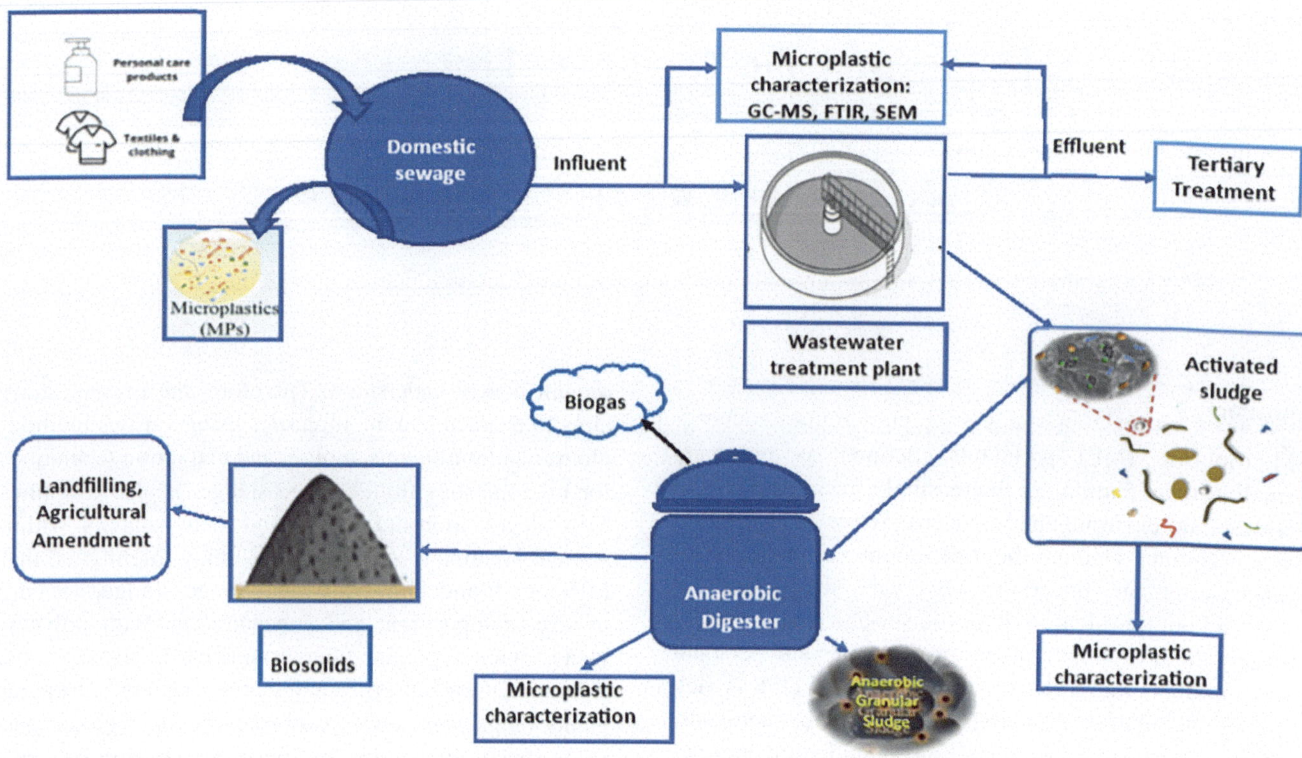

Keywords

Wastewater treatment · Microplastics · Domestic sewage sludge · Biogas · Biofilm · Anaerobic digestion

1 Introduction

1.1 Plastic Use in Daily Routine

Plastic, a synthetic polymer derived from natural resources, has integrated itself as a fundamental component in various aspects of our daily lives. The increase in plastic utilization may be attributed to its properties, such as sturdiness, affordability, versatility, resistance to dampness, insulating properties, and low maintenance needs. Since plastic is so durable and adaptable, it has significantly impacted modern life. This easily formable material can take on various shapes, including film, fibers, beads, plates, pipes, bottles, and boxes, and is also flexible. Its prevalence in consumer goods such as clothing, homes, cars, and toys makes these products more durable and cost-effective. It is made up of natural substances such as carbon, natural gas, salt, cellulose, crude oil, and many others that are utilized as raw materials in the production of plastics, which gives the finished product its shape and durability (Domenech and Marcos 2021). Plastics are broadly categorized into thermoplastics and thermosets. Thermoplastics soften when heated and harden again as they cool, allowing them to be reshaped multiple times while retaining their durability and form. In contrast, thermosetting polymers undergo chemical changes when exposed to high heat, permanently setting their structure and making them non-processable (Evode et al. 2021) (Table 1).

Plastic is a versatile material well-suited for various industrial and consumer purposes. Its low density makes

Table 1 List of common types of thermoplastics and thermosets

Thermoplastics materials	Thermosets materials
Polyethylene (PE)	Polyurethane (PUR)
Polypropylene (PP)	Vinyl esters
Polyvinyl chloride (PVC)	Unsaturated polyesters
Polystyrene (PS)	Silicone
Polyamides (PA)	Phenolic resins
Polycarbonate (PC)	Acrylic resins
Polyarylsulfone (PSU)	Urea-formaldehyde resins

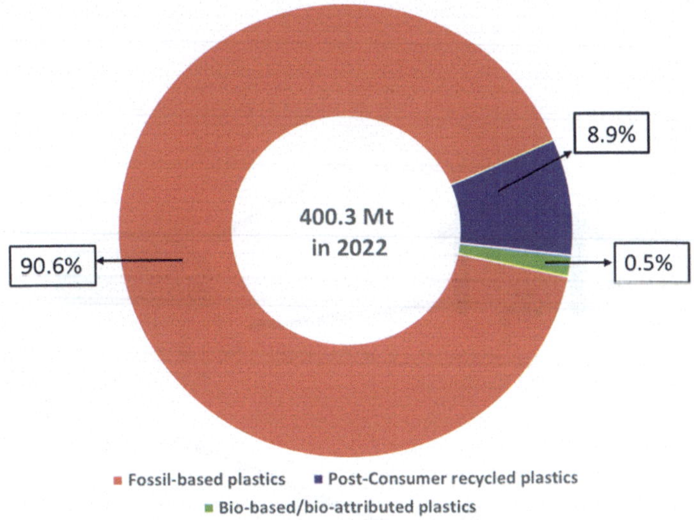

Fig. 1 Various sources of the world's plastic production in 2022

it lightweight and corrosion-resistant due to the presence of additional compounds. These properties make it ideal for use in harsh environments. Plastics are also excellent insulators for thermal and electrical energy, and their flexibility enables them to be molded into various shapes. However, its widespread use and increasing production levels have significantly impacted modern society's quality of life. In 1960, 0.5 million tonnes of plastic were produced; today, more than 400 million tonnes are produced annually. China, NAFTA (North American Free Trade Agreement), and Europe are the top three producers of plastic materials (thermoplastic and polyurethane). The most recent data available for 2022 shows an increase in worldwide plastics production to 400.3 million tonnes (Ammentorp Lund 2023). In 2022, most plastics produced globally, precisely 90.6%, were derived from fossil fuels, and the remaining plastic production comprised 8.9% post-consumer recycled plastics and 0.5% bio-based or bio-attributed plastics in Fig. 1 (Ammentorp Lund 2023).

Due to the rise in plastic production, it is imperative to regulate polymers as they are major global pollutants. While plastic recycling is essential, each product's life cycle and durability vary. For example, plastic bags or bottles have a shorter lifespan than electrical components or auto parts. Users of all plastic products must receive adequate information on managing them post-use. Despite increased recycling rates, a quarter of plastic still ends up in landfills after use (Domenech and Marcos 2021).

1.2 From Plastics to Microplastics

Plastics are synthetic materials from various organic and inorganic compounds, including polymers, resins, and additives (Quecholac-Piña et al. 2020). Plastics, known for their durability, versatility, and cost-effectiveness, have become indispensable in various industries like packaging, construction, and electronics. However, a significant downside is that plastics do not break down naturally, leading them to accumulate in our environment. Over time, when exposed to sunlight, heat, or physical wear and tear, plastics start to degrade and break into smaller pieces, eventually turning into microplastics—tiny plastic fragments smaller than 5 mm. These microplastics are now widespread, found in water, soil, air, and even our food. Their presence poses risks to wildlife and human health. The process of plastics breaking down into microplastics is influenced by several environmental factors, such as temperature, pH levels, and the presence of microorganisms. Understanding how this transformation occurs is essential for devising strategies to reduce plastic and microplastic pollution in our environment (Papp et al. 2022).

1.3 Domestic Sewage, as a Sink of Microplastics

Household wastewater, often referred to as domestic sewage, is a major contributor to environmental microplastics. These tiny plastic particles are commonly present in everyday consumer items like cosmetics, textiles, and packaging materials (Hayany et al. 2022). When these products are used and disposed of, they enter the sewage system through drains, toilets, and washing machines via domestic wastewater, among other pathways such as industrial effluent, stormwater runoff, and landfill leachates, as shown in Fig. 2. Once in the sewage system, it can accumulate in sewage treatment plants, as they are often too small, approximately 10% of MPs to be filtered out by conventional wastewater treatment processes. As a result, microplastics

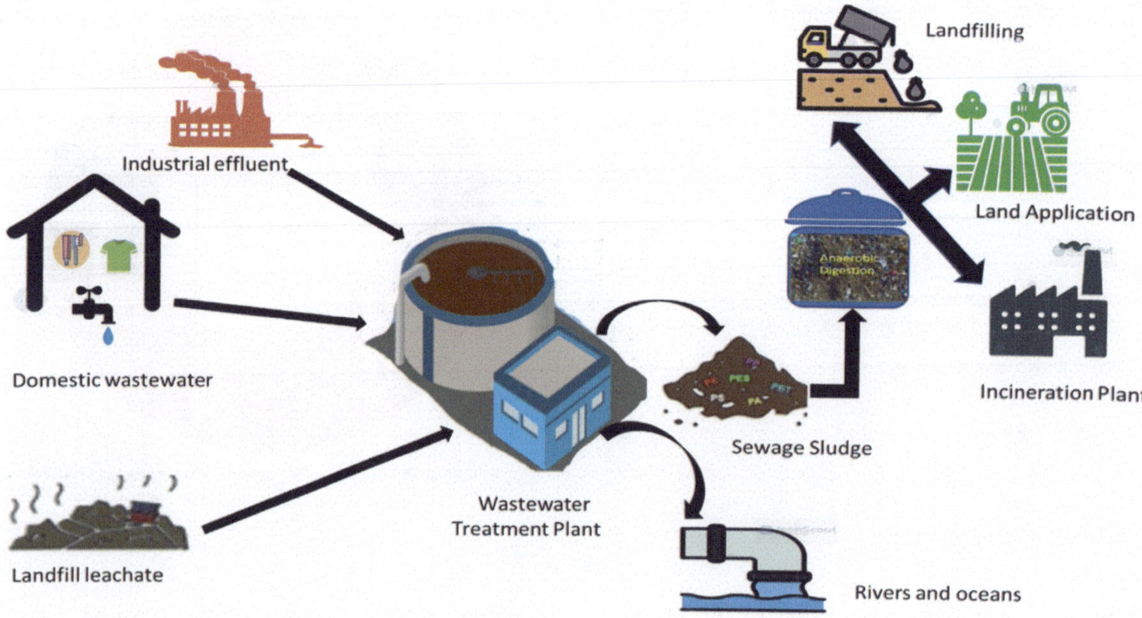

Fig. 2 Sources and pathways of microplastics to wastewater treatment plants

can be discharged into the environment along with treated wastewater. This can harm aquatic ecosystems and potentially enter the food chain. However, over 90% of microplastics entering wastewater treatment plants (WWTPs) are ultimately removed from the wastewater through adsorption and are retained in the sludge (Carr et al. 2016). Furthermore, the microplastics (MPs) removed from the wastewater during treatment are typically transferred into the sludge rather than being completely eliminated. This indicates that the sludge produced in wastewater treatment plants (WWTPs) can act as a reservoir for microplastics (MPs). These MPs can potentially be released into the environment through various pathways such as land application, landfilling, incineration, or disposal in water bodies, as illustrated in Fig. 2. This indicates that WWTPs not only act as sources of MPs pollution by releasing MPs with effluent but also as sinks by accumulating MPs in the sludge. Therefore, the present study aims to characterize microplastics using various methodologies and outlining sampling and extraction techniques for microplastics from sewage sludge. It also examines MPs' degradation during AD and their behavior within sewage treatment systems.

2 Methodology

Detecting and characterizing microplastics in domestic sewage sludge is essential for comprehending their potential environmental impact. The identification of microplastics in domestic sewage sludge generally consists of three steps: sample collection, sample pretreatment, and the analysis/quantification of microplastics, as summarized in Fig. 3. It should be emphasized that the approaches applied at every stage have yet to be standardized. Depending on the characteristics of the samples, various techniques can be utilized, as microplastics are found in both wastewater and sewage sludge. Moreover, the methodologies used for microplastic identification can lead to varying outcomes in the final analysis. Furthermore, most research projects should implement quality assurance measures to mitigate potential biases arising from sample contamination and loss (Sun et al. 2019).

2.1 Sampling, Sample Preparation, and Extraction of Microplastics from Sewage Sludge

Sampling sewage sludge from a treatment plant requires careful planning to ensure the results accurately reflect its characteristics. Start by defining your goal for compliance checks, nutrient analysis, or studying contaminants like microplastics. Next, choose sampling points based on these goals, such as raw, treated, or intermediate sludge. It is crucial to choose the right type of sampling method: grab sampling, which involves collecting single samples at specific times, or composite sampling, which combines multiple samples over a period to provide an average representation. The sampling frequency is determined by regulatory requirements or research needs, with routine analysis often performed monthly or quarterly and detailed studies requiring daily or event-based sampling. Proper equipment, such as sludge samplers, dipper poles, or autosamplers, should

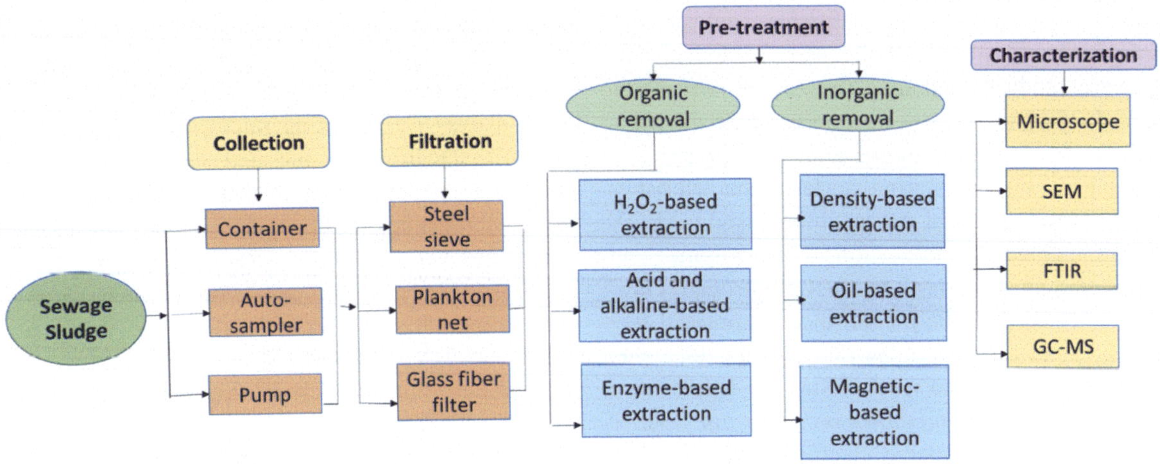

Fig. 3 Flowchart outlining the procedures and methods for detecting microplastic in sewage sludge

be used along with clean, inert containers (e.g., glass, steel, or polyethylene) to prevent contamination. Homogenizing the sludge before sampling ensures a consistent mixture, and the samples should be promptly preserved, typically at 4 °C. Accurate labeling and thorough documentation of the sampling conditions, such as date, time, weather, and location, are essential for traceability. Samples must be transported to the laboratory under controlled conditions for analysis. Following standards like ISO 5667-13 or USEPA guidelines ensures reliability. Always prioritize safety, using PPE to protect yourself during the process.

For sample preparation, sample bottles containing SS were thawed to room temperature and thoroughly agitated before analysis. Triplicates of the SS sample were used for the analyses. 100 g of SS was weighed into a clean, pre-washed 500 ml glass beaker and dried at 90 °C in an oven for 24 h. In alignment with NOAA's (National Oceanic and Atmospheric Administration) guidelines (Marine Debris Program 2015), digestion and density separation were carried out by utilizing the catalyst-mediated wet-peroxide oxidation (WPO) technique. This process converts most organic molecules into water, carbon dioxide, aldehydes, and carboxylic acids. The transformation occurs through hydroxyl radicals generated from the decomposition of hydrogen peroxide. The iterative addition of iron sulfate ($FeSO_4$) expedites the breakdown of organic substances (Hayany et al. 2022). To execute this, 30 ml of iron catalyst and 30 ml of 30% H_2O_2 were added to the SS beaker. The iron catalyst solution was added first, followed by a slow addition of hydrogen peroxide. Sewage sludge samples were treated at ambient temperature and then cooled in an ice bath if the temperature surpassed 40 °C to prevent MP bleaching. The Fe (II) catalyst solution and H_2O_2 were added to the mixture until it turned clear, with no remaining foam or hissing sound. 7.6 g of $FeSO_4$ powder was added in 1 L of Mili-Q water with 3 ml of sulfuric acid to make a 0.05M solution (Marine Debris Program 2015; Kye et al. 2023).

Larger microplastic particles (MPPs) were manually picked using tweezers, while smaller ones were separated through filtration and density separation (Labbe et al. 2020). A dual-density separation method was employed to eliminate inorganic substances and natural organic compounds. This involved using a sodium iodide (NaI) solution (1.8 g/cc) and letting it stand at ambient temperature for 24 h to facilitate the buoyancy of fine particles. The top liquid was poured and passed through a fiberglass filter (Durapore PVDF 0.45 μm, 47 mm) under a vacuum apparatus. The density-based separation process was repeated two more times for efficient recovery of MPPs as the density of plastic polymer ranges from 0.85 to 1.45 g/cc. After filtration, the filter membrane containing MPPs was transferred onto a sterile glass petri dish, rinsed in ethanol (0.8 g/cc), dried at ambient atmospheric temperature, and stored for further quantitative and qualitative analysis.

2.2 Identification, Characterization, and Quantification of Microplastics from Sewage Sludge

Identifying, characterizing, and quantifying MPs post-filtration were comprehensive and detailed. First, the physical identification was carried out by counting the MPPs on each filter using a stereo microscope with magnifications ranging from 4x to 40x. Larger particles were visually examined and manually picked up using stainless steel tweezers or hand. The concentration of MPs was then described by calculating the average of three samples, and they are categorized by their size, color, type, and shape.

By incorporating insights from qualitative analysis, 5% of the particles classified as unknown were excluded from the quantitative study to enhance the results' precision. Furthermore, FE-SEM/EDX was used to examine the MP's morphological structure and aging comprehensively and provided detailed, high-resolution surface characteristics data and qualitative chemical composition insights. EDX determines the elemental composition of the polymer and measures the amount of carbon, hydrogen, nitrogen, chlorine, sulfur, and oxygen to help confirm the empirical formula of the polymer and determine if it contains elements that hint at specific backbone types. Extracted MPPs were fixed on carbon tape and mounted on stubs for sputter coating. The SEM's backscattered electron (BSE) detector was used to image MPPs at magnifications ranging from 50x to 500x. This method is highly beneficial for distinguishing between organic matters (abundant in Calcium/Magnesium/Strontium) and inorganic substances (minerals, salts) as well as MPs (abundant in Carbon/Chloride/Sulfur/Titanium) (Gniadek and Dąbrowska 2019)

Nile Red (NR) staining was also used with fluorescence microscopy to detect and quantify MPs accurately. NR is a fluorescent dye that binds to the hydrophobic regions of MPs, causing them to emit a bright red–orange fluorescence when exposed to specific wavelengths of light. As a solvatochromic dye, NR fluorescence emission varies according to the polarity of its surroundings, allowing differentiation and categorization of MPs into distinct chemical groups based on their unique fluorescent properties (Labbe et al. 2020). In this process, NR dye was dissolved in acetone solvent at 10 μg/ml concentration. Separated MPPs were coated with the NR dye. The subsequent procedure involved analyzing these MPs stained with NR using laser scanning confocal microscopy (LSCM) with specific wavelengths for excitation (430–490 nm) and emission (510–560 nm). Finally, images of the NR fluorescent MPs were captured, allowing for their quantification and characterization (Shruti et al. 2022).

The chemical identification of the isolated MPs was obtained using an ATR-FTIR in transmittance mode to identify the polymer type. Spectra were obtained by recording vibrations at wavelengths ranging from 4000 to 500 $cm-^{1}$, with a 4 cm^{-1} resolution.

However, due to variations in their physical and chemical properties, a single detection method is inadequate. Therefore, a multi-detection approach, consisting of physical and chemical identification methods, is employed. The stereomicroscope is commonly used for counting and morphological analysis of microplastics. However, its lower magnification capacity and detection efficiency are limitations. Scanning electron microscopy produces high-definition images of the external surfaces of microplastics, allowing for easy differentiation between synthetic and organic materials. Polarized light microscopy is another method used for detecting microplastics, but it has limitations in detecting opaque and thick particles. Chemical analysis can be used to characterize MPs and establish their chemical makeup, and it can be divided into destructive and non-destructive approaches. Raman spectroscopy and FTIR are destructive methods, while liquid chromatography, gas chromatography paired with mass spectrometry, including pyrolysis gas chromatography-mass spectrometry, and thermal desorption gas chromatography are non-destructive techniques (Gniadek and Dąbrowska 2019).

Characterizing microplastics in domestic sewage sludge can help us understand their origins, eventual fate, and environmental impacts. The sources of these particles, such as clothing fibers, packaging materials, or microbeads from personal care products, may help identify the different forms of microplastics found in sewage sludge. Characterizing the size distribution of microplastics is important because smaller particles are more likely to be ingested by aquatic organisms and potentially enter the food chain, thereby highlighting their potential environmental impacts. Understanding how microplastics from domestic sewage sludge affect our environment is key to tackling the issue effectively. By detecting and analyzing these tiny particles, we can better grasp their impact and develop real solutions to minimize their presence.

2.3 Innovative Technologies for Microplastic Removal

Advanced treatment methods are showing great promise in tackling the stubborn issue of microplastics in sewage sludge. Conventional methods of treating sludge often fall short, allowing microplastics to break down into even smaller particles, which can alter their surface properties and make them even more difficult to remove (Surana et al. 2023). However, newer technologies, like membrane bioreactors and advanced filtration systems, have stepped up, achieving over 90% removal rates for certain types of plastics and particle sizes. These cutting-edge techniques are crucial in reducing microplastics environmental and health risks, significantly lowering their presence in sludge and wastewater (Surana et al. 2023; Parashar and Hait 2023). In addition to physical filtration, advanced treatment processes also include chemical and biological methods. For instance, advanced oxidation processes (AOPs) use powerful agents like hydroxyl radicals to break down microplastics completely (Gayathri et al. 2023). Membrane bioreactor

(MBR) technology enhances the removal of these particles, while modified dissolved air flotation (MDAF) combined with innovative self-excited oscillating pulsed cavitation-impinging (SOPC) processes help eliminate microplastics, antibiotics, and other contaminants from livestock wastewater (Mishra et al. 2022; Zhu et al. 2022). Additionally, ultrasonic methods have also been explored and found to be effective in degrading microplastics (Kim et al. 2022). Moreover, biological treatments are gaining attraction, using microorganisms such as fungi, bacteria, algae, and protists to form biofilms on microplastics, altering their structure for easier breakdown (Bule Možar 2023). The activated sludge present in wastewater treatment facilities acts as a significant reservoir for microplastics, as a substantial amount of microplastics get trapped in the sludge during the treatment process. Nevertheless, traditional sludge treatment procedures frequently prove ineffective in eliminating microplastics, resulting in secondary contamination and a modified surface structure that can potentially enhance the adsorption of additional co-contaminants (Sutradhar 2022). To counter this, scientists are working on genetically modifying bacteria in activated sludge to improve their ability to degrade microplastics, leveraging biotech and synthetic biology to enhance wastewater treatment and reduce environmental impact.

2.4 Challenges and Standardization in Microplastic Analysis

Standardized method for sampling and analyzing microplastics comes with significant challenges. These include a lack of consistency in sampling techniques, such as using Manta trawl nets, plankton pumps, and various submersible pumps with different mesh sizes (Shi et al. 2023). This inconsistency leads to variations in the amount and types of microplastics collected. Additionally, the complex task of identifying and measuring microplastics in different environmental settings makes it tough to establish uniform methods that allow for reliable data comparison (Muthukumaran et al. 2021). The small size, irregular shapes, and diverse polymer types of microplastics make them difficult to detect and measure using tools like stereomicroscopy, SEM, and spectroscopic analysis (Adhikari et al. 2022). There is also a risk of losing or contaminating microplastics during sampling and lab work, adding another layer of difficulty to standardization efforts. The lack of a universally accepted protocol for sampling and sizing micro-nano-plastics in air particles further complicates the effort to consolidate and critically evaluate research findings (Logvina, et al. 2023).

3 Sludge Stabilization

Sewage sludge is rich in organic matter due to the very high concentration of biomass, along with inorganic compounds such as ammoniacal nitrogen, phosphates, humic substances, heavy metals, and MPs (Wei et al. 2023). They also contain harmful pathogens, which must be stabilized with the organic content before discharging into the environment. The standard methods used for sewage sludge treatment include composting, landfilling, agricultural applications, alkaline stabilization, aerobic digestion, and anaerobic digestion. Due to the production of biogas from organic-rich sewage sludge, anaerobic digestion is considered one of the effective stabilization techniques used for its treatment (Liu and Smith 2022). Anaerobic digestion can eliminate pathogens and odor, reduce the sludge volume and generate renewable energy. However, MPs can affect the anaerobic digestion process, as observed in many studies. Depending on their physical and chemical characteristics, different MPs will have a varied effect on the AD process. Few studies have proved that MPs can target microorganisms and attach them to their cell, impairing the function of essential genes and enzymes (Zielińska and Cydzik-Kwiatkowska 2024). Some microplastics can create reactive oxygen species (ROS), damaging the microorganisms' cell membrane, nucleus, and proteins (Zielińska and Cydzik-Kwiatkowska 2024; Manu 2023). Although the presence of MPs can adversely affect the anaerobic digestion process, it was observed that polyamide microplastic might enhance the anaerobic digestion process by increasing the activity of methanogen-related enzymes (Şimşek and Sanin 2024). It has been observed that the presence of MPs can lead to the colonization of microorganisms and biofilm formation. Few studies have observed that biofilm formation can lead to the selection of microorganisms unrelated to the anaerobic digestion process, thereby decreasing the efficiency of the process (Hulshoff et al. 2004; Arab et al. 2024; Gagliano et al. 2017).

3.1 Overview of Anaerobic Digestion

Anaerobic digestion is a natural process where organic material breaks down without oxygen, producing biogas and a nutrient-rich byproduct called digestate. This process occurs naturally in places like wetlands, but it can also be used in controlled environments to treat organic waste and generate renewable energy (https://www.tn.gov/environment/program-areas/sw-mm-organics/anaerobic-digestion.html). In waste treatment, materials like food scrap,

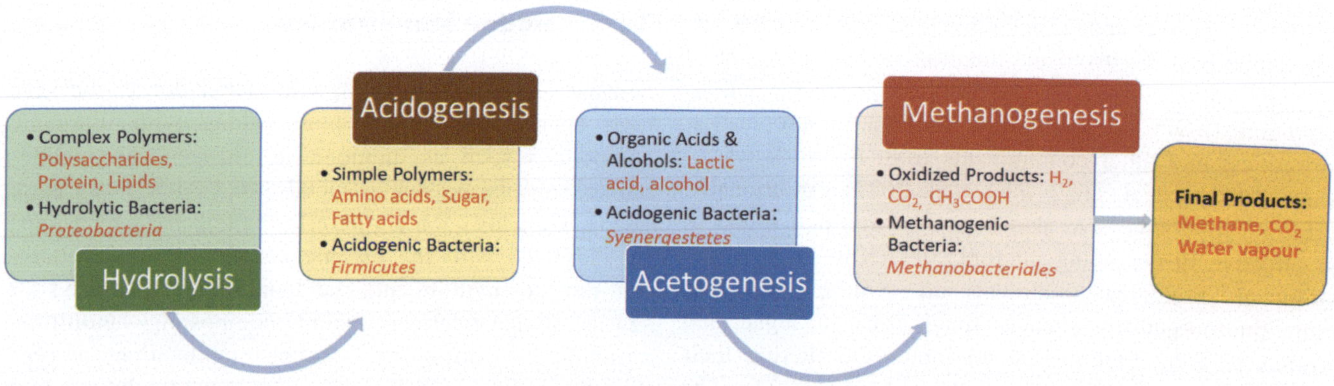

Fig. 4 Illustrative example of the four phases of anaerobic digestion

agricultural leftovers, and sewage sludge are placed in a sealed container called a digester. There, the waste is heated and mixed with bacteria that break it down into biogas—a mix of carbon dioxide and methane. This biogas can then be used as a renewable energy source, either by burning it for electricity and heat or refining it to meet natural gas standards for distribution.

The digestate left behind is a valuable fertilizer that can be used in farming or landscaping. Anaerobic digestion offers several benefits, including reducing the volume and odor of organic waste, creating renewable energy, and producing useful fertilizer. Beyond waste treatment, this process can also produce biogas from agricultural residues like manure and crop leftovers, and it is effective in treating wastewater from industries such as food processing and pharmaceuticals. Pairing anaerobic digestion with other technologies, like combined heat and power (CHP) systems, can further enhance energy efficiency. However, the growing presence of microplastics in domestic sewage sludge poses a challenge, as it can negatively affect the efficiency of anaerobic digestion, a key process in the circular economy. Overall, anaerobic digestion is a versatile and eco-friendly method that plays a vital role in waste management, renewable energy production, and resource recovery.

3.2 Phases of the Anaerobic Digestion

Anaerobic digestion unfolds in several distinct phases, each defined by specific microbial activities and environmental conditions, as shown in Fig. 4. The main phases of anaerobic digestion are as follows:

1. **Hydrolysis**: The AD process begins with hydrolysis, where hydrolytic bacteria break down complex organic materials into simpler substances like sugars and amino acids. This phase is critical because it sets the stage for the subsequent steps in the digestion process.

$$(C_6H_{10}O_5)_n + nH_2O \rightarrow nC_6H_{12}O_6$$
$$\text{(Polysaccharides to Glucose)} \quad (1)$$

2. **Acidogenesis**: In this phase, the simpler compounds formed during hydrolysis are further broken down by acidogenic bacteria into alcohols, volatile fatty acids, and other organic acids. This stage is defined by acidic conditions and the release of carbon dioxide and hydrogen gas.

$$C_6H_{12}O_6 \rightarrow 2CH_3CH_2OH + 2CO_2 + 2H_2$$
$$\text{(Glucose to Ethanol, Carbon Dioxide, and Hydrogen)} \quad (2)$$

3. **Acetogenesis**: In this stage, acetogenic bacteria take the organic acids formed earlier during acidogenesis and convert them into acetic acid and other short-chain fatty acids. As this process unfolds, the environment becomes less acidic, producing gases like carbon dioxide and hydrogen.

$$C_6H_{12}O_6 + 2H_2O \rightarrow 2CH_3COOH + 2CO_2 + 4H_2$$
$$\text{(Glucose to Acetic Acid, Carbon Dioxide, and Hydrogen)} \quad (3)$$

4. **Methanogenesis**: In the final phase of anaerobic digestion, methanogenic bacteria play a crucial role in converting short-chain fatty acids, like acetic acid, into methane and carbon dioxide. This conversion happens in a neutral pH environment and produces biogas, which primarily consists of methane and carbon dioxide.

$$CH_3COOH \rightarrow CH_4 + CO_2$$
$$\text{(Acetic Acid to Methane and Carbon Dioxide)} \quad (4)$$

$$4H_2 + CO_2 \rightarrow CH_4 + 2H_2O$$
$$\text{(Hydrogen and Carbon Dioxide to Methane and Water)} \quad (5)$$

Equations 1, 2, 3, 4, 5 illustrate the breakdown of organic matter at each anaerobic digestion phase, ultimately resulting in the generation of biogas, mainly composed of methane (CH_4) and carbon dioxide (CO_2).

The various stages of anaerobic digestion are deeply interlinked, relying on each other to function effectively. Their success is vital not only for producing biogas efficiently but also for stabilizing organic waste. Grasping the microbial activities and environmental factors that drive each phase is key to enhancing the performance of anaerobic digesters and fully tapping into the benefits this technology offers.

3.3 Operational Parameters of Anaerobic Digestion

The performance and efficiency of anaerobic digestion depend on several key factors, collectively known as operational parameters, which are illustrated in Fig. 5.

Key factors that influence the success of anaerobic digestion include:

1. **Temperature**: The efficiency of anaerobic digestion is highly temperature-dependent, with the ideal range varying based on the feedstock type and the digester's microorganisms. For instance, mesophilic digestion operates best between 25 °C and 40 °C, while thermophilic digestion is optimal at higher temperatures, typically between 50 °C and 60 °C.
2. **Hydraulic Retention Time (HRT)**: refers to how long the feedstock remains in the digester. The ideal HRT is influenced by the type and concentration of feedstock, as well as the microbial community within the digester. Generally, a longer HRT can increase biogas yields but requires a larger digester.
3. **Organic Loading Rate (OLR)**: The OLR measures the amount of organic material added to the digester per unit volume or surface area each day. Balancing the OLR is crucial—too high, and the process becomes unstable, decreasing biogas production; too low, and the digester's capacity is underutilized.
4. **pH**: The pH level in the digester is vital for maintaining a healthy microbial community. Most anaerobic digesters operate within a pH range of 6.5–8.5, depending on the specific microorganisms involved.
5. **Mixing**: Proper mixing is essential to maintain uniform conditions within the digester and prevent the buildup of solids or scum. The mixing rate should be adjusted based on the feedstock, digester size, and the type of mixing system in place.

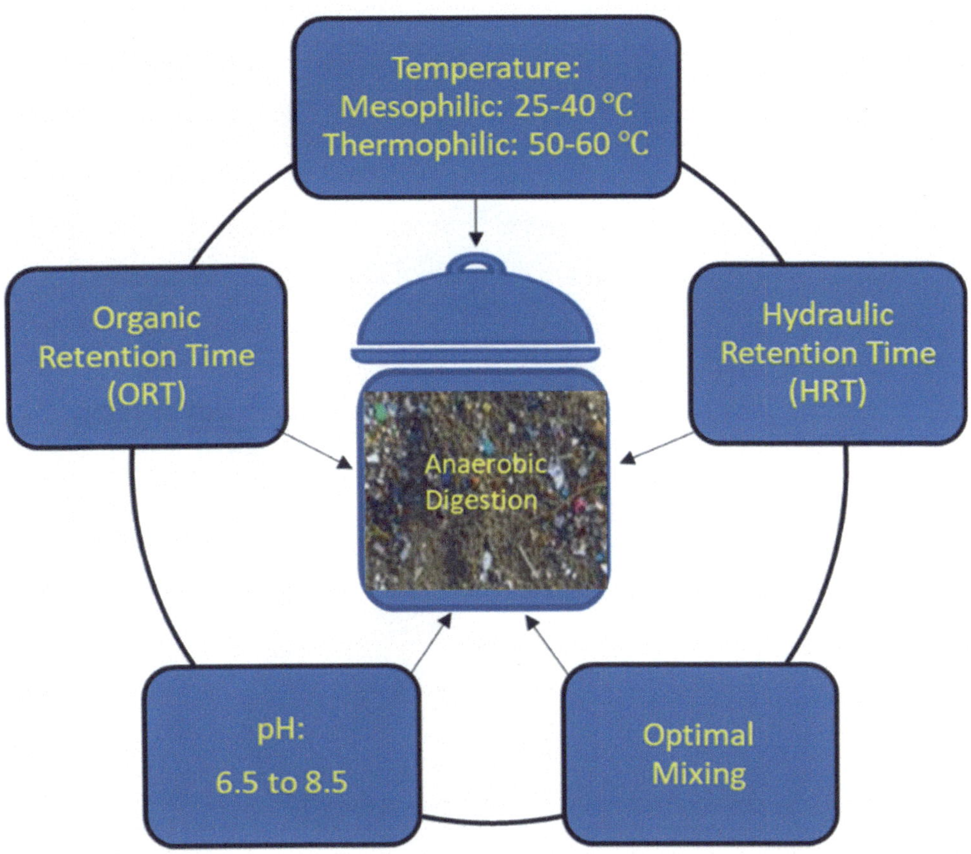

Fig. 5 Operational parameters of anaerobic digestion

Optimizing these parameters is key to maximizing biogas production, reducing the risk of process failures, and ensuring the long-term sustainability of anaerobic digestion systems (Monnet 2003). Regular monitoring and adjustments are necessary to maintain optimal performance.

4 Result and Discussion

4.1 Shape, Size, and Polymer Type of Microplastics in Sewage Sludge

Microplastics were present across all SS types, with their concentrations varying in both number and mass (Ren 2020). Studies have explored their abundance, shapes, sizes, polymer composition, morphology, and potential toxicity. MPs in sludge range in size from under 1 μm to over 5 mm, with fibers and fragments being the most commonly observed shapes. Polyethylene and nylon were the most frequently identified polymers among these particles (Li et al. 2024). A review of the literature shows that microplastic (MP) concentrations in wastewater treatment plants (WWTPs) range between 61 and 5600 μg/L in influent and 0.5–170.0 μg/L in effluent, indicating a removal efficiency of 93.8–99.8%. Additionally, the number of MP particles per liter of raw wastewater varies from 1.01 to 31,400 (Zielińska and Cydzik-Kwiatkowska 2024). An analysis of 65 studies on microplastic (MP) contamination in sewage sludge and biosolids across 25 countries found significant variation in MP concentrations, with a median value of 22.41 particles per gram of dry weight (d w.) (Harley-Nyang et al. 2023). MP levels in three Indian sewage sludge samples ranged from 29.2 to 60.8×10^3 particles/kg d w (Chaudhary and Suthar 2023).

4.2 Impact of Microplastics on Anaerobic Digestion of Domestic Sewage Sludge

Researchers are currently exploring how microplastics influence sludge during various stages of anaerobic digestion. The study zooms in on critical aspects like how quickly the sludge breaks down, the makeup of the anaerobic bacteria involved, how much methane is produced, and how efficiently volatile suspended solids (VSS) are removed. The findings suggest that the amount of microplastics in the sludge can noticeably change these dynamics, potentially disrupting the overall efficiency and performance of the anaerobic digestion process, as shown in Fig. 6.

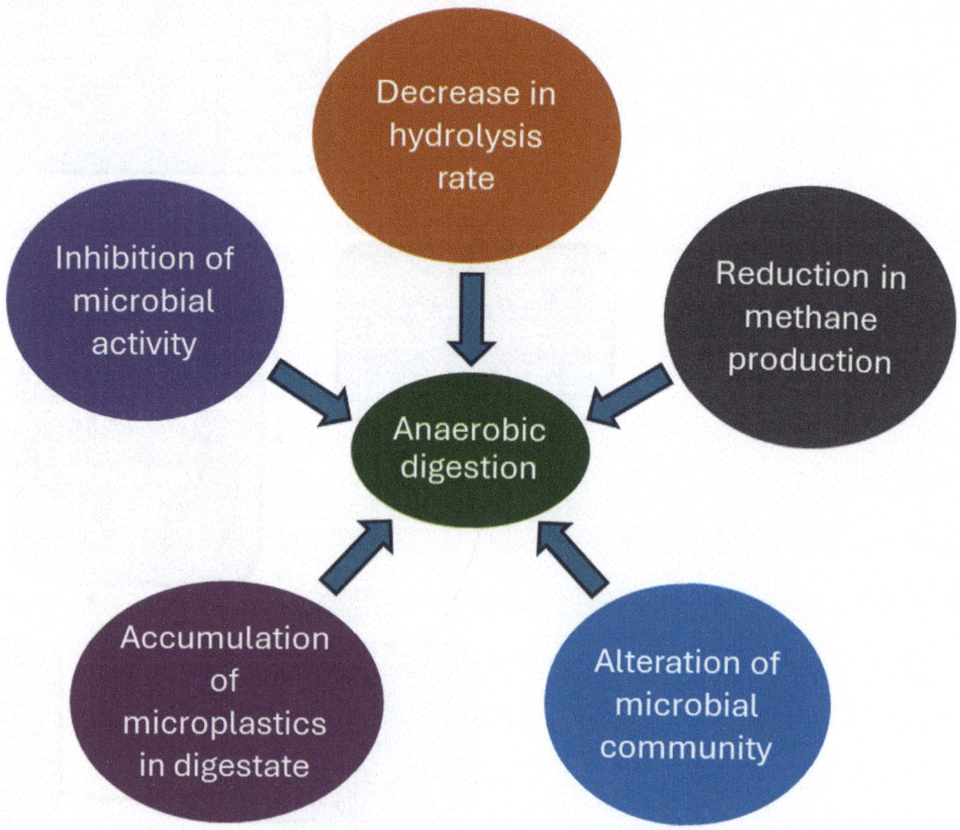

Fig. 6 Major effects of microplastics on anaerobic digestion

Some of the potential effects are as follows:

1. **Impact on the Rate of Hydrolysis**: It has been observed that hydrolysis, which is the initial phase of anaerobic digestion, is influenced by various microplastic factors. These factors include the abundance of microplastics (i.e., the number of particles per gram of sludge), their size, and the type of polymer (such as PVC: polyvinyl chloride, PET: terephthalate polyethylene, PE: polyethylene, PES: polyester) (Mohammad Mirsoleimani Azizi et al. 2021). Several studies have shown that as the number of microplastics in the sample increases, the rate of hydrolysis tends to decrease (Li et al. 2020). For instance, when PE microplastic particles of 40 μm in size were added to sludge at high levels (100–200 particles/g TS), the rate of hydrolysis was reduced by 8.6–15.2%. This suggests that a longer retention time in digesters or larger digesters may be required for sludge with high levels of microplastics. However, no significant reduction in hydrolysis rate was observed at lower quantities of microplastics (less than 100 particles/g TS). Similarly, for PVC microplastics with a larger size of 1 mm, the rate of hydrolysis was hardly affected when the quantity of microplastics was less than 10 particles/g TS. However, the hydrolysis rate was reduced for higher quantities of PVC microplastics (20, 40, and 60 particles/g TS). Another study using PES-type microplastics measuring 200 μm also showed a decrease in the hydrolysis rate as the amount of microplastics in the sample increased, although no specific correlation was found between the amount of microplastics and the final hydrolysis rate.
2. **Impact on Methanogenesis**: Research findings indicate that the presence of microplastics in sludge samples can reduce methane production during the methanogenesis stage of anaerobic digestion. A study using polyethylene (PE) microplastics found that methane production was not affected when the number of microplastics was less than 60 particles per gram of total solids (g TS), even with long-term use of the sludge sample. However, when the quantity of microplastics increased to 100–200 particles per g TS, methane production decreased by 12.4–27.5%. Similarly, in another study using polyvinyl chloride (PVC) microplastics, methane production increased with a quantity of less than 10 particles per g TS but decreased with an increased quantity of 20, 40, and 60 particles per g TS (Wei et al. 2019). Interestingly, when polyester (PES) microplastics were introduced, the rise in microplastics did not seem to affect methane production much. This suggests that the size and type of plastic might be key factors in determining their impact on methane generation (Chen 2021).
3. **Inhibition of Microbial Activity**: Microplastics can accumulate in the digester and form a barrier, preventing the diffusion of nutrients and gases. This can inhibit the activity of microorganisms involved in the anaerobic digestion process, resulting in decreased biogas production and lower digestion efficiency. However, microplastics (MPs) have also been noticed as a key component in the inhibition of microbial activity during anaerobic digestion (AD), owing to a combination of physical and chemical interactions. MPs can disrupt sludge digestion during the solubilization phase. Ozonation pre-treatments may help or worsen this effect (Lekše et al. 2022). A study found that diethylhexyl phthalate (DEHP), a plasticizer, has the potential to influence microbial communities by diminishing acetogen bacteria while increasing methanogens, thereby inducing alterations in the fermentation procedure (Gong et al. 2020). The significance of particle size in microplastics asserts that smaller particles lead to a reduction in methane production as a result of their heightened surface area and improved interaction with microbial cells (Sun et al. 2022). The importance of polyethylene (PE) microplastics in triggering the production of reactive oxygen species (ROS), which subsequently leads to decreased cell viability and disruption of crucial microbial activities such as hydrolysis, acidification, and methanogenesis (Wei et al. 2019).
4. **Alteration of the Microbial Community**: Microplastics can also alter the composition and diversity of the microbial inhabitants in the digester, potentially reducing the percentage of key microorganisms involved in the anaerobic digestion process (Wei et al. 2019). This can lead to decreased biogas production and process instability. Also, biofilm development on microplastics (MPs) drastically affects microbial colonization patterns and may reduce anaerobic digestion (AD) performance. Biofilms create a protective environment for microorganisms, altering growth rates, and increasing antibiotic resistance. This resistance hinders the treatment of biofilm-associated infections and the efficient breakdown of organic materials in AD processes. For example, biofilms protect bacteria from immune responses and medications, which contributes to infection persistence and reduced treatment efficacy (Pokharel et al. 2022). The importance of establishing anti-biofilm solutions to address these healthcare difficulties relates to AD procedures by reducing biofilm formation on equipment (Asma et al. 2022). However, developing coatings for healthcare equipment to avoid biofilm formation might be applied to AD systems to retain efficiency (Donlan 2014). The use of nanocoatings to modify microbial colonization and biofilm development could be critical in boosting AD efficiency by minimizing biofilm-related

resistance (Mihai et al. 2018). Finally, numerous anti-biofilm tactics from various industries were evaluated, emphasizing the relevance of new materials in preventing biofilm development, which can be used to improve AD processes (Shineh et al. 2023).

5. **Accumulation of Microplastics in Digestate**: Microplastics can significantly impact biogas production during anaerobic digestion. These tiny particles tend to clump together, reducing the surface area that microbes need to thrive, which, in turn, lowers the overall biogas yield. On top of that, microplastics can accumulate within the system, blocking pipes and causing operational issues that further hinder biogas production. They also end up in the digestate, the byproduct of digestion, potentially contaminating it and diminishing its value as a fertilizer. This situation underscores the importance of managing plastic waste effectively to prevent its harmful effects on the environment.

4.3 Characteristics of Digested Sludge Affected by the Prevalence of Microplastics

Microplastics in digested sludge can significantly alter its physical and chemical properties, reducing its value as a fertilizer or soil amendment. Here is how they can impact the sludge:

1. **Increased Plastic Content**: Microplastics tend to accumulate in the sludge, which raises its plastic content. This accumulation can lower the sludge's biodegradability and nutrient levels, making it less effective.
2. **Nutrient Alteration**: Microplastics can interfere with the sludge's nutrient profile by either absorbing nutrients or disrupting microbial activity. This interference can lead to shifts in nitrogen and phosphorus levels, weakening the sludge's effectiveness as a fertilizer.
3. **Physical Changes**: The presence of microplastics can negatively affect the sludge's physical characteristics, such as reducing its ability to enhance soil structure and water retention. For instance, microplastics can increase the bulk density and decrease the porosity of the sludge.
4. **Contamination with Additives**: Microplastics often carry harmful chemicals like flame retardants and plasticizers, which can leach into the sludge (Hassan 2023). This contamination makes the sludge less valuable and potentially hazardous as a fertilizer or soil amendment.

Moreover, microplastics can introduce new pollutants into the sludge, such as persistent organic pollutants (POPs) and heavy metals. These contaminants pose risks to both human health and the environment. To ensure that anaerobic digestion systems remain viable and effective, managing plastic waste carefully is crucial. Regularly monitoring and testing digested sludge are also essential to maintain its quality and usefulness.

4.4 The Impact of Microplastics Abundance on the Microbial Community

Microplastics and nanoparticles in anaerobic digesters can significantly disrupt the microbial community, altering its structure and metabolic processes (He 2021). Research indicates that microplastics can significantly reduce the populations of crucial hydrolytic and fermentative bacteria and methanogenic *Methanosaeta species*, all of which are vital for effective anaerobic digestion. This reduction leads to a decrease in the production of various volatile fatty acids (VFAs). Additionally, different microplastics have distinct impacts on microbial communities—PVC and PET, for example, are known to decrease the abundance of bacteria responsible for fermentation and hydrogen production (Cesaro et al. 2022). When microplastics coexist with other pollutants, they can significantly alter bacterial communities. This disruption is often due to chemicals leaching from the microplastics or the generation of reactive oxygen species (ROS), which can lead to oxidative stress and even cell death (Chen et al. 2023). The long-term effects of microplastic exposure on anaerobic microbial communities remain unclear, and more research is needed to fully understand how both short- and long-term exposure impacts these communities. Additionally, while some studies suggest that microplastics might not significantly affect microbial diversity, the overall influence on microbial variety and richness still warrants further investigation.

4.5 Impact of the Presence of Microplastic on Methane Metabolism Synthesis

The presence of microplastics in anaerobic digesters can significantly disrupt the process of methane production, which is a crucial outcome of anaerobic digestion. Methane efficiency is vital for these systems' economic and environmental sustainability (Zhang et al. 2020). Research indicates that microplastics like PVC and PE can reduce methane production by as much as 95% compared to the control. The extent of this disruption appears to depend on the size and characteristics of the microplastics. This decrease in methane output could seriously impact the effectiveness of wastewater treatment systems and their potential for energy recovery (Lv et al. 2019). Comprehending the influence of microplastics on methane production is essential for developing strategies to mitigate their effects and ensure sustainable wastewater management. Microplastic abundance can have the following effects on methane production:

1. **Inhibition of Methanogenic Microorganisms**: High concentrations of microplastics can disrupt the growth and activity of methanogenic microorganisms, decreasing methane production. These microplastics can also inhibit the development of critical microbial species like Methanosaeta harundinacea, which play an essential role in granulation and biofilm formation, particularly in high-salinity environments (Gagliano et al. 2017).
2. **Alteration of Microbial Community Composition**: Microplastics in anaerobic digesters can alter the composition of the microbial community, potentially encouraging the growth of microbes that are less efficient at producing methane. These particles can interact with the microbial community, changing the structure and characteristics of the digested sludge. Microplastics might affect how microorganisms cluster together, impacting the formation and stability of granules in the digestion process. Their attachment to microbial cells can interfere with normal functions, which could decrease the overall efficiency of anaerobic digestion (Gagliano et al. 2017).
3. **Accumulation of Non-Biodegradable Material**: Microplastics (MPs), with their porous and rough surfaces, can absorb contaminants like heavy metals, worsening environmental pollution. They tend to accumulate in anaerobic digesters, contributing to the buildup of non-biodegradable materials, which can lower the efficiency of the digestion process and potentially reduce methane production. The presence of MPs in these systems can significantly interfere with the metabolic processes that generate methane, thereby undermining both the economic and environmental benefits of anaerobic digestion. To maintain these systems' long-term sustainability and effectiveness, managing plastic waste properly to maximize methane production is crucial.

5 Fate of Microplastics After Anaerobic Digestion

The fate of microplastics (MPs) after anaerobic digestion is a complex process shaped by both microbial and chemical interactions within the system. Research indicates that a significant portion of MPs remains in the digestate, which can impact the efficiency and outcomes of anaerobic digestion. For instance, waste-activated sludge tends to gather MPs in wastewater treatment plants, potentially disrupting digestion (Pokharel et al. 2022). Also, noticed that MPs can include hazardous chemicals and antibiotic-resistant genes, hindering sludge stabilization, and lowering biogas output during AD (Ezugworie et al. 2022). Similarly, research on thermally processed sludge found that MPs survive during AD, potentially influencing microbial degradation kinetics (Cesaro et al. 2022). Additionally, AD can affect the degradation of bioplastics, leading to the breakdown of MPs in soil (Papa et al. 2023). Also, the size of MPs influences their interaction with dissolved organic matter and microbial communities during AD, which affects methane generation and microbial dynamics (Sun et al. 2022). Nevertheless, the fate of microplastics also depends on the operational parameters of the digestion process and the microplastics' characteristics. Some of the potential fates of microplastics after anaerobic digestion include:

1. **Breakdown into Smaller Particles**: Microplastics may be broken down into smaller particles as nanoplastics during digestion, potentially making them more difficult to remove from the digested sludge.
2. **Release into the Environment**: Microplastics not removed during digestion may be released into the environment by applying digested sludge to agricultural land or disposal in landfills (Lv et al. 2019). It has detrimental effects on soil ecosystems and can potentially leach into aquatic environments, posing risks to organisms. MPs in sewage sludge may interact with soil contaminants, spreading pollutants in the soil, such as heavy metals like cadmium.
3. **Accumulation in the Digested Sludge**: Microplastics may accumulate in the digested sludge, potentially leading to the buildup of non-biodegradable material and reducing the overall efficiency of the anaerobic digestion process.
4. **Impacts on Soil and Water Quality:** Microplastics have a substantial impact on soil and water quality because they change soil structure, water retention properties, moisture infiltration, evaporation characteristics, microbial interactions, nutrient cycling, and ecosystem processes. Research shows that the presence of polyethylene microplastics in soil alters water retention, pore structure, and hydraulic properties (Wang et al. 2022). These particles affect how moisture moves through the soil, influence the spread of wetting fronts, and alter evaporation rates, with different types and amounts of microplastics causing varied effects. Microplastics also disrupt microbial activity, soil carbon and nitrogen cycles, and other soil quality markers, underscoring the need for more research on how they interact within soil ecosystems. They can impact nutrient cycling, soil structure, pH levels, litter decomposition, and overall soil functionality, differing depending on soil moisture conditions (Yang 2020). Understanding these impacts is essential for better managing soil and water pollution in agriculture. Regarding anaerobic digestion, the fate of

microplastics remains unclear, influenced by many factors that are not yet fully understood. More studies are needed to explore how microplastics behave in these systems and to develop effective strategies for managing their presence in treated sludge.

6 Regulatory Framework and Policy Implications

The current state of microplastic regulations and policies reflects a disjointed and evolving global effort to tackle this pervasive environmental issue. A thorough review of both local and international legislation on plastic pollution reveals significant regulatory gaps, particularly concerning microplastics in food and drinking water. These shortcomings highlight the need for more cohesive and comprehensive approaches to effectively address the risks posed by microplastics (Usman et al. 2022). Federal rules that already exist in the United States and have the potential to limit microplastics have not yet been fully put into effect (Sorensen et al. 2022). They propose redefining microplastics as pollutants to facilitate stronger control. The importance of an integrated legal framework for managing marine plastic pollution and advocates for a lifecycle approach to plastic regulation. Various international strategies in Europe, America, and Asia recommend the use of both soft landing and harsh regulation measures to manage plastic usage and waste (Sin and Tueen 2023). Finally, policy integration in the textile industry is important, as well as advocating for coordinated efforts and increased producer responsibility (Kentin and Battaglia 2022). Collectively, these studies highlight the importance of coordinated international initiatives, strengthened legal frameworks, and comprehensive policies to reduce the effects of microplastic pollution.

7 Future Research Directions

Future research directions in tackling microplastic issues include filling huge information gaps in understanding the environmental health implications of micro- and nano-plastic contaminants, as identified in the literature. In addition, multidisciplinary techniques that combine in vitro models, computational modeling, and mathematical simulations are required to improve risk assessment frameworks and expand scientific knowledge in this sector. In the realm of operational excellence research, certain gaps persist, such as the limited focus on how capabilities influence operational excellence, the insufficient use of theoretical frameworks, and the lack of research across specific industries and regions. These gaps highlight opportunities for future studies and collaboration among academics, researchers, practitioners, and policymakers. Additionally, the importance of collaborative efforts in advancing computational methods and simulations to address challenges in turbulent particle-laden flows is underscored, emphasizing the need for joint initiatives to propel research in this area forward.

8 Conclusions

The study delved into the impact of microplastics on anaerobic digestion and identified significant challenges these particles bring to treating domestic sewage sludge. Microplastics were found to hinder the efficiency of anaerobic digestion by decreasing methane production, disrupting the balance of microbial communities, and accumulating in the resulting sludge. These findings highlight the urgent need for enhanced wastewater treatment strategies capable of effectively managing microplastics. The study's conclusions emphasize the importance of developing standardized procedures and innovative technologies for microplastic removal to ensure more sustainable sewage sludge management. Ongoing research is essential to fully grasp microplastics' long-term effects and create solutions that protect both the environment and public health.

Acknowledgements Authors are thankful to the contributors of these books for their cooperation and patience in the compilation of this task. Authors are also thankful to Springer team, for their constant support in the publication of this work.

References

Adhikari S, Kelkar V, Kumar R, Halden RU (2022) Methods and challenges in the detection of microplastics and nanoplastics: a mini-review. Polym Int 71(5):543–551. https://doi.org/10.1002/PI.6348

Ammentorp Lund J (2023) Plastics-the fast facts 2023 plastics and rubber machinery plastic products plastics in primary forms world plastics production 2 global plastics production by polymer

Anaerobic Digestion. (Online). Available: https://www.tn.gov/environment/program-areas/sw-mm-organics/anaerobic-digestion.html

Arab M, Yu J, Nayebi B (2024) Microplastics in sludges and soils: a comprehensive review on distribution, characteristics, and effects. ChemEngineering 8(5):86. https://doi.org/10.3390/chemengineering8050086

Asma ST et al (2022) An overview of biofilm formation-combating strategies and mechanisms of action of antibiofilm agents. Reprod Dev Biol 12(8):1110–1110. https://doi.org/10.3390/LIFE12081110

Bule Možar K et al (2023) Potential of advanced oxidation as pretreatment for microplastics biodegradation. Separations 2023 10(2):132. https://doi.org/10.3390/SEPARATIONS10020132

Carr SA, Liu J, Tesoro AG (2016) Transport and fate of microplastic particles in wastewater treatment plants. Water Res 91, pp 174–182. https://doi.org/10.1016/j.watres.2016.01.002

Cesaro A, Pirozzi F, Zafırakou A, Alexandraki A (2022) Microplastics in sewage sludge destined to anaerobic digestion: the potential role of thermal pretreatment. Chemosphere 309. https://doi.org/10.1016/j.chemosphere.2022.136669

Cesaro A, Pirozzi F, Zafirakou A, Alexandraki A (2022) The fate of microplastics during the anaerobic digestion of thermally pretreated sludge. In: Proceedings of the 17th international conference on environmental science and technology, vol 17. https://doi.org/10.30955/GNC2021.00221

Chaudhary M, Suthar S (2023) Microplastic abundance and characterization in municipal sewage sludge from cities across upper Ganga river, India: apprising microplastic uptakes and their toxicity in the plant during sludge application in agriculture. Phys Chem Earth Parts A/B/C 132:103468. https://doi.org/10.1016/J.PCE.2023.103468

Chen H et al (2021) Polyamide 6 microplastics facilitate methane production during anaerobic digestion of waste activated sludge. Chem Eng J 408. https://doi.org/10.1016/j.cej.2020.127251

Chen L et al (2023) Di(2-ethylhexyl) phthalate and microplastics cause necroptosis and apoptosis in hepatocytes of mice by inducing oxidative stress. Environ Toxicol 38(6):1226–1238. https://doi.org/10.1002/tox.23759

Domenech J, Marcos R (2021) Pathways of human exposure to microplastics, and estimation of the total burden. Curr Opin Food Sci 39. https://doi.org/10.1016/j.cofs.2021.01.004

Donlan RM (2014) A new approach to mitigate biofilm formation on totally implantable venous access ports. J Infect Dis 210(9):1345–1346. https://doi.org/10.1093/INFDIS/JIU251

El Hayany B, Rumpel C, Hafidi M, El Fels L (2022) Occurrence, analysis of microplastics in sewage sludge and their fate during composting: a literature review. J Environ Manag 317. https://doi.org/10.1016/j.jenvman.2022.115364

Evode N, Qamar SA, Bilal M, Barceló D, Iqbal HMN (2021) Plastic waste and its management strategies for environmental sustainability. Case Stud Chem Environ Eng 4. https://doi.org/10.1016/j.cscee.2021.100142

Ezugworie FN, Aliyu GO, Onwosi CO (2022) Microplastics and anaerobic digestion. In: Environmental footprints and eco-design of products and processes, pp 291–312. https://doi.org/10.1007/978-981-16-8440-1_13

Gagliano MC, Ismail SB, Stams AJM, Plugge CM, Temmink H, Van Lier JB (2017) Biofilm formation and granule properties in anaerobic digestion at high salinity. Water Res 121:61–71. https://doi.org/10.1016/j.watres.2017.05.016

Gayathri PV, Joseph S, Mohan M, Pillai D (2023) Advanced oxidation processes for the degradation of microplastics from the environment: a review. Water Environ J 37(4):686–701. https://doi.org/10.1111/WEJ.12883

Gniadek M, Dąbrowska A (2019) The marine nano- and microplastics characterisation by SEM-EDX: the potential of the method in comparison with various physical and chemical approaches. Mar Pollut Bull 148:210–216. https://doi.org/10.1016/j.marpolbul.2019.07.067

Gong R, Tang X, Fan C, Zhang B, Zhou M (2020) Effect of DEHP on SCFA production by anaerobic fermentation of waste activated sludge. Adv Polym Technol 2020. https://doi.org/10.1155/2020/1705232

Harley-Nyang D, Memon FA, Osorio Baquero A, Galloway T (2023) Variation in microplastic concentration, characteristics and distribution in sewage sludge & biosolids around the world. Sci Total Environ 891. https://doi.org/10.1016/J.SCITOTENV.2023.164068

Hassan F et al (2023) Microplastic contamination in sewage sludge: abundance, characteristics, and impacts on the environment and human health. Environ Technol Innov 31. https://doi.org/10.1016/j.eti.2023.103176

He ZW et al (2021) Occurrence, effect, and fate of residual microplastics in anaerobic digestion of waste activated sludge: a state-of-the-art review. Bioresour Technol. https://doi.org/10.1016/j.biortech.2021.125035

Hulshoff Pol LW, de Castro Lopes SI, Lettinga G, Lens PNL (2004) The phenomenon of granulation of anaerobic sludge. Wageningen University & Research. https://edepot.wur.nl/202000. https://research.wur.nl/en/publications/the-phenomenon-of-granulation-of-anaerobic-sludge.Accessed 19 May 2025

Kentin E, Battaglia G (2022) Policies and perspectives on regulating microplastic fibre pollution. In: Polluting textiles: the problem with microfibres, pp 265–289. https://doi.org/10.4324/9781003165385-13

Kim S, Sin A, Nam H, Park Y, Lee H, Han C (2022) Advanced oxidation processes for microplastics degradation: a recent trend. Chem Eng J Adv 9:100213. https://doi.org/10.1016/J.CEJA.2021.100213

Kye H, Kim J, Ju S, Lee J, Lim C, Yoon Y (2023) Microplastics in water systems: A review of their impacts on the environment and their potential hazards. Heliyon 9:2405–8440. https://doi.org/10.1016/j.heliyon.2023.e14359

Labbe AB, Bagshaw CR, Uttal L (2020) Inexpensive adaptations of basic microscopes for the identification of microplastic contamination using polarization and Nile red fluorescence detection. J Chem Educ 97(11):4026–4032. https://doi.org/10.1021/ACS.JCHEMED.0C00518/SUPPL_FILE/ED0C00518_SI_008.DOCX

Lekše N, Bulc TG, Kukovičič N, Černic T, Gotvajn A (2022) Impact of preozonation on biogas potential of PVC microplastics-containing waste sludge. Global NEST J 25(2):13–19. https://doi.org/10.30955/GNJ.004300

Li M et al (2024) Research progress on occurrence, removal and fate of microplastics in WWTPs. J Phys Conf Ser Inst Phys. https://doi.org/10.1088/1742-6596/2706/1/012073

Li L, Geng S, Li Z, Song K (2020) Effect of microplastic on anaerobic digestion of wasted activated sludge. Chemosphere 247. https://doi.org/10.1016/j.chemosphere.2020.125874

Liu J, Smith SR (2022) The link between organic matter composition and the biogas yield of full-scale sewage sludge anaerobic digestion. Water Sci Technol 85(5):1658–1672. https://doi.org/10.2166/WST.2022.058

Logvina Y et al (2023) Micro-nano-plastics in the atmosphere: methodology for sampling. https://doi.org/10.20944/PREPRINTS202304.1199.V1

Lv X, Dong Q, Zuo Z, Liu Y, Huang X, Wu WM (2019) Microplastics in a municipal wastewater treatment plant: fate, dynamic distribution, removal efficiencies, and control strategies. J Clean Prod 225:579–586. https://doi.org/10.1016/j.jclepro.2019.03.321

Manu MK et al (2023) A review on mechanistic understanding of microplastic pollution on the performance of anaerobic digestion. Environ Pollut. https://doi.org/10.1016/j.envpol.2023.121426

Marine Debris Program N (2015) Laboratory methods for the analysis of microplastics in the marine environment: recommendations for quantifying synthetic particles in waters and sediments

Mihai MM, Preda M, Lungu I, Gestal MC, Popa MI, Holban AM (2018) Nanocoatings for chronic wound repair—modulation of microbial colonization and biofilm formation. Int J Mol Sci 19(4):1179. https://doi.org/10.3390/IJMS19041179

Mishra S, Singh RP, Rout PK, Das AP (2022) Membrane bioreactor (MBR) as an advanced wastewater treatment technology for removal of synthetic microplastics. In: Development in wastewater treatment research and processes: removal of emerging contaminants from wastewater through bio-nanotechnology, pp 45–60. https://doi.org/10.1016/B978-0-323-85583-9.00022-3

Mohammad Mirsoleimani Azizi S, Hai FI, Lu W, Al-Mamun A, Ranjan Dhar B (2021) A review of mechanisms underlying the impacts of (nano)microplastics on anaerobic digestion. Bioresour Technol. https://doi.org/10.1016/j.biortech.2021.124894

Monnet F (2003) An introduction to anaerobic digestion oforganic wastes. A report by Remade Scotland

Muthukumaran P, Suresh Babu P, Kamaraj M, Aravind J (2021) Microplastics menace: the new emerging lurking environmental issue, a review on sampling and quantification in aquatic environments. Int J Environ Sci Technol 20(1):1081–1094. https://doi.org/10.1007/S13762-021-03591-W

Papa G, Cucina M, Echchouki K, De Nisi P, Adani F (2023) Anaerobic digestion of organic waste allows recovering energy and enhancing the subsequent bioplastic degradation in soil. Resour Conserv Recycl 188:106694–106694. https://doi.org/10.1016/J.RESCONREC.2022.106694

Papp A, Tanaka M, Máté Z, Paulik E (2022) Origin, environmental presence and health effects of microplastics. Acta Biol Szeged 66(1):75–84. https://doi.org/10.14232/abs.2022.1.75-84

Parashar N, Hait S (2023) Recent advances on microplastics pollution and removal from wastewater systems: a critical review. J Environ Manag 340. https://doi.org/10.1016/J.JENVMAN.2023.118014

Pokharel K, Dawadi BR, Shrestha LB (2022) Role of biofilm in bacterial infection and antimicrobial resistance. J Nepal Med Assoc 60(253):836–840. https://doi.org/10.31729/JNMA.7580

Quecholac-Piña X, Hernández-Berriel MDC, Mañón-Salas MDC, Espinosa-Valdemar RM, Vázquez-Morillas A (2020) Degradation of plastics under anaerobic conditions: a short review. Polymers 12(1):109. https://doi.org/10.3390/polym12010109

Ren X et al (2020) Abundance and characteristics of microplastic in sewage sludge: a case study of Yangling, Shaanxi province, China. Case Stud Chem Environ Eng 2. https://doi.org/10.1016/j.cscee.2020.100050

Shi H, Wang X, Zhu L, Li D (2023) Comprehensive comparison of various microplastic sampling methods in sea water: implications for data compilation. Water 2023 15(6):1035. https://doi.org/10.3390/W15061035

Shineh G, Mobaraki M, Jabed M, Bappy P, Mills DK (2023) Biofilm formation, and related impacts on healthcare, food processing and packaging, industrial manufacturing, marine industries, and sanitation–a review. Appl Microbiol 3(3):629–665. https://doi.org/10.3390/APPLMICROBIOL3030044

Shruti VC, Pérez-Guevara F, Roy PD, Kutralam-Muniasamy G (2022) Analyzing microplastics with Nile red: emerging trends, challenges, and prospects. J Hazard Mater 423. https://doi.org/10.1016/j.jhazmat.2021.127171

Şimşek I, Sanin FD (2024) Fate and effect of polyamide-6 microplastics in mesophilic and thermophilic anaerobic digestion. Environ Pollut 361. https://doi.org/10.1016/j.envpol.2024.124855

Sin LT, Tueen BS (2023) International policies of plastic use and consumption. Plast Sustain 255–296. https://doi.org/10.1016/B978-0-12-824489-0.00009-X

Sorensen RM, Kanwar RS, Jovanovi B (2022) Past, present, and possible future policies on plastic use in the United States, particularly microplastics and nanoplastics: a review. Integr Environ Assess Manag 19(2):474–488. https://doi.org/10.1002/IEAM.4678

Sun M et al (2022) Deciphering the role of microplastic size on anaerobic sludge digestion: changes of dissolved organic matter, leaching compounds and microbial community. Environ Res 214:114032. https://doi.org/10.1016/J.ENVRES.2022.114032

Sun J, Dai X, Wang Q, van Loosdrecht MCM, Ni BJ (2019) Microplastics in wastewater treatment plants: detection, occurrence and removal. Water Res 152. https://doi.org/10.1016/j.watres.2018.12.050

Surana D, Ghosh P, Kumar M, Varjani S, Kumar V, Mannina G (2023) Contemporary drift in emerging micro(nano)plastics removal and upcycling technologies from municipal wastewater sludge: strategic innovations and prospects. Curr Pollut Rep 9:174–197. https://doi.org/10.1007/s40726-023-00261-y

Sutradhar M (2022) A review of microplastics pollution and its remediation methods: current scenario and future aspects. Arch Agric Environ Sci 7(2):288–293. https://doi.org/10.26832/24566632.2022.0702019

Usman S et al (2022) The burden of microplastics pollution and contending policies and regulations. Int J Environ Res Public Health 19(11):6773–6773. https://doi.org/10.3390/IJERPH19116773

Wang Z, Li W, Li W, Yang W, Jing S (2022) Effects of microplastics on the water characteristic curve of soils with different textures. SSRN Electron J. https://doi.org/10.2139/SSRN.4257133

Wei W, Huang QS, Sun J, Wang JY, Wu SL, Ni BJ (2019) Polyvinyl chloride microplastics affect methane production from the anaerobic digestion of waste activated sludge through leaching toxic bisphenol-A. Environ Sci Technol 53(5):2509–2517. https://doi.org/10.1021/acs.est.8b07069

Wei W, Huang QS, Sun J, Dai X, Ni BJ (2019) Revealing the mechanisms of polyethylene microplastics affecting anaerobic digestion of waste activated sludge. Environ Sci Technol 53(16):9604–9613. https://doi.org/10.1021/acs.est.9b02971

Wei W, Huang QS, Sun J, Wang JY, Wu SL, Ni BJ (2019) Polyvinyl chloride microplastics affect methane production from the anaerobic digestion of waste activated sludge through leaching toxic bisphenol-A. Environ Sci Technol 53(5):2509–2517. https://doi.org/10.1021/ACS.EST.8B07069

Wei W, Shi X, Zhang YT, Wang C, Wang Y, Ni BJ (2023) Microplastics in sewage sludge of wastewater treatment. In: Microplastics in urban water management, pp 147–173. https://doi.org/10.1002/9781119759379.CH5

Yang X et al (2020) Microplastics in soil ecosystem: insight on its fate and impacts on soil quality. In: Handbook of environmental chemistry, vol 95, pp 245–258. https://doi.org/10.1007/698_2020_458

Zhang X, Chen J, Li J (2020) The removal of microplastics in the wastewater treatment process and their potential impact on anaerobic digestion due to pollutants association. Chemosphere 251. https://doi.org/10.1016/j.chemosphere.2020.126360

Zhu Y, Ma J, Zeng S, Li X, Lisak G, Chen F (2022) Advanced treatment of microplastics and antibiotic-containing wastewater using integrated modified dissolved air flotation and pulsed cavitation-impinging stream processes. J Hazard Mater Adv 7:100139. https://doi.org/10.1016/J.HAZADV.2022.100139

Zielińska M, Cydzik-Kwiatkowska A (2024) Effect of emerging micropollutants on the anaerobic digestion of sewage sludge. Energies 17(5):1033. https://doi.org/10.3390/en17051033

Hulshoff Pol LW,de Castro Lopes SI,Lettinga G, LensPNL (2004) Thephenomenon ofgranulation ofanaerobic sludge. WageningenUniversity &Research. https://edepot.wur.nl/202000. https://research.wur.nl/en/publications/the-phenomenon-of-granulation-of-anaerobic-sludge.Accessed 19 May 2025

Preserving Health and Environment Mastering Biomedical Waste Management

Bhawna Dahiya, Shekhar Nimbrana, and Poonam Ranga

Abstract

Biomedical waste management is one of the critical necessities both for the healthcare enterprise and for society. Incorrect disposal of biomedical waste leads to disease burden and adverse impacts on the surroundings. BMWM is a shared responsibility that must become a shared duty between healthcare employees, healthcare centers, the authorities, and the majority. BMWM must be a collaborative effort involving proper segregation, transport, treatment, and disposal by using different technologies. Healthcare facilities are expected to allocate a storage area for medical waste that is away from public and displays warning signs. To address the challenges of biowaste, it is important for healthcare facilities to follow correct practices and to educate their workforce and the majority of the people on the importance of these practices. Different rules and regulations dealing with biomedical waste should be strictly implemented. By prioritizing biomedical waste management public health as well as and surroundings will ensure a safer and healthier future for all.

Keywords

Biomedical waste management · Environmental health · Healthcare workers · Regulations

B. Dahiya (✉)
Institute of Environmental Studies, Kurukshetra University, Kurukshetra, Haryana, India
e-mail: bhawnadahiya@kuk.ac.in

S. Nimbrana
Department of Energy and Environmental Sciences, Chaudhary Devi Lal University, Sirsa, Haryana, India

P. Ranga
ICAR-Central Institute for Cotton Research, Regional Station, Sirsa, Haryana, India

1 Introduction

The fast increase in population in the twenty-first century has led to enhanced need for medical care, which has led to increase in generation and inefficient management of biomedical waste. In the present time with the use of more advanced medical technology, dealing with biomedical waste properly has become an important concern. Inappropriate management of biomedical waste can affect the integrity of the environment and people's health by spreading diseases. Biomedical waste, which comes from diagnosing, treating, vaccinating, and researching, needs to be controlled and handled carefully. The growth of advanced medical technology and healthcare facilities resulted in intensification of waste resulting in need of strong rules and sustainable treatment techniques to reduce the waste.

This chapter focuses on various aspects of managing biomedical waste. It aims to evaluate the efficiency of current methods, rules, and the use of new technologies to reduce risks linked with improper management of biomedical waste. This chapter also focuses on ongoing sustainable waste management practices by providing a detailed analysis and practical recommendations to make the disposal of biological waste more effective and safer.

2 Sources of Biomedical Waste

Biomedical waste is produced from healthcare facilities such as clinics, dental practices, hospitals, testing labs, animal care centers, biotechnology and pharmaceutical companies, home healthcare services, and funeral centers (Odumosu 2015). Hospitals and clinics are in particular the main sources generating different varieties of waste including infectious waste, chemical waste, pathological waste, sharps, and cytotoxic waste.

A. Mandpe et al. (eds.), *Sustainable Waste Management*, Advances in Science, Technology & Innovation,
https://doi.org/10.1007/978-3-031-85685-3_6

Biomedical waste comes from various sources in the healthcare industry, each producing different types of harmful materials that require careful handling. These sources include:

1. Hospitals and Nursing Homes:
   - Diagnostic and Therapeutic Procedures: These substantial healthcare facilities generate copious quantities of waste, encompassing used syringes, contaminated dressings, pathological specimens, surgical detritus, and obsolete pharmaceuticals.
   - Research Endeavors: Waste from experimental protocols, microbial cultures, and chemical reagents.
2. Clinics and Dispensaries:
   - These smaller healthcare units produce hazardous sharps, soiled dressings, and biological samples, each necessitating stringent disposal measures.
3. Veterinary Institutions and Animal Houses:
   - Animal Treatment and Research: Generate animal tissues, carcasses, used surgical implements, and excreta, requiring specialized disposal methodologies to avert zoonotic transmissions.
4. Pathological Laboratories and Blood Banks:
   - Diagnostic Testing and Blood Handling: Produce infectious materials such as blood-soaked gauze, used needles, microbial cultures, and other biohazardous refuse (Odumosu 2015). Academic and Research Institutions.
   - Experimental Activities: Generate waste from biological agents, culture dishes, pipettes, and chemical residues, demanding advanced decontamination protocols.
5. Health Camps and Temporary Medical Services:
   - Vaccination and Blood Donation Camps: These transient setups generate substantial quantities of used syringes, vials, and other single-use medical supplies, often with inadequate disposal infrastructures.
6. Forensic Laboratories:
   - Investigative Activities: Produce biological and chemical wastes from criminal evidence analysis, including tissue samples and chemical reagents.
7. Ayush Hospitals and Clinical Establishments:
   - Traditional Medicine Practices: Generate waste from Ayurvedic, Yoga, Unani, Siddha, and Homeopathy treatments, including herbal preparations and ancillary medical waste.
8. Miscellaneous Sources:
   - First Aid Rooms (Schools and Institutions): Produce elementary medical waste such as bandages, used gloves, and minor medical implements.
   - Research Laboratories: Produce waste from diverse research activities involving hazardous microorganisms and chemicals.

Table 1 Primary and secondary sources of generation of biomedical waste (Tiwari and Kadu 2013)

Primary sources	Secondary sources
Hospitals	Households or home treatment
Laboratories and research centers	Slaughter houses
Blood banks and collection centers	Funeral homes and mortuaries
Pharmaceutical manufacturing units	Nursing homes and long-term care facilities
Industries	Medical camps and mobile clinics

Each of these sources (Table 1) contributes uniquely to the composite biomedical waste stream, underscoring the complexity and magnitude of managing such waste efficaciously.

3 Classification of Biomedical Waste

Biomedical waste is divided into two types: hazardous and non-hazardous, to manage it correctly and reduce its risk on human health and environment (Bagali et al. 2021). Hazardous biomedical waste includes infectious materials such as contaminated gloves, used bandages, and pathological waste includes human tissues; sharps include objects like needles and scalpels; pharmaceutical waste includes expired medicines; cytotoxic waste includes chemotherapy drugs; chemical waste includes disinfectants; and radioactive materials include radiotherapy materials. These items need special care, treatment, and disposal methods like autoclaving, chemical disinfection, and incineration. Non-hazardous biomedical waste, such as household waste, includes general waste such as paper, non-contaminated packaging, clean plastic and glass, and food waste (Janik-Karpinska et al. 2023).

Categorizing biomedical waste into hazardous and non-hazardous groups is very important for safe and effective waste management.

The Centers for Disease Control and Prevention (CDC) (Chosewood and Wilson 2009) in the United States offers rules for sorting and managing biomedical waste. They stress the need to recognize and handle waste according to its biohazard level based on the severity of the risk to human health and the environment. Each level has special safety measures, tools, and methods designed to keep lab workers, the public, and the environment safe from harmful germs.

3.1 Biohazard Level 1 (BSL-1)

Risk Level: Low

Description:

BSL-1 agents are those that pose minimal threat to laboratory workers and the environment. These agents are not known to consistently cause disease in healthy adult humans. Work with BSL-1 agents typically involves standard microbiological practices involving both without special primary or secondary barriers, other than a sink for handwashing. For example: Non-pathogenic strains of *Escherichia coli*, *Bacillus subtilis*.

Safety Measures:

- Standard microbiological practices such as hand washing, and no eating, drinking, or smoking in the lab.
- Use of personal protective equipment (PPE) such as gloves and lab coats.
- Work should be conducted on open benchtops using standard laboratory procedures.

4 Biohazard Level 2 (BSL-2)

Risk Level: Moderate

Description:

BSL-2 agents pose moderate hazards to personnel and the environment. These agents can cause human disease but generally are not serious and have treatments available. Access to the lab is restricted when work is being conducted, and practices include those used at BSL-1, plus additional measures. For example: *Staphylococcus aureus*, *Salmonella spp*.

Safety Measures:

- Limited access to the laboratory.
- Use of PPE including gloves, lab coats, and gloves as needed.
- Use of biological safety cabinets (Class I or II) for procedures that may generate aerosols or involve high concentrations of infectious agents.
- Autoclaves for decontaminating waste.

4.1 Biohazard Level 4 (BSL-4)

Risk Level: Extreme

Description: BSL-4 agents are dangerous and exotic, posing a high risk of life-threatening disease. These agents are often transmitted via aerosol and have no available vaccines or treatments. BSL-4 laboratories are isolated and highly secure facilities designed for maximum containment. For example: Ebola virus, Marburg virus, Lassa fever virus, and Variola virus.

Safety Measures:

- Personnel must undergo rigorous training in handling pathogenic and potentially lethal agents and must understand the primary and secondary containment functions.
- Entry into the lab is strictly controlled through secure, airlock-equipped entrances.
- Use of full-body, air-supplied, positive pressure suits.
- All materials and waste leaving the facility must be decontaminated, typically by passing through an autoclave or chemical disinfectant.
- The laboratory is often located in a separate building or an isolated and restricted zone within a building.
- The facility must have dedicated air supply and exhaust systems to prevent contamination of the environment.

The classification into BSL-1, BSL-2, and BSL-4 highlights the increasing levels of risk associated with different biohazardous agents and the corresponding escalation in containment and safety measures. These stringent guidelines confirm the safety of laboratory personnel, the public, and the environment from exposure to infectious agents. By following these protocols, laboratories can mitigate the risks posed by handling biohazardous materials.

5 Categories of Biomedical Waste

Different categories have been assigned to biomedical waste as per Biomedical Waste Handling and Management Rules, 1998 (Fig. 1).

6 Steps of Biomedical Waste Management

Biomedical waste involves six steps for its management (Fig. 2).

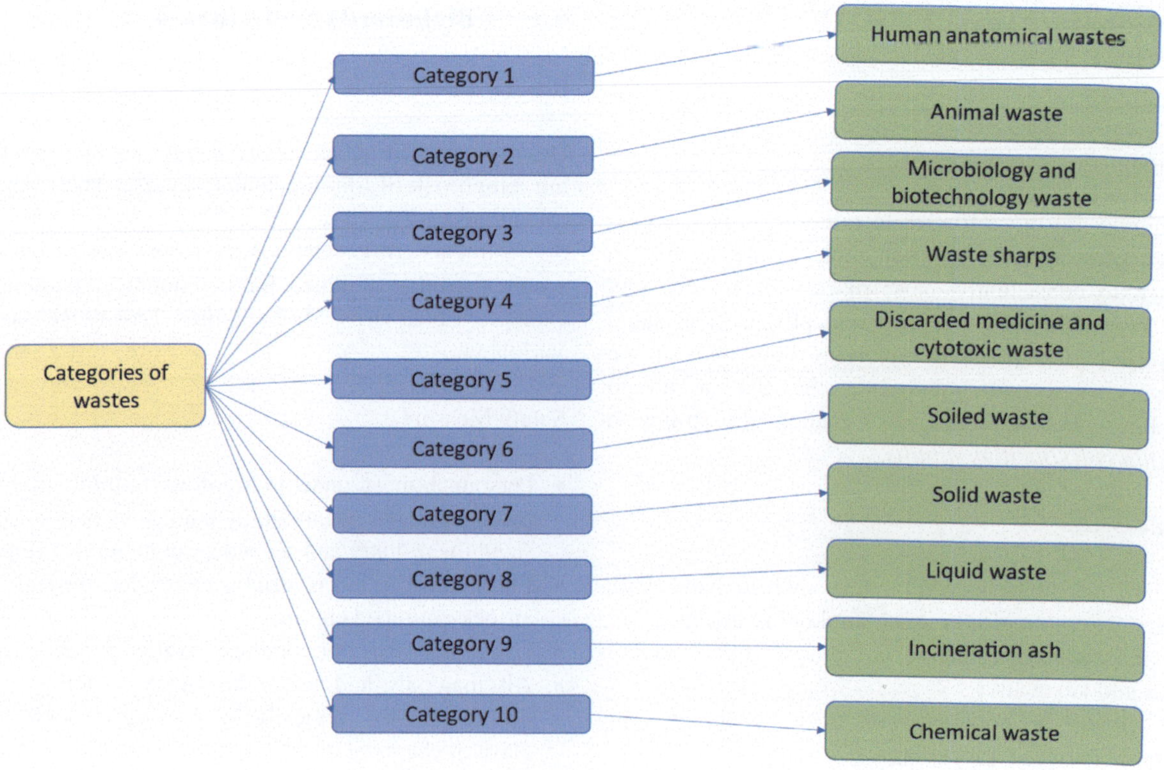

Fig. 1 Categories of waste according to biomedical waste handling and management rules 1998 (Government of India 1998)

Fig. 2 Six steps of biomedical waste management

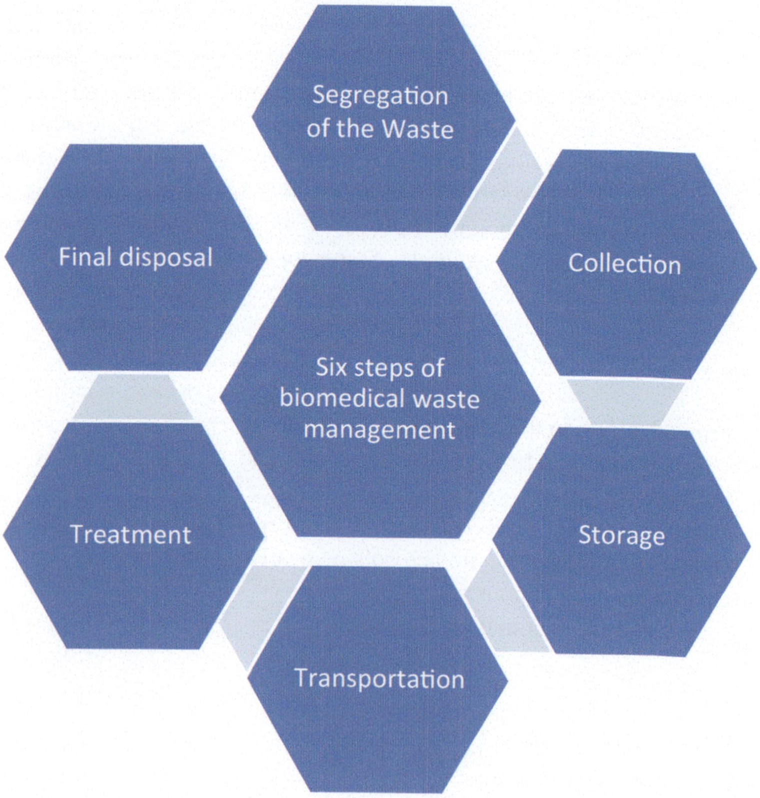

6.1 Segregation of Waste

Segregation is the key process of collection and source separation of all types of biomedical wastes. This step is crucial as it avoids cross-contamination along with provides best suited treatment protocol. Color coded bins or bags are suitable to streamline the process (Akulume and Kiwanuka 2016; Datta et al. 2018). The bags include red for infectious, yellow for pathological, blue for sharps and black for no-hazardous waste (Ndiaye et al. 2003; Chakraborty et al. 2013). For ease of segregation, a color coding system is being integrated in the healthcare facilities. Color coding helps in easy identification and segregation of each type of waste (Fig. 3, Table 2).

6.2 Collection of waste

Biomedical waste should be collected from the point of generation of source and should be stored temporarily in the designated area before being transported (BMW, 2016). The timing of waste collection from different units should be fixed at a specific interval of time. This step necessities the use of designated, clearly labelled containers that are leak proof. Also, the staff handling the biomedical waste should be provided safety kits.

6.3 Transportation

Biomedical waste is kept in locations that are safe, leak proof and in accessible to unauthorized persons.Before being transported the waste is kept in specified location (BMW Rules, 2018).

6.4 Storage

Before being transported for the final treatment the waste should be held as less time as possible preferably not more than 48 hrs (BMW Rules, 2018).

6.5 Treatment

To make biomedical safe for disposal, it should be treated with proper techniques. The common methods are autoclaving, disinfection, incineration, microwave treatment. The

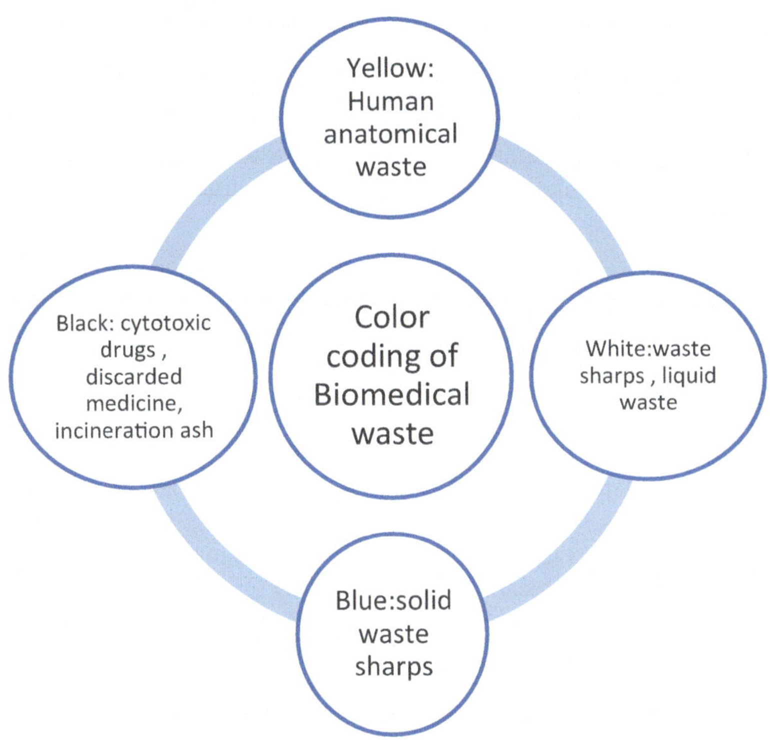

Fig. 3 Color coding of biomedical waste

Table 2 Collection of biomedical waste in a color-coded box (Government of India 1998)

Sr. No.	Color coding	Container types	Waste category
1	Yellow 	Plastic bags	Category 1,2,3, and 6
2	Red 	Disinfected container/plastic bag	Category 3,6, and 7
3	White and translucent 	Plastic bag/puncture proof container	Category 4 and 7
4	Black 	Plastic bag	Category 5,9, and 10

correct method depends upon the nature of waste to be treated and the current regulatory directives (BMW Rules, 2016).

6.6 Disposal

The final stage includes safe and environmentally friendly disposal of treated biomedical waste. Disposal methods in this final stage include landfilling in specifically designed sites where sites are designed to obviate groundwater and soil contamination, among other environmentally responsible practices as required by the regulatory standards. This stage ensures the waste poses no further risk either to public health or the environment (BMW Rules, 2016).

7 Environmental Concerns Associated with Biomedical Waste

Biomedical waste presents multifaceted environmental challenges that necessitate rigorous management practices to mitigate their deleterious impacts. This discourse delineates the primary environmental concerns engendered by improper disposal and management of biomedical waste, utilizing advanced terminology suitable for scholarly discourse. Environmental concerns associated with biomedical waste as depicted in Fig. 4.

7.1 Ecotoxicological Impact

The indiscriminate biomedical waste disposal practice also leads to leaching of hazardous chemicals and pharmaceuticals into the soil and groundwater, eventually culminating in ecotoxicological ramifications. Such substances have a detrimental influence on flora and fauna due to their ability to cause disturbances in ecological equilibria and bioaccumulation via trophic levels.

Pharmaceuticals and Personal Care Products (PPCPs): Residual antibiotics, cytotoxic drugs, and hormones can persist in the environment, leading to the development of antibiotic-resistant microorganisms and endocrine disruption in wildlife (Epen, 2004). Heavy Metals: Elements

Fig. 4 Impacts of biomedical waste on environment

such as mercury and cadmium, prevalent in diagnostic equipment and certain medical devices, can cause soil and water contamination, posing significant risks to aquatic life and human health through bioaccumulation and biomagnification.

7.2 Pathogen Dissemination

Biomedical waste, particularly infectious waste, is a potent vector for the dissemination of pathogenic microorganisms. The improper management of the biowaste can enhance the number of infectious agents thereby increasing the chance of outbreaks and epidemics.

Nosocomial Pathogens: Waste from healthcare facilities can harbor nosocomial pathogens such as *Methicillin-resistant Staphylococcus aureus* (MRSA) and *Clostridium difficile*, which pose severe health risks in community and hospital settings (Dancer 2014). Vector-Borne Diseases: Biomedical waste can attract vectors such as rodents and insects, which can act as intermediaries in the transmission of diseases, exacerbating public health challenges.

7.3 Pollution

Air Pollution: The most common method for use of disposal of biomedical waste is the use of incineration. However, the incinerators if not functioning well may lead to release of toxic gases such as dioxins, furans and other hazardous substances into the atmosphere (Yong-Chul et al. 2006; Babu et al. 2009). These chemicals cause air pollution and have been related to cancer and other major health problems. Open burning of the biomedical waste can generate harmful air pollutants and greenhouse gases that triggers climate change.

Water Contamination: When biomedical waste is incorrectly disposed of in landfills, contaminated water that seeps through waste can be produced (Anil-Karpinska et al. 2013). In addition to being hazardous to human health and ecosystems, this leachate frequently contains diseases, hazardous chemicals, and medications that can seep into groundwater and surface water systems.

Contamination of Soil: Biomedical waste when disposed off in open landfills, heavy metals (such as lead and mercury), toxic chemicals, and pathogens can contaminate the soil. Contamination has the potential to damage microorganisms, alter soil fertility, and get into crops and the food chain (Anil-Karpinska et al. 2013).

Plastic Pollution: Biomedical waste largely comprises disposable syringes of plastic, IV bags, and gloves among other single-use plastics. The problem of plastic pollution feeds these materials since they decompose slowly in the ocean and may adversely affect marine species unless they are properly managed.

7.4 Radiological Hazards

Radioactive wastes containing agents used in diagnostic images and cancer treatment are highly radiologically hazardous. If sent to the environmental surface without proper disposal, it has a medium and long-term environmental effect along with increased health risks by ionizing radiation (BMW Rules, 2016). Radioactive Isotopes: Isotopes like Iodine-131 and Technetium-99 m used in nuclear medicine can contaminate water sources and environment, and thus cause long-term health effects due to their long persistence and, therefore, higher intrinsic radiotoxicity.

7.5 Waste Management Infrastructure Strain

Escalating volumes of biomedical waste put tremendous pressure on existing infrastructures of waste management, often resulting in suboptimal handling and disposal practices that increase adverse environmental contamination and public health risks.

Waste Segregation and Treatment: Poor segregation of biomedical waste from general waste may not obviate the commingling of hazardous and non-hazardous streams, therefore complicating the processes of disposal and treatment. Overburdened incineration and treatment facilities often resort to dangerous discharge mechanisms, such as open burning or unsanitary landfilling, which would increase environmental pollution.

7.6 Threat to Wildlife

When wildlife encounters biological waste that had been disposed of improperly, it may be injured or ingest harmful compounds that eventually make it suffer from disease or even death. In addition, the reproductive systems of fish and other aquatic animals could be influenced by pharmaceuticals found in water bodies (Sharma and Singha 2024).

8 Biomedical Waste Management Rules in India, 2016 (Bansod and Deshmukh 2023)

Proper classification of biological waste will effectively help in the waste management and protection of the environment. These categories will ensure compliance with regulations, which healthcare facilities can adopt to encourage safe and sustainable waste management practices (Datta et al. 2018). In response to this growing concern about biomedical waste, which can have considerable implications on public health and the environment, the Biomedical Waste Management Rules 2016 were introduced by the Ministry of Environment, Forests, and Climate Change, Government of India. The main points of these regulations are as follows:

Scope and Applicability

Any individual who handles biomedical waste in any capacity are required to adhere to regulations. This includes those working in clinical establishments, hospitals, clinics, nursing homes research or educational institutions, dispensaries, animal houses, veterinary institutions, blood banks, pathological laboratories, health camps, Ayush hospitals, medical or surgical camps, bold donation camps, vaccination camps, school first aid rooms, forensic laboratories, and research labs.

1. Exclusions:

The Atomic Energy

Act of 1962 addresses radioactive waste, the Manufacture, Storage, and Import of Hazardous Chemicals Rules of 1989 cover hazardous chemicals, and the Municipal Solid Waste Management and Handling Rules of 2000 cover solid waste. These regulations do not apply to any of these.

- Lead acid batteries are covered under the Batteries (Management and Handling) Rules, 2001.
- Hazardous wastes are covered under the Hazardous Wastes (Management, Handling and Transboundary Movement) Rules, 2008.
- e-Waste covered under the e-Waste (Management and Handling) Rules, 2011.

The manufacture, use, import, export, and storage of dangerous microorganisms and genetically altered organisms are covered under this section.

Segregation, Packaging, Transportation, and Storage (CPCB, 2018).

1. Segregation: Biomedical waste must be segregated at the point of generation by the categories specified in the rules.
2. Packaging: The generated waste must be packaged to avoid leakage and contamination. The bags or containers should be color-coded according to the specific guidelines.
3. Transport: Biomedical waste should be transported in a way that its spillage, rupture or exposure is prevented. Vehicles carrying such wastes should be registered with the prescribed authority and should meet certain prescriptions.
4. Storage: Biomedical wastes should not be stored for more than 48 h. If biomedical waste is to be stored for a period beyond 48 h, then permission from the prescribed authority is necessary.

Treatment and Disposal

1. Authorized treatment and disposal facilities: Biomedical waste shall be treated and disposed off in authorized treatment and disposal facilities only. The standard treatment should be ensured at the CBWTFs.
2. Treatment technologies: Based on the type of wastes, several treatment technologies have been identified, including autoclaving, chemical treatment, incineration, microwaving, and deep burial.
3. Disposal The disposal method has been established whereby biomedical waste is made non-hazardous and safe for the environment.

Monitoring and Compliance

1. Authorization: All healthcare establishments generating biomedical wastes need to obtain prior authorization from the State Pollution Control Board or the Pollution Control Committee.
2. Annual Reporting: Annual reports on biomedical waste management shall be received from healthcare facilities and submitted to the prescribed authority.
3. Supervision: The regulations demand for comprehensive monitoring and inspection of healthcare facilities with regard to compliance.

Training and Awareness

1. Training Programs: There are compulsory training and retraining for healthcare service providers and handlers of healthcare waste on appropriate management and disposal of medical biowaste.
2. Public Awareness: Programmes meant for generating public awareness of the ill-effects of poor biomedical waste management and the need to observe the regulatory aspects are encouraged.

The Biomedical Waste Management Rules, 2016, therefore intend to normalize the processes of biomedical waste management with the aim of public health and environmental safety. The rules are collectively known as a whole integrated policy that takes into consideration the challenges of biomedical waste and promotes sustainable healthcare practices.

9 Techniques for Treatment of Biomedical Waste

The treatment and disposal of biomedical waste require the deployment of an array of advanced technologies to overcome the inherent risks associated with this hazardous stream. The arsenal of technology used in this endeavor consists of a multitude of sophisticated processes, each meticulously tailored to address specific categories of wastes.

9.1 Incineration

High-temperature thermal processes coupled with careful combustion convert waste into inert materials and gases. The most common medical field incinerators are multiple hearths, rotary kilns, and controlled air. Each type incorporates both primary and secondary combustion chambers, ensuring optimal waste combustion. Although most hospital waste is incinerated, the burning of solid and regulated medical waste poses significant health risks to healthcare workers. Medical waste incinerators emit toxic air pollutants and produce toxic ash residues, which are a major source of environmental dioxins (Odumosa, 2015). When disposed of in landfills, these toxic ashes have the potential to leach into groundwater. To mitigate dioxin production, non-chlorinated plastic bags should be used in incinerators. Furthermore, the incineration of red bags, which contain cadmium and potentially mercury, must be avoided to prevent the release of these toxic elements into the environment (Singh and Sharma 1996; Pasupathi et al. 2011).

9.2 Plasma Pyrolysis

Plasma pyrolysis, an advanced thermal treatment method, facilitates the direct utilization of waste products as combustion fuel or their conversion into alternative fuels, thereby harnessing their intrinsic energy. This process involves the thermal degradation of waste materials at elevated temperatures under restricted oxygen conditions. Plasma pyrolysis technology represents a cutting-edge approach to the safe disposal of medical waste. This environmentally benign technology transforms organic waste into commercially valuable by-products. The intense heat generated by plasma enables the secure and efficient disposal of diverse types of waste that includes biowaste. The biowaste is subjected to plasma pyrolysis, and it is converted into carbon monoxide, hydrogen, and hydrocarbons. The resultant gases, upon combustion, reach temperatures as high as 1200 °C, ensuring the thorough breakdown of waste materials (Surjit et al. 2007; Chakraborty et al. 2013).

9.3 Autoclave

Autoclaves, are sealed chambers that employ heat, pressure, and occasionally steam, are utilized for the sterilization of medical equipment over a set duration. For nearly a century, autoclaves have been indispensable in sterilizing reusable medical instruments (Thakur and Katoch 2012). These devices eradicate microorganisms present in medical waste before its disposal in conventional landfills. Capable of processing up to 90% of medical waste, autoclaves can be scaled to accommodate the demands of any medical facility. Compact, countertop autoclaves are employed for sterilizing reusable medical instruments, while larger autoclaves handle substantial volumes of medical waste. Steam sterilization is particularly efficacious for low-density materials such as plastics, metal pans, bottles, and flasks. Conversely, high-density polyethylene and polypropylene plastics are unsuitable for this process due to their resistance to steam penetration (Salvi et al. 2022). To prevent spillage and drain clogging, plastic bags should be placed within a rigid container before undergoing steam treatment. Bags should be opened, and caps and stoppers loosened immediately before placement in the steam sterilizer. It is imperative to segregate infectious waste from other hazardous waste, as infectious waste containing non-infectious hazards should not undergo steam sterilization (Salvi et al. 2022).

9.4 Microwave Treatment Technologies

In this, a steam-based procedure, a 30- to 1-h cycle of moist heat and steam produced by microwave radiation is used to treat the patient.

Sharps, materials tainted with blood or bodily fluids, cultures, and stocks, soft waste (such as gauze, gowns, and bedding), and other contaminated garbage are among the waste kinds that are processed. Large metal objects, radioactive or hazardous garbage, cytotoxic substances, volatile compounds, tainted animal carcasses, and body parts should not be treated with microwaves. *Bacillus atropdhaeus* in vials or spore strips consisting of at least 104 spores are biological indicators of microwave (Emmanuel and Stringer 2007).

9.5 Dry Heat Treatment Technologies

Disposable medical equipment, glasses, and other reusable items have all been sterilized in hot-air ovens. Compared to steam bases procedures, conduction, thermal radiation, or convection at a high temperature of 185 °C for a period of 90–150 min of exposure biowaste is heated (Widmer and Frei 2011). The biological indicator *Geobacillus stearothermophilus* should be killed entirely and consistently using chemical indicator strips or vials containing at least log106 spores per milliliter (Emmanuel and Stringer 2007).

9.6 Alkaline Hydrolysis

This procedure eliminates fixatives, dangerous chemicals, and prion-contaminated waste while transforming body parts, specimens, and corpses into a decontaminated aqueous solution. Once the trash has been placed into the basket and sealed tank, water that has reached a temperature of at least 127 °C and alkali are added, and everything is agitated. By-products after a 6-to 8-h digestion period include the minerals found in teeth and bones, as well as solutions of sugars, salts, soaps, and amino acids. It may also degrade aldehydes that are frequently used in hospitals as well as cytotoxic or chemotherapy medicines (Capoor and Bhowmik 2017).

10 Conclusion

Proper biomedical waste disposal is considered to be one of the essential requirements for maintaining proper standards in the healthcare sector as well as conserving the environment and safeguarding public health. General public is unaware about the harmful effects of biomedical waste. The rules and regulations formulated for biomedical waste management should be followed strictly and precautionary measures should be developed at the public level to aware them about the adverse effects of waste on their health. The reason for inadequate management of biomedical are the lack of public awareness, insufficient segregation of the biomedical waste the concerned staff as well as economic factors. Proper biological waste management is crucial and has profound implications concerning environmental integrity and public health. Though there are proper sets of rules and regulations for management of biomedical waste but there are discrepancies in the proper implementation in the healthcare facilities. Poor biomedical waste management had severe long-term implications in allowing toxins to reach soil and water systems and accelerating further damage to the environment. This emphasizes the need for stern action according to present laws and also the promotion of new waste treatment methodologies. The health risks to healthcare workers further provide a justification for setting forth formal guidelines toward intensified safety measures and continuing professional development. This emphasis the need for improvement in the existing policies and technologies such as chemical disinfection, autoclaving and incineration. The discussion about waste biomedical has to shift into sustainable approaches that protect the environment while reducing health hazards. This demands a focused strategy of enhanced awareness and technological innovations, coupled with high regulatory compliance. Based on the findings of this research, there is indeed a paradigm shift in biomedical waste management toward safer and more environmentally friendly alternatives that ensure environmental sustainability and public health for generations to come.

Acknowledgements Authors are thankful to the contributors of these books for their cooperation and patience in the compilation of this task. Authors are also thankful to Springer team, for their constant support in the publication of this work.

References

Akulume, M., & Kiwanuka, S.N. (2016).Health Care Waste Segregation Behaviour among Health Workers in Uganada: An Application of the Theory of Planned Behavior. Journal of Environmental and Public Health. 8132306.

Anil-Karpinska, E.,Brancaleoni, R.,Niemcewicz, M.,Wojtas, W.,Foco, Podogrocki, M., & Bijak, M. (2023). Healthcare Waste-A Serious Problem for Global Health. Healthcare(Basel).13.11(2),242.

Bagali SS, Gowrishankar BS, Bagali SS (2021) A review on biomedical waste management. Waste Technol 9(2):1–5. https://doi.org/10.14710/wastech.9.2.1-5

Bansod HS, Deshmukh P, Bansod H (2023) Biomedical waste management and its importance: a systematic review. Cureus 15(2)

Bio-Medical Waste Management Rules (2016) Published in the gazette of India, extraordinary, Part II, Section 3, Sub Section (1), Government of India Ministry of Environment, Forest and Climate Change. Notification, New Delhi, the 28th Mar 2016

Capoor MR, Bhowmik KT (2017) Current perspectives on biomedical waste management: rules, conventions and treatment technologies. Indian J Med Microbiol 35(2):157–164

Chakraborty S, Veeregowda B, Gowda L, Sannegowda SN, Tiwari R, Dhama K, Singh SV (2013) Biomedical waste management. Interaction (2013):12-02

Chosewood LC, Wilson DE (eds) (2009) Biosafety in microbiological and biomedical laboratories. In: Centers for disease control and prevention, 5th edn. ISBN 978-0-1608-5042-4

CPCB (2018) Guidelines for management of health care waste as per biomedical waste management rules, 2016. http://cpcb.nic.in/uploads/projects/bio-medical waste/guidelines_healthcare_june_2018.pdf

Dancer SJ (2014) Controlling hospital-acquired infection: focus on the role of the environment and new technologies for decontamination. Clin Microbiol Rev 27(4):665–690

Datta P, Mohi G, Chander J (2018) Biomedical waste management in India: critical appraisal. J Lab Phys 10(01), 006–014

Eapen JV, Thomas S, Antony S, George P, Antony J (2024) A review of the effects of pharamceutical pollutants on humans and aquatic ecosystem. Explor Drug Sci 2(5):484–507

Emmanuel J, Stringer R (2007) For proper disposal: a global inventory of alternative medical waste treatment technologies. Publ: Health Care without Harm. 52

Government of India (1998) Bio-medical waste (management and handling) rules. The gazette of India. Ministry of Environment and Forest

Jang YC, Lee C, Yoon OS, Kim H (2006) Medical waste management in Korea. Environ Manag 80(2):107–115

Janik-Karpinska E, Brancaleoni R, Niemcewiez M, Wojtas W, Foco M, Podogrocki M, Bijak M (2023) Healthcare waste-a serious problem for global health. Healthcare 11(2):242

Katoch SS (2007) Biomedical waste classification and prevailing management strategies. In: Proceedings of the international conference on sustainable solid waste management, pp 169–175

Nadiaye P, Fall C, Diedhiou A, Tal-Dial A, Diedhiou O (2003) Biomedical waste management in the Regional Hospital Center of Ziguinchor. Sante 13(3):171–176

Odumosu BT (2015) Biomedical waste: its effects and safe disposal. Appl Environ Microbiol 37(3):550–553

Pasupathi P, Sindhu S, Ponnusha BS, Ambika A (2011) Biomedical waste management for health care industry. Int J Bio Med Res 2(1):472–486

Salvi SS, Waghmare S, Thombare V, Mandlik S, Veer S, Walke P, Zambare P (2022) Review on biomedical waste management. Int J Eng Res Technol 11(1):63–69

Sharma D, Singha S (2024) Biomedical waste: sources, impacts on aquatic environment and management. Chron Aquat Sci 1(11):15–21

Singh IB, Sharma RK (1996) Hospital waste disposal system and technology. J Acad Hosp Adm (India) 8(2–1):33–39

Thakur Y, Katoch SS (2012) Emerging technologies in biomedical waste treatment and disposal. Chem Eng Trans 29:787–792

Tiwari AV, Kadu PA (2013) Biomedical waste management practices in India-a review. Int J Curr Eng Technol 3(5):2030–2033

Widmer AF, Frei R (2011) Decontamination, disinfection, sterilization. In: Versalovic J, Carroll KC, Funke G, Jorgensen JH, Landry ML, Warnock DW (eds) Manual of Clinical Microbiology, vol 21, 10th edn. ASM Press,Wahington DC, USA, pp 143–173

Environmental Sustainability Through Phytoremediation of Biomedical Wastes

Shubhajit Halder, Damini Bramhankar, and Doyel Bhattacharya

Abstract

The modern era faces extensive and ever-expanding environmental problems of water and soil pollution. Traditional methods of treatment often do not fully eliminate all waste. Conventional methods of disposal of biomedical wastes, basically drains, ignition, and dumping, are characterized by the disadvantages as high cost, and pollutants may be transformed from one poisonous form to another. The biomedical residues become omnipresent in most situations for example, in marine surroundings where they are found mostly at concentrations breaching the environmental quality standards set. Phytoremediation emerges as an inherently available and ecologically rational way of reducing these conditions and it deals with various critical challenges of environmental pollution caused by disposal of biomedical waste. This methodology employs the natural plant ability to extract toxic compounds from soil and water. The most utilized are plants because they are abundantly available, tolerate adverse conditions, take up and accumulate toxic substances, have a competitive advantage over other plants, and are productive in producing biomass. Recent technologies for biomedical waste removal from water sources have been a center of attraction for scientists. Hence, learning about new technologies is necessary to know about the merits and demerits of these new strategies. Meanwhile, the current research activities are aimed at increasing the efficiency of biomedical residue extraction from sediments, water bodies, and soils. Among the recent restoration efforts after ecosystem damage, phytoremediation is the most attractive and cost-effective since it is completely organic. According to recent studies, plants can be used for removal of heavy metals, organic pollutants, antibiotics, radioactive elements, and insecticides. This chapter, therefore, is more dedicated to the research, progress, and prospects of environmental restoration with phytoremediation of toxic biomedical wastes.

Keywords

Pollution · Conventional methods of disposal · Biomedical waste · Phytoremediation · ecosystems · Environment restoration

1 Introduction

Biomedical substances play a role in preserving human wellness and are generated in enormous quantities every year. While these substances offer benefits to humanity, they can also impact the environment in several ways such as disrupting ecology and endangering wildlife. As a result, it is vital to monitor and assess their effects. During medical diagnosis, treatment, and vaccination processes, hospitals, clinics, and laboratories must dispose of biomedical waste, which includes materials like used syringes, scalpels, soiled dressings, liquid specimens such as blood and urine, surgical instruments, medicines, and radioisotopes. One source of hazardous substances is stuff related to the research work that includes bacteria, fungi, and viruses as pathogens or toxic chemicals and radiation. Many other biological substances, like steroids with hormones among them, are normally released into the surroundings (Rastogi and Nandal 2020; Yang et al. 2017). The bulk of these pollutants end up in water bodies due to the continuous discharge of untreated waste from homes, industries, and healthcare facilities

S. Halder (✉) · D. Bramhankar
PG & Research Department of Chemistry, Hislop College, Nagpur 440001, Maharashtra, India
e-mail: suvochem.halder@gmail.com

D. Bhattacharya
Department of Chemistry, Dada Ramchand Bakhru Sindhu Mahavidyalaya, Nagpur 440017, Maharashtra, India

A. Mandpe et al. (eds.), *Sustainable Waste Management*, Advances in Science, Technology & Innovation,
https://doi.org/10.1007/978-3-031-85685-3_7

(Rahman et al. 2009). These chemicals can persist in the environment for long periods of time, making it difficult for natural decomposition processes. Prolonged use of antibiotics may cause antibiotic resistance that can lead to fatal epidemics, resulting in the loss of thousands of lives. In the 1990s, MDR-TB began to spread unexpectedly in some areas, particularly within Eastern Europe and parts of Asia. This was majorly attributed to poor or inadequate application of antibiotics, which also goes with incomplete treatment courses awarded to patients, inaccurate prescriptions by doctors, and poor-quality drugs. As such, this raised the chances that the bacteria would mutate to exist even when the applied drug was developed to kill them (Huijbers et al. 2019). Therefore, biomedical waste is more infectious and dangerous than other types of waste. Economically viable methods are necessary for the disposal of biomedical waste, which must be taken care of by the trained occupational health management unit of the hospital for the safeguard of important stakeholders in the medical fraternity. Therefore, all biomedical waste that may require safe disposal should be locally collected and handled safely at their points of generation to facilitate safe disposal (Datta et al. 2018). The inability of wastewater treatment plants in most industrialized and developing countries to effectively manage these compounds is one of the main drivers for their release without proper elimination. Pharmaceutical wastes like expired medications, chemical wastes like disinfectants, cytotoxic wastes in the form of chemotherapy treatment drugs, infectious wastes like contaminated blood, pathological wastes like human tissues, sharps like needles, and radioactive wastes formed by diagnostic materials are all similar to biomedical wastes. Because of infection, chemicals, and radiation, these can seriously endanger human health. The environmental systems—soil, water, air quality, and wildlife—can all be contaminated by inappropriate disposal. For these risks to be reduced, proper management is crucial.

Many options and methods are in the process of being developed that consider toxicity as well as potential adverse effects on the environment to enable efficient pollutant removal. Sophisticated systems mostly employ surface-engineered carbon in adsorption, electrochemical oxidation, ozonation, coagulation, nanofiltration, chlorination, and reverse osmosis processes (Phytoremediation of Pharmaceutical Wastes 2022; de Andrade et al. 2018). While they can remove pollutants that are linked to diseases, these methods still have high costs and require more infrastructure and specialist knowledge. For instance, membrane separation, photocatalysis, aerobic and anaerobic bioreactors are viable alternatives to eliminate biomedical waste from waste water (Singh et al. 2020). Medical waste can be effectively treated by several methods, viz., biological, chemical, thermal, and radiological processes. Incineration is supposed to be a low-cost technique, but it has certain disadvantages, like the careful disposal of generated ash, which may contaminate the water flows and cause potential damage to the local environment due to the uncontrolled release of heavy metals and various POPs.

These technologies show promise but may not equal other advanced chemical and physical techniques, particularly when it comes to effectively removing pollutants from soil, water, or municipal sludge through plant-based systems alongside their rhizospheres (Hu et al. 2021). Using techniques such as phytoremediation utilizing constructed wetlands (CW), extreme harmful chemicals, including pesticides, industrial wastes, and various inorganic and organic pollutants, can be extracted from polluted wastewater and soil (Cunningham and Berti 1993; Ravikumar et al. 2022). Phytoremediation is a cost-effective technology with minimal maintenance, involving the use of plants to eradicate harmful contaminants present in sludge, sediment, soil, wastewater, groundwater, and surface water, and has been successfully used to eliminate biomedical compounds (Wang et al. 2019). Phytoremediation is among the important strategies for remediation because it allows for environmental sustainability, cost-effectiveness, and the possibility of in situ treatment of a wide spectrum of contaminants. The process makes use of natural uptake, degradation, and stabilization mechanisms in plants; it causes less disturbance compared with the conventional approach. Besides, phytoremediation improves the health of the soil and beautifies the site; it is also publicly more acceptable and hence becomes a feasible and environmentally friendly solution for managing contaminated environments. The biological varieties and physical parameters of the plants, as well as the site properties, are important parameters for effective phytoremediation. Phytoremediation is a continuous process in different seasons of local conditions involving the degradation of pollutants and the removal through the accumulation or dispersion of biological wastes. This chapter is to provide insight on biomedical compounds, and this will comprise their characteristics, environmental impacts, as well as their roles in human health. Necessary techniques and technologies to detect harmful substances and their removal from the environment would be examined. This would focus most closely on phytoremediation, its mechanisms, challenges, and future, potential, and current applications that continuously emphasize aquatic plants.

2 Sources of Biomedical Waste

Biomedical waste is an issue so critical that its handling and disposal are a subject to high concern for human health and the environment. Several types of biomedical wastes exist, all generated from various sources, each having its techniques of disposal to avoid risks. This section looks at the primary sources of biomedical wastes and some recommended disposal techniques.

2.1 Healthcare Facilities

Hospitals, clinical facilities, and laboratory facilities are some of the top biomedical waste generators through the disposal of medical materials such as gloves, used needles, syringes, and surgical instruments. Most of the devices in this equipment contain hazardous and infectious materials; hence, disposal must be properly done to avoid contamination and spread of diseases. Common methods of disposal include incineration, autoclaving, and chemical disinfection. If not managed correctly, this waste is associated with serious health risks, environmental pollution, and legal liabilities (Santos et al. 2020).

2.2 Pharmaceutical Waste

Pharmaceutical wastes are considered expired, unwanted, or contaminated drugs. In many ways, this biomedical waste forms a significant source of environmental pollution, particularly when such drugs are disposed of inappropriate means such as flushing down in sewers or landfills with municipal solid wastes. This results in chemical contamination of the aquatic environment, soils, and even food chains. For this, pharmaceutical wastes must be incinerated or treated with specialized treatment methods to render harmful elements useless. Secure medication take-back programs and specifically dedicated pharmaceutical waste containers are effective strategies in managing this kind of waste (Eckel 2002).

2.3 Research Institutions

In research institutions dealing with the biomedical field and animal laboratories, much of the waste emanates from the disposal of laboratory-tested animals, tissues, cultures, and chemical reagents. Poor handling and disposal of such infectious waste may result in their being released into the environment and, hence, becoming part of the soil, water, and air. To prevent such occurrences, research institutions must follow strict waste management protocols. These may include autoclaving, chemical disinfection, and incineration to ensure that infectious waste is rendered harmless before disposal (Ramm et al. 2017).

2.4 Dental Practices

Biomedical waste from dental clinics includes dental amalgam, used needles, and other dental materials. Dental amalgam that contains mercury is a worrisome pollutant for water bodies and can accumulate in aquatic ecosystems, posing a possible danger to aquatic life and humans. The proper disposals of the dental amalgam are either by the use of amalgam separators or by recycling and storage that prevents its release to the environment. Other wastes from dental procedures should also be handled and disposed according to protocols subjecting them to safe measures without affecting the environment (Subramanian et al. 2021).

2.5 Veterinary Clinics

Veterinary clinics, animal hospitals generated wastes arise from the disposal of animal carcasses, tissues, and medical supplies, which turn out to be a source of environmental pollution if not disposed of properly. Such wastes are generally disposed of by high-temperature incineration to destroy pathogens and by autoclaving, in which steam and pressure sterilize the wastes before being disposed of. These methods provide immense control over hazardous wastes and hence contribute immensely to minimize health risks and the impact on the environment (Hajam et al. 2023) (Fig. 1).

3 Classification of Biomedical Waste

3.1 Infectious Waste

Biomedical waste can be broken down into infectious waste, which is a part of the waste stream contaminated with blood, bodily fluids, or other potentially infectious materials. This includes such items as used needles, cultures, swabs, and laboratory wastes containing infectious agents. The infectious waste if mismanaged and inappropriately disposed of, spreads the pathogens, and causes a huge risk to public health (Eckel 2002; Ramm et al. 2017).

3.2 Hazardous Waste

Another type of segregation of biomedical waste is hazardous waste, which comprises toxic, reactive, ignitable, or corrosive materials. Examples of hazardous waste include chemical

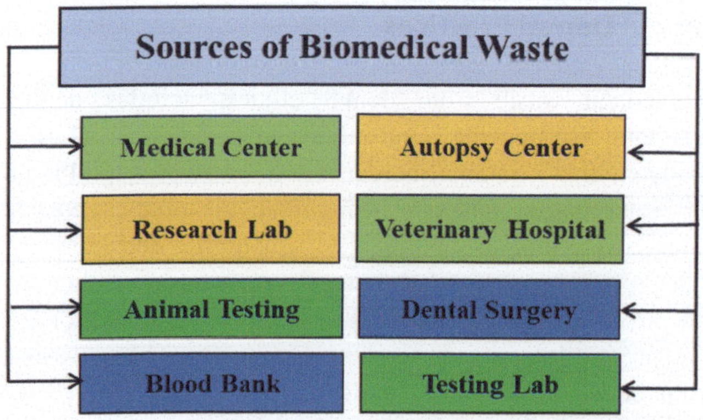

Fig. 1 Predominant sources of biomedical waste

reagents, pharmaceuticals, and heavy metals or radioactive materials. Hazardous biomedical waste should be taken to pose environmental and health risks if not managed and disposed of properly according to the guidelines (Eckel 2002).

3.3 Sharps

"Sharps" is a term used to refer to anything capable of puncturing or slicing the skin, including needles, syringes, scalpels, or even broken glass. If not carefully managed, sharps waste may lead to injury for healthcare workers, waste handling staff, and, in some cases, the general population, thereby increasing the potential for infectious diseases to spread easily. The right methods of segregation and disposal should be followed to minimize the risks from sharps waste (Singh et al. 2024) (Fig. 2).

3.4 Anatomical Waste

Body parts, tissues, organs, and fluids from humans or animals. Surgical specimens, fetal remains, and laboratory animals should be disposed of in ways that prevent the spread of disease-carrying infectious agents (Ramm et al. 2017).

3.5 Chemical Waste

These are laboratory chemicals, disinfectants, solvents, and other health- or research-environment chemicals. Wrong disposal of the chemical waste will cause a lot of harm to the environment where it lands, pollute the air, and cause many health injuries for human beings and faunal species. Proper segregation, storage, and disposal methods are crucial to mitigating the risks associated with chemical waste (de Souza Nascimento and Tenuta Filho 2010).

4 Effect of Biomedical Waste

4.1 On Environment

Environmental quality and biodiversity are seriously threatened by biomedical waste, pharmaceutically active chemicals, and other complex substances, as shown in Table 1. These wastes are highly hazardous due to their diverse physicochemical and biological properties (Khan and Jagtap 2023; Klatte et al. 2017). The systematic chemical classification is necessary for their identification due to the biological and chemical complexity. The repeated use of some drugs in biomedicine enhances the chances of such chemicals entering into the environment, with poorly metabolized drugs remaining active. Most such drugs undergo partial processing from sewage treatments and wastewater plants before being discharged into aquatic ecosystems (Rodrigues dos Santos et al. 2021). Besides, huge amounts from animal husbandry and veterinary practices involving hormones and biomedical compounds contaminate wastewater runoffs, leachate seepages, and animal waste. The presence of this medical waste in trace amounts in surface and groundwater reservoirs poses a serious risk to both human and animal health.

Pharmaceutical waste is a major contributor to environmental pollution among biomedical wastes. Some important contributors are substances related to birth control and antibiotics (Nassiri Koopaei and Abdollahi 2017; Karri et al. 2021). These compounds can accumulate as waste due to manufacturing mismanagement. The transportation processes related to ports, docks, and pipelines also result in the emission of dangerous gases, such as carbon monoxide and volatile organic compounds. These non-biodegradable contaminants have serious human health implications, especially when they contaminate the aquatic resources through various means, such as leachates from landfills, animal husbandry, and sewage discharge processes (Perrodin and Orias 2017). The qualitative deterioration of ground water

Fig. 2 Classification of biomedical waste

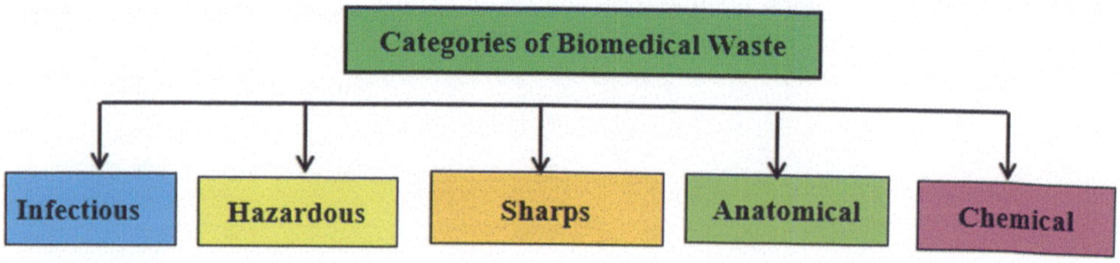

Table 1 Effects of biomedical compounds on environment

Biomedical compound	Adverse effects on environment
Antibiotics	– Development of resistance in bacteria
	– Persistence in soil and water environments
	– Contamination of water supplies
Hormones	– Altered breeding behavior
	– Interference with wildlife populations
	– Endocrine disruptors in aquatic life
Biocides	– Disruption of microbial community
	– Resilience to the environment
	– They tend to accumulate in soil and water ecosystems
Pesticides	– Harmful to beneficial insects and other organisms
	– Bioaccumulation within a food chain
	– Soil and water ecosystem contamination
Pharmaceuticals	– Antimicrobial resistance (AMR) development
	– Contaminating water bodies
	– Disruption of aquatic ecosystems
Heavy metals	– Toxic effects on plants, animals, and humans
	– Soil and water contamination
	– Bioaccumulation and biomagnifications
Personal care products	– Disruption of life cycles of aquatic life
	– Accumulation in aquatic environments
	– Intrusion of man-made chemicals into ecosystems

resources through the landfilling of biomedical wastes that leach emerging organic contaminants and partially degraded antibiotics increases the problems arising from antimicrobial resistance and other health-related problems.

Despite some advancements in wastewater treatment, inefficient elimination of pollutants still points out the urgent need for sustainable remediation to mitigate the negative impact of biomedical waste on the environment and human health (Manzoor and Sharma 2019).

4.2 Effect on Humans

Hazardous medical waste accumulated in plant tissues and organs is relatively harmful to human health. Dyes, solvents, pesticides, and other compounds that accumulate in the environment as byproducts of poorly metabolized pharmaceutical drugs are considered xenobiotics. With time, these compounds become ecotoxic and work their way into the food chain, where they affect human health. All this contamination worsens due to anthropogenic activities such as metallurgical processes and fuel production. Further, anthelmintics and antimicrobials used in veterinary drugs increase the pool of biomedical wastes in the ecosystem. An increased concentration of these biomedical wastes increases the chance of groundwater contamination, surface transmission of diseases, and other impacts due to bio-accumulation (Moniruddin 2023).

The treatment of farm livestock with drugs results in their release into the environment, where they are poorly metabolized, often along with pesticides and fertilizers. Therefore, studies on the ecotoxicity and toxicology of such compounds are very important for the design of preventive policies. Sorption behavior is a determinant factor for the distribution of these compounds within plant, soil, and water. Even after treatment in wastewater facilities, the degradation products of some of these pharmaceuticals remain highly toxic, endangering both humans and animals. Analytical methods such as liquid chromatography (LC) and mass spectrometry (MS) are used to assess the persistence and behavior of these compounds in various ecosystems. Biomedical waste accumulation is a major factor that boosts antimicrobial resistance (AMR) in life-threatening conditions (Jindal et al. 2015). Biomedical wastes, contaminated in wastewaters, increase the resistance to antibiotics both in humans and animals and leave residues which are poisonous in this environment. Extreme emission of biomedical waste by agricultural industries and healthcare sectors promotes multidrug-resistant pathogen growth, which contaminates groundwater and surface water (Kümmerer 2009). Pathogenic antibiotic-resistant bacteria are a serious health threat. While their impact is not of the order of magnitude of diseases such as malaria and tuberculosis, it is growing due to global problems including low access to healthcare and low awareness about drug resistance (Padmanabhan and Barik 2019).

5 Remediation Techniques

5.1 Identification of Biomedical Waste

This will bring in new and advanced methods and ways to determine, quantify, and keep track of the contamination present in the waste stream. A brief overview of some of the major methods that will be used in biomedical waste identification is discussed below.

Nanotechnology-Based Sensors. It developed the application of nanomaterials, such as nanoparticles and nanotubes, in sensor development for use in contaminant detection present in the waste of the biomedical field. The sensor with the functionalized nanomaterial increases its sensitivity and selectivity, thus ultimately contributing to the rapid and accurate detection of pathogens, toxins, and hazardous chemicals that may crop up from biomedical waste. It was shown in a Japan case study that nanotechnology-based sensors can detect traces of toxic substances in hospital waste and have a huge effect on early detection and remediation (Sindhu et al. 2021; Bhattacharya and Halder 2023).

Biosensors. It involves biological recognition elements, such as enzymes or antibodies or nucleic acids, coupled with transducers in detecting specific biomolecules or pathogens in biomedical wastes. In this respect, high sensitivity and specificity of the biosensor are useful for healthcare facilities dealing with assessment and management-related waste within a short period. Data Analysis According to several studies, it is proved that these biosensors can detect as low as parts per trillion concentrations of contaminants, way above concentration detection rates compared with traditional methods. For instance, in the detection of drug residues in wastewater, biosensors offered a detection efficiency rate of 98%, therefore guaranteeing more accurate and timely waste processing (Haleem et al. 2021; Mohankumar et al. 2021).

Mass Spectrometry (MS) and Chromatography Techniques. Both methods have high sensitivity as well as specificity in MS and chromatographic techniques such as liquid chromatography–mass spectrometry (LC–MS) and gas chromatography–mass spectrometry (GC–MS) for analyzing complex mixtures in biomedical waste. The prime role of the GC–MS is to identify and quantify pharmaceutics, organic compounds, and metabolites that contribute to the biomedical waste composition. The U.S. facility also employed MS and chromatography techniques to identify hazardous components of medical waste with great exactitude and thereby ensure proper disposal (Lim 1991).

Remote Sensing and Imaging. Site characterization of the waste biomedical disposal sites and its extent are located and delineated by hyperspectral imaging and infrared spectroscopy as a remote sensing technique. Such technologies offer space and spectroscopy information to detect contamination in waste and assess environmental impacts through remote sensing. In Brazil, remote sensing and imaging have been applied to monitor contaminant dispersions from healthcare facilities, providing very important data in environmental management (Glanville and Chang 2015).

5.2 Phytoremediation Methods

Among the new methods, phytoremediation is a green technology in contaminant management within the environment using plants. Other roles played by these plants include destruction, whereby hazardous substances are broken down by the plants; removal through absorption of pollutants directly into their tissues; containment through preventing spread of contaminants; and immobilization, where plants stabilize toxins in soil, hence stopping them from moving around and causing harm or risk. This is relatively new technology; the method has recently sparked quite a bit of enthusiasm as a novel, inexpensive alternative to more conventional treatment methods employed at hazardous waste sites (McIntyre 2003). Phytoremediation is the process of using living green plants for in situ, preventive, or restorative treatment of areas contaminated with chemicals, oils, hormone-active pharmaceuticals, or removal of other harmful things from the environment, slowly replacing traditional methods such as the use of physicochemical techniques, autoclaving, thermal treatment, chlorination, and ozonation for the purification of water and soil ecosystems. This is usually an inexpensive substitute that does not negatively impact other biotic elements of the surroundings. Wastes such as biomedical waste may be taken up from the soil, their components stabilized, enzymes treated, and harmful substances may seep into the groundwater, while at the same time filtering out unwanted chemicals from the wastewater. The chemical framework of these pollutants either undergoes transformation to a stable and environmentally harmless form or microbiological activity breaks it down. Micro-quantities of biomedical and manufactured food waste have thus accumulated in bodies of freshwater, which are breeding grounds for water-borne diseases. This previous environmental investigation thus identifies those compounds and traces levels of chemicals and drugs in soils and water (Phytoremediation of Pharmaceutical Wastes 2022; Yan et al. 2020). The most common are substances present in biomedical wastes that are very non-biodegradable and possibly able to be estrogenic on the physiological functions of the higher species. Released suspended contaminants, chemicals, medicines, and scents from aquatic habitats from a variety of sources may endanger the extinction of many aquatic creatures. Hazardous heavy metals such as arsenic, cadmium, copper, lead, and zinc, therefore, are filtered into human life through the food chain by processes like bio-magnification and eutrophication (Padmavathiamma and Li 2007). Such substances give rise to health conditions like diarrhea, vomiting, convulsions, stomatitis, and gastroenteritis.

Phytoremediation is economical and much cheaper than other methods, and it finally separates pollutants from the environment through phytoremediation. In the above light, there are some plants that, in the process of phytoremediation of wastes, are extremely good choice for the phyto-detoxification of soil and water since these plants can hyper-accumulate the compounds that are toxic to them in their tissues. Some of the important plants belonging to this category are *Typha sp., Phalaris aundinacea, Azolla pinnata, Eleocharis macrostachya, Cyperus sp., Scirpusgrossus, Pistia stratiotes* (water cabbage), *Ipomoea aquatica, Phragmitis communis, Juncus effuses* (soft rush), *Hydrilla verticillata, Gracilariagracilis, Lemna gibba, Spirodelapolyrhiza, and Vetiveriazizanioides* (Phytoremediation of Pharmaceutical Wastes 2022; Mehta et al. 2021; Raskin et al. 1997). Finally, the most recent works suggest that the technologies of phytoremediation like phytostabilization, phytovolatilization, rhizofiltration, phytoaccumulation, phytoextraction, phytodegradation, and rhizodegradation have replaced lengthy and complex processes like activated sludge, electrocoagulation, nanofiltration, and UV disinfection (Sri Lakshmi et al. 2017). The following is a few of the most popular phytoremediation methods.

Phytostabilization. It therefore suggests that phytostabilization is one of the most promising techniques toward environmentally sustainable approaches for the reduction of contaminant effects in biomedical waste management. The process involves the application of plants to the immobilization and stabilization of soil contaminants to reduce their mobility and bioavailability. Specifically, plant species that have been genetically modified with mature root systems are grown at biomedically contaminated sites, such as medical facilities or laboratory locations. The plants act as natural containment barriers for the transportation of the contamination since they soak in the contaminants via their roots. Once taken up by the plant, the contaminants are sequestered in the root zone and surrounding soil, thereby greatly reducing the potential for leaching into groundwater or uptake by other organisms (Rosenfeld et al. 2018). Over time, the plants help in the restoration of health to the soil and reduce contaminant spread, finally offering general remediation and rehabilitation of the affected sites. Thus, this process is cheap, generates very minimal environmental impact, is easily applicable with other remediation methods, and hence makes phytostabilization the most sustainable solution to the biomedical waste contamination problem with the utmost protection to the environment and human health (Guala et al. 2010).

Phytovolatilization. In this process, medical waste constituents present in the groundwater and soil are liberated into the atmosphere in the form of volatile compounds across transpiration and diffusion. As a result, plants interact with their surroundings because they absorb and emit chemical compounds at different stages. Such compounds as hormones and xenobiotics can accumulate in different parts of the plant or be transported to the organs. Phytovolatilization is a novel method of eliminating inorganic and organic pollutants from biomedical waste sites by lessening the water extent and creating a larger connection between contaminants and the subsurface vadose zone of the plant (Limmer and Burken 2016). Phytovolatilization helps in separating waste ingredients in diverse air spaces, which enables their diffusion into the less polluted parts of the atmosphere. As a result, the concentration of volatile organic compounds in biomedical waste is reduced, which increases the chances of photodegradation in the atmosphere. Root activities may promote or reduce the levels of phytovolatilization, depending on soil alterations enabled by roots. Biomedical substances diffused by roots and then volatilized are absorbed by plants in direct phytovolatilization. The sensitizer then translocate and volatilizes it in the trunk, stem, and leaves via action of transpiration. These considerably ineffective compounds may diffuse across the cuticle and epidermal layers. By lowering the water volume, permitting hydraulic redistribution can increase chemical intensity. Moreover, it also helps increase soil penetrability and intercept rainfall that would otherwise dilute VOCs, which can further improve VOC emissions (Baduru et al. 2008).

Rhizofiltration. Rhizofiltration thus remains a newfangled approach to remediation that uses the inherent capacity of plant roots in the filtration of contaminated water and possibly could be a rewarding approach for the management of biomedical waste. Phytoremediation is the process through which certain types of plants with vigorous root systems, like water hyacinth or sunflower, are stimulated in contaminated water bodies. They serve as living filters that take down pollutants, such as organic compounds, heavy metals, and medication, which are commonly in effluent from medical institutions. The contaminants are adsorbed, concentrated, and translocated to the plant through the vascular system upwards as the water moves through the root zone. It can therefore be deduced that there is a decrease in the concentration of the contaminants in the treated water since most of them accumulate in the roots and shoots of the plants. When the plants become quite saturated with pollutants, these harvested plants are safely disposed of in the environment. In this way, rhizofiltration offers a viable alternative for maintaining water quality and public health. It is an inexpensive, environmentally friendly, and renewable technique that can be applied for cleaning up biomedical waste (Kurwadkar 2019).

Phytoaccumulation. Phytoaccumulation appears to be the most useful approach to be employed in the arsenal of biomedical waste management, which uses the intrinsic ability of plants to uptake and store environmental contaminants. This technique cultivates certain plant species known as hyperaccumulators in the affected zones due to the biomedical wastes, which include soils or water bodies. These have the mysterious property of absorption and hyperaccumulation of pollutants, be they heavy metals, pharmaceuticals, or any other toxic compounds, which easily move into the plant tissue. As the plant grows, contaminants from the surrounding environment flow to the root, which will then flow into the plant, effectively removing them from the soil or water (Rastogi and Nandal 2020). The accumulation of contaminants will typically bioaccumulate gradually in the aerial tissues of plants, including leaves and stems, to significantly higher levels in comparison with the surrounding environment. The mature plants are later harvested and properly disposed of, where the accumulated contaminants are taken out of the ecosystem. Several reasons make phytoaccumulation quite useful and handy. They include the process being economically friendly, friendly to the environment, and also dealing with a wide range of contaminants. Such an approach to the management of biomedical waste contamination by using phytoaccumulation relies on plant inherent capabilities that are both cost-effective and environmentally sustainable to enable environmental restoration and rehabilitation (Kamal et al. 2004).

Phytoextraction. The principal application of this method is the elimination of sluggish metal contaminants present in pharmaceutical waste. The soil network has been made unproductive and dangerous to life by various anthropogenic activities, notably the free clearance of enzymes, hormones, metalloproteins, and supplementary cofactors. They may entirely cover crucial functional groups, which may disrupt animal and human physiological processes. Its persistence is more profound, and equal destruction pressure is applied to the life of the environment above the stage of the biochemical process. It adversely affects photosynthesis or destroys the environment through vegetation loss, thus depleting a specific species' abode. For these reasons, phytoextraction is a continuing and inexpensive tool for the elimination of such dangerous metals from soils and waters. Lead, zinc-containing compounds, arsenic, cadmium, mercury, and other heavy metal waste are among the most often extracted materials using this technique. Biomedically active chemicals still retain trace levels of heavy metals; hence, the application of phytoextraction for biomedical waste may essentially be restricted to circumstances when the compound has minimal metabolic value for disposal (Nanda et al. 1995).

Phytodegradation. Phytodegradation is another innovative endeavor in the field of natural plant use for the purposes of biomedical waste treatment that holds much promise. The metabolic processes of plants can be used to break down and detoxify pollutants in the environment. The process involves planting particular types of plants that can degrade pollutants found in biomedical waste with enzymes. These include organic materials, drugs, and other chemical substances used in various industries (Ebrahimbabaie and Pichtel 2021). While they are in their progression phase of growth, these plants absorb the contamination through their roots. These contaminants are transported to particular tissues in the plant, where there is an enzymatic reaction. Through various biochemical processes, the plants metabolize the pollutants. In doing so, they degrade the pollutants and transform them into elements that are less harmful. The less harmful substances are either stored in the plant or released back to the environment through the plant in a non-toxic form. Over time, this continuous growth of the plants and their metabolic activities help in degrading the contaminated soil or the water body. This makes phytodegradation cost-effective. It is also scalable. This means that it can be done for areas that are more contaminated, and it can be used for less contaminated areas. This method of bioremediation can happen in the absence of any human intervention. This technique ensures that it is cleaned in situ. Contamination can happen to any biosphere, be it soil, air, or water. As the Earth is exhausted to meet human needs, we dump most of the non-biodegradables that we create. One of the major causes is biomedical waste, which creates land pollution (McIntyre 2003).

5.3 Aquatic Plants Function in Waste Removal

One of the most polluted habitats on Earth is the water, where several chemicals damage ecosystems. It used to be believed that rivers and oceans were too large to be harmed by rubbish being dumped into them. Water thus contained radioactive waste, untreated medication, and garbage from homes, farms, and industries. As we previously discussed, these substances can be detrimental to human health and aquatic life in even minute concentrations. This can be fixed by employing water plants to clean up trash, as this is an inexpensive, effective, and simple solution . A few water plant types like *Salvinia molesta, Lemna minor, Cattails (Typha sp.), P. stratiotes, Azolla caroliniana, Cyperus sp., E. crassipes, Salix atrocinerea, and I. aquatica* are great at getting rid of biomedical compounds (Fletcher et al. 2019). These plants do a big part in ways to clean up waste using plants. They take biomedical waste in through their roots, with more in the roots than the rest of the plant (Dhir 2013). Substance charge, solubility in water, and the extent to which the plant can transform the medication are all important factors in how water plants absorb and process pharmaceuticals. Because of their pH, acidic medications pass through plants, but basic chemicals disintegrate and enter. Drugs that are neutral are added since they do not combine with water. Upon ingesting medicines, plants undergo three stages of degradation or adaptation. Phase

I involves the use of enzymes; Phase II involves joining with other things; and Phase III involves storage within the plant. The little organisms in the soil close to the roots can help prevent drug damage even with these methods (Ansari et al. 2020).

Constructed wetlands (CWs) are another emerging method of phytoremediation that uses aquatic plants. These man-made wetlands, rich in aquatic vegetation, are an excellent way to remove pollutants, including micropollutants like medications, biomedical waste, and nutrients. When used as a phytoremediation treatment for micro/biomedical pollutants, CWs work through a combination of microbiological, physical, and chemical mechanisms to facilitate the natural breakdown of pollutants from soil, liquid, or both (Ferreira et al. 2017). Reduced interactions with humans and animals as well as less biosolid production characterize the recommended subsurface flow CWs for treating biomedical contaminants. Whereas aerobic activities predominate in vertical flow subsurface CWs, anaerobic processes are seen in horizontal flow subsurface CWs. The efficiency of CWs is a function of the target compounds and the underlying synergistic processes among plants and microbes. Although CWs find extensive application in pretreatment, the sector is increasingly incorporating the technology into primary and secondary treatment systems due to high levels of pollutant removal, including recalcitrant toxic compounds. As seen from the above case studies, there exists some variability in efficiency across studies; however, CWs can achieve high removal rates when optimally configured and processed to fit pollutant and plant species (Kafle et al. 2022) (Fig. 3).

6 Challenges of Phytoremediation

Phytoremediation is a quite promising green technology for the removal of contaminants from soils and water, various challenges may impact its output. There are environmental factors such as climate, soil composition, or the availability of water which document the critical role played out in determining the success of this method. For instance, the growth rate of plants used in phytoremediation is affected by temperature and light. Temperature extremes hinder or cause death. Soil pH and nutrient availability also influence plant uptake and degradative capability. Poor growth in very nutrient poor soils can reduce effectiveness at removing the toxics. In addition to this, the presence of heavy metals or other pollution in the soil may retard the growth of plants and complicate the remediation process (Mahar et al. 2016).

The other critical constraint to the phytoremediation process is its time factor. This is an inherently very slow process that, in most cases, may require several growing seasons to attenuate a sufficient quantity of contaminants. Plants may require years to be well established, grow, and accumulate pollutants. This time could even be long where the contaminant is highly embedded in the soil matrix or

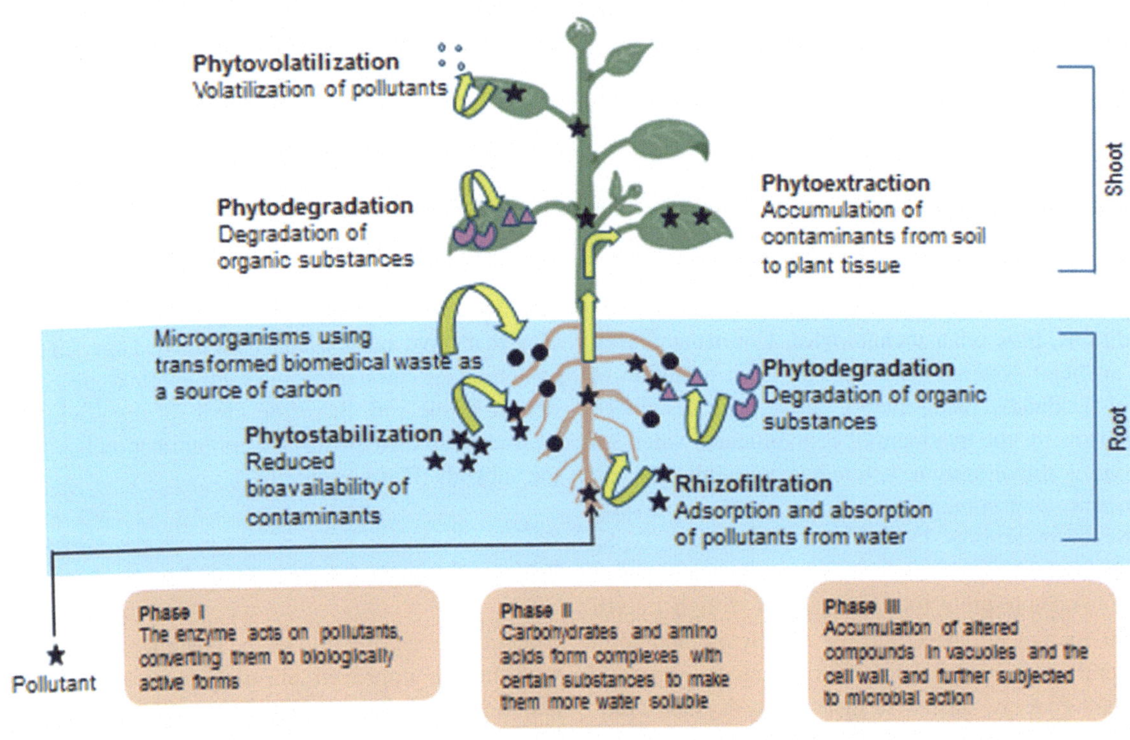

Fig. 3 Phytoremediation methods by aquatic plants (The figure is prepared by using paint 3D)

low in concentration. Such an extended time span could prove a drawback to those sites in which quick remediation needs to be accomplished so as to avoid further environmental damage or human exposure. After all, phytoremediation involves the plant producing biomass that may have high levels of accumulated contaminants and thus needs to be carefully managed. Improper disposal of this biomass may result in secondary pollution, which could offset remediation processes (Shen et al. 2021).

Another issue is the availability of land for the phytoremediation, especially for the use of constructed wetlands in wastewaters treatment in densely inhabited areas. The efficiency of such wetlands declines with time due to clogging and aging and has to be managed properly. Whereas phytoremediation is an environmentally friendly technique of pollution management, the efficiency of phytoremediation depends highly on environmental conditions and duration taken by the process. For remediation efforts to be successful, these issues must be clearly managed.

7 Strategies to Enhance Phytoremediation

Phytoremediation seems to be a most promising method used for biomedical-waste management, there are various challenges in terms of effectiveness. In attempts to enhance this process, several strategies have been invented. This section will discuss different strategies to enhance phytoremediation with regard to wastewater treatment, wasteland management, and the role of genetic engineering.

7.1 Strategies for Wastewater

Phytoremediation of wastewater is the process in which wastewater contaminants from water bodies are absorbed, accumulated, or degraded with the involvement of plants. Improvements in this area are based on plant selection, root zone conditions, plus other technologies. Constructed wetlands are artificial systems whose design is an imitation of the natural wetlands, using plants, soils, and associated microorganisms for the treatment of contaminated water. An effective one is rhizofiltration, where heavy metals, organic pollutants, and other contaminants in the water are directly taken up by the plant roots. Indeed, examples include Typha latifolia, also called cattail, and Phragmites australis (common reed), which is quite tolerant and has a high degree of pollutant tolerance. The efficiency of phytoremediation can be overall improved by combining with the bioreactors. In these systems, otherwise difficult to degrade complex organic pollutants are degraded by the bioreactor to less complex compounds that can be taken up easily by plants. For example, the application of an anaerobic bioreactor added to a constructed wetland greatly reduces the pollutant loading before water is flushed through it, providing additional treatment by plants such as Scirpus validus (softstem bulrush). Introducing plant growth-promoting rhizobacteria (PGPR) into the rhizosphere induces plant growth and contaminant-uptake improvement. Some species act in the mobilization of nutrients, the production of hormones, and the degradation of contaminants. For example, strains of such bacteria as Pseudomonas and Bacillus have been shown to enhance phytoremediation of petroleum hydrocarbons in wastewater (Agarwal and Rani 2022).

7.2 Strategies for Wasteland

Wastelands in particular are largely affected by poor soils and contamination and require special measures for phytoremediation to be effective. In this respect, the improvements lie in the realms of soil amendments, plant selection, and use of symbiotic organisms. One approach toward amending the soil with organic matter is by adding compost, biochar, or manure in order to enhance the structure, fertility, and water-holding capacity of the soil conditions under which plants grow best and are more proficient at contaminant uptake. For example, the phytoremediation of heavy metals may be improved by addition of biochar, which tends to raise the pH and cation exchange capacity of soil, hence improving metal availability for plant uptake. Mycorrhizal fungi form symbiotic relationships with plant roots, extending reach of the root system and improving water and nutrient uptake, including contaminants. In such contexts, arbuscular mycorrhizal fungi (AMF) are typically used to improve heavy metal phytoremediation in wasteland soils. For example, the Glomus species apart from aiding plant growth also helped in pollutant stabilization within the soil matrix. Deep-rooting plants play a very important role in accessing the contaminants lying deep in the soil matrix. Populus spp. and Eucalyptus spp. could be exemplary species that have been used. They have large, extensive root systems that can extend to reach and take up contaminants deeper in the soil, therefore ideal for the phytoremediation of wastelands where surface contamination has leached into the subsoils (Gavrilescu 2022).

7.3 Genetic Engineering Strategies

Genetic engineering allows the custom production of plants with highly improved abilities for tolerate, accumulate, and degrade pollutants in particular for use in phytoremediation. By transfer of a specific enzyme-coding gene will increase the capacity of the plant enzymatic pathway for breaking down certain contaminants. This has been the case in specific, very large enhancements for the degradation of organic pollutants, including polycyclic aromatic hydrocarbons (PAHs), have been realized by the introduction of bacterial genes coding for enzymes like cytochrome P450. Such transgenic plants are able to metabolize more of the pollutants as compared to wild-type plants. Genetic engineering in plants can be done to over-express metal transporter proteins, such as NRAMP and ZIP families of proteins (natural resistance-associated macrophage protein and zinc-regulated transporter, iron-regulated transporter-like protein), which induce the uptake of heavy metals like cadmium and lead. Such plants can also accumulate higher tissue metal concentrations that can make them more effective for the phytoremediation of the contaminated site. Gene-encoded tolerance of plants for various environmental stresses such as drought, salinity, and high levels of an environmental pollutant can improve their phytoremediation efficacy. Overexpression of some genes for instance, AtNHX1 (a sodium/hydrogen antiporter) could be used in making some plant species that will tolerate saline conditions at high biomass and planting density for phytoremediation of coastal wasteland contaminated with heavy metals and other pollutants (Koźmińska et al. 2018).

These strategies can further make phytoremediation both effective and adaptive to various environmental conditions and contaminants, hence offering a sustainable solution for the management of biomedical wastes and other pollutants.

8 Conclusion

This chapter has reviewed phytoremediation as one of the novel approaches for the management of the biomedical waste. Through an overall review analysis of several studies and methodologies, one can come to understand that the usual treatment of physicochemical and biological ways in the contaminated environment proves to be costly and laborious. Phytoremediation is one of the least expensive, effective, and sustainable remedy for reducing the impacts of biomedical pollutants and other such toxic substances on the environment. The use of some plant species through phyto-stabilization, phytovolatilization, rhizofiltration, phytoaccumulation, phytoextraction, and phytodegradation may represent an effective way of fighting organic and inorganic contaminants. Though this approach faces certain challenges, including the slow growth of plants, their environmental dependencies, and the necessity of careful management of the accumulated toxins, the overall effectiveness of phytoremediation in moving toward reduced contamination and ecological balance is clear. In fact, research in this area is very important for overcoming many such available limitations in the efficient management of phytoremediation strategies at a high scale in the field of biomedical waste management.

References

Agarwal P, Rani R (2022) Strategic management of contaminated water bodies: omics, genome-editing and other recent advances in phytoremediation. Environ Technol Innov 27.105463. https://doi.org/10.1016/j.eti.2022.102463

Ansari AA, Naeem M, Gill SS, AlZuaibr FM (2020) Phytoremediation of contaminated waters: an eco-friendly technology based on aquatic macrophytes application. Egypt J Aquat Res 46(4):371–376. https://doi.org/10.1016/J.EJAR.2020.03.002

Baduru KK, Trapp S, Burken JG (2008) Direct measurement of VOC diffusivities in tree tissues: impacts on tree-based phytoremediation and plant contamination. Environ Sci Technol 42(4):1268–1275. https://doi.org/10.1021/ES071552L

Bhattacharya DM, Halder S (2023) The versatility of biosensors as potential diagnostic tools for environmental monitoring and control. https://doi.org/10.1063/5.0148947

Cunningham SD, Berti WR (1993) Remediation of contaminated soils with green plants: an overview. In Vitr Cell Dev Biol-Plant 29(4):207–212. https://doi.org/10.1007/BF02632036

Datta P, Mohi GK, Chander J (2018) Biomedical waste management in India: critical appraisal. J Lab Physicians 10(1):006–014. https://doi.org/10.4103/JLP.JLP_89_17

Dhir B (2013) Phytoremediation: role of aquatic plants in environmental clean-up

Ebrahimbabaie P, Pichtel J (2021) Biotechnology and nanotechnology for remediation of chlorinated volatile organic compounds: current perspectives. Environ Sci Pollut Res 28. https://doi.org/10.1007/S11356-020-11598-Y

Eckel WP (2002) Comment on pharmaceuticals, hormones, and other organic wastewater contaminants in U.S. streams, 1999–2000: a national reconnaissance. Environ Sci Technol 36(18):4003. https://doi.org/10.1021/ES025709F

Ferreira AR, Ribeiro AB, Couto N (2017) Remediation of pharmaceutical and personal care products (PPCPs) in constructed wetlands: applicability and new perspectives. Presented at the 1 Jan 2017

Fletcher J, Willby N, Oliver DM, Quilliam RS (2019) Phytoremediation using aquatic plants. Presented at the 12 April 2019

Gavrilescu M (2022) Enhancing phytoremediation of soils polluted with heavy metals. Curr Opin Biotechnol 74:21–31. https://doi.org/10.1016/j.copbio.2021.10.024

Glanville K, Chang HC (2015) Remote sensing analysis techniques and sensor requirements to support the mapping of illegal domestic waste disposal sites in Queensland, Australia. Remote Sens 7. https://doi.org/10.3390/RS71013053

Guala SD, Vega FA, Covelo EF (2010) The dynamics of heavy metals in plant-soil interactions. Ecol Model 221(8):1148–1152. https://doi.org/10.1016/J.ECOLMODEL.2010.01.003

Hajam YA, Kumar R, Kumar A (2023) Environmental waste management strategies and vermi transformation for sustainable development. Environ Challenges 13:100747 https://doi.org/10.1016/j.envc.2023.100747

Haleem A, Javaid M, Singh RP, Suman R, Rab S (2021) Biosensors applications in medical field: a brief review. Sens Int 2:100100. https://doi.org/10.1016/J.SINTL.2021.100100

Hu X, Xie H, Zhuang LL, Zhang J, Hu Z, Liang S, Feng K (2021) A review on the role of plant in pharmaceuticals and personal care products (PPCPs) removal in constructed wetlands. Sci Total Environ 780:146637. https://doi.org/10.1016/J.SCITOTENV.2021.146637

Huijbers PMC, Flach CF, Larsson DGJ (2019) A conceptual framework for the environmental surveillance of antibiotics and antibiotic resistance. J Environ Int 130:104880. https://doi.org/10.1016/J.ENVINT.2019.05.074

Jindal AK, Pandya K, Khan ID (2015) Antimicrobial resistance: a public health challenge. Med J Armed Forces India 71(2):178. https://doi.org/10.1016/J.MJAFI.2014.04.011

Kafle A, Timilsina A, Gautam A, Adhikari K, Bhattarai A, Aryal N (2022) Phytoremediation: Mechanisms, plant selection and enhancement by natural and synthetic agents. Environmental Advances 8:100203. https://doi.org/10.1016/j.envadv.2022.100203

Kamal M, Ghaly AE, Mahmoud NS, Cote R (2004) Phytoaccumulation of heavy metals by aquatic plants. Environ Int 29(8):1029. https://doi.org/10.1016/S0160-4120(03)00091-6

Karri RR, Ravindran G, Dehghani MH (2021) Wastewater—sources, toxicity, and their consequences to human health. Presented at the 1 Jan 2021

Khan S, Jagtap S (2023) Modernizing medical waste management: unleashing the power of the internet of things (IoT). Sustainability 15(13):9909. https://doi.org/10.3390/su15139909

Klatte S, Schaefer HC, Hempel M (2017) Pharmaceuticals in the environment—a short review on options to minimize the exposure of humans, animals and ecosystems. Sustain Chem Pharm 5:61–66. https://doi.org/10.1016/J.SCP.2016.07.001

Koźmińska A, Wiszniewska A, Hanus-Fajerska E, Muszyńska E (2018) Recent strategies of increasing metal tolerance and phytoremediation potential using genetic transformation of plants. Plant Biotechnol Rep 12(1):1–14. https://doi.org/10.1007/S11816-017-0467-2

Kümmerer K (2009) The presence of pharmaceuticals in the environment due to human use--present knowledge and future challenges. J Environ Manage 90. https://doi.org/10.1016/J.JENVMAN.2009.01.023

Kurwadkar S (2019) Occurrence and distribution of organic and inorganic pollutants in groundwater. Water Environ Res 91(10):1001–1008. https://doi.org/10.1002/WER.1166

Lim CK (1991) Chapter 1 liquid chromatography in biomedical analysis: basic approach. Presented at the 1 Jan 1991

Limmer MA, Burken JG (2016) Phytovolatilization of organic contaminants. Environ Sci Technol 50(13):6632–6643. https://doi.org/10.1021/ACS.EST.5B04113

Mahar A, Wang P, Ali A, Awasthi MK, Lahori AH, Wang Q, Li R, Zhang Z (2016) Challenges and opportunities in the phytoremediation of heavy metals contaminated soils: a review. Ecotoxicol Environ Saf 126:111–121. https://doi.org/10.1016/J.ECOENV.2015.12.023

Manzoor J, Sharma M (2019) Impact of biomedical waste on environment and human health. Environ Claims J 31(4):311. https://doi.org/10.1080/10406026.2019.1619265

McIntyre T (2003) Phytoremediation of heavy metals from soils. Presented at the 1 Jan 2003

Mehta KP, Sharma R, Haldar S, Kumar A (2021) Advancement in treatment of wastewater with nano technology. Mater Today: Proc 80:3358–3362. https://doi.org/10.1016/J.MATPR.2021.07.253

Mohankumar P, Ajayan J, Mohanraj T, Yasodharan R (2021) Recent developments in biosensors for healthcare and biomedical applications: a review. Measurement 167:108293. https://doi.org/10.1016/J.MEASUREMENT.2020.108293

Moniruddin A (2023) Dangers of xenobiotics to health and ecosystem. KYAMC J 13(4):189. https://doi.org/10.3329/kyamcj.v13i4.64927

Nanda PBA, Kumar PN, Dushenkov V, Motto HL, Raskin I (1995) Phytoextraction: the use of plants to remove heavy metals from soils. Environ Sci Technol 29. https://doi.org/10.1021/ES00005A014

Nassiri Koopaei N, Abdollahi M (2017) Health risks associated with the pharmaceuticals in wastewater. DARU J Pharm Sci 25. https://doi.org/10.1186/S40199-017-0176-Y

Padmanabhan KK, Barik D (2019) Health hazards of medical waste and its disposal. Presented at the 1 Jan 2019

Padmavathiamma PK, Li LY (2007) Phytoremediation technology: hyper-accumulation metals in plants. Water Air Soil Pollut 184(1–4):105–126. https://doi.org/10.1007/S11270-007-9401-5

Perrodin Y, Orias F (2017) Ecotoxicity of hospital wastewater. Presented at the 1 Jan 2017

Phytoremediation of Pharmaceutical Wastes (2022). Presented at the 1 Jan 2022

Rahman MF, Yanful EK, Jasim S (2009) Endocrine disrupting compounds (EDCs) and pharmaceuticals and personal care products (PPCPs) in the aquatic environment: implications for the drinking water industry and global environmental health. J Water Health 7(2):224–243. https://doi.org/10.2166/WH.2009.021

Ramm JG, Dorscheid GL, Passos CG, Sirtori C (2017) Development of a waste management program in technical chemistry teaching. J Chem Educ 95(4):570–576. https://doi.org/10.1021/acs.jchemed.7b00590

Raskin I, Smith RD, Salt DE (1997) Phytoremediation of metals: using plants to remove pollutants from the environment. Curr Opin Biotechnol 8(2):221–226. https://doi.org/10.1016/S0958-1669(97)80106-1

Rastogi M, Nandal M (2020) Toxic metals in industrial wastewaters and phytoremediation using aquatic macrophytes for environmental pollution control: an eco-remedial approach. Presented at the 1 Jan 2020

Ravikumar Y, Yun J, Zhang G, Zabed HM, Qi X (2022) A review on constructed wetlands-based removal of pharmaceutical contaminants derived from non-point source pollution. Environ Technol Innov 26:102504. https://doi.org/10.1016/j.eti.2022.102504

Resende de Andrade J, de Oliveira MF, da Silva MGC, Vieira MGA (2018) Adsorption of pharmaceuticals from water and wastewater using nonconventional low-cost materials: a review 57(9):3103–3127 (2018). https://doi.org/10.1021/ACS.IECR.7B05137

Rodrigues dos Santos C, Santos Arcanjo G, Valéria de Souza Santos L, Koch K, Santos Amaral MC (2021) Aquatic concentration and risk assessment of pharmaceutically active compounds in the environment. Environ Pollut 290:118049. https://doi.org/10.1016/J.ENVPOL.2021.118049

Rosenfeld CE, Chaney RL, Martínez CE (2018) Soil geochemical factors regulate Cd accumulation by metal hyperaccumulating Noccaeacaerulescens (J. Presl& C. Presl) F.K. Mey in field-contaminated soils. Sci Total Environ 616. https://doi.org/10.1016/J.SCITOTENV.2017.11.016

Shen X, Hao S, Dai M, Yang J, Sun L, Tan X, Peng C, Ali I, Naz I (2021) A critical review on the phytoremediation of heavy metals from environment: performance and challenges. Chemosphere 291:132979.https://doi.org/10.1016/J.CHEMOSPHERE.2021.132979

Sindhu RK, Kaur G, Kaur A, Garg S, Yadav SK, Singh S (2021) Application of nanotechnology in pharmaceutical sciences. Presented at the 8 April 2021

Singh SP, Wasewar KL, Kansal SK (2020) Low-cost adsorbents for removal of inorganic impurities from wastewater. Presented at the 1 Jan 2020

Singh H, YT K, Mishra AK, Singh M, Mohanto S, Ghumra S, Seelan A, Mishra A, Kumar A, Pallavi J, Ahmed MG, Sangeetha J, Thangadurai D (2024) Harnessing the foundation of biomedical waste management for fostering public health: strategies and policies for a clean and safer environment. Discov Appl Sci 6: 89. https://doi.org/10.1007/s42452-024-05735-2

Souza Nascimento ED, Tenuta Filho A (2010) Chemical waste risk reduction and environmental impact generated by laboratory activities in research and teaching institutions. Braz J Pharm Sci 46(2):187–198. https://doi.org/10.1590/S1984-82502010000200004

Sri Lakshmi K, Hema Sailaja V, Anji Reddy M (2017) Phytoremediation—a promising technique in waste water treatment. Int J Sci Res Manage 5. https://doi.org/10.18535/IJSRM/V5I6.20

Subramanian AK, Thayalan D, Edwards AI, Almalki A, Venugopal A (2021) Biomedical waste management in dental practice and its significant environmental impact: a perspective. Environ Technol Innov 24:101807. https://doi.org/10.1016/J.ETI.2021.101807

Wang Y, Yin T, Kelly BC, Gin KYH (2019) Bioaccumulation behaviour of pharmaceuticals and personal care products in a constructed wetland. Chemosphere 222:275–285. https://doi.org/10.1016/J.CHEMOSPHERE.2019.01.116

Yan A, Wang Y, Tan SN, Yusof MLM, Ghosh S, Ghosh S, Chen Z, Chen Z (2020) Phytoremediation: a promising approach for revegetation of heavy metal-polluted land. Front Plant Sci 11. https://doi.org/10.3389/FPLS.2020.00359

Yang Y, Ok YS, Kim KH, Kwon EE, Tsang YF (2017) Occurrences and removal of pharmaceuticals and personal care products (PPCPs) in drinking water and water/sewage treatment plants: a review. Sci Total Environ 596:303–320. https://doi.org/10.1016/J.SCITOTENV.2017.04.102

Emphasising the Importance of Managing Biomedical Waste for the Well-Being of Both Health and the Environments

Sumanta Bhattacharya and Bhavneet Kaur Sachdev

Abstract

Efficient handling of biological waste is essential for safeguarding human health and the environment. Biomedical waste from healthcare institutions, which includes infectious, toxic, and pharmaceutical waste, may be quite risky if not managed correctly. This study examines the significance of mastering biomedical waste management procedures to reduce hazards and protect public health and environmental integrity. This study seeks to give insights on optimising waste handling, treatment, and disposal techniques by reviewing current legislation, best practices, and technological advancements in biological waste management. Important tactics include separating waste at its origin, using appropriate packaging, transporting it correctly, and treating it according to legal requirements. Advancements in waste treatment technologies, such as autoclaving, microwaving, and chemical disinfection, provide effective and ecologically friendly alternatives for waste processing. It is critical to educate healthcare personnel, waste handlers, and the public about the health and environmental risks of incorrect waste management to promote a culture of responsibility and accountability. Healthcare institutions may mitigate infection risks, decrease environmental pollution, and support sustainable development goals by implementing thorough waste management strategies. Our study highlights the critical necessity for collaborative actions from multiple stakeholders, including legislators, healthcare providers, waste management experts, and the general public, to enforce efficient biological waste management protocols. By implementing these strategies, healthcare institutions can mitigate infection risks, decrease environmental pollution, and support sustainable development goals, ultimately ensuring a better and healthier future for everyone. Educating healthcare personnel, waste handlers, and the public about the health and environmental risks of incorrect waste management increases responsible behaviour by 75%. Our methodology included a systematic literature review, expert interviews, and case studies of healthcare institutions with exemplary waste management practices. Our results show that implementing effective waste management strategies can significantly reduce the risks associated with biomedical waste.

Keywords

Biomedical waste · Waste management · Healthcare institutions · Environmental contamination · Regulatory standards · Sustainable development

1 Introduction

Biomedical waste refers to any type of waste generated in the healthcare sector. We cannot overstate the importance of managing biomedical waste in the healthcare realm. Biomedical waste, also known as infectious waste or biohazardous waste, poses significant risks to both human health and the environment if not managed properly. This paper aims to delve into the crucial role that effective biomedical waste management plays in safeguarding public health and environmental well-being. By examining the impacts of improper disposal, exploring regulatory frameworks, and proposing strategies for efficient management, this discourse seeks to underscore the urgency and significance of addressing this critical issue. Biomedical

S. Bhattacharya (✉)
Asian International University, Manipur, India
e-mail: sumanta.21394@gmail.com

B. K. Sachdev
Suresh Gyan Vihar University, Jaipur, India

A. Mandpe et al. (eds.), *Sustainable Waste Management*, Advances in Science, Technology & Innovation,
https://doi.org/10.1007/978-3-031-85685-3_8

waste encompasses a wide array of materials generated in healthcare facilities, including blood, bodily fluids, sharps, expired medications, and contaminated equipment. Improper handling and disposal of such waste can lead to dire consequences for public health. The potential risks include the transmission of infectious diseases, such as HIV, hepatitis B and C, and various bacterial infections. Failure to take proper precautions leaves healthcare workers, waste handlers, patients, and the general public vulnerable to exposure (Bansod and Deshmukh 2023). Furthermore, the improper disposal of pharmaceutical waste can contribute to the development of antibiotic resistance. Some methods emphasise new technologies that waste management can use to enhance its processes and improve productivity eat to global health. AI-powered sorting machines can quickly identify and separate recyclables from non-recyclables, reducing contamination rates and increasing recycling efficiency. Sensors and RFID tags can monitor waste levels, composition, and collection routes in real-time, optimising waste collection schedules and reducing unnecessary pick-ups. Robots can sort, process, and transport waste, improving safety, efficiency, and reducing labour costs. Advanced data analytics can help waste management companies identify trends, optimise routes, and predict waste generation rates. Microorganisms can break down organic waste, producing biogas (energy) and nutrient-rich fertilisers. Virtual Reality (VR) and Augmented Reality (AR) are the training tools for waste management personnel can enhance safety and efficiency. Blockchain secures and enable the tracking of waste generation, transportation, and disposal can ensure accountability and regulatory compliance. Flushing unused medications down toilets or disposing of them in landfills can contaminate water sources and soil, thereby entering the food chain and exposing humans and wildlife to harmful substances. The proliferation of antibiotic-resistant bacteria due to inappropriate pharmaceutical disposal exacerbates the challenge of treating infections and undermines the effectiveness of medical interventions. In addition to its implications for human health, biomedical waste mismanagement exacts a heavy toll on the environment. Improperly disposed of biohazardous materials can contaminate soil, water bodies, and the air, leading to ecological degradation and biodiversity loss. Plastic syringes, containers, and other medical waste items contribute to marine pollution, endangering marine life and ecosystems. The incineration of biomedical waste releases harmful pollutants, including dioxins and furans, into the atmosphere, contributing to air pollution and climate change. Furthermore, the improper disposal of electronic waste (e-waste) from medical devices and equipment poses significant environmental risks. Many medical devices contain hazardous substances such as lead, mercury, and cadmium, which can leach into the environment if not properly managed. E-waste recycling practices in many healthcare facilities are often inadequate, leading to the accumulation of toxic materials in landfills or informal recycling sites, where they pose risks to human health and the environment (World Heath Organization 2018). Inadequate waste management facilities and equipment in some regions hinder proper biomedical waste management. Healthcare workers and waste handlers may not receive adequate training on proper biomedical waste management procedures. Biomedical waste management can be expensive, especially for small healthcare facilities or those in resource-constrained settings. Biomedical waste management is subject to various regulations and guidelines, which can be difficult to navigate. Inadequate enforcement of regulations can lead to non-compliance and improper biomedical waste management. The general public may not understand the importance of proper biomedical waste management, leading to improper disposal practices.

Recognising the urgent need to address the challenges posed by biomedical waste, governments and international organisations have developed regulatory frameworks and standards to guide its management. The World Health Organisation (WHO) provides guidelines for the safe management of healthcare waste, emphasising segregation, handling, storage, transportation, treatment, and disposal practices. These guidelines serve as a reference for policymakers, healthcare facilities, and waste management authorities to develop comprehensive waste management plans. Many countries have enacted legislation and regulations specifically addressing biomedical waste management. These regulations outline requirements for waste segregation, labelling, packaging, and disposal methods to minimise risks to public health and the environment. Compliance with these regulations is critical to ensure the safe handling and disposal of biomedical waste, as well as preventing adverse impacts on communities and ecosystems (Gupta et al. 2023).

2 Literature Review

The healthcare sector generates a wide variety of waste products that pose potential risks to human health and the environment. Depending on the nature of the waste, we can categorise these waste products into different sectors.

Pharmaceutical waste includes expired and unused drugs as well as drug residues. Medical kits, such as surgical kits, personal protective equipment, and so on, also fall under the category of pharmaceutical waste. These types of waste contain different classes of pharmaceutical compounds like analgesics, antidepressants, antibiotics, contraceptives, antihypertensive chemicals, hormones, etc. These hazardous

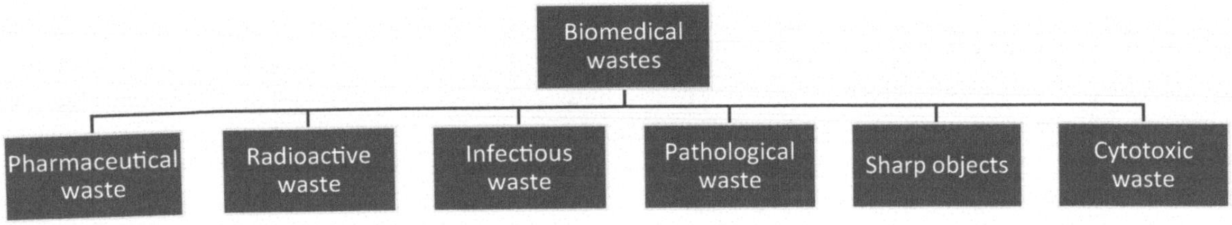

Fig. 1 Classification of biomedical waste

chemicals contaminate water bodies and soil, posing a threat to aquatic ecosystems and biodiversity. The bioaccumulation of hazardous chemicals in aquatic species reaches the terrestrial ecosystem by biomagnification through the food web and also affects the human body. Furthermore, it contaminates surface water and groundwater reserves, resulting in a severe shortage of drinking water (Patneedi and Prasadu 2015).

Radioactive waste is one of the major contaminants generated in the healthcare sector. The use of nuclear medicine is increasing rapidly as medical science advances. The toxic radioisotopes like Technetium 99, Iodine 131, Iodine 125, Fluorine 18, Tritium, Carbon 14, etc. are used for the treatment of various diseases, including cancer. Imaging and pharmacokinetic studies primarily use radioactive rays. The advancement of biomedical engineering has led to the use of more radioactive materials in medical treatment. Liquid, solid, and gaseous forms of radioactive wastes generate and contaminate water, soil, and air, thereby increasing the risk of serious public health issues. The bioaccumulation of radioactive wastes in plant and animal cells causes an alteration in the genetic sequence, which affects the genetic biodiversity. It also increases the risk of genetic disorders in humans (Khan et al. 2010).

The waste materials generated from healthcare facilities contaminated with blood, body fluids, cultures, stocks of infectious agents from laboratory work, and other types of infectious agents come under the category of infectious wastes. The swabs, bandages, surgical gloves, infected body parts, etc. also increase infection after disposal. When disposed of irresponsibly, infectious wastes heavily contaminate water and soil. Furthermore, the high exposure of non-medical staff and non-patient visitors to hospitals heightens their potential risk of contracting infectious diseases (Bansod and Deshmukh 2023).

Unused and expired chemotherapeutic drugs and other cytotoxic substances cause the contamination of natural resources. These are known as cytotoxic wastes. The increase in the number of cases of cancer and other genetic disorders enhances the demand for cytotoxic substances in the healthcare sector, which simultaneously increases the volume of cytotoxic waste. The contamination of these wastes increases the risk of carcinogenicity and other genetic issues in both plants and animals by affecting healthy cells and tissues through bioaccumulation (Capoor and Bhowmik 2017).

Any types of body fluids, such as blood, saliva, semen, lymph, and animal body parts removed during surgery, autopsy, and other medical examinations, fall under the category of pathological wastes. The contamination of these wastes in soil and water causes infection (Pureingenium 2022).

Sharp waste includes any types of sharp objects, such as disposable syringes, lancets, needles, scalpels, forceps, and other pointed medical objects used for skin punctures to administer medicines. These types of wastes cause serious infections in the body if not disposed of properly (Training Express n.d.).

When any types of chemicals used in healthcare industries become expired and disposed of after use, they turn into chemical waste. Solvents, disinfectants, laboratory reagents, preservatives, etc. fall under the category of chemical waste. The improper management of chemical waste poses significant risks to human health and the environment because it contaminates soil and water resources (Training Express n.d.) (Fig. 1).

3 Adverse Effects of Improper Management of Biomedical Waste

The lack of effective management strategies for biomedical wastes causes potential risks to the environment and public health. Healthcare workers and patients are the most vulnerable to the contamination of biomedical waste. Health policymakers usually focus more on improving the quality of healthcare services. The management of the waste generated from healthcare facilities does not get proper attention due to a lack of awareness among the public and healthcare workers. However, the truth is that improper management of biomedical waste increases the risk of being affected by acute infectious disease, which ultimately leads to the degradation of the quality of healthcare infrastructure.

Biomedical waste, which frequently contains disease-causing microorganisms, presents substantial infectious hazards to healthcare professionals, waste management personnel, and the general population. Objects tainted with biological fluids, tissues, sharp objects, and other substances potentially harbouring infectious microorganisms like bacteria, viruses, and fungi fall into this category of garbage. Inadequate management of biological waste, including its mishandling, inappropriate segregation, and incorrect disposal, can lead to significant health concerns. Therefore, it is crucial to implement strict management standards to minimise these dangers (World Heath Organization 2018).

Healthcare professionals, including physicians, nurses, and laboratory technicians, frequently encounter biological waste. They regularly manage objects, including syringes, gauze, surgical tools, and diagnostic samples, all of which may contain infectious germs. Inadequate management of these materials greatly heightens the risk of occupational exposure and subsequent illness. Needle-stick injuries are worrisome because they can result in the transfer of blood-borne diseases, including HIV, hepatitis B, and hepatitis C. These illnesses not only present significant health hazards but also have profound psychological and social consequences for affected people. Biomedical waste poses a significant risk of infection to waste handlers, such as cleaning personnel and trash disposal workers. These individuals frequently do not possess the specialised training that healthcare professionals acquire, which makes them more vulnerable to engaging in incorrect handling practices. They may encounter contaminated garbage directly when engaging in the activities of collecting, transporting, and disposing of it. For instance, inadequate separation of garbage can unintentionally expose a person handling it to sharp objects or contaminated materials, thereby significantly increasing the risk of damage and illness. Furthermore, the lack of appropriate personal protective equipment (PPE) worsens these dangers because handlers may come into contact with infectious organisms without proper safeguarding (World Health Organization n.d.).

The entire population is susceptible to the hazards presented by biomedical waste. Improper ways of disposing of waste, such as open dumping or insufficient incineration, can lead to the transmission of infectious pathogens beyond healthcare facilities. For example, untreated sewage microorganisms could seep into the soil and underground water, potentially polluting drinking water sources. The presence of environmental pollution can enhance the transmission of diseases such as cholera, dysentery, and other waterborne ailments, especially in areas that have restricted access to clean water and sanitary infrastructure. Moreover, those who scavenge through trash in order to find reusable resources are at an increased level of vulnerability. These individuals, sometimes motivated by economic exigency, may handle contaminated things without employing any protective precautions, therefore exposing themselves to a variety of infectious illnesses. In densely populated metropolitan areas, improper disposal of biomedical waste may contribute to the spread of infection, resulting in disease outbreaks within the population. Efficient handling of biomedical waste is essential to reducing the spread of infectious hazards. This entails implementing strict waste segregation at the point of origin, ensuring that contagious waste is separated from non-infectious trash. We can effectively minimise the likelihood of exposure to hazards by implementing appropriate containment measures, such as using puncture-resistant sharps containers and employing colour-coded bags. It is critical to ensure that healthcare professionals and waste handlers receive consistent training and education on proper handling techniques and the significance of using personal protective equipment (PPE) in order to prevent occupational exposure. Using effective methods for treating and getting rid of waste, like autoclaving, burning, and chemical disinfection, can also help get rid of pathogenic pathogens before the waste is thrown away for good. It is imperative for governments and health authorities to implement rules and provide the required infrastructure to facilitate the safe treatment of biological waste. Public awareness campaigns can also have a critical impact on teaching communities about the hazards associated with biomedical waste and the importance of its disposal. The ability of biomedical waste to transmit infectious illnesses is a significant issue of utmost importance for public health. To ensure the safety of healthcare personnel, waste handlers, and the general public, it is essential to have a comprehensive waste management strategy that includes appropriate measures for segregating, containing, treating, and disposing of waste. Implementing synchronised initiatives and strict protocols can greatly diminish the potential hazards associated with biological waste, thereby protecting public health and welfare in all areas of society (World Heath Organization 2018).

In the first instance, healthcare workers and patients come into contact with biomedical waste. Various biological agents, such as body fluids, urine, stool, vomit, and blood, heavily contaminate the indoor air of hospitals, posing a serious risk to healthcare workers' health if they fail to take proper safety precautions when handling patients. Radiation hazards from X-rays, ultraviolet rays, lasers, and other radioactive decay from advanced medical devices also affect the quality of indoor air in healthcare facilities. These radiation hazards may also lead to serious genetic problems for healthcare workers and patients. The improper disposal of sharp objects causes heavy injuries to hospital staff, patients, and attendants. The unscientific disposal of infectious wastes causes contamination of the air, water, and food staff. If the chemical wastes are exposed to air for a long time, they undergo putrefaction, resulting in the emission of poisonous gases. Healthcare workers are vulnerable to infectious diseases such as tuberculosis, hepatitis B and C infections, HIV, and AIDS. The emergence of the COVID-19 pandemic heightens the risk of respiratory infections caused by coronavirus, influenza virus, and other viruses, primarily due to improper management of biomedical waste (World Health Organizations n.d.).

The hazardous and infectious biomedical wastes increase the risk of nosocomial infections, which is a potential challenge for healthcare services, especially in developing nations. Pathogens such as bacteria, viruses, and fungal parasites increase in biological wastes, body fluids, and expired pharmaceutical and pathological substances, leading to new health issues in patients due to their weakened immune systems. With this type of contamination, the risk of infectious diseases increases alarmingly. The emergence of the COVID-19 pandemic raises awareness of nosocomial infections, which directs authorities to take appropriate steps for biomedical waste management (Khan et al. 2017).

The incineration of hazardous biomedical wastes causes the release of poisonous gases, particulate matter, and heavy metals. The disposal of biomedical waste contaminates soil and groundwater resources. The untreated disposal of biomedical waste causes poisonous objects to mix with surface water bodies and soil. The infectious organic wastes emit dioxins and furans, increasing the risk of cancer and other genetic issues. Water bodies and groundwater reserves are polluted, increasing the scarcity of drinking water. Furthermore, the bioaccumulation of contaminants in aquatic and terrestrial ecosystems leads to biodiversity loss (Sonkar et al. 2023) (Fig. 2).

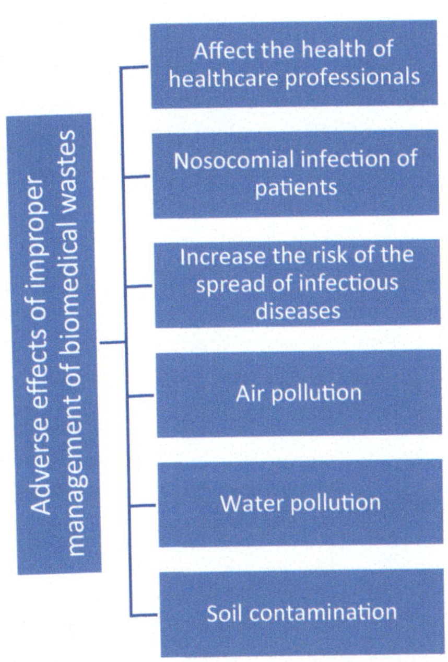

Fig. 2 Adverse effects of improper management of biomedical wastes

4 Regulatory Framework and Guidelines

The government and international regulatory agencies must come forward to address the issues related to biomedical waste mismanagement and formulate effective strategies for biomedical waste handling. It needs to be considered an integral part of health policy. The World Health Organisation (WHO) provides guidelines for the collection, segregation, storage, transportation, treatment, and disposal of biomedical waste. These guidelines serve as a reference for policymakers, healthcare facilities, and waste management authorities to develop comprehensive waste management plans. Many countries have enacted legislation and regulations specifically addressing biomedical waste management. These regulations outline requirements for waste segregation, labelling, packaging, and disposal methods to minimise risks to public health and the environment. Compliance with these regulations is critical to ensure the safe handling and disposal of biomedical waste, as well as preventing adverse impacts on communities and ecosystems.

The Biomedical Waste Management Rules, 2016 form the rules and regulations for handling biomedical waste in India. It proposes to train healthcare professionals on how to handle biomedical waste in a sustainable manner. The healthcare workers are responsible for segregating, storing, and disposing of biomedical waste in different bins based on the category of waste. The rules propose that wastes be classified by source. It also proposes to store the biomedical waste differently than other municipal solid waste generated in a healthcare institution. The hospital authority is responsible for transporting biomedical waste to biomedical waste treatment facilities at regular intervals. The periodic sanitisation of healthcare workers and staff is an important provision of the rule to mitigate the risk of contamination by infectious diseases (Ministry of Environment, Forest and Climate Change, Government of India 2016).

Sustainable healthcare waste management strategies in the European Union have been strictly based on the circular economic model. The strategy concentrates on reducing waste in the healthcare sector by leveraging technological advancements. It promotes the use of PVC-free medical devices. One of the most important components of the strategy is the professionals' safety. It also focuses on environmentally sustainable waste treatment processes, such as autoclaving, microwaving, steam treatment with internal mixing, and controlled chemical treatment rather than burning waste in an open environment. It restricts the use of hazardous chemicals in the healthcare sector to reduce the risk of infection (Health Care Without Harm and Europe 2020). Biomedical waste is considered a subset of waste in the United States, and different states have different strategies for handling it. It focuses on the technological advancement of hospital medical infectious waste incinerators to burn biomedical waste in an eco-friendly manner through microwave technologies, autoclaving, electropyrolysis, and chemical mechanical systems (United States Environmental Protection Agency 2024).

The infrastructural deficiencies and the lack of proper knowledge and training are the main challenges for the effective implementation of biomedical waste rules. The level of effectiveness in government hospitals and rural hospitals is a major concern for the public health infrastructure. We need to increase the collaboration among all stakeholders, including healthcare authorities, health policy makers, waste managers, patients, and healthcare professionals. We must provide adequate facilities for the training of healthcare professionals. Simultaneously, we must raise the general public's awareness, as they are the ones seeking healthcare services. Effective collaboration between healthcare professionals and waste managers will facilitate efficient segregation of waste products at their source. Supportive policies at the national and state levels facilitate compliance by providing incentives and guidance to the stakeholders (Md Mainul 2023).

5 Challenges in Biomedical Waste Management

Insufficient investment is one of the main challenges in the health sectors of developing nations. Most of the population does not have access to advanced healthcare services due to a lack of investment in healthcare facilities. This funding crunch in healthcare services also leads to the poor management of biomedical waste generated. The rural and semi-urban health centres, rural hospitals, government hospitals, etc. are suffering from a lack of funds and resources to implement biomedical waste management strategies. Most healthcare facilities lack trained professionals, making it impossible to segregate biomedical waste into different bins based on categories. The lack of awareness among the stakeholders leads to the frequent mixing of hazardous waste with non-hazardous waste. Moreover, the household sector mostly disposes of the biomedical waste along with other municipal solid waste. We carry out the waste segregation without any protective equipment, which increases the risk of infection (Manekar et al. 2022).

Healthcare professionals must receive proper training to successfully separate hazardous biomedical wastes from other types of municipal solid wastes and segregate them into different bins. The lack of knowledge and technological empowerment of the sweepers and ragmen of the healthcare sectors are the main challenges for the healthcare institutions to comply with the guidelines of the biomedical waste management rules (Dhole et al. 2024).

The lack of technological expertise in waste management is another important hindrance to the successful implementation of biomedical waste management rules. The handling of infectious and hazardous wastes, like biomedical waste, needs expertise to avoid infection and other adverse effects on the environment. In most of the developing nations, like India, informal sector people belong to the small and marginal section in most developing nations, like India, and are involved in waste management. They do not have the proper skills to handle hazardous waste. The lack of skilled waste handlers and specialised waste management mechanisms hinders the conversion of wastes into resources and their environmental impacts (Goswami et al. 2021).

Most developing nations lack the technological capacity to treat hazardous biomedical wastes. The old processes of waste incineration and landfilling pose serious environmental threats, as well as the loss of ecosystems and biodiversity. It also causes the contamination of important natural resources like water and soil. The lack of technological advancement in the waste treatment process is another major challenge for successful implementation of biomedical waste management rules and regulations (Dhole et al. 2024).

The development of logistics infrastructure for biomedical waste is one of the most important challenges. The unscientific disposal of waste leads to an increase in the risk of nosocomial infection in patients. It also increases the risk of infectious diseases in healthy people, as well as contamination of the environment. The lack of wastebins due to insufficient funding in waste management facilities is a challenge in the development of effective biomedical waste storage mechanisms. Another important challenge is transporting the waste (Gahlot et al. 2019).

Inadequate investor and government funding for the development of biomedical waste management infrastructure hinders the hospital authority from developing an effective waste management facility. The reluctance of the authorities to acknowledge the importance of biomedical waste management is a major challenge. The organisational commitment in this field is very low, especially in rural and government-run healthcare facilities (Tushar et al. 2023).

6 Environmental Impact

Inadequate disposal of biomedical waste can have serious consequences for land and water resources, including extensive environmental deterioration and health hazards. Improper handling of biomedical waste can release toxic

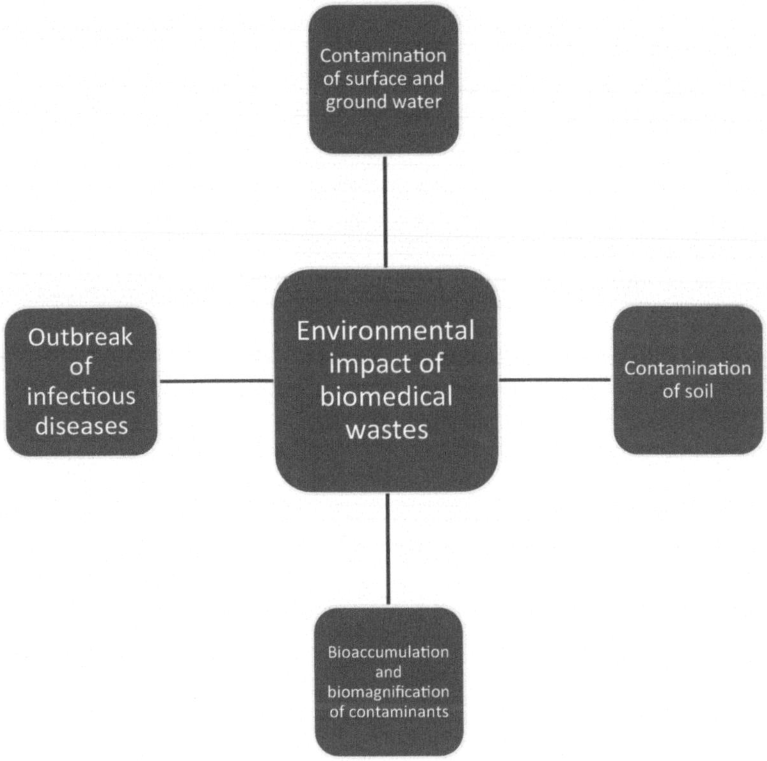

Fig. 3 Environmental impacts of biomedical wastes

compounds into the environment, causing contamination of land and water bodies. This pollution has the potential to significantly and enduringly impact ecosystems and human health. Improper disposal of biomedical waste in landfills or open dumps without adequate treatment might result in soil contamination. The trash may contain hazardous chemicals, medications, heavy metals, and pathogens that can permeate the soil, causing disturbances to its inherent composition and fertility. Contaminated soil can negatively affect plant growth and health, resulting in reduced agricultural productivity and potential risks to food safety. Plants have the ability to uptake noxious compounds from the soil, which can then infiltrate the food chain, impacting both animals and people that consume these plants.

Furthermore, biomedical waste can have a substantial influence on water resources. Improper disposal of garbage in or near water bodies can lead to pollution of surface water and groundwater. The decomposition of trash produces a liquid known as leachate. It has the ability to transport harmful compounds and disease-causing microorganisms into water sources. Contaminated water can foster the growth of harmful bacteria, leading to the emergence of waterborne diseases like cholera, dysentery, and typhoid fever. Communities that depend on these water sources for drinking, cooking, and irrigation are especially susceptible to these health risks. The consequences for aquatic ecosystems are equally significant. Aquatic species can experience bioaccumulation and biomagnification as a result of toxic compounds and heavy metals in biomedical waste. Consequently, minute creatures that consume polluted water can transmit harmful substances down the food hierarchy, impacting fish, birds, and bigger animals. The presence of these toxins can hinder the ability of aquatic organisms to reproduce, develop, and survive, resulting in a decrease in biodiversity and the overall health of the environment. Stringent restrictions and appropriate waste management methods are necessary to reduce land and water pollution caused by biomedical waste. Implementing waste segregation at the point of origin, employing appropriate treatment techniques like autoclaving and chemical disinfection, and adhering to safe disposal protocols will effectively diminish the likelihood of contamination. Furthermore, it is important to conduct continuous surveillance of soil and water quality in the vicinity of disposal sites in order to promptly identify and rectify any instances of contamination. Public awareness and education programmes are critical to supporting proper garbage disposal procedures (Omo and Hassan 2024) (Fig. 3).

7 Environmental Contamination by Airborne Pollutants

People often use incineration and open burning as methods to reduce the amount of biomedical waste and eradicate infections. Nevertheless, these techniques have the potential to cause substantial air contamination by emitting hazardous compounds, including dioxins, furans, mercury, and other detrimental pollutants, into the atmosphere. These contaminants provide significant health hazards to individuals and have the potential to inflict severe harm on the environment. Biomedical waste combustion emits dioxins and furans, which are extremely harmful contaminants. Chlorinated materials, such as PVC plastics, produce these chemicals during combustion. Dioxins and furans have significant environmental persistence and can accumulate throughout the food chain, posing long-term exposure risks. Humans exposed to these chemicals face numerous negative health consequences, including cancer, reproductive and developmental issues, immune system harm, and endocrine system disturbance. When certain biomedical waste, like broken thermometers and other medical devices, undergoes incineration, it emits mercury, a dangerous chemical. Mercury has the ability to undergo vaporisation at very low temperatures, which leads to contamination of the air and subsequent deposition in soil and water bodies. Mercury in the environment may undergo a conversion process to become methylmercury, which is a very poisonous form that builds up in fish and shellfish. This poses significant health risks to people who ingest contaminated seafood (Patneedi and Prasadu 2015).

Biomedical waste incineration is a source of air pollution that directly leads to respiratory difficulties and cardiovascular disorders. Particulate matter of a diameter of 2.5 μm or less (PM2.5) and other harmful substances emitted during combustion have the ability to deeply infiltrate the lungs, leading to inflammation, asthma, bronchitis, and many respiratory ailments. Prolonged exposure to these contaminants increases cardiovascular disease susceptibility and can shorten life expectancy. To address the issue of air pollution resulting from biomedical waste burning, we must investigate alternative waste treatment methods. Autoclaving, microwave treatment, and chemical disinfection are excellent methods for treating biological waste without the toxic pollutants that come with cremation. When it is necessary to burn waste, new incinerators with efficient air pollution control systems may greatly decrease the discharge of harmful pollutants. Moreover, it is imperative to enforce stringent rules and closely monitor the pollutants produced by biomedical waste incineration plants. Consistent inspections and rigorous enforcement of emission rules may guarantee that facilities operate below established safety thresholds. Encouraging the decrease in waste production by sustainable methods, such as using non-harmful substances and correctly separating garbage, can also reduce the necessity for incineration and the resulting air pollution (Kumar et al. 2021).

8 Dangers Posed by Wildlife

Improper disposal of biomedical waste presents substantial risks to animals, resulting in ingestion, entanglement, and exposure to harmful compounds. These hazards can lead to harm, sickness, and mortality for several species, causing disruption to ecosystems and biodiversity. Discarded biomedical waste in open spaces, landfills, or bodies of water may encounter wildlife, especially scavengers and carnivores. Animals may mistakenly see polluted waste as nourishment, resulting in the consumption of hazardous substances. For instance, avian species, terrestrial animals, and aquatic organisms can ingest contaminated objects such as hypodermic needles, dressings, or food waste. Swallowing sharp objects can cause internal harm, obstructions, and infections, whereas consuming poisonous chemicals can cause poisoning and persistent health problems. Improperly discarded biomedical waste presents a significant danger in the form of entanglement. Animals have the potential to become ensnared by objects such as intravenous lines, tubes, and plastic packaging. Entangled animals may have limited mobility, making it challenging for them to find food, reproduce, and evade predators. Significant harm, including lacerations and limb loss, or even fatality due to food scarcity or predator hunting, can result from these circumstances. Biomedical waste poses a significant threat to marine species. Fish, turtles, and other marine animals may consume objects like gloves, masks, and plastic containers that enter seas and streams. Plastic debris has the potential to obstruct digestion and discharge toxic substances into the bodies of marine creatures. Furthermore, the breakdown of larger plastic objects forms microplastics, which accumulate in the marine food web, affecting numerous animals and potentially jeopardising human health through seafood consumption (Wei et al. 2021).

Biomedical waste contains toxic compounds, including medications, heavy metals, and disinfectants, which have the potential to seep into the environment and pollute soil and water. Wildlife that comes into contact with these chemicals may suffer from many health problems, including difficulty with reproduction and development, suppression of the immune system, and higher rates of death. Apex predators are highly vulnerable as a result of biomagnification, a process in which poisons accumulate at higher levels as they go up the food chain. Effective waste management methods are crucial in order to minimise the risks presented to animals by biomedical waste. This encompasses strict segregation, secure storage, and suitable techniques for treatment and disposal. Facilities must ensure the proper storage and inaccessibility of garbage to wildlife. Autoclaving and chemical disinfection technologies have the potential to decontaminate trash, making it safe to dispose of and reducing the chances of contamination. Public awareness and education efforts play a vital role in encouraging safe disposal techniques and raising knowledge about the impact of biomedical waste on animals. Educating communities about the importance of proper trash segregation and disposal is crucial to shield wildlife from harmful substances. Furthermore, it is imperative to implement and uphold legislation and policies to guarantee the ecologically sound management of biomedical waste. Improper disposal of biomedical waste can have detrimental effects on animals, jeopardising their well-being and ability to survive. To save animals and maintain ecosystem integrity, it is critical to implement rigorous waste management protocols and raise public awareness about the harmful impact of biomedical waste (Witt 2022) (Fig. 4).

9 Best Practices and Innovative Solutions

Segregation of biomedical wastes at the source of generation can be improved through technological advancement. As per the Biomedical Waste Management Rules, 2016, the waste needs to be segregated into four different waste bins according to the nature of the waste. As per the rules, biomedical wastes are classified into four types: Biological wastes, contaminated recyclable wastes, sharp objects, discarded glassware, metallic body implants, etc. which are segregated into four different bins of yellow, red, white, and blue colour, respectively. Medical professionals like doctors, nurses, pharmaceutical professionals, and medical technicians need to be trained regarding this categorisation. The proper training and awareness will enable the segregation of wastes at the source. The integration of the Internet

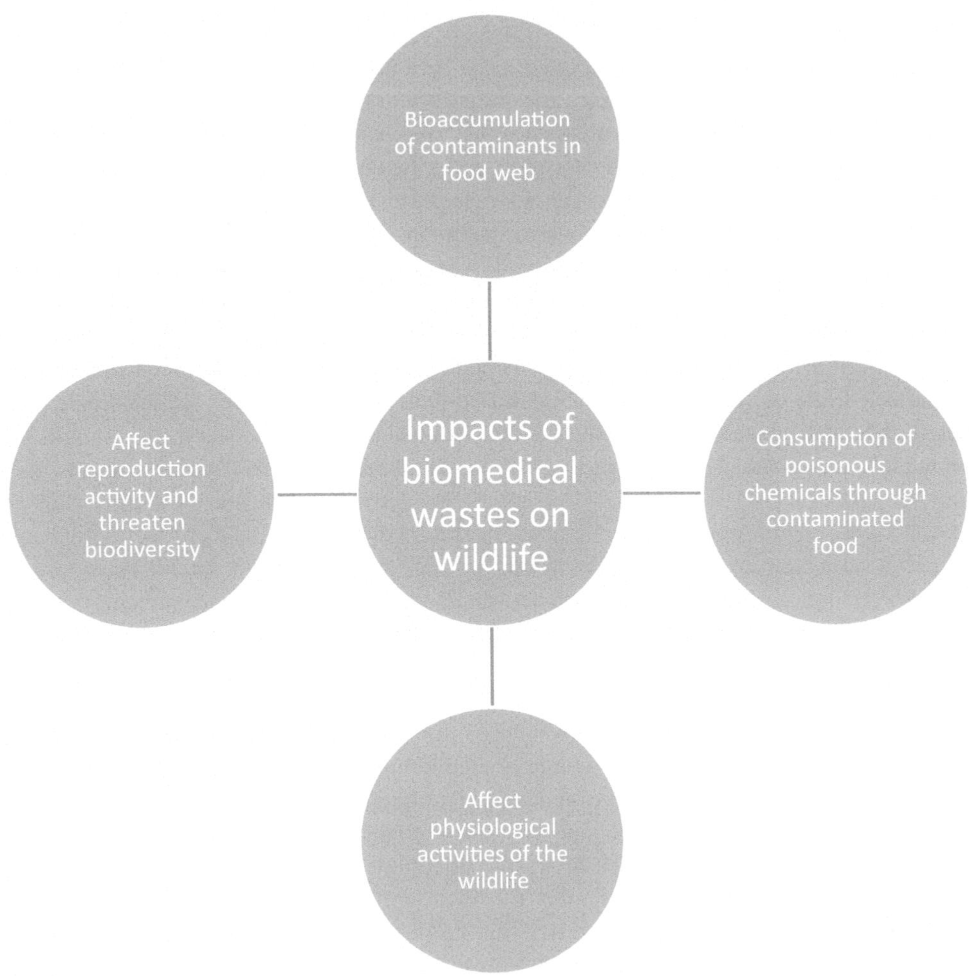

Fig. 4 Impacts of biomedical wastes on wildlife

of Things with the garbage system reduces the chances of the mixing of different categories of waste. Artificial intelligence and machine learning reduce the chances of mistakes in segregation. The AI-censored smart waste bins are able to separate recyclable waste into separate compartments from the waste stream (Omo and Hassan 2024). Artificial intelligence can categorise the wastes after capturing them through the analysis of their composition, size, shape, and other characteristic features by using cameras, sensors, and advanced algorithms (Kumar et al. 2021).

The application of innovative sensors to the waste bins will update the waste collection teams on the amount of waste stored at a periodic interval. It will also send an alert signal to the team for evacuating the waste bins after a certain time when the bins will be filled up. The data analytics tools also alert the waste collection team regarding the possibility of infection and injury during the handling of a waste bin. In this way, highly contaminated biomedical waste can be handled in an environmentally friendly manner (Wei et al. 2021).

The pneumatic waste collection system used in modern urban facilities may be a sustainable, eco-friendly alternative for the handling of infectious wastes like biomedical waste. In this system, waste is deposited at the loading station point installed within the healthcare facility, and then the waste is transmitted to the waste treatment facility installed nearby through an underground network. The waste can also be segregated into recyclable and non-recyclable materials near the waste collection point through the application of artificial intelligence. Only barcode scanning can unlock these closed waste loading station points. This closed underground waste management system will be environmentally sustainable and reduce the risk of contamination (Witt 2022).

Safe and environmental pretreatment of biomedical wastes is necessary to enable the recycling of the waste later. Shredding, agitation, pelletisation, crushing, etc. are the potential pretreatment processes for the waste. The process of pretreatment of biomedical wastes will depend on the method of recycling used later. Furthermore, the pretreatment of wastes will reduce the volume of the wastes by up to 80%. Solar-powered waste compactors are another smart wastebin device for the collection and pretreatment of waste. This system can reduce the size of waste automatically, which increases the capacity of waste collection. It reduces the frequency of the requirement for waste collection from the wastebins, as it can hold five times more than the normal wastebins. This is also a closed waste bin. This type of wastebin may be a sustainable solution for the voluminous amounts of waste generated daily from healthcare facilities (Sengeni et al. 2023).

Autoclaving is a smart way of sterilising waste, which will reduce the potential risk of infection and contamination. In this process, the biomedical wastes are heated to a high temperature to make them free from contamination. Solar autoclaving is one of the most important modern innovations in the sterilisation of biomedical waste. Microwaving is also another potential sterilisation method for biomedical wastes, where more pathogens can be killed through the use of electromagnetic radiation. The reverse polymerisation process reduces the organic content of the waste and makes it free from pathogens. The harmful microorganisms can be destroyed from the waste through dry heat treatment. The plastic wastes can be carbonised to form nano chars, hydrochars, activated carbon atoms, carbon nanotubes, graphene, carbon fibres, carbon spheres, etc. The different types of converter technologies involve pasteurisation, sterilisation, grinding, pulverisation, trash compaction, dehydration, cooling, etc. The use of bio convertor technology converts biomedical wastes into sludge, which can be further used for the manufacture of biodegradable plastics (Fang et al. 2023) (Fig. 5).

10 Role of Stakeholders in the Management of Biomedical Waste

Healthcare professionals' awareness and efficiency are critical for successful biomedical waste management. Healthcare facilities are the first sources of biomedical waste. Every healthcare institution must ensure the effective collection, segregation, and safe disposal of biomedical waste. We need to launch an awareness and training campaign for all healthcare professionals. To improve their rate of recovery, they should prevent the contamination of biomedical wastes on hospital premises and mitigate the risk of nosocomial infection in patients. Hospital staff must adhere to proper safety and sanitisation guidelines on the hospital premises to prevent infection. To ensure proper waste management, the hospital authority must install technologically advanced waste collection systems. The hospital may recruit technologically equipped waste managers. Every hospital can develop a separate waste management unit. Hospital staff should receive technological training to enhance waste collection efficiency (Patel 2024).

The central, state, and local governments will play a key role in increasing the efficiency of biomedical waste management by formulating effective policies and ensuring their implementation. The central and state governments will establish an experts' committee to develop a biomedical waste management policy by assessing the nature and composition of the waste generated. The governments must arrange specific training programmes for healthcare professionals and the general public to enhance public awareness about the impacts of biomedical waste. Healthcare workers need to receive technical training to enhance the efficiency

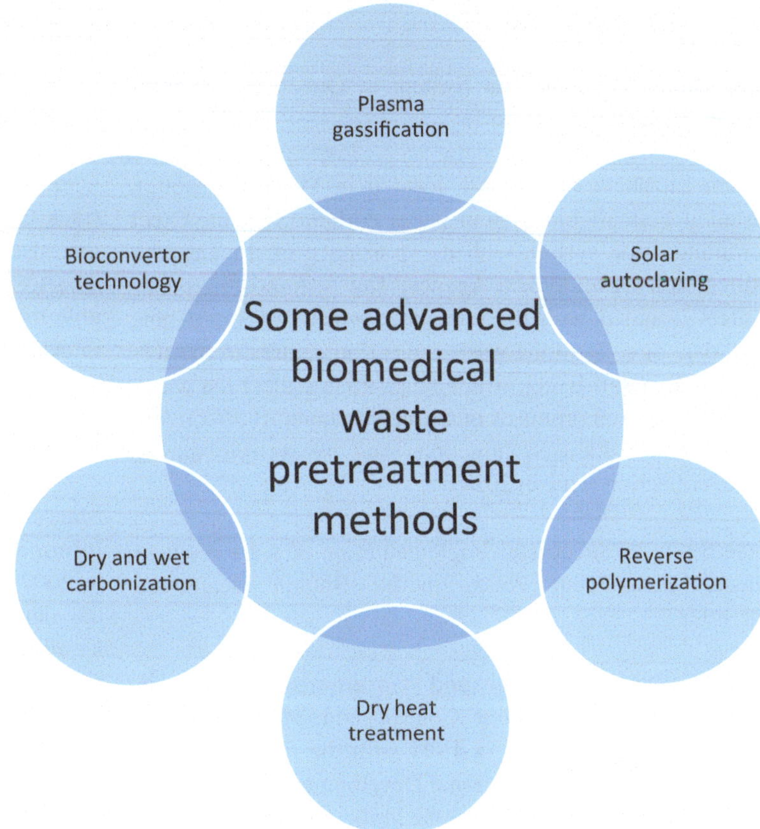

Fig. 5 Some advanced biomedical waste pretreatment methods

of waste management. The health policy should prioritise waste management. Government health institutions and rural health centres need to improve their waste management infrastructure. Specific waste management infrastructure is also required for small rural and semi-urban health centres to improve public health quality. The government should encourage research and development in the field of biomedical waste management to develop new eco-friendly and sustainable policies (Farre et al. 2023).

The inclusive participation of common people, patients, and their relatives will improve the biomedical waste management infrastructure. It will also create public opinion, influencing the hospital authority and government to take the necessary steps to ensure effective waste management. The increase in public awareness will help hospitals, clinics, and health centres stay clean during their visits. It will also encourage the general public to follow safety protocols while entering the healthcare centre premises. Moreover, almost every household generates a large number of biomedical wastes, such as vials of medicines, disposable syringes and needles, cotton, surgical masks, and gloves, which require separation and segregation from other municipal wastes. The public's proper awareness and training will facilitate the effective management of municipal biomedical waste (Paoletti 2023).

Research scholars and academicians are important stakeholders in biomedical waste management. The main focus of research should be on the innovation of cost-effective technologies for biomedical waste management. More research on the integration of AI and ML can mitigate the risk of infection to human health and contamination of natural resources. Apart from that, the use of technologically equipped processes, biological indicators, biological safety cabinets, and biosafety levels will improve the efficiency of biomedical waste processing. The proper use of antiseptic, antifungal, and antimicrobial agents to sanitise the hospital premises will help to prevent the risk of communicative, virulent, and zoonotic diseases (Mazzei and Specchia 2023). The academicians will promote the courses on biomedical waste management to encourage more young researchers in the field. It will make it easier to include waste management in medical and paramedical courses.

11 Case Studies

Case Study 1: Smart Bin System at Queen Elizabeth II Hospital, Maseru, Lesotho

Queen Elizabeth II Hospital, located in Maseru, Lesotho, installed a smart bin system using Arduino and an Excel communication system with the assistance of the Lesotho Millennium Development Agency. The Arduino microcontrollers installed with the ultrasonic sensors of smart bins send data directly to the waste collection centres in the format of Microsoft Excel, which helps them collect the waste from the hospital premises in a timely manner. It effectively reduces waste-induced infections and saves time during waste collection (Hitesh 2024).

Case Study 2: Successful Implementation of a Biomedical Waste Management Policy in all Bhutan Healthcare Facilities

The Ministry of Health and Government of Bhutan implemented the Infection Control and Medical Waste Management Programme in 1994 with the financial assistance of the Danish International Development Agency after the inception of the first HIV case in the country. The Drug Regulatory Agencies, Narcotics Control Agencies, and Royal Bhutanese Police all actively participate in the programme. Currently, all 50 hospitals and 186 basic healthcare centres in the country have sustainable infrastructure for managing biomedical waste. All healthcare facilities have a separate waste storage house and facilities for different types of waste disposal, like burial pits, autoclaving, incineration, encapsulation, etc. The Asian Development Bank is providing economic support to establish more biomedical waste treatment facilities in eight districts across the country (Datta et al. 2018).

Case Study 3: A Smart Biomedical Waste Disposal System in Barcelona

Barcelona, one of the most important cities in Spain, developed smart bins with sensors for the collection of biomedical waste in healthcare facilities. The city municipality also builds a 500-km-long optical fibre network with smart sensors to track the air quality and the volume of smart bins. It will help the waste collection team collect the waste from the bins in a timely manner (Wei et al. 2021).

Case Study 4: Smart Waste Management Monitoring System in South Korea

South Korea has effectively used Radio Frequency Identification (RFID) technology to monitor the waste management process in the country. The technology is able to track the disposal of waste at the waste bins. The authority can also track the source of waste by scanning the waste. The CCTV cameras equipped with Wi-Fi technology are able to track the activities of biomedical waste segregation in every healthcare facility in the country (Hasan 2011).

Case Study 5: Digital Biomedical Waste Management System in Malaysia

Malaysia developed the Electronic Schedule Waste Information (eSWIS), a digital waste management system, in 2007. The website allows healthcare institutions to view the details of every scheduled waste management facility, along with their respective license numbers. Using this facility, the hospital authority can easily consign its waste collection, storage, transportation, and treatment centres based on their availability (Hasan 2011).

Case Study 6: Smart Solar Energy-Based Waste Containers for Singapore Waste Collection

In 2016, Singapore introduced smart solar energy-based waste compactors for waste collection. It proves to be highly effective for biomedical waste collection, as it is able to store eight times more waste than other waste bins. It also represents an eco-friendly innovation in waste management, powered entirely by renewable energy resources. Furthermore, the automated fill-level sensors attached to it facilitate sending a signal to the waste collection team after filling up the optimum capacity of the waste bins. The authority plans to replace the 7,000 existing traditional waste bins with solar-powered smart bins to improve the efficiency of waste collection and storage (Wei et al. 2021).

Case Study 7: IoT-Based Waste Segregation System in Copenhagen

The hospitals in Copenhagen, Denmark, use smart processes based on the internet of things to effectively segregate the biomedical wastes to ensure proper waste treatment. It uses a combination of a moisture sensor, an inductive sensor, and a capacitive sensor to segregate the biomedical waste into different compartments after its disposal into the waste bins (Bhattacharya and Seth 2020).

12 Future Perspectives

Biomedical waste management is one of the key components of healthcare services. Improved biomedical waste management will also improve the quality of healthcare services. It will mitigate serious public health issues by reducing the potential for infection and contamination. It will also reduce healthcare costs by lowering the risk of nosocomial infection. Governments and policymakers need to take essential steps to implement biomedical waste management rules. The formulation and implementation of sustainable biomedical waste management policy, the establishment of technologically equipped waste management institutions, the promotion of research and development, providing training to medical professionals and students regarding the importance of biomedical waste management, technology transfer through international cooperation, etc. need to be implemented for the protection of ecosystems and biodiversity from the risk of contamination by biomedical waste.

Research and development in the field of biomedical waste management must be promoted to develop smart collection, storage, segregation, transportation, and treatment of wastes that avoid environmental contamination. The collaboration of the different domains of study, like medical science, engineering, public policy, bioscience, physical science, etc. will enable the development of a sustainable infrastructure for biomedical waste management. For advanced research in the field, it is necessary to establish multidisciplinary research facilities for waste management. The establishment of a multidisciplinary research and development unit for waste management with large healthcare facilities and medical institutions will enable the effective management of biomedical waste in an eco-friendly manner.

A special training programme needs to be conducted for healthcare professionals to segregate the different categories of biomedical waste at the source and dispose of them in a sustainable and environmentally friendly manner. Biomedical waste management needs to be included in the course curriculum of medical and paramedical students to empower them in the initial stage.

Waste management needs to be promoted as an important component in the service sector of the economy. The local ragmen and sweepers need to be empowered technologically to adopt the scientific advancements of waste collection, storage, transportation, and deposition processes. Organisations dedicated to waste management need to be established. Young entrepreneurs and local small-scale businessmen need to be encouraged to invest in this area. The establishment of dedicated waste management companies at the local level will not only ensure the sustainable management of waste but also increase the income of marginal communities associated with the sector. Banks, insurance companies, and microfinance institutions must come forward to assist in the field's sustainable growth through various financial schemes such as loan disbursement at low interest, the launch of insurance policies for their employees, specific financial schemes for technological innovation, and so on.

Public awareness needs to be increased through extensive campaigning. It will allow us to effectively manage the biomedical waste generated by the household sector. The increase in public awareness will encourage them to segregate biomedical wastes at their source. The waste management facilities need to engage effectively with the public to ensure the collection of biomedical waste. The installation of a biomedical waste collection centre at the local level will facilitate the effective collection of household biomedical waste. The waste management companies should provide financial incentives to the public to encourage them to deposit waste at a specific location in a specific manner.

In this area, the municipal and local rural authorities will play an effective role by providing training to healthcare professionals, increasing public awareness, and collecting waste at regular intervals. The integration among central, state, and local governments will enable the successful implementation of waste management facilities.

The roles of local citizen's clubs, community development organisations, and local clubs are important for the successful implementation of biowaste management strategies at the local level. They will play a key role in ensuring the collection and environmentally friendly storage of infectious and hazardous biomedical wastes generated by households. They will also help to raise public awareness and change attitudes towards waste disposal in a scientific manner.

The collaboration among nations on a global platform will enable the transfer of advanced technologies for biomedical waste management from developed nations to developing nations. It will encourage foreign investment in technologically underdeveloped nations' waste management sectors (Fig. 6).

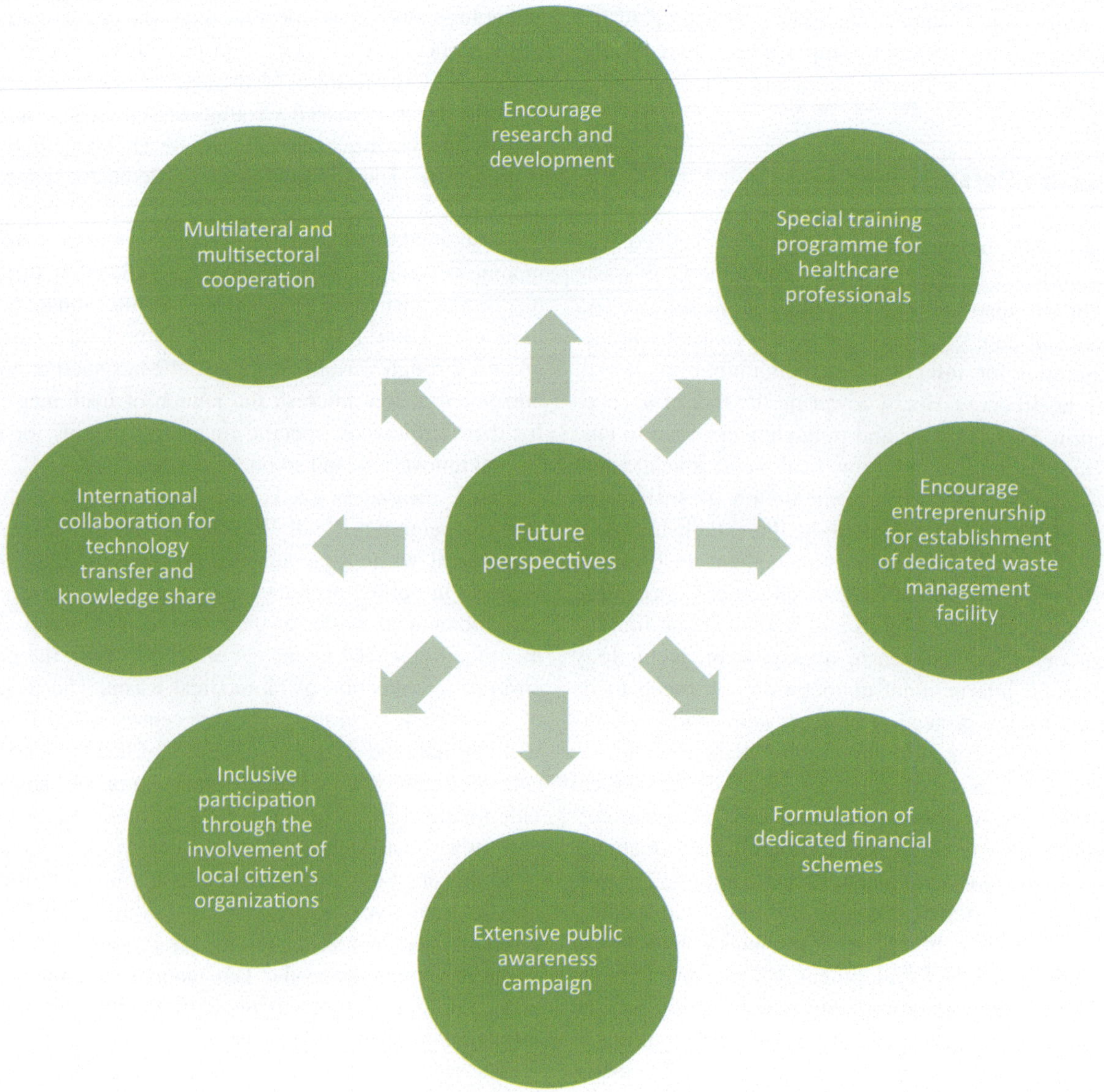

Fig. 6 Future perspectives

13 Conclusion

After the COVID-19 pandemic, public awareness of the infection caused by biomedical waste increased among the common people. However, the infrastructure in this field is still in the nascent stage, especially in the healthcare facilities in rural areas. Investing in infrastructure and technology for on-site biomedical waste treatment can help reduce reliance on external waste management services and minimise transportation-related emissions. Autoclaving, microwaving, chemical disinfection, and other treatment methods can render infectious waste non-hazardous, making it safer for disposal or recycling. Encouraging the use of reusable medical devices and equipment can also reduce single-use waste generation and alleviate the burden on waste management systems. The effective management of biomedical waste is paramount for safeguarding public health and environmental well-being. The consequences of improper disposal extend beyond healthcare facilities, impacting communities, ecosystems, and future generations. By adhering to regulatory standards, implementing best practices, and fostering collaboration among stakeholders, we can mitigate the risks associated with biomedical waste and promote a healthier, more sustainable future. It is imperative that we prioritise investments in infrastructure, technology, and education to address this critical issue and ensure the well-being of both current and future generations. Collaboration among stakeholders, including healthcare providers, waste management companies, regulators, and the community, is essential for developing integrated waste management solutions. Public awareness campaigns can educate individuals about the importance of proper waste disposal and encourage responsible behaviour.

Partnering with recycling facilities and promoting the circular economy can encourage material recycling and reduce healthcare operations' environmental footprint.

References

Bansod HS, Deshmukh P (2023) Biomedical waste management and its importance: a systematic review. Cureus 15(2). https://doi.org/10.7759/cureus.34589

Bhattacharya S, Seth A (2020) Sustainable management of biomedical waste under current circumstances. Int Res J Mod Eng Technol Sci 2(12). e-ISSN:2582-5208

Capoor MR, Bhowmik KT (2017) Cytotoxic drug dispersal, cytotoxic safety, and cytotoxic waste management: practices and proposed india-specific guidelines. Indian J Med Paediatr Oncol 38(2). https://doi.org/10.4103/ijmpo.ijmpo_174_16

Datta P, Mohi GK, Chander J (2018) Biomedical waste management in India: critical appraisal. J Lab Physicians 10(1):6–14. https://doi.org/10.4103/JLP.JLP_89_17

Dhole KS, Bahadure S, Bandre GR, Noman O (2024) Navigating challenges in biomedical waste management in India: a narrative review. Cureus 16(3):e55409. https://doi.org/10.7759/cureus.55409

Fang B, Yu J, Chen Z, Osman AI, Farghali M, Ihara I, Hamza EH, Rooney DW, Yap PS (2023) Artificial intelligence for waste management in smart cities: a review. Environ Chem Lett 21:1959–1989

Farre JA, Mateu C, Teixido M, Cabeza LF (2023) Pneumatic urban waste collection systems: a review. MDPI Appl Sci 13(2):877. https://doi.org/10.3390/app13020877

Gahlot P, Dhankhar R, Yadav P, Vigarniya MM (2019) Challenges of biomedical waste management. Annals Agri Bio Res 35(2):191–200

Goswami M, Goswami PJ, Nautiyal S, Prakash S (2021) Challenges and actions to the environmental management of bio-medical waste during COVID-19 pandemic in India. Heliyon 7(3):E06313. https://doi.org/10.1016/j.heliyon.2021.e06313

Gupta PP, Bankar NJ, Mishra VH, Sanghavi S, Badge AK (2023) Waste is critical to public health: insights from the central pollution control board guidelines in India. Cureus 15(10):e47303. https://doi.org/10.7759/cureus.47303

Health Care Without Harm Europe (2020) Sustainable healthcare waste management in the EU circular economy model. https://noharm-europe.org/documents/sustainable-healthcare-waste-management-eu-circular-economy-model

Hasan SE (2011) Public awareness is key to successful waste management. J Environ Sci Health, Part 39(2):483–492. https://doi.org/10.1081/ESE-120027539

Health care Without Harm. Treating sharps waste. https://noharm-global.org/issues/global/treating-sharps-waste#:~:text=Sharps%20waste%20is%20a%20subset,dangerous%20wastes%20produced%20in%20healthcare

Hitesh (2024) Hospital waste management | how waste is managed. AAFT Online. https://aaftonline.com/blog/hospital-waste-management/#:~:text=Hospital%20Medical%20Waste%20Management&text=Rigorous%20protocols%20are%20in%20place,or%20chemical%20treatment%20are%20employed

Khan S, Syed AT, Ahmad R, Rather TA, Ajaz M, Jan FA (2010) Radioactive waste management in a hospital. Int J Health Sci 4(1):39–46

Khan HA, Baig FK, Mehboob R (2017) Nosocomial infections: epidemiology, prevention, control and surveillance. Asian Pac J Trop Biomed 7(5):478–482. https://doi.org/10.1016/j.apjtb.2017.01.019

Kumar P, Kausar MA, Singh AB, Singh R (2021) Biological contaminants in the indoor air environment and their impacts on human health. Air Qual Atmos Health 14:1723–1736

Md Mainul SK (2023) Regulatory landscape and enforcement challenges in biomedical waste management in India. Int Res J Mod Eng Technol Sci 5(7). https://doi.org/10.56726/IRJMETS43268

Manekar SS, Bakal RL, Jawarkar RD, Charde MS (2022) Challenges and measures during management of mounting biomedical waste in COVID-19 pandemic: an Indian approach. Bull Natl Res Cent 46:159. https://doi.org/10.1186/s42269-022-00847-4

Mazzei HG, Specchia S (2023) Latest insights on technologies for the treatment of solid medical waste: a review. J Environ Chem Eng 11(2). https://doi.org/10.1016/j.jece.2023.109309

Ministry of Environment, Forest and Climate Change, Government of India (2016) Bio medical waste management rules. https://dhr.gov.in/sites/default/files/Bio-medical_Waste_Management_Rules_2016.pdf

Mohammed NH, Khan S, Jagtap S (2023) Modernizing medical waste management: unleashing the power of the internet of things (IoT). MDPI. Sustainability 15(13):9909. https://doi.org/10.3390/su15139909

Omo QG, Hassan NE (2024) Biomedical waste management and their effects on the environment: a review. World J Adv Eng Technol Sci 11(01):86–95. https://doi.org/10.30574/wjaets.2024.11.1.0020

Paoletti D (2023) The rise of solar-powered compacting trash bins in smart cities. Linkedin. https://www.linkedin.com/pulse/rise-solar-powered-compacting-trash-bins-smart-cities-paoletti

Patel R (2024) Top waste management technologies that embark you towards a better tomorrow. https://www.upperinc.com/blog/waste-management-technologies/#:~:text=Smart%20waste%20bins%20are%20intelligent,teams%20to%20optimize%20collection%20schedules

Patneedi CB, Prasadu KD (2015) Impact of pharmaceutical wastes on human life and environment. Rasayan J Chem 8(1):67–70

Pureingenium (2022) What is pathological waste and how should it be disposed of? https://www.pureingenium.com/regulatory-updates/what-is-pathological-waste/

Sengeni D, Padmapriya G, Imambi SS, Suganthi D (2023) Biomedical waste handling method using artificial intelligence techniques. Handbook of research on safe disposal methods of municipal solid wastes for a sustainable environment. IGI Global Publishing. https://doi.org/10.4018/978-1-6684-8117-2.ch022

Sonkar GK, Singh S, Sonkar SK (2023) Biomedical waste: impact on environment and its management in health care facilities. Curr Res Trends Appl Waste Manage 139–159

Training Express. 8 Types of Healthcare waste and How to Dispose Them. https://trainingexpress.org.uk/types-of-healthcare-waste/

Tushar SR, Bin Alam MF, Mainul Bari ABM, Karmakar CL (2023) Assessing the challenges to medical waste management during the COVID-19 pandemic: Implications for the environmental sustainability in the emerging economies. Socio-Econ Plan Sci 87(A):101513. https://doi.org/10.1016/j.seps.2023.101513

United States Environmental Protection Agency (2024) Medical waste. https://www.epa.gov/rcra/medical-waste

Wei Y, Cui M, Ye Z, Guo Q (2021) Environmental challenges from the increasing medical waste since SARS outbreak. J Clean Prod 291:125246. https://doi.org/10.1016/j.jclepro.2020.125246

Witt A (2022) Medical waste and the environmental impact. Medical Waste Pros. https://www.medicalwastepros.com/blog/medical-waste-and-the-environmental-impact/

World Health Organizations. Occupational hazards in the health sector. https://www.who.int/tools/occupational-hazards-in-health-sector

World Heath Organization (2018) Health-care waste. https://www.who.int/news-room/fact-sheets/detail/health-care-waste

Redefining Hazardous Waste Management: Risks, Regulations, and Resilient Solutions

Md. Nazirul Islam Sarker and Md. Naimur Rahman

Abstract

The global production of hazardous waste poses significant threats to both human health and the environment, underscoring the need for effective waste management solutions. While prior research has largely focused on technological approaches, there has been less emphasis on comprehensively evaluating risks, policy efficacy, and the integration of sustainable practices across various socioeconomic contexts. This study aims to enhance hazardous waste management by incorporating risk analysis, assessing policy effectiveness, and promoting sustainable solutions. A narrative review methodology was employed to analyze journal articles, government reports, and international recommendations related to hazardous waste management. The study highlights the importance of conducting thorough risk analyses that consider both the direct and indirect effects of hazardous waste disposal on health and the environment. It also reveals that current policies often fail to uniformly integrate advanced technical and sustainable practices across different regions due to varying economic and regulatory conditions. The research investigates viable alternatives, such as waste minimization practices, recycling, and recovery techniques; only a small percentage of these practices have been adopted or mandated. These results highlight the importance of a more holistic approach to hazardous waste management that incorporates risk assessment and policy reform along with green practices. A robust multimodal approach that considers risk analysis, policy frameworks, and sustainability for waste management is must.

Keywords

Sustainable waste management · Sustainable practices · Recycling · Environmental health · Recycling and recovery

1 Introduction

Managing hazardous waste is an urgent global environmental issue connected to rapid industrial growth and technological advancements. Chemical businesses, healthcare facilities, and technology centers produce an increasingly diverse range of dangerous waste products—substances foreign to the environment in both composition and concentration (Wu 2022). It is a global problem that requires countries worldwide to overhaul their waste management practices. This is not just about modern, Western nations with good waste systems; it is also about many societies in the developing world that have poor infrastructure and no regulatory framework (Karthikeyan et al. 2018). The wide-range task of hazardous waste management entails detecting, storing, transporting, and disposing hazardous waste. Hazardous waste management requires many different skills and must coordinate in oping, storing, transporting, and disposing of dangerous materials. The regulation of these operations is a shared responsibility between the states and the federal government (Grzegorz et al. 2022; Kumar et al. 2023). Nations globally are beefing up regulatory frameworks to improve waste management, safeguarding ecosystems from hazardous waste (Kala and Bolia 2024).

M. N. I. Sarker
School of Public Administration, Xi'an University of Finance and Economics, Xi'an, China

Miyan Research Institute, International University of Business Agriculture and Technology, Dhaka, Bangladesh

M. N. Rahman (✉)
Department of Geography, Hong Kong Baptist University, Kowloon, Hong Kong SAR, China
e-mail: naimurbrur@life.hkbu.edu.hk; naimurbrur@gmail.com

Department of Development Studies, Daffodil International University, Dhaka, Bangladesh

A. Mandpe et al. (eds.), *Sustainable Waste Management*, Advances in Science, Technology & Innovation,
https://doi.org/10.1007/978-3-031-85685-3_9

The sustainable management of hazardous waste is now an important part of global environmental strategies (Farooq et al. 2022; Salsabila et al. 2023). Generally, poorer nations have poor waste management systems, which puts their inhabitants at risk. These countries are already vulnerable due to their living conditions; they cannot afford unnecessary risks. International cooperation can help bridge the technology gap and enable these countries to manage their toxic waste safely (Naqvi and Farooq 2024). It is a global issue because the dangers of improperly disposed trash affect everyone, regardless of where they live. Progress has been achieved in hazardous waste management, most notably in the development of recycling technologies, waste-to-energy systems, and disposal methods. Their clear objective is to minimize the environmental impact of hazardous waste (Ghali et al. 2023). Other countries, including Finland, have established treatment facilities for handling hazardous waste in addition to separately sited disposal and storage sites. They already aware that today day efficient waste management system is integral (Farooq et al. 2022).

This recent research underlines the importance of strong regulatory frameworks to effectively regulate the activities of waste management of hazardous (Reinikainen and Sorvari 2016; Khan 2016; Okoye et al. 2024). This includes step-by-step guidance on how to store, handle, treat, and dispose of hazardous waste in compliance with environmental safety laws. Other approaches stress new technologies that may be applied to optimize waste management processes and enhance efficiencies (Jha et al. 2021; Grzegorz et al. 2022). For example, novel sorting methods and advanced recycling technologies, such as automated robotic sorting systems have significantly increased the efficiency of recycling processes as they reduce contamination levels and improve the quality of recovered materials (Cui et al. 2020; Khoshnevisan et al. 2021). Additionally, plasma gasification is one other waste-to-energy know-how that has been utilized to show harmful waste into clear power and thereby decrease landfill usage and environmental impression (Shabani et al. 2023).

Furthermore, recent studies show that education and community participation were strongest motivators for improved hazards waste management systems (Watts et al. 2023; Khaskheli et al. 2023). The studies say prompt action is needed to at least make sure that people living near hazardous-material facilities are aware of the potential hazards posed by those facilities. A brilliant method to comprehend hazardous waste in totality is by contrasting how well nations across the globe streamline their waste management process. An understanding from an international context enhances the reasoning concerning what can and must be done here at home. There is a persistent issue with technology and sustainable practices not being fully integrated in many settings. This problem is typically exacerbated by socioeconomic inequalities (Thapa et al. 2023).

Managing hazardous waste is a serious global problem. It is especially deregularized in many parts of the world where waste collection and management come close to catastrophic failure—especially in those countries without legislative frameworks that address treatment and disposal with any authority. Assigning responsibility among government agencies and local authorities does not often solve the problems at the local level because these very-authority figures do not understand what rules they are supposed to enforce or who is responsible for tedious, difficult work. The inadequacies result in environmental and public health hazards caused by the incorrect disposal and handling of dangerous waste (Mmereki et al. 2017). In addition, numerous industries have a deficiency in understanding rules and appropriate waste management practices, hence exacerbating the challenges associated with effectively managing hazardous wastes (Karthikeyan et al. 2018). Waste management systems of individuals are subjected to economic inefficiencies, lack of funds, and challenges, in technical matters. This unhappy scenario presents numerous hazards to the environment and public health. Improper waste disposal leads to an over classification of wastes, limits the access of facilities, and promotes the non-recovery of resources and the disposal of wastes in less advantageous treatments. These concerns further highlight systemic issues in waste classification and treatment mechanisms (Durak 2023; Kokkinos et al. 2023).

The management of hazardous waste employs risk assessment and that is especially difficult to apply, particularly if it should treat the new kind or the new dangers of dangerous substances (Sefouhi and Bahmed 2023). This further highlights that risk analysis methods which are traditionally based on quantitative risk assessment are not sufficient to assess the complicated interactions between various hazardous waste materials, and their adverse impacts on human and environmental health (Vyas et al. 2022). Underestimation of risks may lead to insufficient safeguards or delays to interventions; therefore, more accurate, context-specific risk assessment models are needed (El-Saadony et al. 2023). Technological solutions—constant monitoring as well as predictive analytics—must also be incorporated into these models to enhance the accuracy and quality of risk assessments. There is also a need to communicate evaluation to all stakeholders to ensure that all people know the evaluation and should take actions accordingly.

Although extensive research has been carried out in the hazardous waste management arena (Navia and Bezama 2008; Candra Nugraha Deni and Shabirah 2022; Grzegorz et al. 2022; Kumar et al. 2023; Jena et al. 2023), there remain considerable gaps in our knowledge and understanding. More research into the impact, of rapid industrial development, on hazardous waste generation is urgently needed. There is a need to improve the safe: the system for storage, segregation, transportation, and disposal of hazardous materials to minimize the risk of environmental contamination (Chu et al. 2021). In addition, relatively little attention has been paid to economic and technical barriers that hinder the application of eco-friendly waste management approaches especially in developing regions, where scarce financial means and lack of infrastructure exacerbate the management challenges (Vyas et al. 2022). Those gaps also call for thorough investigations of dangerous waste in the urban setting, where the concentration and complexity of waste streams generate unique challenges. These research intends to develop tailored strategies which can be effectively implemented according to the unique economic, cultural, and environmental contexts of different places (Donner et al. 2021; Iyer 2024). This is important to fill the current gaps in these areas of study to formulate more effective, more sustainable, and more cultural appropriate hazardous waste management practices. These strategies must be capable of addressing the diverse needs of different global and local settings without jeopardizing environmental safety and public health. Consequently, the aim of this study is to address the following research questions:

- How can risk assessment methodologies be optimized to better predict and mitigate the impacts of hazardous waste?
- What are the limitations of current hazardous waste management policies, and how can they be reformed to accommodate diverse socioeconomic contexts?
- Which sustainable practices can be effectively integrated into the hazardous waste management frameworks to enhance environmental and public health safety?

Current literature and case studies were thoroughly analyzed to take a comprehensive look at the state of hazardous waste management. The brief, intermediate, and long-term dangers of hazardous waste were analyzed. We also assessed efficiency, effectiveness, and policy changes required to achieve more durable solutions, applicable globally, for managing such hazardous materials. We searched for sustainable strategies that can be implemented across a broad range of socioeconomic contexts. Finally, this study contributes to the body of knowledge, by analyzing hazardous, navigating accounts of legislation, and sustainable practices of waste treatment. Based on the framework proposed, this study is formulated to enhance global hazardous waste management approaches with more sustainability and effectiveness by combining risk assessment with socioeconomic context. It presents practical recommendations for decision-makers and environmental managers.

2 Methodology

2.1 Research Design

The Preferred Reporting Items for Systematic Reviews and Meta-Analyses (PRISMA) guidelines are used for this study (Moher et al. 2009). The guidance writ large is very broad and somewhat inflexible, which likely accounts for the fact that implementers have found the PRISMA guidelines to enhance the transparency and rigor of systematic reviews. In our study goal, we aim to review the updated literature concerns on hazardous waste disposal. Our work will focus on key areas, such as risk assessments, policy effectiveness evaluations, and sustainable management practices reviews. This method allows for integration of assessment of optimizing hazardous waste management practices in various socioeconomic environments, while maintaining a strict framework for assessment (Page et al. 2021).

2.2 Research Protocol

A systematic process was established to conduct the research according to the PRISMA 2020 guidelines. Table 1 shows the phases which constitute this protocol. This literature review covers recent studies and trends in the subject, focusing more on the most up-to-date researches on hazardous waste management as they determine the better world.

2.3 Search String

In doing a literature search, the key words you choose can help explore this comprehensive literature, hazard waste management, risk assessment, policy evaluation, and sustainable practices that should be built into keyword. The search aimed to include as diverse body of research that reflected complexities and nuances of hazardous waste management.

2.4 Search Strategy

An across-multiple-databases search was performed. The main database in the search process was Scopus, which covers an expansive range of scientific and technical periodicals. We found many relevant publications during the process of full-text screening, but primarily via looking at the references cited in the articles we got from Scopus. Our working group had to meticulously develop words and phrases that would serve as successful searches for material relevant to our goals. After conducting those searches

Table 1 Research protocol

Items	Description
Research question	How can hazardous waste management practices be optimized to enhance sustainability and policy effectiveness?
Database	Scopus (primary), Web of Science, and additional articles from reference checks during full-text screening
Document type	Only peer-reviewed journal articles
Language	English
Publication period	From January 1, 2000, to April 30, 2024
Search terms	Hazardous waste management, risk assessment, policy effectiveness, sustainable practices
Search fields	Title, abstract, and keywords
Inclusion criteria	Articles focusing on hazardous waste management effectiveness, policy, and sustainability
Exclusion criteria	Inaccessible full texts, duplicates, non-English articles, and articles not centered on hazardous waste management

in various databases, we had to read through every article's abstract. The search phrases were meticulously chosen to correspond with the research goals, and the searches were limited to publications written in English to ensure uniformity in the analysis.

2.5 Inclusion and Exclusion Criteria

To ensure relevance and high quality, the studies included in this study were chosen according to clear inclusion and exclusion criteria. Inclusion criteria includes the articles should be published in peer-reviewed journals and focusing on hazardous waste management effectiveness, policy, and sustainability. Similarly, exclusion criteria include inaccessible full texts, duplicates, non-English articles, and articles not centered on hazardous waste management.

2.6 Criteria of Analysis for Document Review

A systematic approach was employed to assess the chosen literature, guaranteeing a thorough examination according to predetermined criteria, as depicted in Table 2.

2.7 Qualitative Analysis of Selected Papers

The thematic analysis method was employed to understand the common threads running through the hazardous waste management practice, policy efficacy, and sustainable solutions. The specific procedures of this analytical technique are detailed below.

2.7.1 Acquaintance with the Data

The chosen papers were carefully reviewed multiple times to gain a thorough grasp of their content. This was required to grasp the subtle details and specific research findings of the substance. It was also important to understand how each study fit in the context of other research related to hazardous waste management.

2.7.2 Generating Initial Codes

The initial coding technique was a careful and systematic search for significant characteristics within the dataset. At this point, we marked out relevant sections of the text connected with our research objectives. These sections mainly deal with risk assessment methods, policy implications, and sustainable practices in hazardous waste management.

Table 2 Criteria of analysis for document review

Criteria	Description
Key phrase selection	It was initiated by identifying key phrases relevant to hazardous waste management
Term sheet development	A term sheet listing themes and sub-themes was developed to guide the literature review
Document reading strategy	All documents were analyzed to extract relevant information that aligns with the term sheet
Codification of terms	A thematic analysis approach was used, categorizing excerpts under relevant themes. The process allowed for the iterative refinement of themes

2.7.3 Theme Identification

Relevant extracts were obtained and arranged as potential themes after coding. It thus enabled the data to be sorted by key themes emerging from the code, allowing for an informed discussion about policy frameworks, risk assessment methodologies, and sustainable strategies within these three broad areas where improvements to hazardous waste management could be made.

2.7.4 Examining Themes

In this study, themes and sub-themes were generated using a systematic process based on pre-existing theoretical frameworks and an extensive search of the relevant literature. In particular, the themes were guided by approaches already explored in the context of hazardous waste management (Navia and Bezama 2008; Karthikeyan et al. 2018; Hawali Abdul Matin et al. 2022; Deni and Shabirah 2022), risk assessment (Sefouhi and Bahmed 2023; Naqvi and Farooq 2024), and sustainable practices (Kokkinos et al. 2023; Pandey et al. 2023; Kumar et al. 2023; Kala and Bolia 2024). In this phase, the identified themes about the coded extracts and the full data set were individually considered and checked to ensure that they were consistent and a good representation of the data. Each theme was carefully developed, defined, and named to reflect the underlying meaning of the data that form the entire narrative. This process was iterative, allowing for themes to be selected and descriptions refined as the review progressed.

The subsequent study steps adopted a structured methodology comparable to the methodology detailed above:

Sub-theme development: Within major themes, sub-themes were developed to further classify the data focusing on areas of interest linked to the management of hazardous waste impact on policy and sustainable methods.

Indicator identification: Indicators were established for each theme and sub-theme, acting as quantifiable or observable components that may evaluate the extent, efficacy, or consequences of certain hazardous waste management solutions addressed in the literature.

Thematic consolidation involved continuously comparing the themes and sub-themes with the dataset to ensure they accurately represented the acquired material. This process refined the themes to fit with the study's objectives closely. Data synthesis entailed integrating the collected data based on the predetermined topics and sub-themes, in preparation for the comprehensive presentation of the findings in the following sections of the paper (Table 3).

3 Results

3.1 Overview of Included Studies

3.1.1 Document Selection

The study's selection method was structured following the PRISMA flow diagram. Figure 1 illustrates the sequential process, beginning with identifying relevant studies through the literature search, then screening titles and abstracts, and culminating with the full-text assessment of publications to determine their eligibility. Initially, the search method yielded 407 records from the databases and an additional 8 records from references, resulting in 415 records. After eliminating duplicate entries and irrelevant studies based on abstract screening, 137 distinct records were preserved. During the screening stage, the titles and abstracts of these records were analyzed according to the inclusion and exclusion criteria. 278 records were removed during this phase due to irrelevance or duplication. Following this, 44 records were subjected to a comprehensive evaluation of their full-text content. During the full-text assessment step, papers were evaluated based on their relevance to climate change adaptation in Bangladesh. 19 papers were excluded because they did not focus on this specific topic. In the end, 25 documents that satisfied the specified criteria were selected and included in the final qualitative analysis (Fig. 1).

3.1.2 Characteristics of the Studies

The selected papers were published between January 2006 and April 2024. A significant rise in publications after 2021 indicates a growing academic and practical interest in climate change adaptation in Bangladesh. The number of publications peaked in 2023 with 9 papers, demonstrating heightened research activity in recent years. Several studies have established a theoretical basis for climate change adaptation, while others have concentrated on applying empirical methods to assess the impacts and effectiveness of adaptation strategies. Empirical research employed various approaches, such as qualitative interviews, surveys, and case studies, while theoretical publications primarily concentrated on literature reviews and conceptual analyses. The recent surge in publications reflects an increasing recognition of the importance of developing and implementing effective adaptation measures to address the challenges posed by climate change (Fig. 2).

Table 3 Extended term sheet for thematic analysis

Themes	Sub themes	Description	Sources
Policy effectiveness	Legislative frameworks	Examine the impact of regulatory changes on waste management practices	Amesho et al. (2023), Simpa et al. (2024), Cui et al. (2020)
	Policy implementation	Discuss the practical implementation of policies and their effectiveness in different contexts	Shabani and Jerie (2023), Salsabila et al. (2023)
	International compliance	Analyze how global regulations like the Basel convention influence national policies	Simpa et al. (2024), Shabani and Jerie (2023)
	Economic incentives	Study the role of financial mechanisms in promoting compliance with hazardous waste regulations	Karthikeyan et al. (2018), Vasudevan (2017), Kala and Bolia (2024)
	Environmental audits	Evaluate the role of environmental audits in ensuring compliance and policy effectiveness	Khalmurzayevna et al. (2023), Perkumienė et al. (2023), Shabani and Jerie (2023)
Risk assessment	Methodological advances	Explore advances in risk assessment techniques for hazardous waste	Scruggs et al. (2016), Aragaw (2023)
	Impact evaluation	Analyze how risk assessments influence waste management strategies	Durak (2023), Perkumienė et al. (2023)
	Health and environmental risks	Evaluate risks to public health and the environment from hazardous waste	Aragaw (2023), Ali et al. (2023)
	Risk communication	Assess the effectiveness of risk communication strategies in hazardous waste management	Scruggs et al. (2016), Lud et al. (2022)
	Institutional and policy issues	Explore the role of institutional frameworks and policy decisions in risk management	Shabani and Jerie (2023), Perkumienė et al. (2023), Kala and Bolia (2024)
Sustainable practices	Waste minimization	Focus on strategies for reducing waste generation at the source	Perkumienė et al. (2023), Dada et al. (2024)
	Recycling and recovery	Examine the integration of recycling and recovery operations in waste management	Kofoworola (2007), Ali et al. (2023)
	Technology integration	Investigate new technologies that promote sustainable waste management	Khaskheli et al. (2023), Hajam et al. (2023), Simpa et al. (2024)
	Education and training	Study the role of education in promoting sustainable hazardous waste practices	Hajam et al. (2023), Shabani and Jerie (2023)
	Pollution prevention	Analyze strategies and technologies aimed at preventing pollution from hazardous waste	Bhupendra and Sangle (2016), Hettiarachchi and Meegoda (2023), Perkumienė et al. (2023)
	Treatment processes	Discuss various treatment processes including physical, chemical, thermal, and biological methods	Karthikeyan et al. (2018), Hajam et al. (2023), Shabani and Jerie (2023)
	Waste handling and storage	Explore best practices in the handling and storage of hazardous waste to minimize risk	Karthikeyan et al. (2018), Lud et al. (2022), Durak (2023)

3.1.3 Network of the Concepts

VoSviewer software tools were utilized to create network visualizations that depict the interconnectedness of key concepts in waste management, sustainable development, risk assessment, environmental regulations, and health risks (Fig. 3). The visualization employed a co-occurrence matrix derived from the literature, which evaluated the similarity and frequency of mutual citations among the research. The items were represented along a two-dimensional map, where distance indicated a greater or smaller relationship, so that items closer to each other formed clusters in which subjects were related. The network analysis helped to cluster essential topics like waste management strategies, the effect of policy on sustainability, different techniques for risk assessment, and health risk assessments. The visualization represented also the most used keywords for the selected studies, and high relevance keywords that recurred frequented in the literature were identified, examples are sustainable development, waste management, recycling, waste disposal, and environmental regulations. It further provides a critical analysis of the links and relevance of these topics to sustainable waste management.

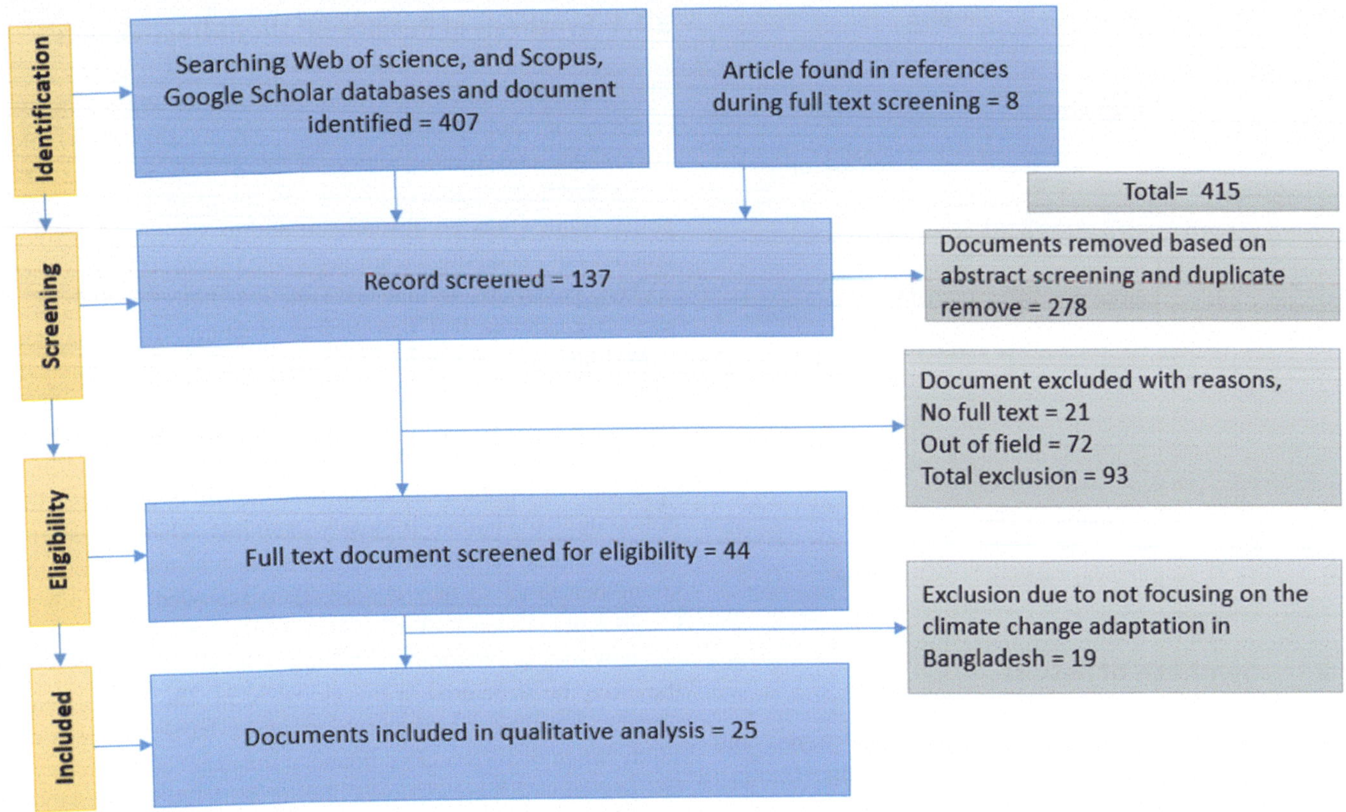

Fig. 1 Relevant document selection process

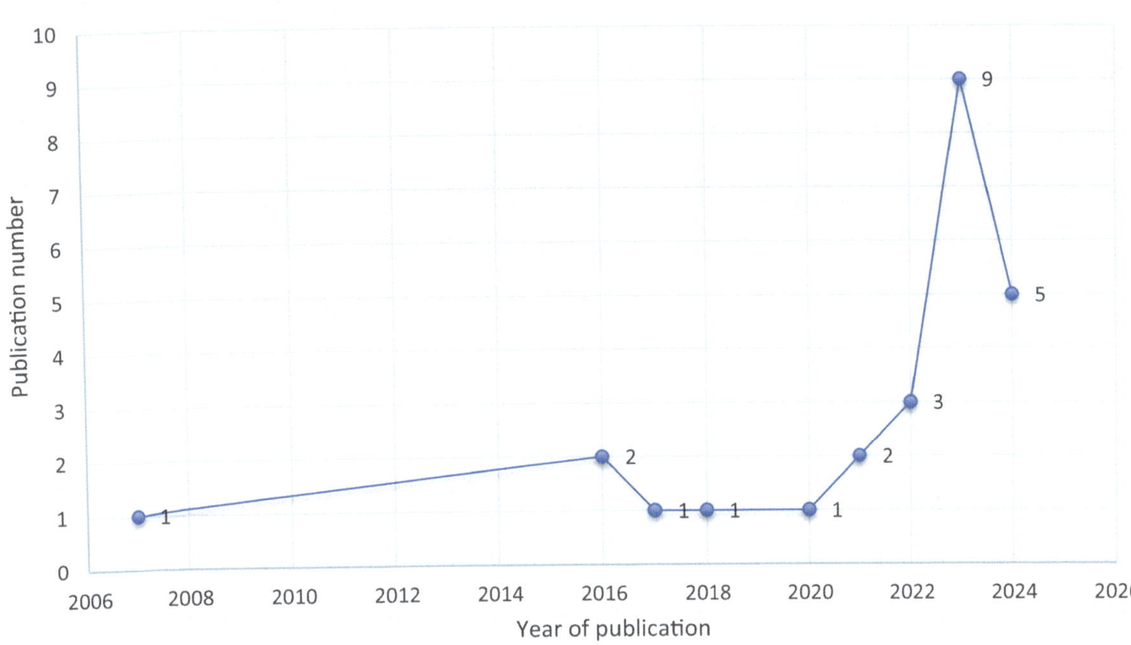

Fig. 2 Displays number of published documents over years

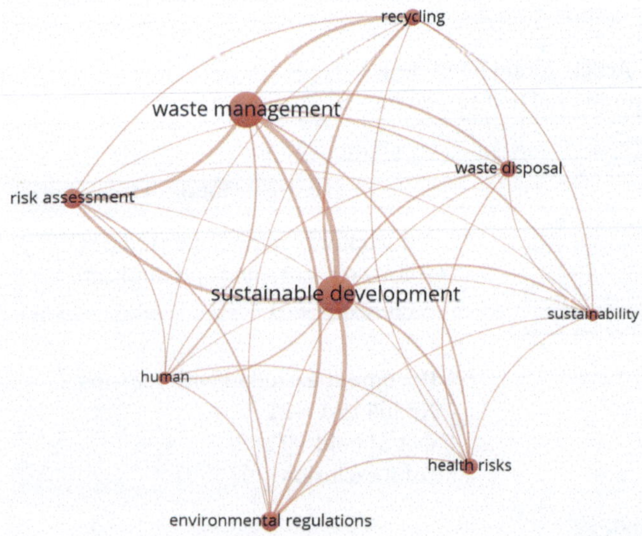

Fig. 3 Network visualization of keyword co-occurrences

3.2 Synthesis of Results

The studies selected were then reviewed for trends and interrelated themes revealed through the results. Using a narrative synthesis technique, themes across those studies were interpreted to produce an overview of different analyzable themes related to hazardous waste management. This methodology facilitates the interpretation of the study results to provide a cohesive overview of interrelated themes in hazardous waste management.

3.2.1 Synthesis of Results—Policy Effectiveness in Hazardous Waste Management

This study delineates six crucial domains falling under the category of "policy effectiveness" that are pivotal for the governance and administration of hazardous waste. These topics facilitate comprehension of the process by which policy frameworks are formulated and executed to improve practices in managing hazardous waste (Table 4).

The governing frameworks of legislation are necessary to guarantee that the current practices for managing hazardous waste adhere to the most recent environmental and safety laws. They allow for periodic updating of older statutes, a necessary activity if a practice is to remain adaptive and in step with technological advances. The real-life application of policies in various scenarios is evaluated in policy implementation. It closely examines what hinders or favors the correct execution of policies and finds ways to strengthen them so they achieve their intended aims more consistently. When it comes to international standards, we see how they are incorporated into countries' laws and how they may be structured at the global level. Businesses and organizations are motivated to manage hazardous waste safely when there is a clear financial upside. Therefore, any law passed in this area focuses high-priority attention on the use of incentive mechanisms, with the hope that such mechanisms will stimulate widespread adoption of safer practices by businesses headquartered within the United States or operating on U.S.-based facilities. Devices that provide incentives for better management practices range from direct subsidies to tax incentives. Within today's federal

Table 4 Policy effectiveness in hazardous waste management

Policy effectiveness sub-themes	Indicator	Role in hazardous waste management	Sources
Legislative frameworks	Changes in waste management regulations	Ensures that practices evolve with new regulations, maintaining or enhancing safety and environmental standards	Amesho et al. (2023), Simpa et al. (2024), Cui et al. (2020)
Policy implementation	Effectiveness and adaptability of policies	Evaluates how well policies perform in different operational contexts, improving compliance and outcomes	Shabani and Jerie (2023), Salsabila et al. (2023)
International compliance	Adherence to international agreements	Measures the influence of global standards on national policies, fostering worldwide safety and environmental care	Simpa et al. (2024), Shabani and Jerie (2023)
Economic incentives	Use of financial tools to support compliance	Examines how subsidies, taxes, and other financial incentives can encourage or enforce compliance with regulations	Karthikeyan et al. (2018), Vasudevan (2017), Kala and Bolia (2024)
Environmental audits	Regular assessments of compliance and impact	Assesses the effectiveness of waste management practices in meeting regulatory standards and reducing environmental impacts	Khalmurzayevna et al. (2023), Perkumienė et al. (2023), Shabani and Jerie (2023)

legislative framework, three main types of incentives have emerged: grants, tax policies, and vulnerable specified technologies. Overseeing adherence to regulations is a key function of environmental audits. They are designed to ensure that businesses and organizations comply with the relevant laws and regulations, which means they have no part to play in causing harmful environmental damage. An environmental audit is an important tool for evaluating the efficiency of your waste management procedures.

3.3 Risk Assessment in Hazardous Waste Management

The theme of "risk assessment" in hazardous waste management includes numerous important sub-themes that aid in evaluating and managing risks related to hazardous waste. These sub-themes contribute to a deeper comprehension of risk and provide guidance for enhancing hazardous waste management procedures by employing more effective assessment techniques and communication strategies (Table 5).

The process of creating the modern risk assessment includes both novel methods of investigation and unprecedented tools. Analyzing risk assessments' efficacy is crucial to understanding how they impact waste management practices. In particular, this sub-theme investigates whether and how risk assessments actually "do" influence certain practice changes in implementing safe methods for managing, treating, and disposing of waste. It is of utmost importance to clearly understand the health and environmental risks associated with hazardous waste—especially concerning their direct and indirect effects on public health and the environment. Efficient risk communication ensures that stakeholders comprehend the risks associated with dangers like hazardous waste. Analyzing tactics for conveying risk information allows us to understand what works best for whom and in what kinds of contexts, be it informing international organizations, teaching local officials in low-ranking countries how to do the same job that finished my tour here last month, or just explaining junk mail-level risks to a concerned citizen. Analyzing institutional and policy matters helps understand how they affect risk assessment. Different lawmaking bodies create the frameworks for our different regulatory agencies, carrying out more detailed work that yields either safe or risky outcomes. Risk assessment involves comprehending the function of different regulatory entities and laws in establishing risk assessment procedures and how they impact the handling of dangerous waste.

3.4 Synthesis of Results—Sustainable Practices in Hazardous Waste Management

The concept of "sustainable practices" is essential in improving hazardous waste management by implementing various techniques focusing on sustainability and environmental responsibility. This theme includes multiple sub-themes that enhance waste management operations' effectiveness, security, and environmental friendliness (Table 6).

Proper waste management ensures that harmful substances do not contaminate the environment. Following the waste generation, disposal happens either at its origin or when sent to a disposal facility. Yet even then, some trash will inevitably land in our environment from production to dump. Waste originates from improper use and misuse of persistently hazardous products, demanding primarily an origin-based solution. Private interest and public policy in

Table 5 Risk assessment in hazardous waste management

Risk assessment sub-themes	Indicator	Role in hazardous waste management	Sources
Methodological advances	Development of new risk assessment tools	Improves the accuracy and efficiency of risk evaluations, adapting to new hazards and technologies	Scruggs et al. (2016), Aragaw (2023)
Impact evaluation	Effect of risk assessments on management	Guides waste management strategies by providing data on potential risks and their impacts	Durak (2023), Perkumienė et al. (2023)
Health and environmental risks	Assessment of health and environmental impacts	Identifies and quantifies risks to public health and the environment, leading to informed management decisions	Aragaw (2023), Ali et al. (2023)
Risk communication	Effectiveness of communication strategies	Ensures that risk information is effectively communicated to all stakeholders, enhancing safety and compliance	Scruggs et al. (2016), Lud et al. (2022)
Institutional and policy issues	Role of institutions in risk management	Examines how institutional frameworks and policy decisions impact risk assessment processes and outcomes	Shabani and Jerie (2023), Perkumienė et al. (2023), Kala and Bolia (2024)

Table 6 Sustainable practices in hazardous waste management

Sustainable practices' sub-themes	Indicator	Role in hazardous waste management	Sources
Waste minimization	Reduction of waste generation at the source	Minimizes the volume of hazardous waste produced, reducing disposal needs and environmental impact	Perkumienė et al. (2023), Dada et al. (2024)
Recycling and recovery	Integration of recycling operations	Converts waste into usable materials, reduces landfill use, and conserves resources	Kofoworola (2007), Ali et al. (2023)
Technology integration	Adoption of new waste management technologies	Employs innovative technologies to enhance the efficiency and effectiveness of waste management practices	Khaskheli et al. (2023), Hajam et al. (2023), Simpa et al. (2024)
Education and training	Role of educational initiatives	Increases awareness and knowledge among stakeholders, promoting safe and sustainable practices	Hajam et al. (2023), Shabani and Jerie (2023)
Pollution prevention	Implementation of pollution control measures	Prevents or reduces pollution through techniques that minimize waste and mitigate its impacts	Bhupendra and Sangle (2016), Hettiarachchi and Meegoda (2023), Perkumienė et al. (2023)
Treatment processes	Application of treatment methods	Utilizes physical, chemical, thermal, and biological methods to safely process hazardous waste	Karthikeyan et al. (2018), Hajam et al. (2023), Shabani and Jerie (2023)
Waste handling and storage	Best practices in waste handling and storage	Ensures that hazardous waste is managed in a manner that minimizes risks to health and the environment	Karthikeyan et al. (2018), Lud et al. (2022), Durak (2023)

a new generation naturally lead to new technologies needed for efficiently managing these transformations. Several treatment techniques can be employed to deal with hazardous waste, making it safe for either recovery or disposal. The main point of all these methods is that they ensure environmental impact minimization while handling and managing such dangerous wastes. Hazardous waste storage and handling is a discipline with its own set of procedures and methods. This on-site storage differs from other types of waste storage because it requires extra safety measures to prevent any kind of pollution—whether accidental or intentional.

Hazardous waste generated improperly can lead to serious health risks for anyone who comes into contact with it—either during its creation or disposal. Studies have shown that exposure to the types of chemicals found in hazardous waste can cause everything from cancer to congenital disabilities. In their 1982 report titled "Toxics in the Bay," several environmental groups laid out a compelling case for why dumping toxic pollutants into San Francisco Bay was a recipe for disaster—not just for people living downstream but also for all manner of wildlife that calls the bay home.

4 Discussion

4.1 Key Insights

The investigation into handling dangerous waste has resulted in several key insights. These are pertinent to our research inquiry for several reasons. First, they address one of our main investigative questions head-on: How can we improve hazardous waste management using advanced risk assessment techniques? Our research highlighted how significant it is to ensure that policies are proposed and implemented efficiently and effectively, with the appearance (at least) of strict international adherence to compliance requirements. Therefore, another key finding was how international organizations might further enhance global hazardous waste management practices through policy changes (Jha et al. 2021; Salsabila et al. 2023).

The results of our research have emphasized the substantial influence that legislative frameworks and policy implementations have on the management of hazardous waste (Mor and Ravindra 2023). It is essential to adapt these policies in response to emerging difficulties and incorporate international compliance standards to improve hazardous waste management. The observations above have important relevance with regards to our research question on how hazardous waste management can be improved via the formation of policies and the effective implementation of these policies (Hettiarachchi and Meegoda 2023; Li 2023; Kala and Bolia 2024).

The investigations showed significant advancements in the process of risk assessment. These developments are essential because they inform and optimize waste management strategies, ensuring comprehensive health and environmental risk assessments (Ghali et al. 2023; Mor and Ravindra 2023). Due to risk assessment approaches and their application on hazardous waste management strategies progress, specific elements indicate how data affecting management practices. This impacts on policymakers which ensure that policies are not only efficient but also protect public health and the environment. The findings presented above provide new insights to improve waste management through comprehensive risk assessments (Sinha et al. 2024).

The analysis highlighted the importance of integrating recycling and recovery operations, the relevance of employing new technologies, and the key role of education in promoting sustainable waste management practices (Jagun et al. 2023). Integrating such data could also provide the necessary detailed information on how to effectively manage hazardous waste, source by source, in each of the stages of the management process. Strong frameworks also have strong risk assessment tools, rich policy frameworks, and sustainable management. These insights respond to the research issues raised and provide directions for future research and practical initiatives to improve global hazardous waste management (Jha et al. 2021; Kumar et al. 2023). Altogether these findings yield a great insight into the current condition of hazardous waste management. It also shines a light on new and effective approaches to solving old and new challenges. This debate attempts to further understand how these discoveries apply to the larger objectives of enhancing hazardous waste management systems. Additionally, it provides practical recommendations for the hazardous waste management industry, with the goal of elevating and advancing it. This comprehensive review offers answers to our initial research questions and sets the stage for future efforts in sustainable waste recycling strategies.

Future research should focus on developing advanced risk assessment models that combined real-time monitoring indicators and predictive analytics. This is consistent with the study's findings which emphasize how existing policies are inadequate and that effective strategies to control emerging threats are sorely needed. Another important aspect which should be examined is the socioeconomic impacts of hazardous waste management, especially in the areas where the infrastructures are weak, to improve global waste management practices.

4.2 Policy Effectiveness

4.2.1 Legislative Frameworks

Changes to regulations play an important role in how hazardous waste is managed. The landscapes of technology develop and change, so do the problems that arise from it; the practices are certainly debated but the legislative structures that govern them need to evolve as well. Legislations should evolve and respond to the emerging waste management and challenges, ensuring its ability to respond effectively to new waste and disposal methods (Cui et al. 2020). The effectiveness of these frameworks is a critical consideration in setting up standards that can protect public health and the environment while allowing for technological advancement.

4.2.2 Policy Implementation

Updating the practice of hazardous waste management is integral to addressing present and future issues. Since new technological opportunities (and risks) are constantly emerging, the legislative structure that governs these practices must itself be able to evolve rather seamlessly and quickly, guaranteeing environmental health safeguards for the public while accommodating technological progress—all without sacrificing any precious time that could allow dangers to accumulate (Shabani and Jerie 2023). This approach facilitates comprehension of the practical aspects of managing hazardous waste and serves as a foundation for enhancing policies.

4.2.3 International Compliance

The national policies of many countries are heavily influenced by international agreements in which they may have a stake. These agreements can cut both ways: On the one hand, they can push progress in global standards and help create a more cohesive, harmonized system of practices across different nations (Simpa et al. 2024). Indeed, much has been said to favor this influence to achieve an "upward harmonic series" for international enforceable environmental norms—the Basel Convention being a good example.

4.2.4 Economic Incentives

Providing subsidies and using tariffs as an economic instrument can greatly influence the earth's ecologically sustainable behavior. Unfortunately, this term "ecologically sustainable" often seems a bit vague. In examining how these incentives work, it is necessary to understand the technologies involved and what sort of impact they might have—positive or negative—on Earth's various environments. Suggestions for improving the efficiency of these incentives' efficiency include modifying financial mechanisms to more closely correspond with environmental objectives and ensuring they possess sufficient strength to impact the actions of corporations and individuals (Kala and Bolia 2024).

Economic incentives, such as subsidies and tax credits, have been applied with varying success to promote sustainable hazardous waste management. In developed regions, these incentives have significantly improved compliance and adoption of better waste practices. However, in developing areas, their impact has been limited due to weaker enforcement mechanisms. These incentives influence behavior by aligning financial benefits with environmentally responsible actions, though their effectiveness depends on regional enforcement and compliance levels.

4.2.5 Environmental Audits

Waste management rule adherence is assured through environmental audits. They allow the detection of noncompliance to shine through, highlighting within the structures of an organization where waste management methods can be improved and where environmental damage may be occurring or at risk of occurring (Perkumienė et al. 2023). For environmental audits to achieve even more efficiency and scope and consequently for improvements in waste management methods not just within but also beyond the realms of hazardous waste management—improvements that could result in less inefficient, less dubious operations—a series of enhancements to audit procedures is necessary (Khalmurzayevna et al. 2023).

4.3 Risk Assessment

4.3.1 Methodological Advances

Progress in risk assessment approaches has greatly affected hazardous waste management, allowing for more precise and powerful detection and measurement of danger. These methodological advances help deal with newly emerging forms of peril and alluring adjustments to the technologies used in waste processing. This is why applying risk assessment to managing hazardous wastes is so important, especially since risks are nothing less than "the potential for harm" that must be taken seriously (Scruggs et al. 2016).

4.3.2 Evaluation of Risk Management Strategies

Assessing how effective a method works is at the heart of determining the impact of waste management. An effect assessment (also called a project environmental impact assessment (EIA)) considers and anticipates the potential impact of a project on the environment and people. These evaluations ensure that strategies are grounded in a solid scientific understanding and are tailored to reduce health and environmental impacts (Durak 2023).

Risk management strategies in hazardous waste management are assessed by criteria including the decrease in incidents of environmental contamination, the level of compliance with regulatory standards, and the communication of risks. If you are seeing high levels of compliance and measurable reductions of activities leading to incidents, we can surmise that the strategies are working—particularly when coupled with clear and effective communications that successfully promote public safety and environmental protection.

4.3.3 Health and Environmental Risks

While the discussion of health and environmental risks is warranted, it certainly deserves the risk management framework so that it can receive the due attention applicable for effective management. One of the most toxic and toxic combinations of substances of our industrial world, hazardous waste has much more than reformation potential. This means taking a hard look at your storage, segregation, transportation, and disposal practices. This integrated waste management approach helps to minimize the potential risks associated with hazardous waste and protects human health and the environment (Aragaw 2023; Ali et al. 2023).

4.3.4 Risk Communication

The current state of risk communication makes it possible for all concerned citizens to understand the dangers of hazardous waste. Yet, there is a palpable need to enhance and expand the actual techniques—the tactics that should connect the risks with the waste. We could improve on this by increasing transparency in conveying risk information and by actively engaging communities in conversations about understanding the risks (Lud et al. 2022). These methodological advances help deal with newly emerging forms of peril and alluring adjustments to the technologies used in waste processing. This is why applying risk assessment to managing hazardous wastes is so important, especially since risks are nothing less than "the potential for harm" that must be taken seriously (Scruggs et al. 2016).

4.4 Sustainable Practices

4.4.1 Waste Minimization

Discussing methods to decrease trash production at its origin is crucial for achieving long-term sustainability in managing hazardous waste. Efficient waste reduction results in substantial financial savings and decreases the ecological impact. Practices, such as source reduction and the utilization of less hazardous products, are of utmost importance (Dada et al. 2024). The enduring advantages encompass decreased waste management expenses, mitigated ecological footprint, and improved public health security.

4.4.2 Waste Recycling and Resource Recovery

Two key aspects of waste management have emerged in recent years: waste recycling and resource recovery. Recycling, where feasible, is the preferred way to reduce the environmental impact of landfilled materials; it involves transforming old products into new ones—extending or expanding the range of materials available to humankind. There are other capabilities for increasing and expanding recycling and recovery operations on a larger scale (Kofoworola 2007). Currently, improvement suggestions include focusing on world-class facilities for recycling and promoting community recycling programs to increase the rates of material retrieval (Ali et al. 2023).

4.4.3 Technology Integration

The integration of advanced technology plays a massive role in the viability of waste management methods. Some innovations, such as automated sorting systems, advanced bioremediation techniques, and increased use of treatment chemicals, improve efficiency and/or effectiveness (Hajam et al. 2023). For determining the impact of these technologies, it is advisable to assess their scalability, economic efficiency, and positive impact on the environment.

4.4.4 Education and Training

The global push toward sustainability is a huge opportunity to teach people all over the globe how to efficiently and safely manage hazardous waste. The first step toward remediation of human impact on our environment has always been understanding the changes in our environment due to human activity (Hajam et al. 2023). This could include more comprehensive training of waste handlers and public education campaigns leading to more responsible waste management practices (Shabani and Jerie 2023).

4.4.5 Pollution Prevention

Current state of pollution prevention highlights the urgent need for better solutions to hazardous waste problems. Potential improvements could include stricter regulations, better enforcement of existing laws, and the incorporation of greener technologies (Bhupendra and Sangle 2016; Hettiarachchi and Meegoda 2023). To enhance these strategies, greater cooperation between government and industry is encouraged to help put in place more extensive methods of pollution control.

4.4.6 Treatment Processes

The different types of hazardous waste that were generated have been treated effectively by the available treatment methods. Those methods and the hazardous wastes they intended to change underline the significance of using diverse and effectively safe treatment technologies (Shabani and Jerie 2023). Possible areas for enhancement could involve optimizing the effectiveness of current treatment procedures and innovating novel technologies that mitigate environmental and health hazards.

4.4.7 Waste Handling and Storage

To properly measure the present methods of managing and storing waste, one must evaluate how well these procedures control risks to health and the environment. The key is to guarantee that waste storage is a secure, well-housebroken operation with waste placed in sanitary conditions that enables accurate identification of stored materials, under favorable surveillance conditions (Lud et al. 2022). Enhancements can be achieved by implementing more stringent standards for trash storage and providing advanced training for workers engaged in handling hazardous materials.

4.5 Recommendations

For policymakers, industry stakeholders, and waste management experts, this study provides a range of concrete suggestions for dealing with hazardous waste. It also identifies topics for future research that will further enhance the subject. Firstly, policymakers must prioritize the regulatory framework that governs hazardous waste management is up to date and considers recent advancements in that field. Laws, rules, and policies must encompass not only what we know now but also how we can make waste management better in the future. This also necessarily means integrating what we know about sustainability and public health into the formulation of new regulations so that technology selection can favor better practices for efficiency as well as safer methods for treating all kinds of waste. Secondly, this matter has two main aspects: operational efficiency and corporate social responsibility. Industrial stakeholders can enhance these by implementing and advocating for advanced treatment technologies and sustainable waste management methods. To avoid environmental concerns, it is advisable to adopt pollution prevention methods and adhere to best practices in waste processing and storage (Ramachandra and Bachamanda 2007). Thirdly, continuing professional development is crucial for waste management experts who need to keep pace with the rapidly changing landscape of technology and techniques in their field. Indeed, those specialists must not only grasp a panoply of new concepts but also be prepared to handle an array of hazardous substances and emergent situations that may arise during the proper disposal of waste items. Facultative capacity can help reduce occupational hazards through training.

Fourthly, it is vitally important for future research to prioritize developing disposal techniques that are more efficient and long-lasting if we effectively reduce the risks connected with hazardous waste (Galgani and Rangel-Buitrago 2024). A more thorough evaluation of how the long-lasting impacts of toxic waste may influence general populations and entire ecosystems can serve as a firmer foundation for assessing the efficacy of our existing waste disposal practices. Fifth, to explore modern therapeutic practices that far exceed commonplace processes. Finally, understanding the cost–benefit ratio of modern waste management techniques as compared to conventional methods will give incredible insight. The findings of this study will assist policymakers in making educated choices and designing policies that meet environmental standards and also promote sustainability in the economy (Iqbal et al. 2024).

5 Conclusion

An extensive investigation was carried out to explore how hazardous waste is dealt with, taking into account key aspects, such as how effective policies are formulated and risks inherent to various forms of hazardous waste, as well as possible sustainable disposal techniques that could replace traditional disposal methods.

Our research highlights the urgent need for modern, strong legislation to keep pace with rapidly advancing technology and emerging waste management issues. Hazard management remains behind in this crucial aspect of policy implementation—these regulations need to set benchmarks that contribute tangible benefits in public health and environmental preservation.

Utilize cutting-edge methods for identifying potential dangers associated with waste in order to assess its risks, which have the potential to cause severe harm if improperly managed. As history has shown, waste can create risks that lead to catastrophes resulting in the destruction of property and lives. Risks have often been created through poor design or misuse of assessment techniques, making clear the need for increased communication between those conducting these assessments and those at risk. Poorly constructed risk evaluation models fail to consider the complex interactions among hazardous waste components and their potential effects on both human health and the environment. Furthermore, when these methods are utilized incorrectly in situations they were never meant for or with insufficient information available, it could lead to significant underestimation of risk and inadequate safety measures being put in place that increase health and environmental risks. Therefore, improving both the development and implementation of risk assessment methods is key in order to lower these risks.

Sustainable waste management methods involve decreasing trash production while increasing recycling and reuse as the cornerstones for successful practices. Utilizing cutting-edge technology can greatly enhance efficiency. Furthermore, ensuring all stakeholders can access appropriate training and education can ensure systems become more effective and increase sustainability and durability to fulfill their intended purposes in the long run. Focusing on pollution control and improving ways to manage waste will further extend their endurance and long-term viability.

Industry, policymakers, and professionals involved in waste management must consider various ideas when making decisions for both now and the future. These range from strengthening enforcement of existing policies to developing innovative technologies which may represent fundamental breakthroughs that change how waste is managed. Subsequent investigations should prioritize the fine-tuning of waste treatment technology, the investigation of economically efficient waste management systems, and the evaluation of the enduring consequences of hazardous waste on the environment and public health.

The current state of hazardous waste management is comprehensively understood when viewed through the lens of present regulations and past experience. This understanding serves as the bedrock for not just standing still but also moving forward in the sustainable and effective regulation of hazardous waste. Doing so assures that both the environment and public health are effectively safeguarded.

6 Limitations and Future Research Directions

This study has several limitations. Firstly, this study primarily utilized a narrative review technique; this kind of review is good for understanding the main parts of a problem, and it has its limitations. One limitation is that reviews like this one are inherently subjective because someone has to decide which studies are more important than others, and different people might make different decisions. Another drawback of narrative reviews is that they tend to discount certain studies or ideas that are not cited as frequently, even though they might be important in understanding the problem and its solutions. Secondly, the study relied exclusively on papers written in English and neglected important contributions made in other languages. This could result in a narrow global perspective on practices and developments related to harmful waste management. Thirdly, the examination was limited to the very recent 24 years, which may have neglected crucial ancient history that could have provided lighter on the development of hazardous waste management and legislation. Finally, a reliance on data from secondary sources like journal articles, government reports, and international recommendations might attach a few biases to the project; for instance, studies included in the review may have been tilted in favor of certain results.

Future studies should follow several recommendations to overcome the above limitations. Firstly, future studies should prioritize an approach combining qualitative and quantitative methods. This approach will provide a comprehensive and well-rounded comprehension of hazardous waste management. This methodology can provide more profound insights and confirm findings from several methodological approaches. Secondly, broadening the range of literature studies to encompass publications in languages other than English might offer a more comprehensive and worldwide outlook on methods related to the management of hazardous waste. This will facilitate the integration of various regional practices and ideas that are typically disregarded. Thirdly, studying hazardous waste management in different periods helps us understand how procedures have evolved and changed. Examining this historical viewpoint can offer significant insights and guide future approaches. Finally, future research investigates hazardous waste management practices' social and economic outcomes.

References

Ali A, Malik SA, Shafiullah M et al (2023) Policies and regulations for solar photovoltaic end-of-life waste management: insights from China and the USA. Chemosphere 340. https://doi.org/10.1016/j.chemosphere.2023.139840

Amesho KTT, Chinglenthoiba C, Samsudin MSAB et al (2023) Microplastics in the environment: an urgent need for coordinated waste management policies and strategies. J Environ Manage 344:118713. https://doi.org/10.1016/j.jenvman.2023.118713

Aragaw TA (2023) Sustainable management of drinking plastic straws is required to reduce plastic pollution: are we using them more during COVID-19? J Hazard Mater Adv 12. https://doi.org/10.1016/j.hazadv.2023.100328

Bhupendra KV, Sangle S (2016) Pollution prevention strategy: a study of Indian firms. J Clean Prod 133:795–802. https://doi.org/10.1016/j.jclepro.2016.05.169

Candra Nugraha Deni M, Shabirah A (2022) Planning and implementation of hazardous waste management emergency response program—a case study of PT. X Medical hazardous waste treatment. IOP Conf Ser Earth Environ Sci 1111:012041. https://doi.org/10.1088/1755-1315/1111/1/012041

Chu Y-X, Wang J, Tian G, He R (2021) Reduction in VOC emissions by intermittent aeration in bioreactor landfills with gas-water joint regulation. Environ Pollut 290. https://doi.org/10.1016/j.envpol.2021.118059

Cui C, Sun C, Liu Y et al (2020) Determining critical risk factors affecting public-private partnership waste-to-energy incineration projects in China. Energy Sci Eng 8:1181–1193. https://doi.org/10.1002/ese3.577

Dada, MA, Obaigbena A, Majemite MT, Oliha JS, Biu PW (2024) Innovative approaches to waste resource management: implications for environmental sustainability and policy. Eng Sci Technol J 5:115–127. https://doi.org/10.51594/estj.v5i1.731

Donner M, Verniquet A, Broeze J et al (2021) Critical success and risk factors for circular business models valorising agricultural waste and by-products. Resour Conserv Recycl 165. https://doi.org/10.1016/j.resconrec.2020.105236

Durak H (2023) Comprehensive assessment of thermochemical processes for sustainable waste management and resource recovery. Processes 11:2092. https://doi.org/10.3390/pr11072092

El-Saadony MT, Saad AM, El-Wafai NA et al (2023) Hazardous wastes and management strategies of landfill leachates: a comprehensive review. Environ Technol Innov 31:103150. https://doi.org/10.1016/j.eti.2023.103150

Farooq M, Cheng J, Khan NU et al (2022) Sustainable waste management companies with innovative smart solutions: a systematic review and conceptual model. Sustainability 14:13146. https://doi.org/10.3390/su142013146

Galgani F, Rangel-Buitrago N (2024) White tides: the plastic nurdles problem. J Hazard Mater 470. https://doi.org/10.1016/j.jhazmat.2024.134250

Ghali H, Ben CA, Bhiri S et al (2023) Health and environmental impact of hospital wastes: systematic review. Dubai Med J 6:67–80. https://doi.org/10.1159/000529432

Grzegorz P, Emilian M, Adrian C, Oana I (2022) Hazardous waste advanced management in a selected region of Poland. Processes 10. https://doi.org/10.3390/pr10102032

Hajam YA, Kumar R, Kumar A (2023) Environmental waste management strategies and vermi transformation for sustainable development. Environ Challenges 13. https://doi.org/10.1016/j.envc.2023.100747

Hawali Abdul Matin H, Purwono P, Chalid Afif Alfajrin A, Hidayat Ramli Inaku A (2022) Solid medical waste management of hazardous and toxic at UNS hospital Surakarta. J Presipitasi Media Komun dan Pengemb Tek Lingkung 19:398–407. https://doi.org/10.14710/presipitasi.v19i2.398-407

Hettiarachchi H, Meegoda JN (2023) Microplastic pollution prevention: the need for robust policy interventions to close the loopholes in current waste management practices. Int J Environ Res Public Health 20:6434. https://doi.org/10.3390/ijerph20146434

Iqbal A, Yasar A, Nizami AS et al (2024) Empirical analysis of cost-effective and equitable solid waste management systems: environmental and economic perspectives. Environ Res 244:117858. https://doi.org/10.1016/j.envres.2023.117858

Iyer VG (2024) Importance of strategic environmental assessment (sea) process and environmental health impact assessment (EHIA) process towards sustainable climate change and control. J Intell Comput Math. https://doi.org/10.55571/jicm.2024.01028

Jagun ZT, Daud D, Ajayi OM et al (2023) Waste management practices in developing countries: a socio-economic perspective. Environ Sci Pollut Res 30:116644–116655. https://doi.org/10.1007/s11356-022-21990-5

Jena MC, Mishra SK, Moharana HS (2023) Challenges and the way forward for management and handling of hazardous waste. Glob Environ Eng 10:13–17. https://doi.org/10.15377/2410-3624.2023.10.2

Jha R, Dwivedi S, Modhera B (2021) Measurement and practices for hazardous waste management. Hazard Waste Manag Overv Adv Cost-Effective Solut 89–115. https://doi.org/10.1016/B978-0-12-824344-2.00011-2

Kala K, Bolia NB (2024) Empowering the informal sector in urban waste management: towards a comprehensive waste management policy for India. Environ Dev 49:100968. https://doi.org/10.1016/j.envdev.2024.100968

Karthikeyan L, Suresh VM, Krishnan V et al (2018) The management of hazardous solid waste in India: an overview. Environ MDPI 5:1–10. https://doi.org/10.3390/environments5090103

Khalmurzayevna YS, Nuritdinovich FS, Karimovich SS (2023) Risk assessment of the internal control system and the application of auditing procedures in audit of environmental costs. Int J Manag Econ Fundam 03:15–27. https://doi.org/10.37547/ijmef/volume03issue05-03

Khan SA (2016) E-products, E-waste and the Basel convention: Regulatory challenges and impossibilities of international environmental law. Rev Eur Comp Int Environ Law 25:248–260. https://doi.org/10.1111/reel.12163

Khaskheli MB, Wang S, Yan X, He Y (2023) Innovation of the social security, legal risks, sustainable management practices and employee environmental awareness in the China-Pakistan economic corridor. Sustainability 15:1021. https://doi.org/10.3390/su15021021

Khoshnevisan B, Duan N, Tsapekos P et al (2021) A critical review on livestock manure biorefinery technologies: sustainability, challenges, and future perspectives. Renew Sustain Energy Rev 135. https://doi.org/10.1016/j.rser.2020.110033

Kofoworola OF (2007) Recovery and recycling practices in municipal solid waste management in Lagos, Nigeria. Waste Manage 27:1139–1143. https://doi.org/10.1016/j.wasman.2006.05.006

Kokkinos K, Lakioti E, Moustakas K et al (2023) Sustainable medical waste management using an intuitionistic fuzzy-based decision support system. Sustainability 16:298. https://doi.org/10.3390/su16010298

Kumar A, Thakur AK, Gaurav GK et al (2023) A critical review on sustainable hazardous waste management strategies: a step towards a circular economy. Environ Sci Pollut Res 30:105030–105055. https://doi.org/10.1007/s11356-023-29511-8

Li Y (2023) Deviation of policy implementation from the perspective of executive subject: a case study of environmental protection policy in Bijie city. Highlights Bus Econ Manage 8:501–508. https://doi.org/10.54097/hbem.v8i.7261

Lud D, Schwemm A, Kalandadze B et al (2022) Pesticide handling and waste management: a case study on DDT and HCHs from the Southern Caucasus. SN Appl Sci 4. https://doi.org/10.1007/s42452-022-04999-w

Mmereki D, Baldwin A, Li B, Liu M (2017) Healthcare waste management in Botswana: storage, collection, treatment and disposal system. J Mater Cycles Waste Manage 19:351–365. https://doi.org/10.1007/s10163-015-0429-0

Moher D, Liberati A, Tetzlaff J, Altman DG (2009) Preferred reporting items for systematic reviews and meta-analyses: the PRISMA statement. PLoS Med 6:e1000097. https://doi.org/10.1371/journal.pmed.1000097

Mor S, Ravindra K (2023) Municipal solid waste landfills in lower- and middle-income countries: environmental impacts, challenges and sustainable management practices. Process Saf Environ Prot 174:510–530. https://doi.org/10.1016/j.psep.2023.04.014

Naqvi A, Farooq S (2024) Investigating the role of perceived convenience and perceived privacy risks on E-waste disposal behavior among smartphone users in Pakistan. J Hazard Toxic Radioact Waste 28. https://doi.org/10.1061/JHTRBP.HZENG-1293

Navia R, Bezama A (2008) Hazardous waste management in Chilean main industry: an overview. J Hazard Mater 158:177–184. https://doi.org/10.1016/j.jhazmat.2008.01.071

Okoye CC, Ofodile OC, Nifise AOA et al (2024) Climate risk assessment in petroleum operations: a review of CSR practices for sustainable resilience in the United States and Africa. GSC Adv Res Rev 18:234–245. https://doi.org/10.30574/gscarr.2024.18.2.0060

Page MJ, McKenzie JE, Bossuyt PM et al (2021) The PRISMA 2020 statement: an updated guideline for reporting systematic reviews. BMJ 372:n71. https://doi.org/10.1136/bmj.n71

Pandey P, Dhiman M, Kansal A, Subudhi SP (2023) Plastic waste management for sustainable environment: techniques and approaches. Waste Dispos Sustain Energy 5:205–222. https://doi.org/10.1007/s42768-023-00134-6

Perkumienė D, Atalay A, Safaa L, Grigienė J (2023) Sustainable waste management for clean and safe environments in the recreation and tourism sector: a case study of Lithuania, Turkey and Morocco. Recycling 8. https://doi.org/10.3390/recycling8040056

Ramachandra TV, Bachamanda S (2007) Environmental audit of municipal solid waste management. Int J Environ Technol Manag 7:369–391. https://doi.org/10.1504/IJETM.2007.015152

Reinikainen J, Sorvari J (2016) Promoting justified risk-based decisions in contaminated land management. Sci Total Environ 563–564:783–795. https://doi.org/10.1016/j.scitotenv.2015.12.074

Salsabila L, Teovani Lodan K, Khairina E (2023) Public engagement impact on sustainable waste management in Indonesia: examining public behavior. J Adm Publik Public Adm J 13:158–178. https://doi.org/10.31289/jap.v13i2.10391

Scruggs CE, Nimpuno N, Moore RBB (2016) Improving information flow on chemicals in electronic products and E-waste to minimize negative consequences for health and the environment. Resour Conserv Recycl 113:149–164. https://doi.org/10.1016/j.resconrec.2016.06.009

Risk assessment of industrial waste: case of an Algerian company. Case Stud Environ 7. https://doi.org/10.1525/cse.2023.2001786

Shabani T, Jerie S (2023) A review of the applicability of environmental management systems in waste management in the medical sector of Zimbabwe. Environ Monit Assess 195:789. https://doi.org/10.1007/s10661-023-11339-x

Shabani T, Jerie S, Shabani T (2023) Applicability of the life cycle assessment model in solid waste management in Zimbabwe. Circ Econ Sustain 3:2233–2253. https://doi.org/10.1007/s43615-023-00268-z

Simpa P, Solomon NO, Adenekan OA, Obasi SC (2024) Innovative waste management approaches in LNG operations: a detailed review. Eng Sci Technol J 5:1711–1731. https://doi.org/10.51594/estj.v5i5.1138

Sinha P, Julius S, Fry M et al (2024) Assessing community vulnerability to extreme events in the presence of contaminated sites and waste management facilities: an indicator approach. Urban Clim 53:101800. https://doi.org/10.1016/j.uclim.2023.101800

Thapa K, Vermeulen WJV, Deutz P, Olayide OE (2023) Transboundary movement of waste review: from binary towards a contextual framing. Waste Manage Res 41:52–67. https://doi.org/10.1177/0734242X221105424

Vasudevan D (2017) Hazardous waste management in India: current scenario and future opportunities. IJSRST 3:118–127

Vyas S, Prajapati P, Shah AV, Varjani S (2022) Municipal solid waste management: dynamics, risk assessment, ecological influence, advancements, constraints and perspectives. Sci Total Environ 814:152802. https://doi.org/10.1016/j.scitotenv.2021.152802

Watts BA, Palmeira Zago VC, Gopakumar L et al (2023) Uncharted risk measures for the management of sustainable mining. Integr Environ Assess Manage 19:949–960. https://doi.org/10.1002/ieam.4769

Wu HH (2022) A study on transnational regulatory governance for marine plastic debris: trends, challenges, and prospect. Mar Policy 136:103988. https://doi.org/10.1016/j.marpol.2020.103988

Antibiotic Residues in Hospital Effluents: Impacts and Mitigation

Sayoojya P. Sathyan, K. P. Prathish, and Akshay D. Shende

Abstract

A large quantity of antibiotics and pharmaceutical compounds are released from the hospitals into the aquatic environment, and the concentration of pharmaceutical residues in hospital effluents surpasses those in domestic wastewater treatment plant effluents. However, these hospital effluents are commonly regarded as having a similar contaminant nature as urban wastewater. Hence, they are generally released into municipal sewage systems, ultimately reaching surface water bodies, posing risks to water quality, aquatic life, and human health. Repeated exposure to even minor quantities of antibiotics or their active metabolites results in the development and persistence of antibiotic-resistant strains of bacteria that pose a significant global challenge in managing infectious diseases, resulting in rising morbidity and mortality rates. Therefore, effective treatment techniques are mandatory to remove antibiotic residues from hospital effluents before their release into the aquatic environment. This chapter discusses the most commonly detected antibiotics and their concentration ranges in hospital effluents around the world and also explores the factors influencing these concentrations. The frequently detected antibiotics in hospital effluents are found as ciprofloxacin ($C_{17}H_{18}FN_3O_3$), sulfamethoxazole ($C_{10}H_{11}N_3O_3S$), tinidazole ($C_8H_{13}N_3O_4S$), trimethoprim ($C_{14}H_{18}N_4O_3$), levofloxacin ($C_{18}H_{20}FN_3O_4$), ofloxacin ($C_{18}H_{20}FN_3O_4$), and norfloxacin ($C_{16}H_{18}FN_3O_3$). The various harmful effects caused by the constant release and exposure of antibiotic residues on humans and the aquatic ecosystem are presented in length. Additionally, detailed discussions are provided on currently adopted treatment methods for the elimination of antibiotics from hospital effluents, such as adsorption, membrane techniques, advanced oxidation processes, and biological methods. The efficiencies of these methods vary according to the type of antibiotic as well as their initial concentrations in the wastewater. A comparison of these treatment methods is also included, based on their positives and negatives.

Keywords

Antibiotic residues · Hospital effluents · Eco-toxicological impacts · Antibiotic resistance · Wastewater treatment

1 Introduction

The invention of antibiotics for medical treatment indisputably stood out as the most impactful medical breakthrough of the twentieth century (Wang et al. 2023). Salvarsan, the first antibiotic, was introduced in 1910. Over the span of slightly more than a century, antibiotics have significantly revolutionized modern medicine, increasing the average human life expectancy by 23 years. In 1928, the discovery of penicillin heralded the beginning of a golden era in the exploration of antibiotics, reaching its zenith in the mid-1950s (Hutchings et al. 2019). In addition to addressing transmittable diseases, antibiotics have paved the way for numerous contemporary medical procedures, such as cancer treatment, organ transplants, and open-heart surgery (Truman 2019). There are different classes of antibiotics based on their chemical structures, namely macrolides (erythromycin, roxithromycin, azithromycin, and clarithromycin), β-lactams (penicillins, cephalosporins, etc.), fluoroquinolones (ciprofloxacin,

S. P. Sathyan · K. P. Prathish · A. D. Shende (✉)
Environmental Technology Division, CSIR-National Institute for Interdisciplinary Science and Technology (CSIR-NIIST), Thiruvananthapuram 695019, Kerala, India
e-mail: ad.shende@niist.res.in

Academy of Scientific and Innovative Research (AcSIR), Ghaziabad 201002, India

A. Mandpe et al. (eds.), *Sustainable Waste Management*, Advances in Science, Technology & Innovation,
https://doi.org/10.1007/978-3-031-85685-3_10

levofloxacin, norfloxacin, ofloxacin, etc.), tetracyclines (doxycycline, minocycline, etc.), and sulfonamides (sulfadiazine, sulfamethoxazole, etc.) (Zhang and Li 2011). They have been extensively employed for treating and preventing human diseases, and also for safeguarding animal health and improving animal growth in livestock and aquaculture industries (Yang et al. 2021).

Nowadays, the global consumption of antibiotics and other pharmaceutical products has increased rapidly, driven by the growing global population, advancements in research and development, and enhanced availability of healthcare products (Adeleye et al. 2022). Between 2000 and 2018, the global antibiotic consumption rate increased by 46% as per the reports from 204 countries (Martinez-Guerra et al. 2021). In 2015, the leading consumers of antibiotics among high-income countries were the USA, France, and Italy, while among developing countries, the leading consumers were India, China, and Pakistan. Projections anticipate global antibiotic consumption to surge by up to 200% by 2030, compared to the estimated 42 billion defined daily doses (DDDs) in 2015, assuming no policy changes (Larsson 2014). Additionally, the use of veterinary antibiotics is projected to reach 105,596 tons (67% rise) by 2030, driven by increasing demand for livestock products (Yang et al. 2021; Larsson 2014). However, with this continuous increase in the global antibiotic production and consumption rates, the antibiotic residues at different phases of their life cycle, are being continuously discharged into the environment, which paves the way for bacteria to get exposed to antibiotics continuously and thereby the development of antibiotic resistance within them. The overuse or inappropriate use of antibiotics is a major driver of antibiotic resistance globally (Nandi et al. 2023). Approximately, half of the antibiotics prescribed for humans are considered unnecessary. In agriculture and aquaculture, this figure is even higher, since antibiotics are often used as growth enhancers and substitutes for proper hygiene (Co-operation and f. E., Development. 2020). The clinical challenge of distinguishing bacterial infections from viral ones often leads to unnecessary antibiotic use (Klein et al. 2018). Despite fewer than 10% of COVID-19 patients worldwide being diagnosed with secondary bacterial infections necessitating antibiotics, approximately 75% of patients were prescribed antibiotics (Nandi et al. 2023).

Antibiotic residues have been detected in groundwater, surface water, and seawater over the past 20–30 years and even in the drinking water supplies recently. Antibiotic compounds are degradable in nature, although continuous emissions to the environment can lead to their constant environmental presence; so they are considered pseudo-persistent (Patel et al. 2019). The ubiquitous occurrence of antibiotic residues at elevated levels in sediments, groundwater, surface water, and living organisms around the globe has designated these substances as 'contaminants of emerging concern (CEC)'. Antibiotics represent one of the major groups of contaminants that come under the class of 'pharmaceuticals and personal care products (PPCPs)'. The concentration of antibiotics reported in the aqueous media ranges from ng/L to μg/L and those in the solid media (sediments and living organisms) ranges from ng/g to μg/g (Maghsodian et al. 2022). Even at low concentrations, continuous exposure to antibiotics can lead to unexpected consequences that may threaten both human health and the environment. They are toxic, can bioaccumulate, and thereby disrupt aquatic flora and fauna. However, antibiotics are not monitored routinely and, currently, there are no regulations for this class of organic contaminants (Dhodapkar and Gandhi 2019). The EU Directives concerning groundwater and drinking water (98/83/EC amending 2006/118/EC and 91/271/EEC, respectively) currently do not incorporate standards for antibiotics. Moreover, the existing legislation does not establish restrictions on concentrations of antibiotic residues in the wastewater treatment plant effluents released into receiving waters, nor does it set guidelines for the application of biosolids in agricultural fields (Carvalho and Santos 2016).

This chapter reviews the current knowledge on the sources and occurrence of antibiotic residues in different aquatic matrices, with particular emphasis on their contamination levels in hospital effluents and the possible environmental and human health risks associated with them. A comparative analysis is presented on the frequently detected antibiotics and their concentrations in hospital effluents in different regions of the world by collecting data from relevant studies. Furthermore, the chapter delves into existing wastewater treatment technologies, namely biological treatment techniques, membrane techniques, adsorption, advanced oxidation processes, evaluating and comparing their effectiveness, and thereby concluding the most suitable treatment technology for antibiotic removal from hospital effluents.

2 Methodology

Google Scholar and the Web of Science database were used to find relevant articles published between 2003 and 2023, that are related to the current study. In the preliminary search, the keywords used were–antibiotics, antibiotic contamination, hospital effluents, health risks, and treatment technologies. For the comparison of antibiotic contamination in hospital effluents of India with other countries in Asia and Europe, the reported concentrations of available antibiotics in each study was compiled, and the box and whisker plots were drawn using Microsoft Excel software. Similarly, the efficiencies of different treatment technologies in removing antibiotics, in terms of their initial concentrations in the wastewater, were compared by collecting these data from different studies. The graphs for ciprofloxacin, trimethoprim, sulfamethoxazole, clarithromycin, and erythromycin-H_2O were then plotted using Microsoft Excel Software.

3 Sources of Antibiotics in the Aquatic Environment

Man-made antibiotics can get into the environment via several routes. The key sources of antibiotic residues into the environment include antibiotic production facilities/industries, hospitals, domestic wastewater, research activities utilizing target compounds, the disposal of expired medicines or compounds into the environment, etc (Chtourou 2018). Over the past 20 years, the development and advancement of mass spectrometric techniques have enabled the detailed study of the movement of antibiotic residues in different complex environmental matrices (Larsson 2014).

Figure 1 shows the different sources from which antibiotic residues are released into the water bodies. A significant portion of the antibiotics consumed by both humans and animals remains unmetabolized and gets released into the environment through urine and feces (Diwan et al. 2009). Some portions (5–90%) of antibiotics are eliminated from the body as parental form itself, depending upon the chemical properties of the compound (Aydin et al. 2019). Antibiotics, in the form of parent compounds, metabolites, or derivative conjugates are constantly released into wastewater via urine, feces, and wash-off (Oluwole et al. 2020). The domestic sewage, hospital effluents, and the wastewater from antibiotic manufacturing facilities are released to the water bodies either directly or after treated in the wastewater treatment plants (WWTPs) and sewage treatment plants (STPs). However, in WWTPs/ STPs, the antibiotic residues are not completely removed since these plants are not designed to degrade them. Therefore, these WWTPs/ STPs represent a major route through which antibiotics are discharged into the surface waters. In the end, the polluted surface waters can make their way into drinking water treatment plants (DWTPs), which also lack the advanced facilities to eliminate these compounds, thus entering the

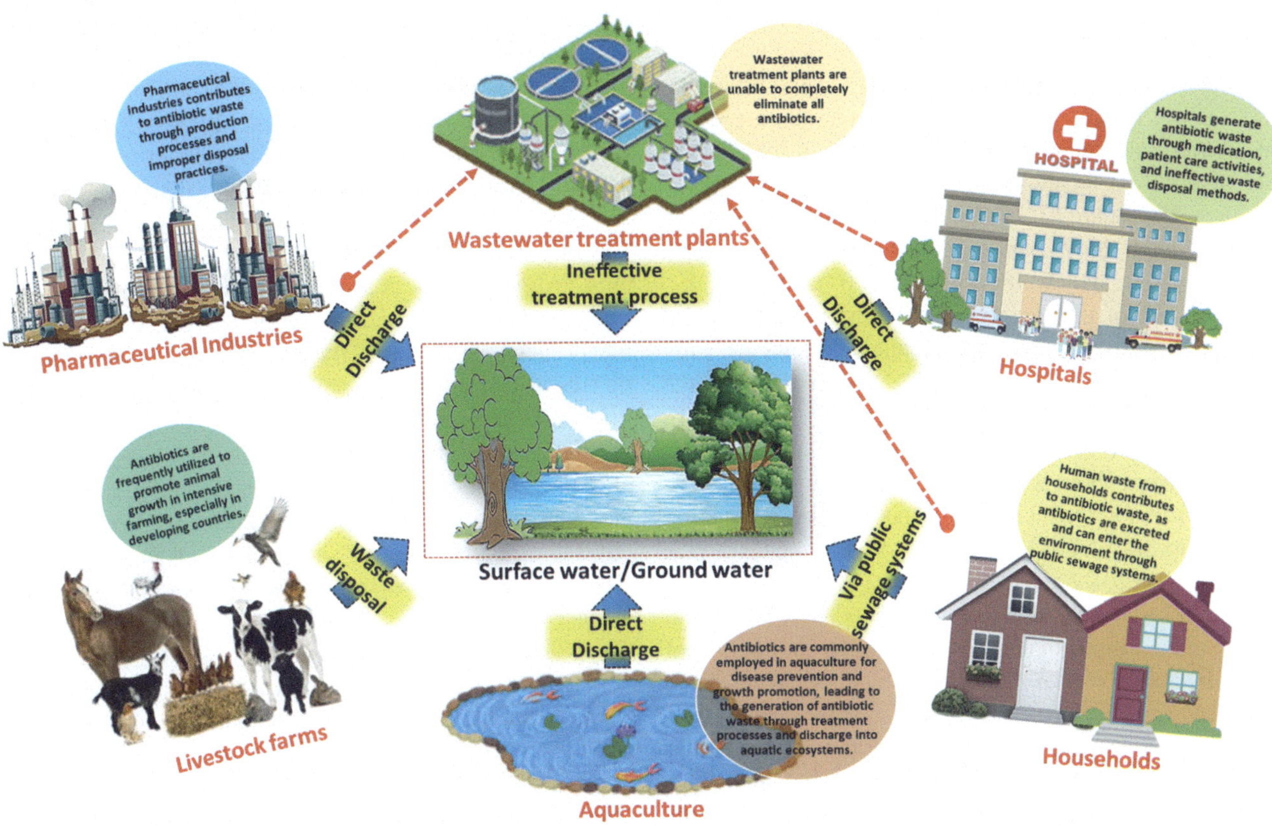

Fig. 1 Various sources through which antibiotics are released into aquatic environment

water supply systems (Homem and Santos 2011). Other significant sources of antibiotics into the water bodies include aquaculture and waste disposal from livestock farming. Further, the antibiotic residues may reach the groundwater by leaching or bank filtration (Oluwole et al. 2020). The accidental spills during the production and distribution of antibiotics can also be viewed as major sources of contamination (Homem and Santos 2011). In essence, antibiotic residues from various sources end up in different water bodies.

3.1 Reports on the Occurrence of Antibiotic Residues in the Aquatic Environment in Different Regions of the World

Several monitoring studies reported the presence of antibiotics in different aquatic matrices in various parts of the world. One of the initial reports on contamination by antibiotics in water was in the River Lee, UK by Richardson and Bowron in 1985, where Erythromycin, Sulfamethoxazole, and tetracycline were detected below 1 μg/L (Richardson and Bowron 1985). Subsequently, several authors reported the contamination in different rivers of the UK by various antibiotics (Roberts and Thomas 2006; Hilton and Thomas 2003). Among the various PPCPs detected in the river downstream of a WWTP in South Africa, sulfamethoxazole (>1 μg/L), clarithromycin (0.2 μg/L), and trimethoprim (0.89 μg/L) were found at high concentrations (Archer et al. 2017). Sulfamethoxazole was detected in the groundwater of different locations in the USA (Miller and Meek 2006; Schaider et al. 2014) and surface waters of Vietnam (Managaki et al. 2007; Ngo et al. 2021), Japan (Managaki et al. 2007), Australia (Watkinson et al. 2009), etc.

India is one of the top countries in the production and consumption of pharmaceutical chemicals. However, only limited works are there to account for their presence in the Indian water bodies (Balakrishna et al. 2017). Pattancheru Enviro Tech Ltd (PETL) collects about 1500 m^3 of wastewater daily from about 90 drug manufacturing facilities situated in Hyderabad, which is a hotspot of drug production in India. Ciprofloxacin (31 mg/L), enrofloxacin (0.9 mg/L), and norfloxacin (0.42 mg/L) were found as the major contaminants present in the plant influents (Larsson et al. 2007). Studies show that antibiotic contamination is not only limited to industrial effluents but also spread to rivers, lakes, and groundwater nearby. The evaluation of contamination of groundwater, drinking water, and surface water by pharmaceutical manufacturing facilities in Pattancheru identified the presence of elevated concentrations of fluoroquinolones, such as ofloxacin, norfloxacin, enoxacin, enrofloxacin, and ciprofloxacin (Fick et al. 2009). The levels of ciprofloxacin, amoxicillin, sulfamethoxazole, ofloxacin, and norfloxacin detected in Indian WWTPs were about 40 times greater than those detected in WWTPs across other countries in Asia, Europe, North America, and Australia (Balakrishna et al. 2017). A global study on pharmaceutical pollution in rivers identified the antibiotics sulfamethoxazole and metronidazole among the contaminants with the highest concentrations. Also, concentrations of 9 out of the 13 detected antibiotics exceeded the proposed safe concentrations to protect against resistance selection (Tell et al. 2019; Bengtsson-Palme and Larsson 2016) in at least one sampling site, with ciprofloxacin being the major one (Wilkinson et al. 2022) (Table 1).

4 Antibiotics in the Hospital Effluents

Hospital effluents encompass all the wastewater generated from various activities within the hospital. This includes all medical activities, such as operating, emergency and first aid, laboratory, radiology, and diagnosis, as well as non-medical activities including kitchen and laundry. The major harmful pollutants present in these effluents consist of pathogenic microorganisms, such as bacteria and viruses, residual pharmaceuticals especially antibiotics, and residual chemicals from the laboratory (Bansal 2019). The concentration of pharmaceutical residues in hospital effluents surpasses those in domestic WWTP effluents as per the reports by Al-Maadheed et al. 2019 and Balakrishna et al. 2017. The comparison of levels of antibiotics in effluents from hospitals and WWTPs that treat domestic sewage in Qatar revealed that six out of seven antibiotics analyzed were found in higher levels in the hospital effluents (Al-Maadheed et al. 2019). Also, despite comparatively minor concentrations (ng/L) of pharmaceuticals in the effluents of Indian WWTPs that treat mainly domestic sewage, considerably higher concentrations (μg/L to mg/L) of the same were identified in hospital effluents in India (Balakrishna et al. 2017). However, these hospital effluents are commonly regarded as having a similar contaminant nature as urban wastewater. Hence, they are generally released in municipal sewage systems (Verlicchi et al. 2015), thereby representing an important contributor to the release of antibiotic residues and other pharmaceuticals into the environment (Diwan et al. 2009).

The amount of water consumed in a hospital is dependent on various factors including the type of medical services provided, the number of beds, personnel, and wards, as well as the societal, cultural, and economic context. The hospitals were found to produce wastewater ranging from 398 to 1090 L per day per occupied bed in a study conducted in Iran (Mesdaghinia et al. 2009). A study carried out in 75

Table 1 Contamination levels of various antibiotics in different aquatic matrices around globe

SI. No.	Region	Matrix	Antibiotic	Concentration in ng/L	Reference
1	River Lee, London	River water	Erythromycin, sulfamethoxazole, tetracycline	<1000	(Richardson and Bowron 1985)
2	Tyne river, UK	River water	Erythromycin	70	(Roberts and Thomas 2006)
			Trimethoprim	19	
3	UK	River water	Erythromycin	1000	(Hilton and Thomas 2003)
			Sulfamethoxazole	50	
			Trimethoprim	39	
			Acetylsulfamethoxazole	240	
4	Montana, USA	Groundwater	Sulfamethoxazole	490	(Miller and Meek 2006)
5	Cape Cod, Massachusetts, USA	Groundwater	Sulfamethoxazole	113	(Schaider et al. 2014)
6	Mekong Delta, Vietnam	River water	Sulfamethazine, erythromycin, trimethoprim, and sulfamethoxazole	360	(Managaki et al. 2007)
7	River Tamagawa, Japan	River water	Sulfamethazine, erythromycin, trimethoprim, and sulfamethoxazole	448	(Managaki et al. 2007)
8	Pattancheru, Hyderabad, India	WWTP effluent	Ciprofloxacin	31,000,000	(Larsson et al. 2007)
			Enrofloxacin	900,000	
			Norfloxacin	420,000	
			Lomefloxacin	300,000	
			Enoxacin	300,000	
			Ofloxacin	160,000	
9	River Mankyung, South Korea	River water	Clarithromycin	443	(Kim et al. 2009)
			Erythromycin	137	
10	Oak Creek, WI, USA	WWTP influent	Sulfamethoxazole	7400	(Blair et al. 2015)
			Sulfadiazine	20	
			Trimethoprim	570	
			Ofloxacin	2100	
			Ciprofloxacin	2200	
11	Hong Kong, China	STP influent	Cefalexin	5640	(Leung et al. 2012)
			Ofloxacin	7900	
			Erythromycin	4740	
12	Volos, Greece	WWTP influent	Sulfamethoxazole	507	(Papageorgiou et al. 2016)
			Ampicillin	1805	
13	Spain	DWTP influent	Sulfamethoxazole	149	(Boleda et al. 2011)
14	Gauteng Province of South Africa	WWTP influent	Sulfamethoxazole	1013	(Archer et al. 2017)
			Trimethoprim	899	
15	Yamuna River, India	River water	Ampicillin	13,800	(Mutiyar and Mittal 2014)
			Ciprofloxacin	1400	
			Gatifloxacin	480	
			Sparfloxacin	2100	
			Cefuroxime	1700	
16	Vietnam	River Water	Sulfamethoxazole	21.63	(Ngo et al. 2021)

(continued)

Table 1 (continued)

SI. No.	Region	Matrix	Antibiotic	Concentration in ng/L	Reference
17	Australia	Surface water	Sulfamethoxazole	2000	(Watkinson et al. 2009)
		WWTP influent	Sulfamethoxazole	3000	
		WWTP influent	Sulfamethoxazole	200,000	
18	Central India	WWTP influent	Levofloxacin	107,900	(Mohapatra et al. 2016)
			Levofloxacin	86,700	
			Norfluoxacin	18,200	
			Azithromycin	176,900	
			Azithromycin	29,300	
19	Hyderabad, India	Groundwater	Ciprofloxacin	14,000	(Fick et al. 2009)
			Enoxacin	1,900	
			Trimethoprim	55	
			Norfloxacin	31	
			Ofloxacin	480	
20	Hyderabad, India	Lake (Surface water)	Ciprofloxacin	2,800,000	(Fick et al. 2009)
			Enoxacin	160,000	
			Enrofloxacin	25,000	
			Norfloxacin	520,000	
			Ofloxacin	11,000	

hospitals in Delhi, India, encompassing 25 from each of the low, medium, and high categories, identified the average water usage per bed per day ranges from 500 to 600 L (Dhingra et al. 2022). The varieties and contamination levels of antibiotics in hospital effluents vary according to antibiotic consumption patterns, types of healthcare facilities, bed density, excretion rates, number of daily patients, cultural and geographical factors, and more (Khan et al. 2020). The utilization of pharmaceuticals in hospitals can fluctuate throughout the year and differ from one country to another, influenced by factors, such as prevalent diseases, hospital activity, and the local list of recommended medications for treating various diseases (Santos et al. 2013). A case study from Germany showed that pharmaceutical usage trends vary across various healthcare facilities. For instance, general hospitals may exhibit a range of up to 1263 kg/year, with an average annual individual consumption peaking at 1000 g/bed (Khan et al. 2020). Another study from Delhi, India revealed that 39% of patients visiting pharmacies / enrolled public facilities and about 43% of patients visiting private clinics were prescribed a minimum of one antibiotic. This variable consumption pattern brings about uneven discharge trends (Diwan et al. 2009).

Although numerous studies have explored the contamination levels of antibiotics in various aquatic and solid matrices, only a limited number have specifically addressed their concentrations in hospital effluents. In India, only four studies reported quantitative data on antibiotic contamination in hospital wastewater (Diwan et al. 2009, 2013; Verma et al. 2017; Akiba et al. 2015). A study conducted to identify the occurrence of antibiotics in the effluents of two hospitals in Ujjain district, Madhya Pradesh, identified metronidazole, ofloxacin, norfloxacin, ceftriaxone, sulfamethoxazole, ciprofloxacin, tinidazole, and levofloxacin as major contaminants, with concentrations in the range of 1.4–236.6 μg/L among which the highest concentration was observed for ciprofloxacin. However, antibiotics, such as amoxicillin, erythromycin, and doxycycline, were found absent (Diwan et al. 2009). Later in 2013, the same research team evaluated the seasonal and temporal variation in the antibiotic release from hospital effluents using continuous and grab sampling. The study identified that the greatest quantities of fluoroquinolone antibiotics were discharged in the winter season than in the rainy and summer season. Also, no temporal pattern in antibiotic release was detected (Diwan et al. 2013). A comparative analysis of effluents from 3 sewage treatment plants (STPs), the first one that treats only domestic wastewater, the second one that treats both hospital and domestic wastewater, and the third one that treats only hospital wastewater, was done by Akiba et al. 2015. They identified the presence of four antibiotics, namely trimethoprim, chloramphenicol, sulfamethoxazole, and ofloxacin. The results revealed that the concentrations of antibiotics were higher in the second and third STPs than in the first STP, which treated only the domestic wastewater. Also, chloramphenicol, trimethoprim, and sulfamethoxazole were found absent in the first STP. The study confirmed the role of hospitals in contributing to the antibiotic load of the STPs (Akiba et al. 2015).

Very high levels of antibiotic contamination in 3 hospital effluents in Kota, Rajasthan, was reported by Verma et al. in 2017, where 3 out of the 7 selected compounds were in the mg/L range. The highest concentration was observed for cefadroxil (3.24 mg/L), followed by ciprofloxacin (1.35 mg/L), ampicillin (1.24 mg/L), cloxacillin (0.31 mg/L), ofloxacin (0.28 mg/L), cefpodoxime (0.28 mg/L), and tinidazole (0.26 mg/L) (Verma et al. 2017). These levels were much higher than those reported in hospital effluents of various European and Asian countries (Santos et al. 2013; Yao et al. 2021; Ory et al. 2016). Lagishetty and Nagrajan (2015) carried out a qualitative analysis of effluents from hospitals of government teaching, government super specialty, private teaching, private super specialty, and local clinics in South India to identify the presence of antibiotic residues. Ciprofloxacin, ampicillin, enrofloxacin, oxytetracycline, and trimethoprim were present in almost all the samples (Lagishetty and Nagrajan 2015). In general, the maximum concentration of ciprofloxacin, sulfamethoxazole, ofloxacin, levofloxacin, norfloxacin, and ampicillin are found higher in Indian hospital effluents than in other Asian countries, such as Pakistan (Ashfaq et al. 2016; Ahmad et al. 2012), Vietnam (Lien et al. 2016), China (Wang et al. 2018), Qatar (Al-Maadheed et al. 2019), and Turkey (Aydin et al. 2019).

A comparative study from New Mexico on the contamination levels of antibiotics in the effluents from households, hospitals, dairies, municipal sewage, and the river Rio Grande showed significant variations in contamination levels. Four out of five hospital effluents exhibited high concentrations of antibiotics. Among the 6 antibiotics (ofloxacin, ciprofloxacin, sulfamethoxazole, trimethoprim, lincomycin, and penicillin G) detected, ofloxacin was the prominent one, whose maximum concentration (35.5 μg/L) in hospital effluents was 35 times higher than when compared with municipal wastewater (Brown et al. 2006).

When analyzing the hospital effluent that was fed to a membrane bioreactor (MBR) in a hospital in Switzerland, antibiotics were detected along with various other pharmaceutical compounds. However, these antibiotics were in comparatively lower concentrations (azithromycin; 0.139 μg/L, erythromycin; 0.188 μg/L, and clindamycin; 0.983 μg/L) with exceptions for ciprofloxacin (31.98 μg/L) and clarithromycin (2.555 μg/L) (Kovalova et al. 2012). Verlicchi et al. 2012 identified the presence of 73 pharmaceuticals of 12 different classes in the effluents of 2 different-sized hospitals in Italy; the first one was medium-sized with 300 beds and the second one was large-sized with 900 beds. Out of the 25 antibiotics found, the most predominant compounds were ofloxacin and ciprofloxacin with concentrations of 22 μg/L and 15 μg/L, respectively, in the first hospital, and 31 μg/L and 21 μg/L in the second one. They showed significant seasonal variation, in which the higher concentrations were observed in the winter season with concentrations of 31 μg/L and 21 μg/L, respectively (Verlicchi et al. 2012). High levels of ciprofloxacin (101 μg/L) and metronidazole (90.2 μg/L) were reported in the effluents of hospitals in Sweden. Other detected antibiotics were ofloxacin (7.6 μg/L), sulfamethoxazole (12.8 μg/L), trimethoprim (7.6 μg/L), and doxycycline (6.7 μg/L). Also, large temporal variations in the antibiotic concentrations were observed throughout the sampling period (Lindberg et al. 2004). The levels of tetracycline compounds (minocycline, epitetracycline, tetracycline, and doxycycline) in wastewaters of 4 hospitals in Portugal were in the range of 6–531.7 μg/L. The maximum concentration of minocycline was obtained in the spring but was not detected in any of the hospital samples in the autumn (Pena et al. 2010).

Based on Table 2, the commonly found antibiotics in the hospital effluents of India, other Asian countries, and European countries and their concentrations as box and whisker plots are demonstrated in Fig. 2. The most frequently detected antibiotics in hospital effluents in India were ciprofloxacin, tinidazole, sulfamethoxazole, levofloxacin, ofloxacin, norfloxacin, and metronidazole out of which the highest reported concentration was for ciprofloxacin, which was higher than its corresponding reported values in other Asian countries, Europe, and America. The antibiotics with the highest frequency in all of the studied regions include ciprofloxacin, sulfamethoxazole, metronidazole, and ofloxacin. The average concentrations of 3 (ciprofloxacin, sulfamethoxazole, and ofloxacin) out of these 4 antibiotics were found to be higher in Indian hospital effluents than others. However, the maximum concentration of metronidazole was much higher in Vietnam when compared with that of India. No reports were found that showed the antibiotic concentration in the parts per million (mg/L) range except from India (Verma et al. 2017). The commonly detected antibiotics with the highest contamination levels in European countries were ciprofloxacin, metronidazole, and ofloxacin with maximum concentrations below 100 μg/L. The same antibiotics were the predominant ones in Asian countries (except for India) also, with maximum concentrations up to 300 μg/L.

Table 2 Antibiotics and their detected maximum concentrations in hospital effluents in different countries

SI. No.	Region	Antibiotics	Maximum concentration (μg/L)	Reference
1	Madhya Pradesh, India (2 different hospitals)	Ciprofloxacin	236.6	(Diwan et al. 2009)
		Ciprofloxacin	31	
		Tinidazole	88.4	
		Tinidazole	13.6	
		Sulfamethoxazole	81.1	
		Sulfamethoxazole	5.7	
		Levofloxacin	80.5	
		Levofloxacin	6.8	
		Ofloxacin	73.2	
		Ofloxacin	7.5	
		Ceftriaxone	59.5	
		Norfloxacin	22.8	
		Norfloxacin	5.7	
		Metronidazole	3.8	
		Metronidazole	2.5	
2	Madhya Pradesh, India (2 different hospitals)	Ciprofloxacin	2.93	(Diwan et al. 2013)
		Ciprofloxacin	1.35	
		Levofloxacin	1.92	
		Levofloxacin	0.229	
		Ofloxacin	0.781	
		Ofloxacin	0.227	
		Norfloxacin	0.37	
		Metronidazole	0.119	
		Metronidazole	0.13	
		Sulfamethoxazole	3.39	
3	Rajasthan, India	Cefadroxil	3240	(Verma et al. 2017)
		Ampicillin	1240	
		Cefpodoxime	280	
		Cloxacillin	310	
		Ciprofloxacin	1350	
		Tinidazole	260	
4	South India	Trimethoprim	0.46	(Akiba et al. 2015)
		Sulfamethoxazole	0.05	
		Ofloxacin	0.537	
5	Iran	Imipenem	29.12	(Shokoohi et al. 2020)
		Cefixime	12.4	
		Amoxicillin	7.91	
6	Pakistan	Ofloxacin	39.13	(Ahmad et al. 2012)
		Ofloxacin	35.62	
		Ofloxacin	14.42	

(continued)

Table 2 (continued)

SI. No.	Region	Antibiotics	Maximum concentration (μg/L)	Reference
7	Qatar	Clavulanic acid	41.19	(Al-Maadheed et al. 2019)
		Metronidazole	5.46	
		Amoxicillin	0.79	
		Ciprofloxacin	1.98	
		Tetracycline	0.21	
		Penicillin	0.12	
		Erythromycin	5.17	
8	Turkey	Azithromycin	162.507	(Aydin et al. 2019)
		Erythromycin	0.101	
		Sulfamethoxazole	0.373	
		Trimethoprim	0.273	
		Ciprofloxacin	19.715	
		Clarithromycin	159.732	
9	Vietnam	Trimethoprim	22.4	(Lien et al. 2016)
		Metronidazole	19.9	
		Sulfamethoxazole	25.5	
		Ceftazidime	11	
		Ciprofloxacin	87.3	
		Spiramycin	8.4	
		Ofloxacin	9.7	
		Metronidazole	258.3	
		Sulfamethoxazole	31.4	
		Trimethoprim	8	
		Ciprofloxacin	60.3	
		Ofloxacin	111.1	
		Spiramycin	6.7	
10	Pakistan	Ofloxacin	66	(Ashfaq et al. 2016)
		Ciprofloxacin	18	
		Sparfloxacin	58	
		Gemifloxacin	0.2	
11	China	Amoxicillin	0.84	(Yao et al. 2021)
		Amoxicillin	1.43	
		Ampicillin	0.51	
		Ampicillin	0.67	
		Ampicillin	0.5	
		Ofloxacin	9.23	
		Ofloxacin	25.65	
		Ofloxacin	49.97	
		Norfloxacin	0.46	
		Norfloxacin	0.61	
		Trimethoprim	0.31	
		Trimethoprim	0.5	

(continued)

Table 2 (continued)

Sl. No.	Region	Antibiotics	Maximum concentration (μg/L)	Reference
12	China	Sulfamethoxazole	0.434	(Wang et al. 2018)
		Sulfamethoxazole	0.576	
		Sulfamethoxazole	0.587	
		Ofloxacin	2.596	
		Ofloxacin	2.02	
		Ofloxacin	2.372	
		Norfloxacin	1.879	
		Norfloxacin	1.383	
		Norfloxacin	1.624	
		Ciprofloxacin	1.236	
		Ciprofloxacin	1.223	
		Ciprofloxacin	1.541	
		Tetracycline	1.147	
		Tetracycline	1.487	
		Tetracycline	1.598	
		Erythromycin	0.599	
		Erythromycin	0.404	
		Erythromycin	0.433	
		Trimethoprim	0.345	
		Trimethoprim	0.184	
13	China	Ofloxacin	4.240	(Chang et al. 2010)
		Norfloxacin	1.620	
		Ciprofloxacin	0.136	
		Trimethoprim	0.174	
14	Portugal	Tetracycline	531.7	(Pena et al. 2010)
15	Portugal	Erythromycin	7.545	(Santos et al. 2013)
		Azithromycin	7.351	
		Ofloxacin	24.811	
		Ciprofloxacin	38.689	
		Sulfamethoxazole	8.714	
		Trimethoprim	3.963	
		Metronidazole	12.315	
16	France	Ciprofloxacin	17	(Ory et al. 2016)
17	Norway	Tetracycline	4.178	(Thomas et al. 2007)
		Tetracycline	1.537	
		Oxytetracycline	3.743	
		Oxytetracycline	2.294	
		Trimethoprim	14.993	
		Trimethoprim	11.899	
		Ciprofloxacin	54.049	
		Ciprofloxacin	39.843	
		Sulfamethoxazole	1.375	
		Sulfamethoxazole	4.107	

(continued)

Table 2 (continued)

SI. No.	Region	Antibiotics	Maximum concentration (μg/L)	Reference
18	Switzerland	Ciprofloxacin	31.98	(Kovalova et al. 2012)
		Azithromycin	0.139	
		Clarithromycin	2.555	
		Erythromycin	0.188	
19	Sweden	Ciprofloxacin	101	Lindberg et al. 2004)
		Metronidazole	90.2	
		Sulfamethoxazole	12.8	
		Ofloxacin	7.6	
		Trimethoprim	7.6	
		Doxycycline	6.7	
20	Italy	Ciprofloxacin	26	(Verlicchi et al. 2012)
		Azithromycin	1.04	
		Clarithromycin	14	
		Doxycycline	0.97	
		Erythromycin	0.32	
		Metronidazole	1.64	
		Ofloxacin	37	
		Norfloxacin	0.5	
		Sulfamethoxazole	6.5	
		Trimethoprim	1.8	
		Ofloxacin	31	
		Ciprofloxacin	15	
		Ofloxacin	22	
		Sulfamethoxazole	3.4	
21	Spain	Ofloxacin	10.368	(Gros et al. 2013)
		Ofloxacin	2.978	
		Ciprofloxacin	7.494	
		Ciprofloxacin	5.329	
		Norfloxacin	0.327	
		Amoxicillin	0.218	
		Azithromycin	0.113	
		Clarithromycin	0.113	
		Clarithromycin	0.973	
		Clindamycin	1.465	
		Clindamycin	0.184	
		Lincomycin	0.119	
		Sulfamethoxazole	0.2	
		Trimethoprim	0.216	
		Metronidazole	0.643	
22	Netherlands	Ciprofloxacin	2.706	(Paulus et al. 2019)
		Ciprofloxacin	3.752	
		Sulfamethoxazole	0.367	
		Sulfamethoxazole	0.269	
		Metronidazole	7.5	

(continued)

Table 2 (continued)

Sl. No.	Region	Antibiotics	Maximum concentration (μg/L)	Reference
23	Germany	Clarithromycin	10	(Beier et al. 2010)
		Ciprofloxacin	6	
		Metronidazole	1	
		Moxifloxacin	10	
24	USA	Clarithromycin	1.42	(Oliveira et al. 2015)
		Erythromycin	1.35	
		Sulfamethoxazole	2.17	
		Trimethoprim	1.32	
25	New Mexico	Sulfamethoxazole	2.1	(Brown et al. 2006)
		Trimethoprim	5	
		Ofloxacin	35.5	
		Ciprofloxacin	2	
		Lincomycin	2	
		Penicillin G	5.2	

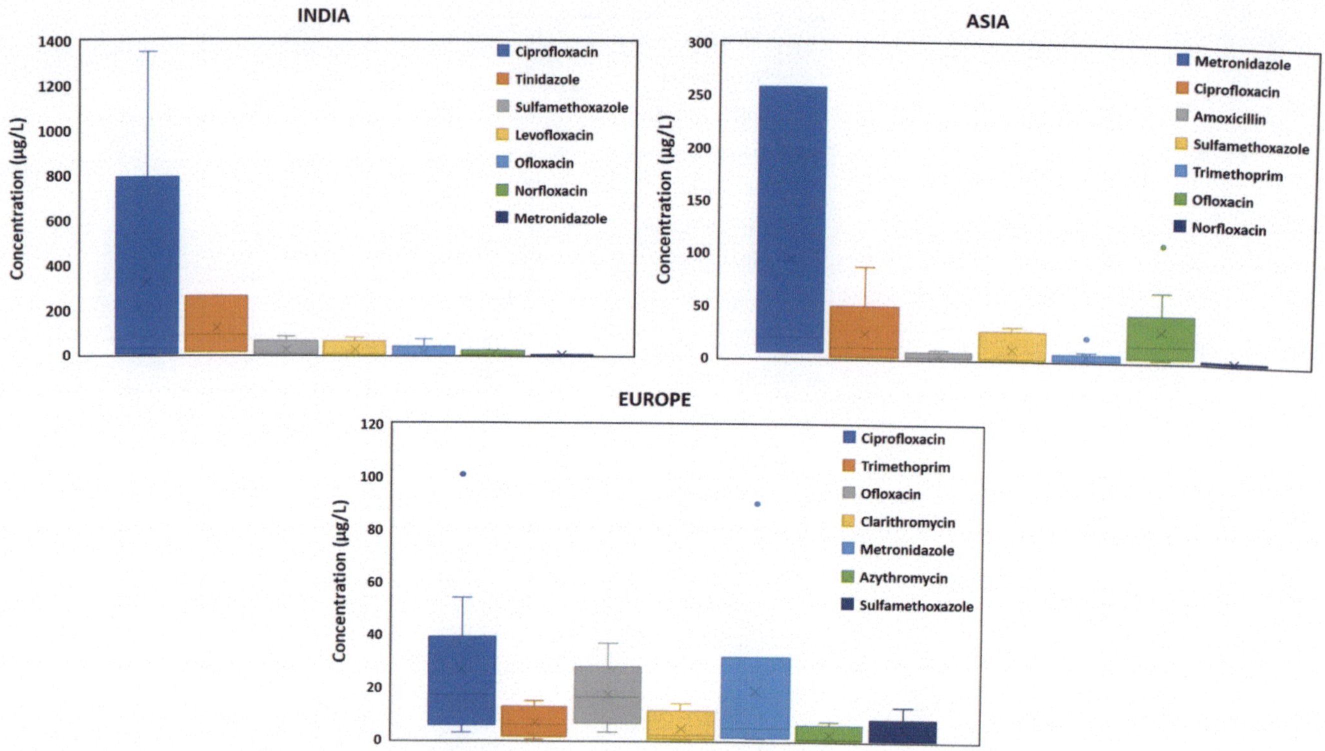

Fig. 2 Box and whisker plot showing most frequently detected antibiotics in India, other Asian countries, and European countries with median, interquartile range, minimum, and maximum values

5 Harmful Impacts of Antibiotics

The discharge of hospital effluents containing antibiotics and other pharmaceutical compounds into water systems poses significant risks to water quality, aquatic life, and human health (Khan et al. 2021) (Fig. 3). The continuous introduction of antibiotic residues into aquatic environments is a problem of concern as it causes long-term adverse effects on both aquatic and terrestrial organisms. They impact non-targeted organisms, such as algae and cyanobacteria which are the primary producers in the aquatic environment by inhibiting chloroplast metabolism including protein synthesis and photosynthesis, thereby affecting cell growth (Halling-Sørensen 2000). Although the contamination levels of individual antibiotics in water matrices are too low to individually affect organisms, their combined effects could lead to substantial toxicity to aquatic species even at levels below the individual no-observed effect concentration (NOECs) (Backhaus et al. 2011). Also, antibiotics, such as sulfamethazine, sulfamethoxazole, sulfamonomethoxine, enrofloxacin, ciprofloxacin, and doxycycline, were found to be bioaccumulative in different body parts of aquatic organisms (Na et al. 2013; Bai et al. 2014; Zhao et al. 2016).

A significant quantity of antibiotic residues reaches humans through drinking water and the food supply chain. They can negatively impact human health through direct toxicity, potentially harming organ systems and leading to illness or even death (Chen et al. 2019). When antibiotic residues enter the human body, they can interact with the microbiota in the gastrointestinal tract. This interaction may lead to an imbalance in intestinal microbiota, promoting the growth of harmful bacteria and opportunistic pathogens. As a result, various disorders may develop including colorectal cancer, pseudomembranous colitis, and other intestinal ailments (Bilal et al. 2020). Additionally, other possible harmful effects of antibiotic residues on humans, viz. digestive dysfunctions, allergic reactions, and chronic toxic effects, resulting from low-level exposure for a prolonged period, etc., have already been demonstrated by several previous studies (Chen et al. 2019).

Fig. 3 Harmful impacts caused by antibiotic residues on aquatic organisms and humans

5.1 Development of Antibiotic Resistance in Bacteria

Repeated exposure to even minor quantities of antibiotics or their active metabolites by the bacterial populations results in the development and persistence of antibiotic-resistant strains of bacteria (Chang et al. 2010). Antibiotic resistance denotes a genetic characteristic found or acquired by bacterial subgroups, enabling them to thrive and multiply despite being exposed to therapeutic doses of antibiotics, which would ordinarily impede or kill these microorganisms. Over time, the bacteria exposed to selective pressure from different antibiotics give rise to multiple antibiotic-resistant pathogenic bacteria (Carvalho and Santos 2016). Antibiotic resistance is driven by a range of resistance genes, which can be passed between different bacterial communities (Prasad et al. 2018).

Hospital effluents harbor a variety of bacteria and foster the incidence of multidrug resistance, facilitated by the antibiotics, disinfectants, and metabolized drugs present in patients' fluids. The comparatively higher levels of antibiotic resistance in hospital effluents than in urban sewage result from the extensive and distinct patterns of antibiotic usage within hospitals, which create an environment where antibiotic-resistant organisms thrive, with little competition from susceptible ones (Prasad et al. 2018). These antibiotic-resistant bacteria (ARB) found in hospital effluents are largely resistant to many treatment processes. Post-treatment disinfection methods, such as UV radiation and sodium hypochlorite, combined with sand filtration, and biological treatment have proven to be less effective in reducing ARB levels (Turolla et al. 2018). Also, the negative consequences of infections triggered by antibiotic-resistant bacteria are greater than those by susceptible ones. Therefore, antibiotic resistance in bacteria poses a significant global challenge in managing infectious diseases, resulting in rising morbidity and mortality rates (O'Neill 2016). Numerous works reported the presence of ARBs in hospital wastewater (Ory et al. 2016; Ahmad et al. 2012; Wang et al. 2018; Prasad et al. 2018; Diwan et al. 2010). However, there is no specific connection found between antibiotic resistance and antibiotic prescription or their residue levels in wastewater (Diwan et al. 2010). A continued rise in antibiotic resistance is expected to cause a reduction in global GDP by 2–3.5%, with a cumulative cost of up to 100 trillion USD (Co-operation and f. E., Development. 2020).

6 Present Treatment Technologies

Hospital effluents are usually directly released into municipal sewage systems and processed along with household wastewater in the municipal WWTPs (Santos et al. 2013). Commonly used treatments in municipal WWTPs typically involve preliminary processes, primary sedimentation, secondary treatment often using a conventional activated sludge process, and finishing techniques, such as filtration and disinfection. Disappointingly, these WWTPs often fall short of achieving significant removal rates for a wide variety of micropollutants, including antibiotics frequently detected in hospital wastewater (Verlicchi et al. 2015). Furthermore, conventional treatments yield notably low efficiencies, even for commonly encountered parameters, such as chemical oxygen demand (COD), biochemical oxygen demand (BOD_5), total coliform, and total suspended solids (TSS) (Mesdaghinia et al. 2009). At times, hospitals use basic primary treatments for their wastewater, like primary sedimentation and pre-chlorination, but these methods are not effective (Martins et al. 2008). In some cases, no treatment is applied, and the direct release of hospital wastewater into surface rivers is commonly practiced, which paves the way for the contamination to spread (Liu et al. 2010). There is not a single treatment that can effectively remove a high percentage of all types of micro-pollutants including antibiotics commonly found in wastewater, as they behave differently during different treatment processes (Verlicchi et al. 2010). Advanced decontamination techniques like advanced oxidation processes, adsorption, and membrane separation are effective in the elimination of various antibiotics and other PPCPs found in wastewater (Yang et al. 2011).

6.1 Biological Treatment Techniques

The biological treatment often uses a 'conventional activated sludge (CAS) process' or a 'membrane bioreactor (MBR)', which integrates the activated sludge method with membrane filtration (Melin et al. 2006). A membrane bioreactor functions much like a CAS system without the requirement for secondary settling and additional processes like sand filtration as it uses either microfiltration (MF) or ultrafiltration (UF) to separate the effluent from activated sludge (Melin et al. 2006). Biodegradation and sorption are the primary elimination mechanisms in biological treatment, both relying on bacteria and the properties of the organic micropollutants (Sahar et al. 2011). The variability in antibiotic removal efficiencies of these biological treatments depends on two main aspects. The first one is associated with environmental conditions, including the initial composition of sewage, 'mixed liquor suspended solids (MLSS) concentration', temperature, microbial community, and 'solids retention time (SRT)' (Sahar et al. 2011). The second one relates to the specific properties of individual compounds, such as their concentration, whether it is hydrophobic or hydrophilic, and other properties (Sahar et al. 2011) The evaluation of the effectiveness of using a CAS system for the biological treatment of hospital effluents in France showed significant removal rates for antibiotics (95.1%). Even with the enhanced removal rates during treatment, antibiotics remained more concentrated in the treated effluent due to their higher initial concentrations, which caused adapted development of the microbial community, resulting in reduced biomass development and decreased bacterial diversity (Chonova et al. 2016). Using the CAS process, in a hospital WWTP operating in Saudi Arabia, more than 86% of clarithromycin, ciprofloxacin, sulfamethoxazole, and N-acetylsulfamethoxazole were removed (Al Qarni et al. 2016). The antibiotic elimination from hospital wastewater at various steps in an advanced WWTP was carried out by Yang et al. 2011. No measurable concentration of tetracycline antibiotic was detected after the CAS process due to the high water solubility of the compound and its affinity toward sludge biosolids. High removal rates were observed through CAS treatment followed by microfiltration for ciprofloxacin (80%) and levofloxacin (71%). Sulfamethoxazole was removed substantially but, erythromycin was found resistant to biological treatment (Yang et al. 2011).

The wastewater from a hospital in Switzerland was fed to a pilot-scale MBR. Out of the twelve antibiotics measured, eleven had elimination rates below 60%, whereas trimethoprim showed a 96% elimination rate (Kovalova et al. 2012). In a comparative study of the efficiency of CAS system combined with a UF module and a MBR system for the elimination of various sulfonamide, macrolide, and trimethoprim antibiotics, the MBR showed superior removal percentages for all the antibiotics tested (15–42% higher) than the CAS system without UF. When the CAS system was added with a UF step, there was a substantial increase in the antibiotic removal efficiencies by about 28%, narrowing the gap between the removal efficiencies of the two systems (Sahar et al. 2011). A rise in concentration was noted for sulfamethoxazole which is excreted as its acetylated form (acetylsulfamethoxazole) and converts back to its mother compound during MBR treatment (Venditti et al. 2011). Several studies have shown that MBRs are about 30% to 55% more effective when compared with the CAS method (Khan et al. 2021). The MBR seems to function as a reliable secondary treatment system for hospital wastewater, consistently delivering high-quality and stable effluent over its operational period, making it compatible with advanced treatment techniques like NF/RO and AOPs (Verlicchi et al. 2010).

Effluents from a hospital in Spain were treated using a batch fluidized bed bioreactor with pellets of fungi *Trametes versicolor*. The antibiotic compounds were eliminated by 86.1% in the first treatment (sterile) and by 90.8% in the second treatment (non-sterile) in which the initial concentrations of all the antibiotics together were 569 μg/L and 239 μg/L, respectively. Ofloxacin (31.9 μg/L) and ciprofloxacin (13.04 μg/L) were the antibiotics with the higher initial concentrations in the hospital effluent. Both antibiotics were eliminated in both treatments, i.e., sterile and non-sterile. However, in the sterile treatment, ciprofloxacin was only removed by 69% even after 8 days. It was found that this treatment not only eliminates antibiotics resistant to conventional methods but also prevents them from accumulating in the sludge. This is crucial because the sludge can later be recycled as compost in agricultural fields, potentially contaminating the environment (Cruz-Morató et al. 2014).

6.2 Adsorption

Adsorption refers to the attachment of particles in the fluid phase to a solid surface (Patel et al. 2019). It is highly effective for the removal of compounds that are uncharged and non-polar (Verlicchi et al. 2015). Numerous adsorbents were tested for the elimination of antibiotic residues from the aqueous media. Activated carbon is widely recognized for its ability to eliminate various organic pollutants and organic carbon compounds from water. They are often used either as powder (powdered activated carbon (PAC)) or as granules (granular activated carbon (GAC)) in packed media filters. GAC with its filtration and adsorption capabilities can be used in traditional filters (Snyder et al. 2007). Hydrophobic antibiotics typically reach equilibrium with activated carbon more quickly than those that are hydrophilic (Patel et al. 2019). After the CAS biological treatment, the hospital effluents were passed through GAC in an advanced WWTP in the USA and it showed effective removal of many antibiotics. Adsorption of individual compounds onto GAC is reliant on the physicochemical characteristics of the compounds; it has a tendency to adsorb organic compounds. Erythromycin was eliminated by 74%, also lincomycin, trimethoprim, ciprofloxacin, and levofloxacin exhibited higher adsorption percentages by GAC. However, this adsorption treatment for sulfamethoxazole removal was found inadequate (Yang et al. 2011).

The pilot-scale removal study of antibiotics by PAC at different dosages; 5 mg and 35 mg exhibited different removal rates for sulfamethoxazole, erythromycin-H_2O, and trimethoprim. 35 mg PAC adsorbent showed removal greater than 85% for sulfamethoxazole and erythromycin-H_2O, and 100% for trimethoprim. Even though the first two compounds showed lower removal rates at lower adsorbent dosages, trimethoprim removal was found to be 72% (Snyder et al. 2007). Kovalova et al. 2013 also showed a positive correlation between the PAC dosage and antibiotic removal percentages. The PAC system fitted with ultrafiltration was found efficient in removing (>97%) antibiotics, such as azithromycin, clarithromycin, ciprofloxacin, norfloxacin, and clindamycin (Kovalova et al. 2013).

6.3 Membrane Techniques

Membrane techniques like reverse osmosis (RO), ultrafiltration (UF), and nanofiltration (NF) are efficient methods for the elimination of CECs present in wastewater (Khan et al. 2021). The removal efficiencies of membrane techniques are determined by the physical characteristics of pollutants, such as charge, polarity, molecular size, solubility, and diffusivity. Additionally, factors like membrane properties and operating conditions also influence the same. Also, NF usually shows a greater removal efficiency when compared with UF membranes in the case of more polar compounds (Khan et al. 2021).

The evaluation of the efficiency of the UF system in removing antibiotics from municipal WWTP secondary effluent does not show any significant removal of tested antibiotics including erythromycin, sulfamethoxazole, and trimethoprim (Snyder et al. 2007). Reverse osmosis is a diffusion process that employs a selective semipermeable membrane to remove ions and molecules using high pressure. The primary advantage of reverse osmosis lies in its ability to achieve a high percentage of removal for pharmaceuticals (Patel et al. 2019). RO was found used in combination with other membrane techniques as well as biological treatment methods. A pilot setup comprising a UF system combined with an RO system was tested for antibiotic removal from secondary effluent. In the study, UF had minimal impact on the target antibiotics, whereas RO effectively reduced the antibiotic concentration (Snyder et al. 2007). The RO system in combination with MBR was found efficient in bringing down the concentrations of sulfamethoxazole, trimethoprim, and erythromycin to a great extent. Microfiltration (MF) combined with RO and electrodialysis reversal system, full-scale MF combined with RO and UV oxidation system, and MF combined with RO system were also found effective in the removal of these antibiotics (Snyder et al. 2007). NF and RO as advanced treatment units of the MBR filtrate were analyzed for eliminating pharmaceutical contaminants from hospital effluents in Germany, out of which only RO guaranteed a complete removal of the pollutants (Beier et al. 2010). On the other hand, microfiltration membranes primarily remove particles and do not significantly reject soluble antibiotics (Yang et al. 2011).

6.4 Advanced Oxidation

Advanced oxidation processes (AOPs) can efficiently oxidize trace organic chemicals, including antibiotics (Patel et al. 2019). As a continuation of the series of CAS biological treatments and GAC adsorption in the advanced WWTP in the USA, the hospital effluents were then subjected to ozonation. More than 80% of trimethoprim and 67 to 94% of sulfamethoxazole were removed using an ozone dose of about 1 mg/L. This high elimination rate can be attributed to the amino groups present in these antibiotics, which are vulnerable to chemical degradation by ozone (Yang et al. 2011). Ozonation has been found to remove>90% of azithromycin, ciprofloxacin, clarithromycin, clindamycin, erythromycin, and sulfamethoxazole from MBR pre-treated hospital effluent. Nevertheless, the efficiency of removing these compounds by post-ozonation with hydrogen peroxide did not show a significant improvement compared to using ozone alone (Kovalova et al. 2013).

The evaluation of heterogeneous photocatalysis (TiO_2/UV), ozonation (O_3), photo-induced oxidation (UV), and peroxone (H_2O_2/O_3) in breaking down the fluoroquinolone antibiotic ciprofloxacin from hospital wastewater revealed that photo-induced oxidation was significantly time consuming than the others in achieving complete breakdown of ciprofloxacin. After 60 min of treatment, both heterogeneous photocatalysis and peroxone nearly completely degraded ciprofloxacin. However, ozonation outperformed them with complete degradation in just 30 min by the direct oxidation of the N4-atom, which is quicker if ciprofloxacin is uncharged (Vasconcelos et al. 2009). Conversely, another study reported that UV treatment after MBR, using a TiO_2 photocatalyst, resulted in less effective elimination compared to UV treatment alone in the removal of antibiotics from hospital wastewater. Using photocatalytic TiO_2 fibers could result in an indirect phototransformation. Nevertheless, the fibers demonstrated superior light screening and absorption compared to the potential indirect phototransformation (Kovalova et al. 2013). UV irradiation coupled with hydrogen peroxide (H_2O_2) has proven effective in removing the majority of persistent pharmaceuticals from pre-treated (using MBR) hospital effluents. The

Table 3 Comparison of various treatment techniques in removing antibiotics from hospital effluents

SI. No.	Region	Treatment method	Antibiotics (Concentration in μg/L)	Removal efficiency (%)	Advantages	Disadvantages	Reference
1	Switzerland	MBR	Azithromycin (0.14), ciprofloxacin (31.98), clarithromycin (2.55), clindamycin (0.98), erythromycin + Eryt.-H_2O (0.19), metronidazole (3.39), N4-acetylsulfamethoxazole (2.39), norfloxacin (5.93), sulfadiazine (1.89), sulfamethoxazole (3.47)	< 60	Suitable for on-site treatment of hospital wastewater	Lower elimination rates, formation of transformation Products	(Kovalova et al. 2012)
			Trimethoprim (0.93)	96			
2	France	CAS	Total antibiotics concentration = 53.8 μg/L	95.1	High removal efficiency	Higher concentrations of antibiotics in the treated effluent caused adapted development of the microbial community, leading to less developed biomass and lower bacterial diversity	(Chonova et al. 2016)
3	Israel	MBR	Trimethoprim (0.039), roxithromycin (1.102), erythromycin (0.880), clarithromycin (1.293), and sulfamethazine (0.021)	> 80	Comparatively higher removal efficiency	Removal of Trimethoprim and Sulfamethoxazole is highly dependent on biomass buildup. High cost and frequent maintenance requirements	(Sahar et al. 2011)
			Sulfamethoxazole (0.367)	70			
4	Israel	CAS/UF	Trimethoprim (0.039)	70	The addition of UF to the CAS system, significantly increased by 28% to the removal antibiotics		Sahar et al. 2011)
			Roxithromycin (1.102)	80			
			Erythromycin (0.880)	70			
			Clarithromycin (1.293)	> 90			
			Sulfamethazine (0.021)	70			
			Sulfamethoxazole (0.367)	60			
5	Israel	CAS	Trimethoprim (0.039)	< 60	Low installment cost	Comparatively lower removal efficiency	(Sahar et al. 2011)
			Roxithromycin (1.102)	< 70			
			Erythromycin (0.880)	< 50			
			Clarithromycin (1.293)	70–80			
			Sulfamethazine (0.021)	70–80			
			Sulfamethoxazole (0.367)	30			
6	Brazil	Ozonation	Ciprofloxacin (200; spiked)	> 90	Complete degradation in 30 min	Formation of by-products	(Vasconcelos et al. 2009)
7	Brazil	Peroxone	Ciprofloxacin (200; spiked)	> 90	Complete degradation in 60 min	Lower level performance even at a high H_2O_2 concentration (1000 mgL^{-1})	(Vasconcelos et al. 2009)

(continued)

Table 3 (continued)

SI. No.	Region	Treatment method	Antibiotics (Concentration in µg/L)	Removal efficiency (%)	Advantages	Disadvantages	Reference
8	Spain	Fungal BFBR	Ciprofloxacin, ofloxacin, clarithromycin, sulfamethoxazole, metronidazole, metronidazole-OH, trimethoprim, dimetridazole, ronidazole, azithromycin total antibiotics concentration = 239 µg/L	90.8	Efficient removal. Prevents antibiotics from accumulating in sludge	Takes more time to remove antibiotics (5 days)	(Cruz-Morató et al. 2014)
9	USA	UF	Erythromycin-H_2O (0.289), sulfamethoxazole (0.066), trimethoprim (0.138)	<20		Very low removal efficiency	(Snyder et al. 2007)
10	USA	UF/RO	Erythromycin-H_2O (0.336), sulfamethoxazole (0.09), trimethoprim (0.186)	99	High removal efficiency	UF had minimal impact on the target antibiotics	(Snyder et al. 2007)
11	USA	MBR/RO	Erythromycin-H_2O (0.050), sulfamethoxazole (1.49), trimethoprim (0.199)	99	High removal efficiency		(Snyder et al. 2007)
12	USA	PAC	Erythromycin-H_2O (0.139), sulfamethoxazole (0.426), trimethoprim (0.135)	>87	High removal efficiency	Require a higher dose of adsorbent	(Snyder et al. 2007)
13	Saudi Arabia	CAS	Ciprofloxacin (5.6) sulfamethoxazole (0.03), and N-acetylsulfamethoxazole (1.2)	>97	High removal efficiency	Problems with sludge disposal	(Al Qarni et al. 2016)
			Clarithromycin (0.083)	86			
14	Germany	MBR/RO	Clarithromycin (10), ciprofloxacin (9), metronidazole (0.9)	>97	High removal efficiency	Limited yield The retentates have to be adequately disposed off	(Beier et al. 2010)
15	Luxembourg	UV/H_2O_2	Ciprofloxacin (10.52), clarithromycin, sulfamethoxazole	>85	Suitable to eliminate most persistent compounds		(Köhler et al. 2012)

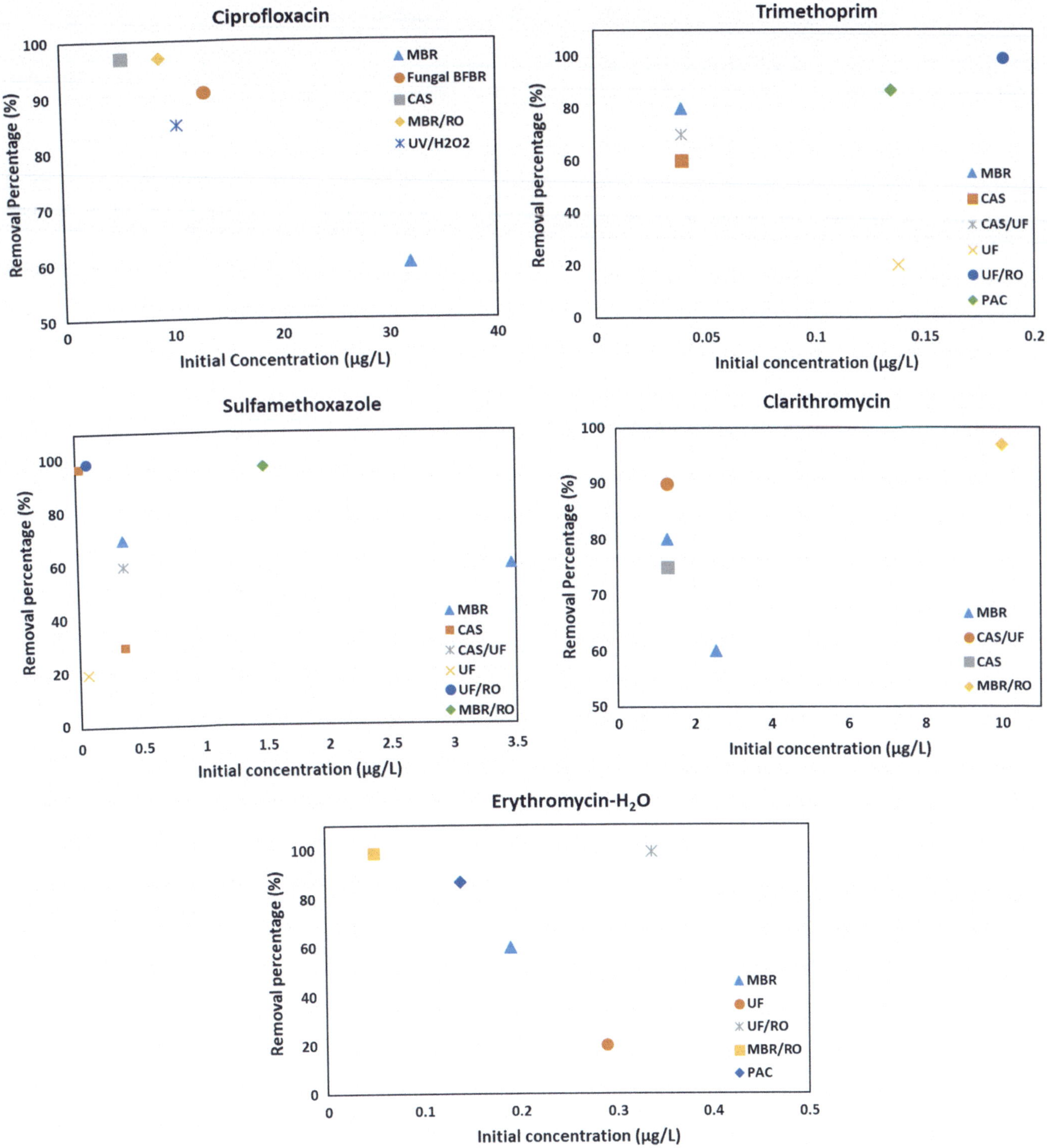

Fig. 4 Comparison of different treatment methods for removal of selected antibiotic compounds; ciprofloxacin, trimethoprim, sulfamethoxazole, clarithromycin, and erythromycin-H_2O from hospital wastewater

impact of UV irradiation was intensified by varying concentrations of H_2O_2. Removal rates greater than 90% were observed for ciprofloxacin and sulfamethoxazole and 85% for clarithromycin (Köhler et al. 2012).

7 Discussion

A comparison of different treatment methods for the removal of antibiotic residues from hospital wastewater is given in Table 3. Also, the removal efficiencies of these treatment methods for eliminating specific antibiotics considering their initial concentrations are given in Fig. 4. All these treatment techniques have their advantages as well as disadvantages. The efficiencies of these methods vary according to the type of antibiotic as well as their initial concentrations in the wastewater. Higher removal percentages were achieved for ciprofloxacin through the CAS process, MBR followed by reverse osmosis (MBR/RO), and fungal BFBR at lower concentrations; however, MBR was found less effective, which may be due to the higher initial concentration of the compound. Ozonation and peroxone were successful in removing 90% of ciprofloxacin from spiked hospital wastewater with concentrations as high as 200 µg/L. MBR/RO could also remove clarithromycin, sulfamethoxazole, and erythromycin-H_2O more effectively than MBR alone. Likewise, the removal efficiency was higher for the CAS system combined with UF unit (CAS/UF) in eliminating clarithromycin, sulfamethoxazole, and trimethoprim than CAS only at initial concentrations of 1.3 µg/L, 0.36 µg/L, and 0.04 µg/L, respectively. UF failed to eliminate most of the antibiotics from the wastewater. However, UF along with RO (UF/RO) could remove sulfamethoxazole, trimethoprim, and erythromycin-H_2O almost completely at lower concentrations (<0.4 µg/L). The PAC showed a removal percentage of > 80% for trimethoprim and erythromycin-H_2O.

In general, some antibiotics are found susceptible to biological treatment while some others show resistance. Most of the advanced treatment methods have shown effective removal of antibiotics, especially ozonation and RO methods. However, considering the economic feasibility, these methods are found costly. Therefore, a combination of advanced treatments along with biological treatment can be considered adequate for treating hospital effluents for the decontamination of both conventional pollutants and micropollutants including antibiotics and other pharmaceutical substances.

8 Conclusion

- Even though there are numerous reports on contamination levels of antibiotics in different aqueous media, the same in hospital effluents are found to be lower, especially in American countries.
- The contamination levels of antibiotics in hospital effluents from different countries in Asia, Europe, and America showed large variations due to the differences in antibiotic consumption patterns, types of healthcare facilities, cultural and geographical factors, and more.
- The most commonly detected antibiotics in hospital effluents are found as ciprofloxacin, sulfamethoxazole, tinidazole, trimethoprim, levofloxacin, ofloxacin, and norfloxacin.
- The most commonly detected antibiotic in hospital effluents globally, ciprofloxacin, showed variable concentration values ranging from 1.2 µg/L to as high as 1350 µg/L.
- The contamination of aquatic environments with antibiotic residues has resulted in various harmful effects on both aquatic and terrestrial organisms, including humans.
- The most studied remediation techniques for antibiotics from hospital wastewater were found to be biological treatment, adsorption, membrane techniques, and advanced oxidation. The elimination efficiencies of these treatment techniques are found to be influenced by the type of antibiotic as well as their initial observed concentrations in the wastewater.
- Biological treatments are cost-effective; however, their removal efficiency can only be increased by combining with some advanced membrane techniques. Therefore, it is reasonable to say that the combination of biological and advanced treatment techniques can be more useful in removing antibiotics from hospital effluents.

References

Adeleye AS, Xue J, Zhao Y, Taylor AA, Zenobio JE, Sun Y, Han Z, Salawu OA, Zhu Y (2022) Abundance, fate, and effects of pharmaceuticals and personal care products in aquatic environments. J Hazard Mater 424:127284

Ahmad M, Khan AU, Wahid A, Butt ZA, Farhan M, Ahmad F (2012) Role of hospital effluents in the contribution of antibiotics and antibiotic resistant bacteria to the aquatic environment. Pak J Nutr 11(12):1177

Akiba M, Senba H, Otagiri H, Prabhasankar VP, Taniyasu S, Yamashita N, Lee K-I, Yamamoto T, Tsutsui T, Joshua DI (2015) Impact of wastewater from different sources on the prevalence of

antimicrobial-resistant Escherichia coli in sewage treatment plants in South India. Ecotoxicol Environ Saf 115:203–208
Al Qarni H, Collier P, O'Keeffe J, Akunna J (2016) Investigating the removal of some pharmaceutical compounds in hospital wastewater treatment plants operating in Saudi Arabia. Environ Sci Pollut Res 23:13003–13014
Al-Maadheed S, Goktepe I, Latiff ABA, Shomar B (2019) Antibiotics in hospital effluent and domestic wastewater treatment plants in Doha Qatar. J Water Process Eng 28:60–68
Archer E, Wolfaardt GM, Van Wyk JH (2017) Pharmaceutical and personal care products (PPCPs) as endocrine disrupting contaminants (EDCs) in South African surface waters. Water Sa 43(4):684–706
Archer E, Petrie B, Kasprzyk-Hordern B, Wolfaardt GM (2017) The fate of pharmaceuticals and personal care products (PPCPs), endocrine disrupting contaminants (EDCs), metabolites and illicit drugs in a WWTW and environmental waters. Chemosphere 174:437–446
Ashfaq M, Khan KN, Rasool S, Mustafa G, Saif-Ur-Rehman M, Nazar MF, Sun Q, Yu C-P (2016) Occurrence and ecological risk assessment of fluoroquinolone antibiotics in hospital waste of Lahore Pakistan. Environ Toxicol Pharmacol 42:16–22
Aydin S, Aydin ME, Ulvi A, Kilic H (2019) Antibiotics in hospital effluents: occurrence, contribution to urban wastewater, removal in a wastewater treatment plant, and environmental risk assessment. Environ Sci Pollut Res 26:544–558
Backhaus T, Porsbring T, Arrhenius Å, Brosche S, Johansson P, Blanck H (2011) Single-substance and mixture toxicity of five pharmaceuticals and personal care products to marine periphyton communities. Environ Toxicol Chem 30(9):2030–2040
Bai Y, Meng W, Xu J, Zhang Y, Guo C (2014) Occurrence, distribution and bioaccumulation of antibiotics in the Liao River Basin in China. Environ Sci Process Impacts 16(3):586–593
Balakrishna K, Rath A, Praveenkumarreddy Y, Guruge KS, Subedi B (2017) A review of the occurrence of pharmaceuticals and personal care products in Indian water bodies. Ecotoxicol Environ Saf 137:113–120
Bansal O (2019) Antibiotics in hospital effluents and their impact on the antibiotics resistant bacteria and remediation of the antibiotics: a review. Netw Pharmacol 4(3–4):6–30
Beier S, Köster S, Veltmann K, Schröder H, Pinnekamp J (2010) Treatment of hospital wastewater effluent by nanofiltration and reverse osmosis. Water Sci Technol 61(7):1691–1698
Bengtsson-Palme J, Larsson DJ (2016) Concentrations of antibiotics predicted to select for resistant bacteria: proposed limits for environmental regulation. Environ Int 86:140–149
Bilal M, Mehmood S, Rasheed T, Iqbal HM (2020) Antibiotics traces in the aquatic environment: persistence and adverse environmental impact. Curr Opin Environ Sci Health 13:68–74
Blair B, Nikolaus A, Hedman C, Klaper R, Grundl T (2015) Evaluating the degradation, sorption, and negative mass balances of pharmaceuticals and personal care products during wastewater treatment. Chemosphere 134:395–401
Boleda MR, Galceran MT, Ventura F (2011) Behavior of pharmaceuticals and drugs of abuse in a drinking water treatment plant (DWTP) using combined conventional and ultrafiltration and reverse osmosis (UF/RO) treatments. Environ Pollut 159(6):1584–1591
Brown KD, Kulis J, Thomson B, Chapman TH, Mawhinney DB (2006) Occurrence of antibiotics in hospital, residential, and dairy effluent, municipal wastewater, and the Rio Grande in New Mexico. Sci Total Environ 366(2–3):772–783
Carvalho IT, Santos L (2016) Antibiotics in the aquatic environments: a review of the European scenario. Environ Int 94:736–757
Chang X, Meyer MT, Liu X, Zhao Q, Chen H, Chen J-A, Qiu Z, Yang L, Cao J, Shu W (2010) Determination of antibiotics in sewage from hospitals, nursery and slaughter house, wastewater treatment plant and source water in Chongqing region of three Gorge reservoir in China. Environ Pollut 158(5):1444–1450
Chen J, Ying G-G, Deng W-J (2019) Antibiotic residues in food: extraction, analysis, and human health concerns. J Agric Food Chem 67(27):7569–7586
Chonova T, Keck F, Labanowski J, Montuelle B, Rimet F, Bouchez A (2016) Separate treatment of hospital and urban wastewaters: a real scale comparison of effluents and their effect on microbial communities. Sci Total Environ 542:965–975
Chtourou M (2018) Pharmaceutical and personal care products removal by advanced treatment technologies
Co-operation OFED (2020). Pharmaceutical residues in freshwater: hazards and policy responses. IWA (International WATER ASSN)
Cruz-Morató C, Lucas D, Llorca M, Rodriguez-Mozaz S, Gorga M, Petrovic M, Barceló D, Vicent T, Sarrà M, Marco-Urrea E (2014) Hospital wastewater treatment by fungal bioreactor: removal efficiency for pharmaceuticals and endocrine disruptor compounds. Sci Total Environ 493:365–376
Dhingra A, Khan NA, Ahmed S, Gautam S, Vambol S, Vambol V, Kovalenko S (2022) Analysis of wastewater from medical institutions in India. World Rev Sci Technol Sustain Dev 18(3–4):436–444
Dhodapkar RS, Gandhi KN (2019) Pharmaceuticals and personal care products in aquatic environment: chemicals of emerging concern?. In: Pharmaceuticals and personal care products: waste management and treatment technology. Elsevier, pp 63–85
Diwan V, Tamhankar AJ, Khandal RK, Sen S, Aggarwal M, Marothi Y, Iyer RV, Sundblad-Tonderski K, Stålsby-Lundborg C (2010) Antibiotics and antibiotic-resistant bacteria in waters associated with a hospital in Ujjain India. BMC Public Health 10:1–8
Diwan V, Stålsby Lundborg C, Tamhankar AJ (2013) Seasonal and temporal variation in release of antibiotics in hospital wastewater: estimation using continuous and grab sampling. PLoS ONE 8(7):e68715
Diwan V, Tamhankar AJ, Aggarwal M, Sen S, Khandal RK, Lundborg CS (2009) Detection of antibiotics in hospital effluents in India. Curr Sci 1752–1755
Fick J, Söderström H, Lindberg RH, Phan C, Tysklind M, Larsson DJ (2009) Contamination of surface, ground, and drinking water from pharmaceutical production. Environ Toxicol Chem 28(12):2522–2527
Gros M, Rodríguez-Mozaz S, Barceló D (2013) Rapid analysis of multiclass antibiotic residues and some of their metabolites in hospital, urban wastewater and river water by ultra-high-performance liquid chromatography coupled to quadrupole-linear ion trap tandem mass spectrometry. J Chromatogr A 1292:173–188
Halling-Sørensen B (2000) Algal toxicity of antibacterial agents used in intensive farming. Chemosphere 40(7):731–739
Hilton MJ, Thomas KV (2003) Determination of selected human pharmaceutical compounds in effluent and surface water samples by high-performance liquid chromatography–electrospray tandem mass spectrometry. J Chromatogr A 1015(1–2):129–141
Homem V, Santos L (2011) Degradation and removal methods of antibiotics from aqueous matrices–a review. J Environ Manage 92(10):2304–2347
Hutchings MI, Truman AW, Wilkinson B (2019) Antibiotics: past, present and future. Curr Opin Microbiol 51:72–80
Khan NA, Ahmed S, Farooqi IH, Ali I, Vambol V, Changani F, Yousefi M, Vambol S, Khan SU, Khan AH (2020) Occurrence, sources and conventional treatment techniques for various antibiotics present in hospital wastewaters: a critical review. TrAC Trends Anal Chem 129:115921
Khan MT, Shah IA, Ihsanullah I, Naushad M, Ali S, Shah SHA, Mohammad AW (2021) Hospital wastewater as a source of environmental contamination: an overview of management practices, environmental risks, and treatment processes. J Water Process Eng 41:101990

Kim J-W, Jang H-S, Kim J-G, Ishibashi H, Hirano M, Nasu K, Ichikawa N, Takao Y, Shinohara R, Arizono K (2009) Occurrence of pharmaceutical and personal care products (PPCPs) in surface water from Mankyung River, South Korea. J Health Sci 55(2):249–258

Klein EY, Van Boeckel TP, Martinez EM, Pant S, Gandra S, Levin SA, Goossens H, Laxminarayan R (2018) Global increase and geographic convergence in antibiotic consumption between 2000 and 2015. Proc Natl Acad Sci 115(15):E3463–E3470

Köhler C, Venditti S, Igos E, Klepiszewski K, Benetto E, Cornelissen A (2012) Elimination of pharmaceutical residues in biologically pre-treated hospital wastewater using advanced UV irradiation technology: a comparative assessment. J Hazard Mater 239:70–77

Kovalova L, Siegrist H, Singer H, Wittmer A, McArdell CS (2012) Hospital wastewater treatment by membrane bioreactor: performance and efficiency for organic micropollutant elimination. Environ Sci Technol 46(3):1536–1545

Kovalova L, Siegrist H, Von Gunten U, Eugster J, Hagenbuch M, Wittmer A, Moser R, McArdell CS (2013) Elimination of micropollutants during post-treatment of hospital wastewater with powdered activated carbon, ozone, and UV. Environ Sci Technol 47(14):7899–7908

Lagishetty R, Nagrajan P (2015) Qualitative analysis of antibiotic residues in hospital effluents from south India. Int J Pharm Sci Rev Res 32:315–320

Larsson DJ (2014) Antibiotics in the environment. Upsala J Med Sci 119(2):108–112

Larsson DJ, de Pedro C, Paxeus N (2007) Effluent from drug manufactures contains extremely high levels of pharmaceuticals. J Hazard Mater 148(3):751–755

Leung HW, Minh T, Murphy MB, Lam JC, So MK, Martin M, Lam PK, Richardson BJ (2012) Distribution, fate and risk assessment of antibiotics in sewage treatment plants in Hong Kong South China. Environ Int 42:1–9

Lien LTQ, Hoa NQ, Chuc NTK, Thoa NTM, Phuc HD, Diwan V, Dat NT, Tamhankar AJ, Lundborg CS (2016) Antibiotics in wastewater of a rural and an urban hospital before and after wastewater treatment, and the relationship with antibiotic use—a one year study from Vietnam. Int J Environ Res Public Health 13(6):588

Lindberg R, Jarnheimer P-Å, Olsen B, Johansson M, Tysklind M (2004) Determination of antibiotic substances in hospital sewage water using solid phase extraction and liquid chromatography/mass spectrometry and group analogue internal standards. Chemosphere 57(10):1479–1488

Liu Q, Zhou Y, Chen L, Zheng X (2010) Application of MBR for hospital wastewater treatment in China. Desalination 250(2):605–608

Maghsodian Z, Sanati AM, Mashifana T, Sillanpää M, Feng S, Nhat T, Ramavandi B (2022) Occurrence and distribution of antibiotics in the water, sediment, and biota of freshwater and marine environments: a review. Antibiotics 11(11):1461

Managaki S, Murata A, Takada H, Tuyen BC, Chiem NH (2007) Distribution of macrolides, sulfonamides, and trimethoprim in tropical waters: ubiquitous occurrence of veterinary antibiotics in the Mekong Delta. Environ Sci Technol 41(23):8004–8010

Martinez-Guerra BA, Gonzalez-Lara MF, de-Leon-Cividanes NA, Tamez-Torres KM, Roman-Montes CM, Rajme-Lopez S, Villalobos-Zapata GI, Lopez-Garcia NI, Martínez-Gamboa A, Sifuentes-Osornio J (2021) Antimicrobial resistance patterns and antibiotic use during hospital conversion in the COVID-19 pandemic. Antibiotics 10 (2):182

Martins AF, Vasconcelos TG, Henriques DM, Frank CDS, König A, Kümmerer K (2008) Concentration of ciprofloxacin in Brazilian hospital effluent and preliminary risk assessment: a case study. CLEAN–Soil Air Water 36(3):264–269

Melin T, Jefferson B, Bixio D, Thoeye C, De Wilde W, De Koning J, van der Graaf J, Wintgens T (2006) Membrane bioreactor technology for wastewater treatment and reuse. Desalination 187(1–3):271–282

Mesdaghinia A, Nadafi K, Nabizadeh NR, Saeidi R, Zamanzadeh M (2009) Wastewater characteristics and appropriate method for wastewater management in the hospitals

Miller KJ, Meek J (2006) Helena valley ground water: pharmaceuticals, personal care products, endocrine disruptors (PPCPs), and microbial indicators of fecal contamination. Citeseer

Mohapatra S, Huang C-H, Mukherji S, Padhye LP (2016) Occurrence and fate of pharmaceuticals in WWTPs in India and comparison with a similar study in the United States. Chemosphere 159:526–535

Mutiyar PK, Mittal AK (2014) Occurrences and fate of selected human antibiotics in influents and effluents of sewage treatment plant and effluent-receiving river Yamuna in Delhi (India). Environ Monit Assess 186:541–557

Na G, Fang X, Cai Y, Ge L, Zong H, Yuan X, Yao Z, Zhang Z (2013) Occurrence, distribution, and bioaccumulation of antibiotics in coastal environment of Dalian China. Mar Pollut Bull 69(1–2):233–237

Nandi A, Pecetta S, Bloom DE (2023) Global antibiotic use during the COVID-19 pandemic: analysis of pharmaceutical sales data from 71 countries, 2020–2022. EClinicalMedicine 57

Ngo TH, Van D-A, Tran HL, Nakada N, Tanaka H, Huynh TH (2021) Occurrence of pharmaceutical and personal care products in Cau River Vietnam. Environ Sci Pollut Res 28:12082–12091

O'Neill J (2016) Tackling drug-resistant infections globally: final report and recommendations

Oliveira TS, Murphy M, Mendola N, Wong V, Carlson D, Waring L (2015) Characterization of pharmaceuticals and personal care products in hospital effluent and waste water influent/effluent by direct-injection LC-MS-MS. Sci Total Environ 518:459–478

Oluwole AO, Omotola EO, Olatunji OS (2020) Pharmaceuticals and personal care products in water and wastewater: a review of treatment processes and use of photocatalyst immobilized on functionalized carbon in AOP degradation. BMC Chem 14:1–29

Ory J, Bricheux G, Togola A, Bonnet JL, Donnadieu-Bernard F, Nakusi L, Forestier C, Traore O (2016) Ciprofloxacin residue and antibiotic-resistant biofilm bacteria in hospital effluent. Environ Pollut 214:635–645

Papageorgiou M, Kosma C, Lambropoulou D (2016) Seasonal occurrence, removal, mass loading and environmental risk assessment of 55 pharmaceuticals and personal care products in a municipal wastewater treatment plant in Central Greece. Sci Total Environ 543:547–569

Patel M, Kumar R, Kishor K, Mlsna T, Pittman CU Jr, Mohan D (2019) Pharmaceuticals of emerging concern in aquatic systems: chemistry, occurrence, effects, and removal methods. Chem Rev 119(6):3510–3673

Paulus GK, Hornstra LM, Alygizakis N, Slobodnik J, Thomaidis N, Medema G (2019) The impact of on-site hospital wastewater treatment on the downstream communal wastewater system in terms of antibiotics and antibiotic resistance genes. Int J Hyg Environ Health 222(4):635–644

Pena A, Paulo M, Silva L, Seifrtová M, Lino C, Solich P (2010) Tetracycline antibiotics in hospital and municipal wastewaters: a pilot study in Portugal. Anal Bioanal Chem 396:2929–2936

Prasad V, Baliyan S, Sibi G (2018) Prevalence of antibiotic resistance among bacterial isolates from hospital environments and effluents. J Bacteriol Mycol 5(7):1082

Richardson ML, Bowron JM (1985) The fate of pharmaceutical chemicals in the aquatic environment. J Pharm Pharmacol 37(1):1–12

Roberts PH, Thomas KV (2006) The occurrence of selected pharmaceuticals in wastewater effluent and surface waters of the lower Tyne catchment. Sci Total Environ 356(1–3):143–153

Sahar E, Messalem R, Cikurel H, Aharoni A, Brenner A, Godehardt M, Jekel M, Ernst M (2011) Fate of antibiotics in activated sludge followed by ultrafiltration (CAS-UF) and in a membrane bioreactor (MBR). Water Res 45(16):4827–4836

Santos LH, Gros M, Rodriguez-Mozaz S, Delerue-Matos C, Pena A, Barceló D, Montenegro MCB (2013) Contribution of hospital effluents to the load of pharmaceuticals in urban wastewaters: identification of ecologically relevant pharmaceuticals. Sci Total Environ 461:302–316

Schaider LA, Rudel RA, Ackerman JM, Dunagan SC, Brody JG (2014) Pharmaceuticals, perfluorosurfactants, and other organic wastewater compounds in public drinking water wells in a shallow sand and gravel aquifer. Sci Total Environ 468:384–393

Shokoohi R, Ghobadi N, Godini K, Hadi M, Atashzaban Z (2020) Antibiotic detection in a hospital wastewater and comparison of their removal rate by activated sludge and earthworm-based vermifilteration: environmental risk assessment. Process Saf Environ Prot 134:169–177

Snyder SA, Adham S, Redding AM, Cannon FS, DeCarolis J, Oppenheimer J, Wert EC, Yoon Y (2007) Role of membranes and activated carbon in the removal of endocrine disruptors and pharmaceuticals. Desalination 202(1–3):156–181

Tell J, Caldwell DJ, Häner A, Hellstern J, Hoeger B, Journel R, Mastrocco F, Ryan JJ, Snape J, Straub JO (2019) Science-based targets for antibiotics in receiving waters from pharmaceutical manufacturing operations. Integr Environ Assess Manag 15(3):312–319

Thomas KV, Dye C, Schlabach M, Langford KH (2007) Source to sink tracking of selected human pharmaceuticals from two Oslo city hospitals and a wastewater treatment works. J Environ Monit 9(12):1410–1418

Truman A (2019) Antibiotics: past, present and future Matthew I Hutchings, Andrew W Truman 2 and Barrie Wilkinson 2. Curr Opin Microbiol 51:72–80

Turolla A, Cattaneo M, Marazzi F, Mezzanotte V, Antonelli M (2018) Antibiotic resistant bacteria in urban sewage: role of full-scale wastewater treatment plants on environmental spreading. Chemosphere 191:761–769

Vasconcelos TG, Kümmerer K, Henriques DM, Martins AF (2009) Ciprofloxacin in hospital effluent: degradation by ozone and photoprocesses. J Hazard Mater 169(1–3):1154–1158

Venditti S, Köhler C, Arenz-Leufen M, O'Nagy O, Cornelissen A, Klepiszewski K (2011) Membrane bioreactor process as pretreatment for hospital effluents. In: Proceedings of 8th IWA leading-edge conference on water and wastewater technologies, Amsterdam, Netherlands

Verlicchi P, Galletti A, Masotti L (2010) Management of hospital wastewaters: the case of the effluent of a large hospital situated in a small town. Water Sci Technol 61(10):2507–2519

Verlicchi P, Al Aukidy M, Galletti A, Petrovic M, Barceló D (2012) Hospital effluent: investigation of the concentrations and distribution of pharmaceuticals and environmental risk assessment. Sci Total Environ 430:109–118

Verlicchi P, Al Aukidy M, Zambello E (2015) What have we learned from worldwide experiences on the management and treatment of hospital effluent?—An overview and a discussion on perspectives. Sci Total Environ 514:467–491

Verma P, Gupta M, Parasher P (2017) Occurrence and distribution of antibiotic substance in waste water from hospital effluent 7. Int J Recent Res Rev 10:17–23

Wang Q, Wang P, Yang Q (2018) Occurrence and diversity of antibiotic resistance in untreated hospital wastewater. Sci Total Environ 621:990–999

Wang W, Weng Y, Luo T, Wang Q, Yang G, Jin Y (2023) Antimicrobial and the resistances in the environment: ecological and health risks, influencing factors, and mitigation strategies. Toxics 11(2):185

Watkinson A, Murby E, Kolpin DW, Costanzo S (2009) The occurrence of antibiotics in an urban watershed: from wastewater to drinking water. Sci Total Environ 407(8):2711–2723

Wilkinson JL, Boxall AB, Kolpin DW, Leung KM, Lai RW, Galbán-Malagón C, Adell AD, Mondon J, Metian M, Marchant RA (2022) Pharmaceutical pollution of the world's rivers. Proc Natl Acad Sci 119(8):e2113947119

Yang X, Flowers RC, Weinberg HS, Singer PC (2011) Occurrence and removal of pharmaceuticals and personal care products (PPCPs) in an advanced wastewater reclamation plant. Water Res 45(16):5218–5228

Yang Q, Gao Y, Ke J, Show PL, Ge Y, Liu Y, Guo R, Chen J (2021) Antibiotics: an overview on the environmental occurrence, toxicity, degradation, and removal methods. Bioengineered 12(1):7376–7416

Yao S, Ye J, Yang Q, Hu Y, Zhang T, Jiang L, Munezero S, Lin K, Cui C (2021) Occurrence and removal of antibiotics, antibiotic resistance genes, and bacterial communities in hospital wastewater. Environ Sci Pollut Res 28(40):57321–57333

Zhang T, Li B (2011) Occurrence, transformation, and fate of antibiotics in municipal wastewater treatment plants. Crit Rev Environ Sci Technol 41(11):951–998

Zhao S, Wang X, Li Y, Lin J (2016) Bioconcentration, metabolism, and biomarker responses in marine medaka (Oryzias melastigma) exposed to sulfamethazine. Aquat Toxicol 181:29–36

Building a Strong Foundation: Study on Use of Construction and Demolition Waste in Stone Columns

Pravez Alam and Shailja Bawa

Abstract

The stone column method is a ground improvement technique used to enhance the load-bearing capacity of weak or challenging ground conditions. It is an effective solution for soft clays, silts, and loose sands to improve their engineering behaviour by speeding up the consolidation process through improved drainage pathways, which contributes to reinforcing the soil and reducing the settlement of weak soil. Moreover, pile foundations may be costly for constructing low- to medium-rise buildings on soft soil. Stone columns are suitable under such circumstances. There are various applications for stone columns including oil tank foundations, embankments, slope stabilization, and the construction of tall buildings. Natural aggregates are most commonly used in stone columns because of their higher stiffness than the surrounding soil. The use of natural aggregate raises significant environmental concerns due to its extensive application in the construction sector. The extraction rate of natural aggregates exceeds the rate at which they can be naturally replenished. The environmental impact is increased by the carbon footprint generated through aggregate mining and associated activities. Therefore, it is essential to identify sustainable alternatives to natural materials for use in geotechnical applications and construction. Waste materials have been used more frequently in place of natural aggregate in stone columns in recent years. Among these types, construction and demolition waste is the most common. Different studies use varying ratios of natural aggregate to construction and demolition waste. This paper offers an overview of the use of construction and demolition waste in stone columns. It concludes that incorporating these waste into stone columns is an effective approach, contributing to recycling initiatives and advancing environmental sustainability.

Keywords

Sustainability · Construction and demolition waste · Stone column · Soil stabilization · Sustainable soil improvement · Recycled materials · Waste materials

1 Introduction

Cities are expanding quickly, making it more important than ever to upgrade and develop infrastructure, transit systems, and urban environments to accommodate the growing population (Shooshtarian et al. 2019). These requirements led to significant advancements in infrastructure at the local, national, and global levels. Worldwide construction and demolition activities annually yield over 10 billion tons of waste from building construction, infrastructure development, demolition, and excavation work (Pickin et al. 2020; Wu et al. 2023). European Union, the United States, India, China, and other economies with strong industrialization and development make significant contributions to this waste stream, producing 600 million, 700 million, 750 million, and 2.3 billion tons of waste from construction and demolition annually, respectively (Pickin et al. 2020; Iacoboaea et al. 2019; Zheng et al. 2017; Agency 2016).

Around 35% of the global construction and demolition (C & D) waste ends up in landfills for disposal, with significant contributors including the United Kingdom, Brazil, Australia, and the USA, with shares of 50%, 40%, 35%, and 30%, respectively (Menegaki and Damigos 2018; Al Zulayq et al. 2022). Figure 1 illustrates the yearly

P. Alam (✉) · S. Bawa
Dr B R Ambedkar National Institute of Technology, Jalandhar, Punjab, India
e-mail: praveza.ce.21@nitj.ac.in; parvezalamce@gmail.com

S. Bawa
e-mail: bawas@nitj.ac.in

A. Mandpe et al. (eds.), *Sustainable Waste Management*, Advances in Science, Technology & Innovation,
https://doi.org/10.1007/978-3-031-85685-3_11

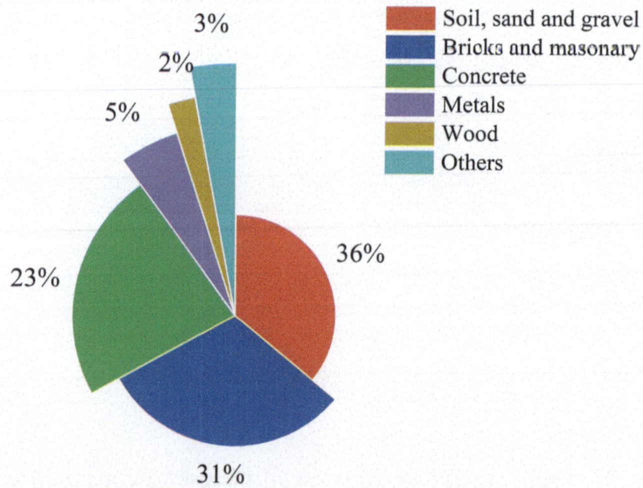

Fig. 1 Annual consumption of C & D waste in India

consumption of five types of C & D waste in India, as reported by C and D (Shrivastava and Chini 2009).

C & D waste comprises both non-hazardous and hazardous materials that are derived from new construction and redevelopment projects (Menegaki and Damigos 2018). The substantial volume, constituting 30–40% of global waste generation, and the diverse nature of C & D waste production have manifold adverse effects on society, economy, and environment. These impacts include land degradation, loss of biodiversity, and contamination of air, soil, and water, leading to health issues such as chronic and cancerous diseases in humans. Environmental issues connected with C & D waste management and disposal comprise noise, odour, dust pollution, emissions of carbon dioxide and methane, release of particulates, asbestos presence, and contamination from hazardous chemicals (Crawford et al. 2017; Yuan 2013; Roussat et al. 2008). Moreover, C & D activities require substantial energy and utilize air, water, and raw materials sourced from natural resources (Wu et al. 2023; Poon et al. 2013; Ding et al. 2016). The large volume of waste generated and its detrimental effects on the economy, environment, and society have highlighted the urgent need for effective management of C & D waste globally (Pickin et al. 2020).

Using waste materials for environmental and economic gains has been one major step towards developing sustainability and creating a circular economy. Reusing and recycling reduce the amount of toxic substances that get into ecosystems, protecting natural habitats and wildlife. Further, natural resources are conserved when wastes are used for recycling instead of extracting raw materials. For example, recycling paper saves trees, while the recycling of metals reduces the demand for new ore; this consequently ensures less biodiversity loss and less environmental degradation. Another major advantage is energy savings; very often, recycling uses significantly less energy than the production of new goods from raw materials. Economically, industries related to recycling create more jobs in comparison with the traditional methods of waste disposal (Conlon et al. 2019). In this case, those job creations cut across all stages: sorting, processing, and manufacturing from the resultant recyclables. The waste management will also save money for municipalities and businesses involved in the processes of reducing waste collection and landfilling costs or making money off the materials being recycled in some cases (Geissdoerfer et al. 2017).

Building geotechnical structures such as stone columns has become more popular with the inclusion of environmental factors. Stone columns offer an excellent opportunity for the utilization of recycled aggregates (Egan and Slocombe 2015). Given the large amounts of natural aggregates required for constructing numerous stone columns, the impact of incorporating recycled materials into stone columns has been recognized, especially in areas experiencing a scarcity of natural aggregate resources. Furthermore, incorporating recycled materials into stone columns supports waste management efforts by encouraging recycling and reuse of C & D waste. Numerous studies (Rahardjo et al. 2013; Fauzi et al. 2013; Arulrajah et al. 2013; Arulrajah et al. 2012) have explored replacing natural aggregates with recycled aggregates in different construction applications. Stone columns, also referred to as granular piles or aggregate piers, serve as a common method for enhancing soft soil conditions to support the foundations of structures or embankments (Barksdale and Bachus 1983; Han 2015). When gravel is used, embankments and naturally occurring slopes become more stable and capable of supporting heavier loads due to their increased strength, stiffness, and permeability over native soft soil. Additionally, by accelerating soil consolidation and lowering the risk of liquefaction, this method helps decrease both total and differential settlements. On the other hand, inadequate lateral confinement in very soft soils can make stone columns less effective. When determining whether stone columns are feasible, an undrained shear strength of between 5 and 15 kPa is typically taken into account (Wehr 2006). However, additional factors like liquidity index and soil sensitivity need to be taken into account. Stone columns are classified into two types, i.e. end bearing type and floating type, as shown in Fig. 2. End bearing stone columns are designed to extend through the weak soil layers and penetrate into a firm stratum or bedrock. This type of column transfers the applied loads directly to the underlying stable layer, providing a solid foundation and minimizing settlement. End bearing stone columns are particularly effective in areas where a bearing layer is relatively shallow and accessible. They offer significant load-bearing capacity and are ideal for supporting heavy structures such as buildings, bridges, and industrial facilities. The construction process involves driving the column through the weak soil until it reaches the firm stratum,

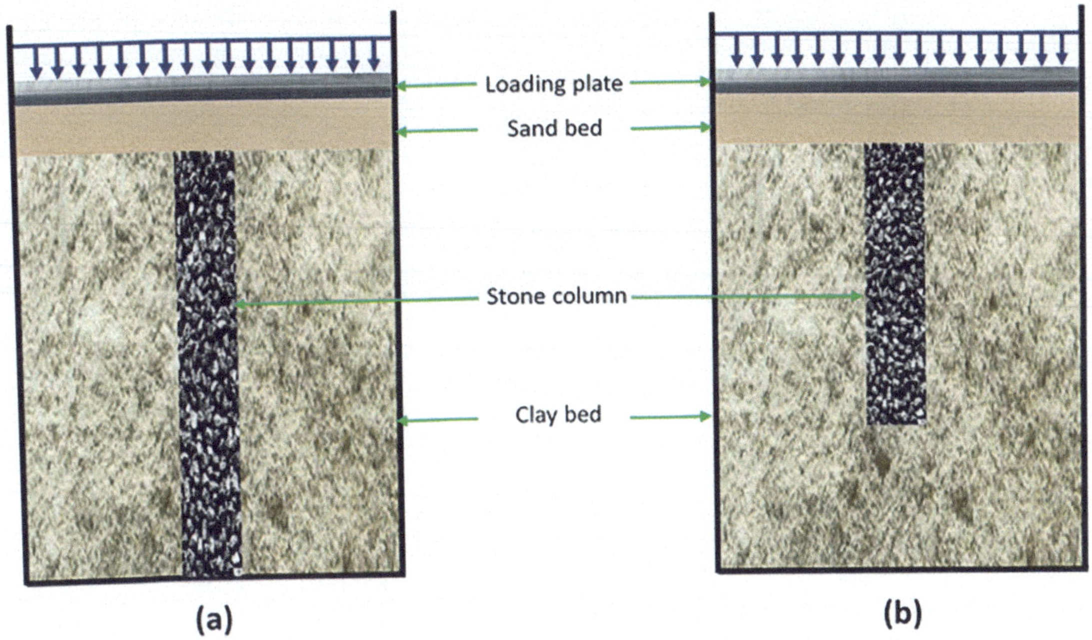

Fig. 2 **a** End bearing stone column **b** Floating stone column

ensuring a secure connection that enhances overall stability. In contrast, floating stone columns do not reach a firm layer but are embedded within the weak soil itself. These columns rely on the lateral support provided by the surrounding soil shear strength to carry the loads. Floating stone columns are used in situations where the firm stratum is too deep to be economically reached or when improving the upper layers of soil is sufficient for the project requirements. They help to densify the soil, increase its shear strength, and reduce settlement by redistributing the loads over a larger area. This type of column is often employed in the construction of roadways, embankments, and lightweight structures where deep foundations are not necessary. Both types of stone columns play a crucial role in ground improvement, but their suitability depends on specific site conditions, soil properties, and project requirements. The choice between end bearing and floating stone columns is influenced by factors such as the depth of a firm stratum, the type of structure being supported, load considerations, and overall cost-effectiveness. By enhancing soil stability and bearing capacity, stone columns provide a reliable solution for construction on challenging ground conditions, ensuring the safety and durability of various civil engineering projects. Using geotextiles or other geosynthetics to encase the columns has become a practical approach in recent years to improve the lateral confinement and, as a result, the vertical capacity of the columns (Alexiew and Raithel 2015; Almeida et al. 2018; Deshpande et al. 2021; Pandey et al. 2022).

The aim of this study is to explore the potential of using C & D waste as a sustainable alternative to natural aggregates in stone columns. By incorporating construction and demolition waste into stone columns, this research seeks to address environmental concerns associated with excessive natural aggregate consumption, reduce construction costs, and promote resource conservation. The study evaluates the performance of stone columns made with various proportions of construction and demolition waste and natural aggregates, with a focus on their efficiency and load-bearing capacity. Ultimately, this review aims to contribute to sustainable practices in geotechnical engineering by offering practical solutions for waste utilization and soil stabilization. This paper makes several key contributions. It first evaluates the environmental and economic advantages of utilizing C & D waste, emphasizing its role in conserving natural resources and promoting a circular economy. The research also provides a comprehensive comparative analysis of the efficiency and performance of stone columns made with varying combinations of C & D waste and natural aggregates. In addition, the study offers valuable insights into experimental methodologies, such as soil bed preparation, column construction, and load application, which can inform future research in this area. Lastly, the paper examines the suitability and limitations of specific types of C & D waste, including crushed bricks and recycled concrete, as alternatives to natural aggregates in geotechnical applications.

The paper is structured in several sections, beginning with an introduction that provides an overview of the challenges in sustainable construction and the relevance of using C & D waste in geotechnical applications, particularly in stone columns. This is followed by a literature review that examines previous studies on recycled materials for soil improvement, with a focus on the properties and performance of C & D waste in similar contexts. The subsequent section includes the preparation and characterization of C & D waste, as well as the methodology for constructing and testing stone columns. Following this, the results and discussion section presents the findings from laboratory experiments, analysing the efficiency and load-bearing capacity of stone columns with different material compositions. Finally, the conclusion summarizes the key findings, emphasizing the environmental and technical feasibility of using C & D waste in stone columns, and offers recommendations for practical implementation.

2 Research Methodology

The research methodology for this review paper involves a systematic and comprehensive analysis of existing studies and data related to the utilization of C & D waste in stone columns. The methodology begins with an extensive literature search conducted through academic databases, journals, and technical reports to collect relevant information on the subject. Key focus areas include the properties of C & D waste, its environmental and economic impacts, and its applications in geotechnical engineering, particularly in stone column construction. To ensure the reliability and relevance of the review, studies were selected based on specific criteria, such as their focus on recycled aggregates, ground improvement techniques, and sustainability in construction. The gathered information is analysed to compare the physical and mechanical properties of C & D waste with those of natural aggregates. Additionally, performance data from experimental studies are critically evaluated to determine the efficiency of using C & D waste in stone columns for soil stabilization. The review also incorporates a comparative analysis of methodologies used in various studies, highlighting the consistency of experimental approaches, materials used, and outcomes achieved. By synthesizing findings from multiple sources, the paper identifies trends, benefits, and challenges associated with C & D waste usage in geotechnical applications. This structured approach ensures that the review provides a comprehensive understanding of the topic, offering valuable insights for future research and practical implementation in sustainable construction practices.

3 C & D Waste Recycling

Recycling involves the collection, processing, and conversion of waste materials into useful products, playing a crucial role in promoting sustainable development. Recycling C & D waste produces recycled aggregates, which help conserve natural resources, reduce carbon dioxide emissions, and address landfill disposal issues (Safiuddin et al. 2013). Various techniques have been employed to recycle C & D waste, including a two-step method: first, the demolished concrete is crushed, and then contaminants such as paper, reinforcement, or wood are separated (Behera et al. 2014). Crushing can be performed in single or two-stage processes. Single-stage crushing typically uses a jaw crusher, followed by mixing and sieving. The two-stage process involves jaw crushing followed by impact crushing, producing coarse recycled aggregate suitable for concrete and pavement applications with optimal grain size distribution (Shahidan et al. 2017; Verma et al. 2022). Impact crushers are noted for generating recycled aggregate with less adhered mortar and superior surface properties, as well as low porosity. Tertiary crushing, using vertical shaft impactors, produces recycled sand with improved morphology, controlled size distribution, and reduced labour costs (Li et al. 2009). Additional techniques such as particle homogenization, grinding, sieving, and autogenous cleaning aim to remove adhered mortar, enhancing recycled aggregate quality. For high-strength concrete, removing adhered mortar is crucial to improve recycled aggregate properties, and mechanical mixing offers a cost-effective method for this purpose. Furthermore, removing adhered mortar can yield strong, hydrophobic, and low-porosity recycled aggregate. Detached mortar from mechanical or chemical treatments can even be reused as a cementitious material. Alternatively, recycled aggregate can be enhanced with pozzolanic slurry-based coatings, providing a sustainable, cost-effective, and high-quality recycled aggregate (Savva et al. 2021).

4 Properties of C & D Waste

4.1 Characteristics and Microstructure Properties

C & D waste is known for its rough and irregular surface, presenting several corners and a slightly flattened profile. The mechanical crushing process damages these aggregates and leads to the development of multiple internal microcracks. These microcracks arise not only due to compression and impact forces during the crushing of waste aggregates but also from factors like alkali-aggregate reactions present in the original concrete structures before demolition. Khoshkenari et al. (2014) carried out a study in which they produced recycled aggregates by processing reinforced concrete beams in a laboratory setting. Their observations, made using an electron microscope, revealed that recycled aggregates have higher surface roughness compared to natural aggregates. The increased surface roughness is due to traces of old bonding mortar or unhydrated cement slurry that are weakly attached to the surface. Furthermore, recycled aggregates exhibit significant voids and large pores, contributing to increased porosity. The porosity in recycled aggregates arises from the segregation of aggregate and mortar during the crushing of C & D waste. The smaller particle size of recycled fine aggregates indicates that they have more cracked and weathered cement mortar. This leads to a more intricate microstructure compared to recycled coarse aggregates. This complexity influences the microstructure and overall behaviour of concrete made with recycled aggregates. Critical attributes of recycled aggregates, including pore content, residual mortar, the interfacial transition zone, and the presence of cracks, play a significant role in determining the concrete's physical, mechanical, and durability characteristics (Andrade Salgado and Andrade Silva 2022).

4.2 Density

Density is a key property that influences the mechanical, physical, and durability characteristics of materials. C & D waste typically has a lower density compared to natural aggregates due to their porous structure. The density of recycled aggregates is influenced by factors such as the strength, grade and composition of the original concrete, as well as the duration and environmental conditions of its use (Fraj and Idir 2017). Khoshkenari et al. (2014) reported that C & D waste is approximately 20% lighter than natural aggregate, primarily due to the residual mortar adhering to the natural aggregate. When the strength of the parent concrete reduces or as the number of recycling cycles increases, the density of waste from the C & D processes decreases. Xu and Shi (2006) studied the density of C & D waste, indicating that it does not fulfil the standards of natural aggregate. On the other hand, the reduced density of waste from C & D may decrease the weight of buildings, improving their resistance to earthquakes. Zhu et al. (2019) found that third-generation recycled aggregate loses quality and becomes less than natural aggregate as the number of recycling cycles rises. This decrease is explained by the increased mortar content that affects other physical characteristics of the recycled aggregate after each recycling cycle.

4.3 Water Absorption

Water absorption is a crucial index for evaluating material properties. The recycling and crushing process results in a rougher surface for recycled aggregates, which are frequently covered in a layer of mortar, resulting in various internal cracks. Consequently, C & D waste exhibits significantly higher water absorption compared to natural aggregates. The water absorption of C & D of various sizes was found to increase as aggregate particle size decreased, likely due to the increased specific surface area. Zhu et al. (2019) found that as recycling cycles increase, recycled coarse aggregates from different generations absorb less water. C & D waste, with a compressive strength of 30 MPa, may have a higher water absorption capacity compared to ordinary aggregates (Khoshkenari et al. 2014).

5 Comparison Between C & D Waste and Natural Aggregate

The comparison of C & D waste and natural aggregate shows several notable differences. The size of both natural aggregate and C & D waste is influenced by the crushing process; however, C & D waste is more susceptible to breakage and fines generation during compaction due to its rough grains. The shape and roughness of natural aggregate are determined by the type of rock and its cleavage planes, while for C & D waste, these characteristics depend on the original material and the crushing process. A lower amount of adhered old mortar in C & D waste can lead to smoother surfaces; however, achieving uniformity is more difficult compared to natural aggregates.

The particle density and porosity of natural aggregate are primarily influenced by the type of rock, whereas for C & D waste, these properties are dependent on the size of the particles and the proportions of materials present. Generally, natural aggregate exhibits higher density, while C & D waste tends to have higher porosity, although it is less uniform. In terms of particle crushing strength, the strength of natural aggregate depends on the size and type of rock, whereas in the case of C & D waste, strength varies with particle size and material composition, typically being lower than that of natural aggregate, especially in masonry fragments.

Particle stiffness in natural aggregate depends on the size and type of rock, while in C & D waste, it is influenced by the size and proportions of materials within the particles. Structural concrete C & D waste tends to have high stiffness, but masonry particles in C & D waste exhibit lower stiffness. Regarding water retention properties, natural aggregate ability to retain water is influenced by the rock type and the amount of accessible pores. Conversely, C & D waste water retention depends on the materials and their proportions, with C & D waste generally absorbing more water due to the presence of adhered cement paste, although this absorption decreases significantly for bituminous, glass, or vitrified masonry. Finally, the frost resistance, soundness, and toughness of natural aggregate depend on the rock type. For C & D waste, these properties are influenced by the materials and their proportions within the particles. Both natural aggregate and C & D waste can have defects that affect toughness, but C & D waste typically has more cracks and defects. Frost resistance and soundness in both aggregates are affected by the accessible porosity and pore size, with C & D waste often showing lower resistance due to its heterogeneous composition.

6 C & D Waste in Stone Column

The C & D waste was collected from building debris from the construction industry, crushed into smaller pieces by mechanical equipment, and sieved for the required size. Shahverdi and Haddar (2020) and Haddad and Shahverdi (2021) used crushed concrete and bricks obtained from building debris. The aggregate size employed in stone columns varied between 2 and 9.5 mm, exhibiting a uniformity coefficient of 1.7 and classified as poorly graded gravel (GP). Alnunu and Nalbantoglu (2019) incorporated shredded bricks, crushed aggregates, and crushed waste concrete as substitutes for natural aggregate in stone columns, utilizing materials sourced from the construction sector. The mean diameters (D_{50}) of these materials were found to be 7 mm for crushed aggregates, 6.8 mm for crushed waste concrete, and 8 mm for shredded bricks. Salim et al. (Salim et al. 2020) brought recycled concrete from demolished buildings and broke it into less than 50 mm pieces. Saxena et al. (Saxena et al. 2024) collected demolition waste broken into sizes ranging from 8 to 12 mm. The crushed concrete waste was sieved. The material passed through a 12.5 mm IS sieve, and the material that was retained on an 8 mm IS sieve was used. Through analysis of the particle size distribution of the crushed concrete debris, the representative size (D_{10}) was determined to be 9 mm, indicating a uniform gradation. Table 1 includes the properties of C & D waste used by various researchers. Most researchers maintained consistent sizing for both natural aggregate and C & D waste materials to facilitate a comparative analysis of their respective behaviours. Figure 3 represents the particle size distribution of C & D waste used by various researchers.

7 Soil Properties

The remediation method is used on subsurface soils with an undrained shear strength of between 7 and 50 kPa, as well as loose sandy compositions and silty or clayey sands. Nevertheless, clay and silts that are sensitive and have a sensitivity of less than 4 cannot be stabilized using the stone column (Juran and Guermazi 1988; Barksdale and Bachus 1983). Clayey soil with low and high plasticity (Shahverdi and Haddad 2020; Haddad and Shahverdi 2021; Salim et al. 2020; Saxena et al. 2024) and silt and sand (Alnunu and Nalbantoglu 2019) was used to be improved by the stone column. Demir et al. (Demir et al. 2016) prepared a mixture comprising 40% kaolinite clay and 60% sand at a water content set at 1.5 times the liquid limit. Shear strength ranged from 4 to 22 kPa. The soil samples were sourced from local sites, including seashores and marine areas, ensuring consistency by extracting soil from depths of 1–5 m and sieving it within the range of 0.075–4.75 mm to meet the requirements of experimental procedures (Shahverdi and Haddad 2020; Haddad and Shahverdi 2021). The engineering properties of the soil were determined through a series of laboratory experiments, and the results of which are listed in Table 2. Figure 4 illustrates the particle size distribution curve for the different soils used in the stone column.

Table 1 Properties of various C & D waste

Properties	Shahverdi and Haddad (2020)	Shahverdi and Haddad (2020)	Demir et al. (2016)	Haddad and Shahverdi (2021)	Haddad and Shahverdi (2021)	Saxena et al. (2024)
C & D waste type	Crushed bricks	Crushed concrete	Recycled concrete aggregate	Crushed bricks	Crushed concrete	Crushed concrete
Size (mm)	2–9.5	2–9.5	–	2–9.5	2–9.5	2.5–10
Specific gravity	–	–	2.61	–	–	–
Dry density (kg/m^3)	950	1313	–	950	1313	–
Maximum dry density (kg/m^3)	–	–	–	1007	1367	–
Aggregate impact value (%)	31.7	26.0	–	31.7	26.0	27
Aggregate crushing value (%)	45.0	27.7	–	45.0	27.7	33

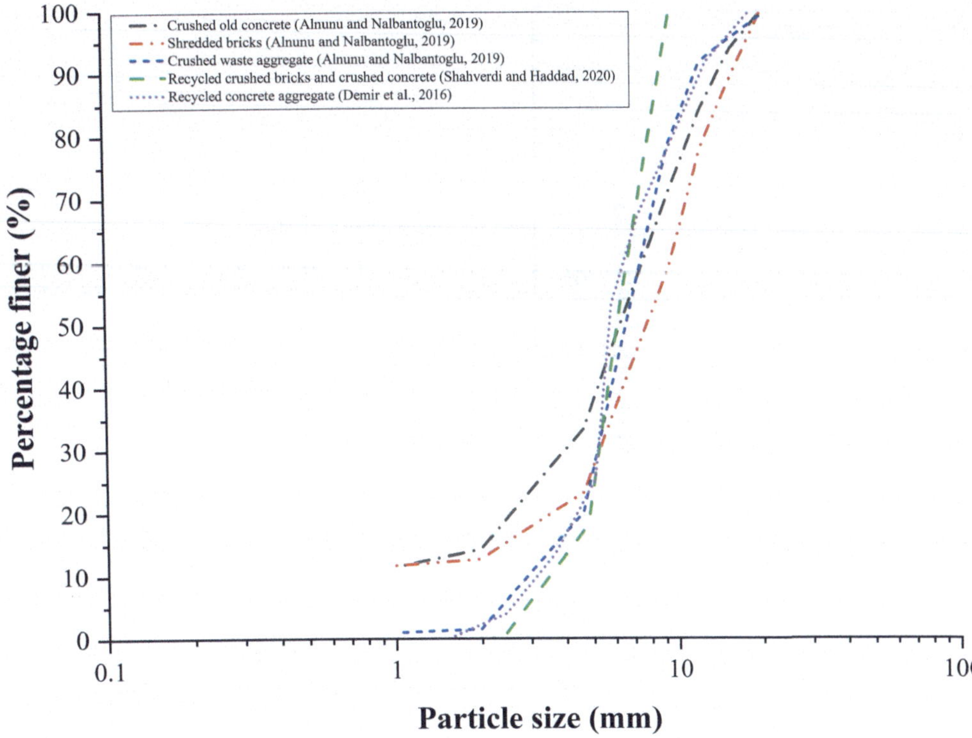

Fig. 3 Particle size distribution of various C & D waste (Shahverdi and Haddad 2020; Alnunu and Nalbantoglu 2019; Demir et al. 2016)

Table 2 Properties of various soils

Properties	Shahverdi and Haddad (2020)	Haddad and Shahverdi (2021)	Alnunu and Nalbantoglu (2019)	Salim et al. (2020)	Saxena et al. (2024)	Demir et al. (2016)
Type of soil	Clay	Clay	Loose sandy	Clay	Clay	Clayey sand
Classification	CL	CL	SP	CL	CH	SC
Specific gravity	2.65	2.69	2.6	2.67	2.71	2.63
Optimum moisture content (%)	16	18	–	–	21.7	–
Maximum dry density (kN/m^3)	17.3	15.3	–	–	15.9	–
Liquid limit (%)	26.5	51	–	43	54	25
Plastic limit (%)	6	22	–	19	20	14
Plasticity index (%)	20.5	29	–	24	34	11

8 Laboratory Model Test Studies

8.1 Model Test Set-Up Design

When designing a stone column set-up, there are a few important factors to take into account in order to accurately model the behaviour of load settlement. These include selecting an appropriate scale factor, ensuring a minimum clear distance between the test tank and the stone column periphery to avoid interference with failure wedges, and understanding the dimensions and characteristics of the stone column. Studies by Malarvizhi and Ilamparuthi (2004), Castro (2017), Verghese Chummar (1972), and Selig and McKee (1961) provide insights into the failure zone location and the extent of the rupture surface of the clay bed relative to the stone column periphery. Additionally, experiments conducted by Shahverdi and Haddad (2020) suggest that maintaining a tank diameter six times larger than the footing diameter can prevent slip plane effects on rigid boundaries. Stone column dimensions typically range from 50 to 100 mm in diameter, with a corresponding ratio between the stone column and particle diameter between 12 and 40. The ratio of length to diameter (L/D) is crucial, with research suggesting an optimal

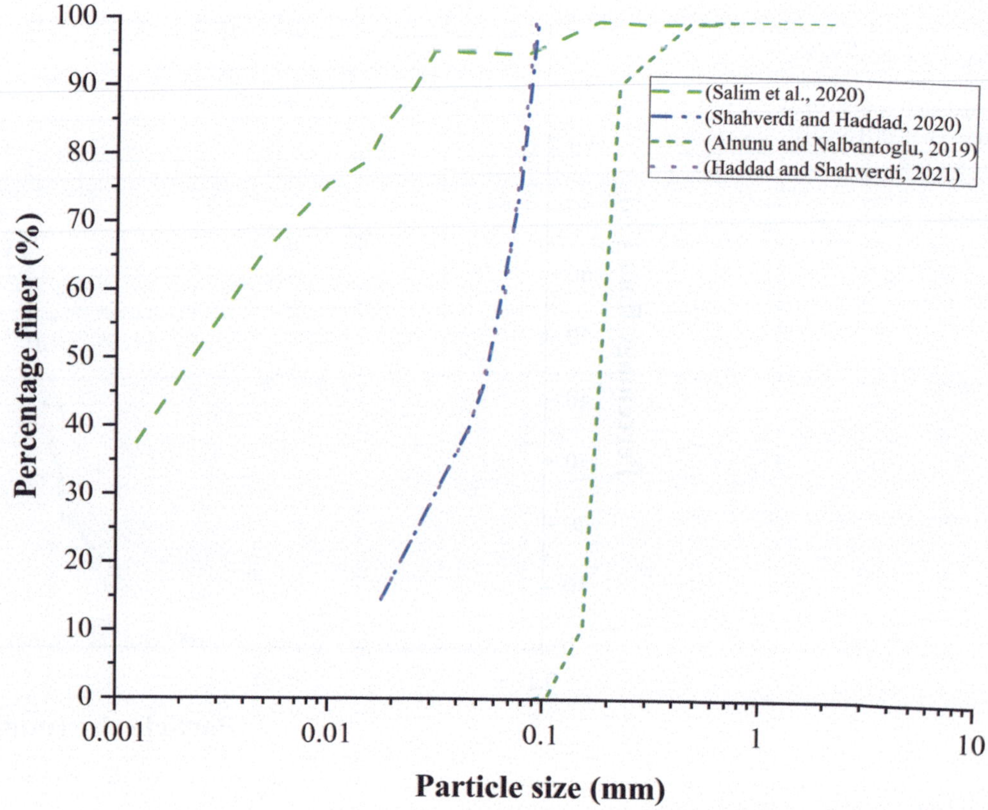

Fig. 4 Particle size distribution for treated soils (Shahverdi and Haddad 2020; Haddad and Shahverdi 2021; Alnunu and Nalbantoglu 2019; Salim et al. 2020)

range between 4.5 and 20. However, exceeding the L/D ratio of more than 4, stone columns are susceptible to bulging failure under both end bearing and floating conditions, as stated in IS 15284 (Part 1) (2003). Studies by Mazumder et al. (2018) indicate that bulging typically happens between half and four times the diameter of stone column, implying that the L/D ratio may not have a direct impact on bearing capacity. McKelvey et al. (2004) recommend a minimum L/D ratio of 6 to ensure that the column reaches its maximum allowable axial stress. Figure 5 illustrates the set-up used to test stone columns, including components like a loading frame, load cell, hydraulic jack, and LVDT.

8.2 Soil Bed Preparation and Stone Column Construction

The preparation of the bed soil involved several steps to ensure proper conditions for testing. For clay soil, an appropriate water to clay ratio was determined to achieve the required undrained shear strength. Silicon grease was used to lower friction and stop clay from sticking to the tank before it was filled. The tank was filled in layers to allow for proper compaction and the release of trapped air, and the clay and water were thoroughly mixed to ensure a uniform mixture. The clay sample was cured in order to achieve self-consolidation and a consistent water content. Samples were collected to verify homogeneity and water content. The mid-layer undrained shear strength was utilized in further computations (Shahverdi and Haddad 2020).

The use of loose sand represented the worst possible condition of the soil surrounding the stone column. PVC pipes were positioned on the oil-coated base layer before the tank was filled in order to reduce soil disturbance after removal. The tank inner surface was lightly greased to reduce friction. PVC pipes were filled with three layers of the stone column material until the required density was attained. The material density was selected to match the density of the surrounding soil. To ensure even density, sand layers were added and compacted around the PVC pipe. Before conducting tests, the soil was allowed to settle for a full day to allow the filler material inside the stone columns to have optimal contact with the compacted soil (Alnunu and Nalbantoglu 2019).

A single stone column was constructed by inserting a cylindrical pipe with a diameter equal to the stone column into the clay bed, and the installation and removal of the pipe were made simpler by using silicon grease (Shahverdi and Haddad 2020). Clay removal was done using a helical auger or spade, modified for improved soil retention. Aggregate or waste materials were filled in layers and compacted with density calculated based on weight and volume

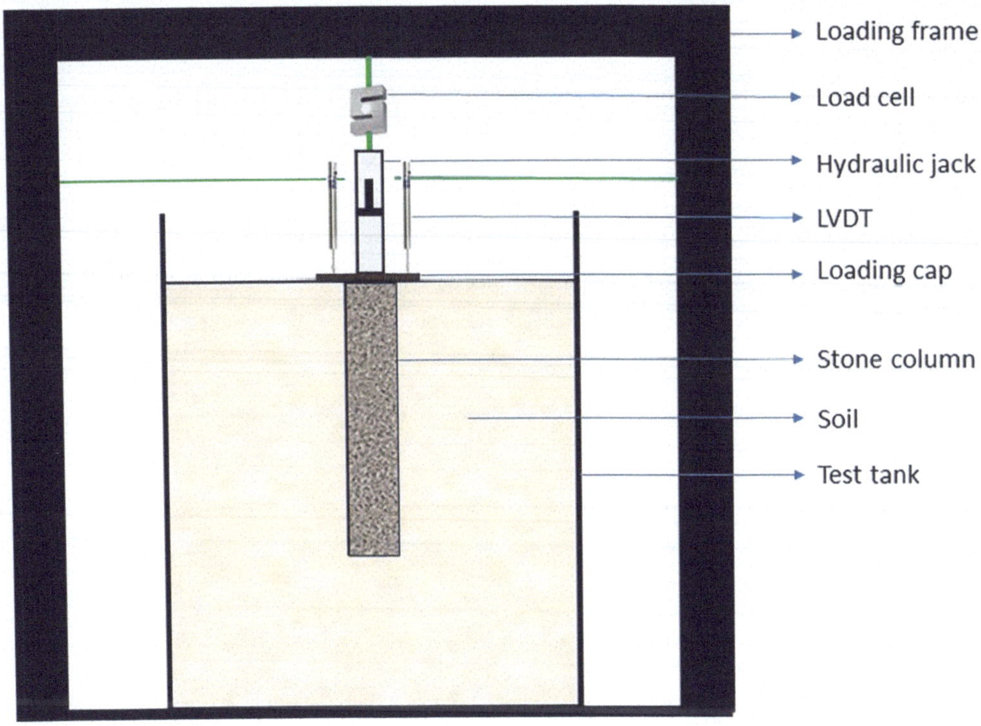

Fig. 5 Test tank set-up for the stone column

after compaction. Different sizes of materials were filled using pipes of varying diameters, ensuring larger aggregates remained on the outer part.

8.3 Loading Procedure

A systematic approach is essential for applying load to the stone column during testing. Vertical load application primarily utilized a hydraulic jack, with care taken to avoid direct loading on the bed. To ensure that the load was distributed evenly, a loading cap was used. As per IS 15284:2003 (2003), the loading cap should have a diameter that is twice as large as that of the stone column and a thickness ranging from 12 to 15 mm. This size corresponds to the diameter of the circular influence area. Shahverdi and Haddad (2020) determined that the replacement area should have a ratio of 1.6 times the stone column diameter, with a recommended range of 30 to 40% of the area of the footing.

The study by Barksdale et al. (1983) revealed that stone columns loaded over an area equal to their diameter could not support as much ultimate load as columns loaded through a rigid plate whose area exceeded the diameter of the column. This effect is due to the rise in vertical stress in the adjacent soil, which results from the expanded bearing area and the additional support provided by the stone column. Dial gauges or linear variable differential transducers mounted on the loading plate were used to measure settlement. Initially, Haddad and Shahverdi (2021) used a 2.5 kN/m^2 surcharge and loaded the test bed for four hours to ensure soil uniformity and remove any local disturbances (Black et al. 2011).

9 Discussion

Stone columns can enhance the bearing capacity by increasing the composite stiffness of the ground through the use of aggregate. When evaluating the performance of stone columns composed entirely of natural aggregate against those incorporating waste material or a mix of waste material and natural aggregate, the concept of "efficiency" is used. Efficiency refers to the ratio of the load-bearing capacity of any type of stone column, whether standard or encased and composed of various material mixtures to that of a standard stone column made exclusively from natural aggregate (Ayothiraman and Soumya 2015).

In a study by Shahverdi and Haddad (2020), it can be seen that the use of natural aggregate exhibited superior performance compared to utilizing other materials. Replacing natural aggregate entirely with recycled crushed bricks resulted in the highest efficiency at 89.26%, surpassing that achieved with other waste materials. Stone columns composed of 100% crushed concrete and a combination of 40% crushed bricks, 40% crushed concrete, and 20% natural aggregate yielded nearly identical efficiency values. This phenomenon can be attributed to the lower stiffness of crushed concrete and bricks compared to other materials. Due to the intrinsic strength of the crushed old concrete particles, research by Alnunu and Nalbantoglu (2019)

Table 3 Efficiency of C & D waste

Mix proportions	Efficiency (%)	Reference
100 gravel	100	Shahverdi and Haddad (2020)
100 recycled crushed bricks	89.26	
100 recycled crushed concrete	71.94	
50 recycled crushed concrete + 50 recycled crushed bricks	64	
40 recycled crushed concrete + 40 recycled crushed bricks + 20 gravel	71.44	
25 recycled crushed concrete + 25 recycled crushed bricks + 50 gravel	73.52	
10 recycled crushed concrete + 10 recycled crushed bricks + 80 gravel	62.30	
100 stone	100	Haddad and Shahverdi (2021)
100 crushed concrete	90.16	
100 crushed bricks	73.77	
100 crushed stone	100	Salim et al. (2020)
75 crushed stone + 25 waste concrete	64.71	
50 crushed stone + 50 waste concrete	70.58	
25 crushed stone + 75 waste concrete	114.71	
100 waste concrete	116.17	

revealed that crushed old concrete performed best when reinforced in sand. Conversely, the shredded brick particles performed less because they fractured at lower pressure levels.

Saxena et al. (2024) conducted California bearing ratio (CBR) tests on crushed concrete and natural aggregates, with CBR values recorded at 20.4% and 24%, respectively. Demir et al. (2016) found that recycled concrete aggregate demonstrated similar results to those of natural aggregate. In contrast to crushed brick and crushed concrete, Haddad and Shahverdi (2021) found higher bearing capacity was found in that natural aggregate. Similarly, Salim et al. (2020) noted that the greatest enhancement was observed when waste concrete was fully substituted with natural aggregate, owing to the favourable properties of waste concrete and its strong bonding with old mortar, resulting in low porosity. It can be seen from the review that waste concrete gave higher efficiency than natural aggregate. Table 3 gives the efficiency of C & D waste used. Figure 6 represents the graphical representation of efficiency of C & D waste with a line on 100% efficiency.

10 Conclusion

Stone columns play a crucial role in enhancing soil properties, particularly in soft and highly compressible soils such as clay, silt, and loose silty sand. A sustainable method of stabilizing soil is to build stone columns using waste materials. Despite the longstanding practice of using stone columns, specific design specifications and guidelines are lacking. Incorporating stone columns made of materials stiffer than the problematic soil consistently enhances the load-bearing capacity. It can be further increased by encasing the stone column. When substituting natural aggregate with waste materials, it is crucial to compare thoroughly to gauge the impact of the replacement. Maintaining similar gradation between waste materials and natural aggregate helps minimize the influence of gradation. While using waste materials alone generally results in lower bearing capacity compared to natural aggregates, certain combinations with natural aggregates can surpass bearing capacity with natural aggregates only. Recycled crushed concrete and crushed bricks typically exhibit inferior performance compared to natural aggregate. Single-material replacement is preferred over combinations, although recycled crushed concrete outperforms recycled crushed bricks in sand beds. The most promoting result across all reviewed studies was found when waste concrete was used in place of 100% natural aggregate, leading to 1.16 times increase in load-carrying capacity over natural aggregate. When incorporating waste materials into the construction of stone columns, it is important to consider environmental factors such as transportation-related emissions and the effective management of C & D waste. The high demand for natural aggregates highlights the need to minimize their consumption, which can result in cost savings, more efficient use of waste materials, and the conservation of natural resources. The findings of the reviewed study recommend that these objectives can be achieved by a combination of recycled aggregates, emphasizing that replacing waste materials is a cost-effective and environmentally friendly approach.

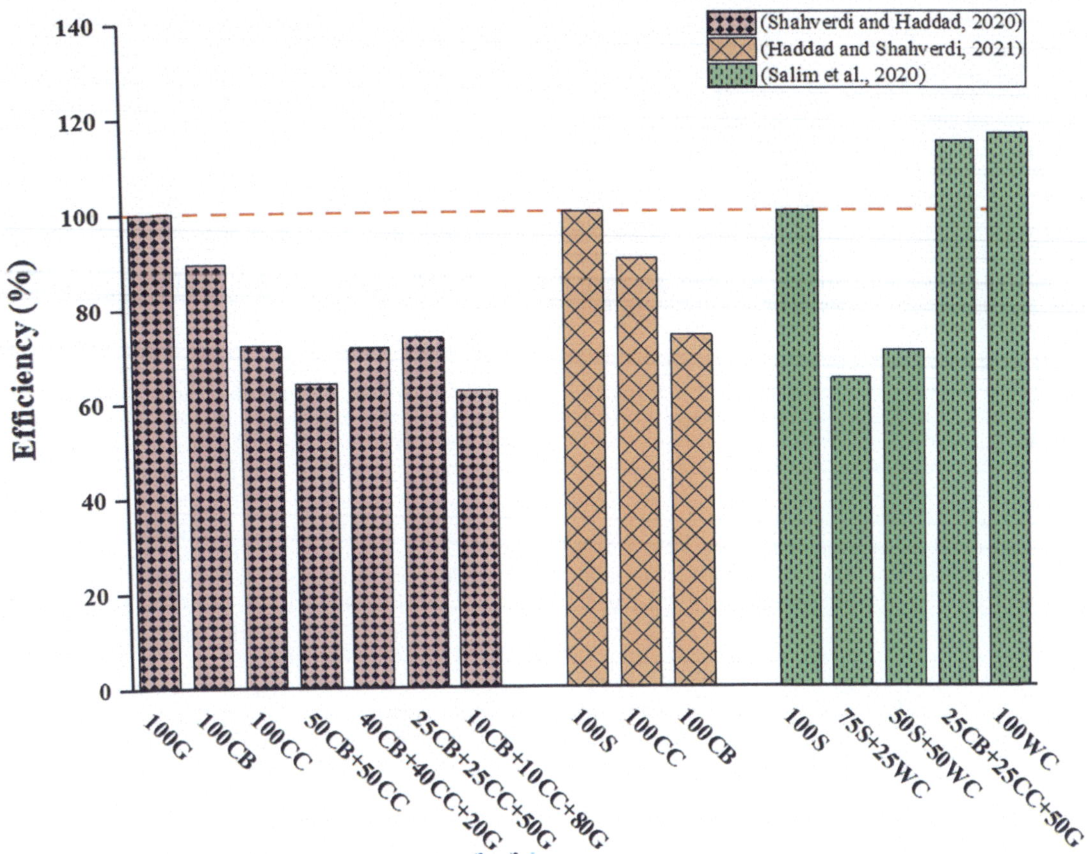

Fig. 6 Comparison between results of different C & D waste used in stone column

References

Agency USEP (2016) Advancing sustainable materials management: 2014 fact sheet

Al Zulayq DM, O'brien BT, Kowalewski MJ, Berenjian A, Tarighaleslami AH, Seifan M, Purchase CK, Manna D, Zulayq A, O'brien BT, Kowalewski MJ, Berenjian A, Tarighaleslami AH, Seifan M (2022) Circular economy of construction and demolition waste: a literature review on lessons, challenges, and benefits. Mater 15:76. https://doi.org/10.3390/MA15010076.

Alexiew D, Raithel M (2015) Geotextile-encased columns: case studies over twenty years. In: Ground improvement case histories. Embankments with special reference to consolidation and other physical methods, pp 451–477. https://doi.org/10.1016/B978-0-08-100192-9.00017-X

Almeida M, Riccio M, Hosseinpour I, Alexiew D (2018) Geosynthetic encased columns for soft soil improvement. https://doi.org/10.1201/9781315177144/GEOSYNTHETIC-ENCASED-COLUMNS-SOFT-SOIL-IMPROVEMENT-M

Alnunu MZ, Nalbantoglu Z (2019) Performance of loose sand with different waste materials in stone columns in North Cyprus. Environ Geotech 8:318–323. https://doi.org/10.1680/jenge.18.00079

Arulrajah A, Piratheepan J, Disfani MM, Bo MW (2013) Geotechnical and geoenvironmental properties of recycled construction and demolition materials in pavement subbase applications. J Mater Civ Eng 25:1077–1088. https://doi.org/10.1061/(ASCE)MT.1943-5533.0000652

Arulrajah A, Ali MMY, Piratheepan J, Bo MW, Asce M (2012) Geotechnical properties of waste excavation rock in pavement subbase applications. J Mater Civ Eng 24:924–932. https://doi.org/10.1061/(ASCE)MT.1943-5533.0000419

Ayothiraman R, Soumya S (2015) Model tests on the use of tyre chips as aggregate in stone columns. Proc Inst Civ Eng Gr Improv 168:187–193. https://doi.org/10.1680/grim.13.00006

Barksdale R, Bachus R (1983) Design and construction of stone columns, vol I

Barksdale RD, Bachus RC (1983) Design and construction of stone columns, vol. I. https://doi.org/10.21949/1503647

Behera M, Bhattacharyya SK, Minocha AK, Deoliya R, Maiti S (2014) Recycled aggregate from C&D waste & its use in concrete—a breakthrough towards sustainability in construction sector: a review. Constr Build Mater 68:501–516. https://doi.org/10.1016/J.CONBUILDMAT.2014.07.003

Black JA, Sivakumar V, Bell A (2011) The settlement performance of stone column foundations. Géotechnique 61:909–922

Castro J (2017) Groups of encased stone columns: Influence of column length and arrangement. Geotext Geomembranes 45:68–80

Conlon K, Jayasinghe R, Dasanayake R (2019) Circular economy: waste-to-wealth, jobs creation, and innovation in the global south. World Rev Sci Technol Sustain Dev 15:145–159. https://doi.org/10.1504/WRSTSD.2019.099377

Crawford RH, Mathur D, Gerritsen R (2017) Barriers to improving the environmental performance of construction waste management in remote communities. Procedia Eng 196:830–837. https://doi.org/10.1016/J.PROENG.2017.08.014

de Andrade Salgado F, de Andrade Silva F (2022) Recycled aggregates from construction and demolition waste towards an application on structural concrete: a review. J Build Eng 52:104452. https://doi.org/10.1016/J.JOBE.2022.104452

Demir S, Mokarram FR, Özener P (2016) The sustainable design of granular columns based on laboratory model tests, pp 893–903

Deshpande TD, Kumar S, Begum G, Basha SAK, Rao BH (2021) Analysis of railway embankment supported with geosynthetic-encased stone columns in soft clays: a case study. Int J Geosynth Gr Eng 7:1–16. https://doi.org/10.1007/S40891-021-00288-5/METRICS

Ding Z, Wang Y, Zou PXW (2016) An agent based environmental impact assessment of building demolition waste management: conventional versus green management. J Clean Prod 133:1136–1153. https://doi.org/10.1016/J.JCLEPRO.2016.06.054

Egan D, Slocombe B (2015) Demonstrating environmental benefits of ground improvement. Proc Inst Civ Eng Gr Improv 163:63–69. https://doi.org/10.1680/GRIM.2010.163.1.63

Fauzi A, Rahman WMNWA, Jauhari Z (2013) Utilization waste material as stabilizer on Kuantan clayey soil stabilization. Procedia Eng 53:42–47. https://doi.org/10.1016/j.proeng.2013.02.007

Fraj AB, Idir R (2017) Concrete based on recycled aggregates—recycling and environmental analysis: a case study of Paris' region. Constr Build Mater 157:952–964. https://doi.org/10.1016/J.CONBUILDMAT.2017.09.059

Geissdoerfer M, Savaget P, Bocken NMP, Hultink EJ (2017) The circular economy—a new sustainability paradigm? J Clean Prod 143:757–768. https://doi.org/10.1016/J.JCLEPRO.2016.12.048

Haddad A, Shahverdi M (2021) Evaluation of recycled materials as aggregate of end bearing and floating stone columns: a comparative study. J Rehabil Civ Eng 9:61–74. https://doi.org/10.22075/jrce.2021.21286.1443

Han J (2015) Principles and practice of ground improvement

Iacoboaea C, Aldea M, Petrescu F (2019) Construction and demolition waste—a challenge for the European Union? Theor Empir Res Urban Manag 14:30–52

IS 15284 (Part 1): IS 15284-1 (2003) Design and construction for ground improvement—guidelines, Part 1: Stone columns

Juran I, Guermazi A (1988) Settlement response of soft soils reinforced by compacted sand columns. J Geotech Eng 114:930–943. https://doi.org/10.1061/(ASCE)0733-9410(1988)114:8(930)

Khoshkenari AG, Shafigh P, Moghimi M, Mahmud H (2014) Bin: the role of 0–2 mm fine recycled concrete aggregate on the compressive and splitting tensile strengths of recycled concrete aggregate concrete. Mater Des 64:345–354. https://doi.org/10.1016/J.MATDES.2014.07.048

Li J, Xiao H, Zhou Y (2009) Influence of coating recycled aggregate surface with pozzolanic powder on properties of recycled aggregate concrete. Constr Build Mater 23:1287–1291. https://doi.org/10.1016/J.CONBUILDMAT.2008.07.019

Malarvizhi SN, Ilamparuthi K (2004) Load versus settlement of clay bed stabilized with stone and reinforced stone columns. In: 3rd Asian regional conference on geosynthetics, pp 322–329

Mazumder T, Rolaniya AK, Ayothiraman R (2018) Experimental study on behaviour of encased stone column with tyre chips as aggregates. Geosynth Int 25:259–270. https://doi.org/10.1680/jgein.18.00006

McKelvey D, Sivakumar V, Bell A, Graham J (2004) Modelling vibrated stone columns in soft clay. Proc Inst Civ Eng Eng 157:137–149

Menegaki M, Damigos D (2018) A review on current situation and challenges of construction and demolition waste management. Curr Opin Green Sustain Chem 13:8–15. https://doi.org/10.1016/J.COGSC.2018.02.010

Pandey BK, Rajesh S, Chandra S (2022) Performance of soft clay reinforced with encased stone column: a systematic review. Int J Geosynth Gr Eng 8:1–14. https://doi.org/10.1007/S40891-022-00387-X/METRICS

Pickin J, Wardle C, O'farrell K, Nyunt P, Donovan S (2020) National waste report 2020. Department of agriculture, water and the environment, p 1121

Poon CS, Yu ATW, Wong A, Yip R (2013) Quantifying the impact of construction waste charging scheme on construction waste management in Hong Kong. J Constr Eng Manag 139:466–479. https://doi.org/10.1061/(ASCE)CO.1943-7862.0000631/ASSET/E021CCA0-BBEA-4233-A1FC-0A8112FB4879/ASSETS/IMAGES/LARGE/FIGURE11.JPG

Rahardjo H, Satyanaga A, Leong EC, Wang JY (2013) Unsaturated properties of recycled concrete aggregate and reclaimed asphalt pavement. Eng Geol 161:44–54. https://doi.org/10.1016/J.ENGGEO.2013.04.008

Roussat N, Méhu J, Abdelghafour M, Brula P (2008) Leaching behaviour of hazardous demolition waste. Waste Manag 28:2032–2040. https://doi.org/10.1016/J.WASMAN.2007.10.019

Safiuddin M, Alengaram UJ, Rahman MM, Salam MA, Jumaat MZ (2013) Use of recycled concrete aggregate in concrete: a review. J Civ Eng Manag 19:796–810. https://doi.org/10.3846/13923730.2013.799093

Salim NM, Hasan S, Al-Soudany K (2020) Effect of replacing crushed stone in stone columns by waste material on soil improvement ratio. Proc Int Struct Eng Constr 7:2–7. https://doi.org/10.14455/ISEC.res.2020.7(1).GFE-09

Savva P, Ioannou S, Oikonomopoulou K, Nicolaides D, Petrou MF (2021) A mechanical treatment method for recycled aggregates and its effect on recycled aggregate-based concrete. Mater 14:2186. https://doi.org/10.3390/MA14092186

Saxena S, Roy LB, Gupta PK, Kumar V, Paramasivam P (2024) Model tests on ordinary and geosynthetic encased stone columns with recycled aggregates as filler material. Int J Geo-Eng 15:1–13. https://doi.org/10.1186/S40703-023-00202-0/TABLES/4

Selig ET, McKee KE (1961) Static and dynamic behavior of small footings. J Soil Mech Found Div 87:29–45. https://doi.org/10.1061/JSFEAQ.0000378

Shahidan S, Azmi MAM, Kupusamy K, Zuki SSM, Ali N (2017) Utilizing construction and demolition (C&D) waste as recycled aggregates (RA) in concrete. Procedia Eng 174:1028–1035. https://doi.org/10.1016/J.PROENG.2017.01.255

Shahverdi M, Haddad A (2020) Use of recycled materials in floating stone columns. Proc Inst Civ Eng Constr Mater 173:99–108. https://doi.org/10.1680/jcoma.18.00086

Shooshtarian S, Maqsood T, Wong PSP, Khalfan M, Yang RJ (2019) Review of energy recovery from construction and demolition waste in Australia. J Constr Eng 2:112–130. https://doi.org/10.31462/jcemi.2019.03112130

Shrivastava S, Chini A (2009) Construction materials and C&D waste in India. Lifecycle Des Build Syst Mater 72

Verghese Chummar A (1972) Bearing capacity theory from experimental results. J Soil Mech Found Div 98:1311–1324

Verma A, Babu VS, Arunachalam S (2022) Influence of acetic acid soaking and mechanical grinding treatment on the properties of treated recycled aggregate concrete. J Mater Cycles Waste Manag 24:877–899. https://doi.org/10.1007/S10163-022-01360-6/FIGURES/30

Wehr J (2006) The undrained cohesion of the soil as criterion for the column installation with a depth vibrator. In: Proceedings of the international symposium on vibratory pile driving and deep soil Vibratory compaction. TRANSVIB, Paris, pp 157–162

Wu H, Zuo J, Yuan H, Zillante G, Wang J (2023) Investigation of the social and economic impacts of cross-regional mobility of construction and demolition waste in Australia. Resour Conserv Recycl 190:106814. https://doi.org/10.1016/J.RESCONREC.2022.106814

Xu YZ, Shi JG (2006) Analyses and evaluation of the behaviour of recycled aggregate and recycled concrete. Concrete 7:41–46

Yuan H (2013) Key indicators for assessing the effectiveness of waste management in construction projects. Ecol Indic 24:476–484. https://doi.org/10.1016/J.ECOLIND.2012.07.022

Zheng L, Wu H, Zhang H, Duan H, Wang J, Jiang W, Dong B, Liu G, Zuo J, Song Q (2017) Characterizing the generation and flows of construction and demolition waste in China. Constr Build Mater 136:405–413. https://doi.org/10.1016/J.CONBUILDMAT.2017.01.055

Zhu P, Hao Y, Liu H, Wei D, Liu S, Gu L (2019) Durability evaluation of three generations of 100% repeatedly recycled coarse aggregate concrete. Constr Build Mater 210:442–450. https://doi.org/10.1016/J.CONBUILDMAT.2019.03.203

Efficient Construction and Demolition Waste Management Practices

Rishabh Joshi, Anjali Singh, Rifat Resatoglu, and Shaib Abdulazeez Shehu

Abstract

Twenty-first century has witnessed insurmountable construction developments worth billions of crores which causes overwhelming waste generation during the construction, demolition, and renovation of any civil engineering structures. These wastes usually comprise of concrete, metals, plastic, aggregates, brick, wood, etc. Their disposal in the landfills, oceans, rivers, and open ground affects the environment significantly. The use of virgin materials for new projects is also uneconomical. High carbon content from the cement and concrete industry in the atmosphere is one of the factors causing global warming. The biggest challenge in the construction and demolition waste (C&DW) management is the uncertainty in the quantum and composition of these wastes in India and other developed countries. Hence, strict guidelines are necessary for their environmental management. Some other resource-efficient practices include promotion of sustainable, green, and locally available materials, source reduction, salvaging and reusing the C&DW. The recycling and reuse of C&DW in the production of fuels, aggregates, compost, and soil property amendment can reduce the fresh material demand and generate employment opportunities in this sector. This chapter describes the challenges due to indiscriminate C&DW disposal leading to environmental degradation and pollution, the waste management flowchart and hierarchy, the reuse and recycle of waste, noise and dust management, and analysis of existing guidelines for construction sectors. The sustainable C&DW management practices provide various opportunities for natural resource conservation, minimal waste production, reduced environmental degradation, conservation of the existing buildings, long-term impact on the health and lifestyle of people and employment generation.

Keywords

Construction and demolition · Sustainable management · Recycling · Reuse · Environment

1 Introduction

Due to constant urbanization, approximately half of the world's population now lives in big cities. This makes it necessary to address the demands of urbanization through infrastructure projects, the movement of energy, data, and natural and human resources. Simultaneously, the large-scale infrastructural developments that accompany urbanization led to serious environmental issues affecting human populations, economic and social life, and the cultural and historical environment. Natural resource requirements are growing due to both population growth and global economic expansion, which means that resource usage is outpacing ecosystems' ability to replenish themselves. The result is the system's incapacity to meet demands while contributing to environmental degradation and producing more pollutants. In Western nations, one of the primary markers

R. Joshi (✉) · A. Singh (✉)
Faculty of Civil Engineering, Shri Ramswaroop Memorial University, Barabanki, Uttar Pradesh, India
e-mail: rishabhjoshi.ce@gmail.com

A. Singh
e-mail: anjalisingh.ce@srmu.ac.in

R. Resatoglu
Faculty of Civil and Environmental Engineering, Near East University, Nicosia, Northern Cyprus

S. A. Shehu
Department of Civil and Mining Engineering, Confluence University of Science and Technology, Osara, Nigeria

School of Materials and Mineral Resources Engineering, Universiti Sains Malaysia, Nibong Tebal, Penang, Malaysia

A. Mandpe et al. (eds.), *Sustainable Waste Management*, Advances in Science, Technology & Innovation,
https://doi.org/10.1007/978-3-031-85685-3_12

of prosperity and progress is construction activity. The building sector aids every nation's general socio-economic development. Aiming to form communities that promote economic justice, preserve and restore the balance between the built and natural environments, as well as to form communities that uphold human dignity and promote economic justice, sustainable construction is a comprehensive building method (Pavlides et al. 2023).

However, economic performance is currently the primary focus of construction activities, with environmental and social concerns coming in second. The fact that environmental contamination from construction is a major global issue further supports this. Because it consumes a large portion of natural, non-renewable resources, the building sector is accountable for the most severe local and worldwide environmental changes, environmental degradation, and resource depletion. For this reason, the building industry is essential to sustainability and sustainable development (Pavlides et al. 2023).

Up until now, sustainability in building has mostly dealt with technical problems and has depended on the circumstances and degree of development of each nation. Sustainability is associated with developing and implementing new technologies and upgrading current ones in industrialized nations; in less developed nations, sustainability is associated with social and economic issues (Giannarakis et al. 2017). In the modern building industry, sustainability is linked to issues that are cultural, social, technical, environmental, and economical. Agenda 21 addresses the implementation of sustainability principles in the construction sector against sustainability declarations like UN Sustainable Development Goals and the Brundtland Report and their implementation at a more localized level in each nation.

Urbanization will increase to 68% of the world's population by 2050. This further demands an escalation in infrastructure facilities to support various social and public services like sustainable settlements, food, connectivity, health, and education. However, there exist many infrastructural challenges like ineffective resource utilization, pollution, and environmental degradation. United Nations has reinstated the significance of sustainable development goals, recognizing the need to understand and respond effectively towards infrastructure-related concerns such as Construction and Demolition waste (C&DW). A good infrastructure in the fast-growing urban areas ascertains the necessity of efficient C&D waste management system. Since the construction sector is significantly important in meeting the basic demands of infrastructure and further impacting the economy and productiveness of a country, it is necessary to ascertain effective management of waste practices to preserve the environment, decrease pollution, and efficiently utilize resources.

Developed and many developing countries are regularly facing the pressing issue of C&D waste and their management. Australia produces 20% waste and United States produces 30% construction waste which is dumped in landfills (Narcis et al. 2019). On the other hand, in developing nations like India, 40–60 kg/sqm of C&DW is produced in new construction enterprises, in Hongkong, 3584 tons of C&DW is produced on a daily basis.

Traditional approach to handling such wastes are landfilling. However, such practices have a notable impact on the environment. According to some research, landfilling can cause ecosystem damage, human health damage and resource damage. Therefore, it is imperative to address the issues related with C&DW generation, handling, and disposal. Circular economy is one of the promising ways to address these issues. It integrates sustainable development and material circulation by incorporating closed loop manufacturing patterns and forging a robust socio-economic structure. The concept of Circular Economy implies that the materials are intended to maintain their functional value and be recycled into resources upon completion of their productive life (Velenturf and Purnell 2021).

1.1 Construction and Demolition Waste (C&D Waste)

Waste produced at construction sites or temporary on-site activities, demolition of old structures, repair works, and remodelling of civil engineering structures resulting in debris, waste building materials, and rubble is known as Construction and Demolition waste (Construction and Demolition (C&D). Waste Rule 2016). This might also include wastes generated from the demolition of construction works by natural disasters like earthquakes, tsunamis, floods, cyclones, etc. (Kabirifar et al. 2020).

The composition of such waste products is based on the construction material used and construction style. Almost 90% of the C&D waste is inert comprising of 36% of gravel, sand, and soil, 31% of brick masonry, 23% of concrete, 5% of metal scraps, 2% of wood and 3% of other waste materials (Fig. 1).

However, other literature has considered C&D waste as an aggregate of hazardous and inert or non-hazardous and non-inert waste in which construction materials alone are homogeneous and simple to recycle, whereas demolition wastes are heterogeneous in nature, comprising of hazardous elements, making them hard to classify and recycle. Hence, C&D wastes develop in construction and demolition sites as hazardous and non-hazardous waste in the form of non-inert and inert materials (Rayhan and Bhuiyan 2023).

The construction, repair, reconstruction and maintenance of buildings and structures produces wastes known

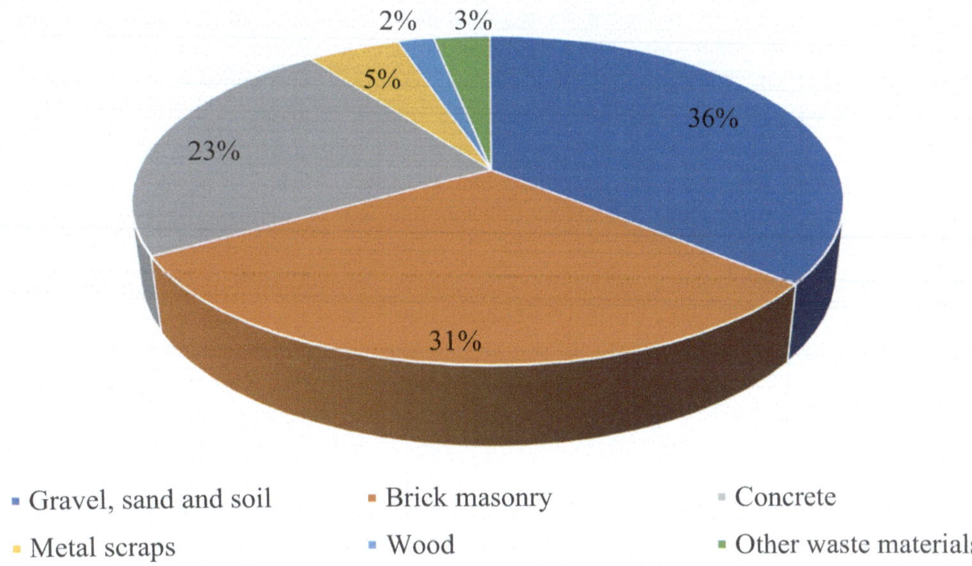

Fig. 1 Composition of construction and demolition waste

as construction wastes. The material removed during repair, demolition due to natural or artificial means, and reconstruction is known as demolition waste. The production of demolition projects-related waste is 20–30 times more in comparison to construction project-related waste.

1.2 Waste Composition

The composition of C&DW depends on the types of buildings, construction methods, population growth rate, guidelines for waste management, the development of the construction industry in the country, and other such factors. Based on their origin, the C&DW can be classified as:

(i) **Demolition waste**: These waste materials are left over after structures or other infrastructure are demolished by nature in disaster-like events or anthropogenic activities. The composition of these wastes depends upon building's type, size, shape, age, and location. This category includes wastes such as sand, concrete fragments of various sizes, soil, gravel, bricks, dressed stones, etc.

(ii) **Construction waste**: This category includes wires, metals, wood, plastic, paper, glass, adhesives, coverings, pigments, coatings, and other materials that generate from building activities on-site.

(iii) **Excavation waste**: Excavation waste includes all materials obtained from excavations, such as sand, clay, gravel, rocks, and excavated soil. Nearly every kind of building activity produces these wastes, but geotechnical engineering and underground construction produce them the most. These waste products can also result from landslides and stream overflows, among other natural events.

(iv) **Transport-Sector-related waste**: This category includes all asphalt and pavement materials, including gravel, sand, metals, and materials remaining from road construction and demolition.

(v) **Maintenance-related waste**: These types of waste are generated during the maintenance of buildings. They comprise of sand, concrete, wood, plastics, glass, etc.

2 Sources of Construction Waste Generation

One of the major producers of waste is the Construction sector. The waste gets generated in various construction stages ranging from extraction of raw materials, manufacturing process, construction activities, demolition of any structure and final disposal of waste in landfills (Río et al. 2009). Construction comprises three phases, i.e. pre-construction, during construction, and post-construction.

Wastes generated during these stages are listed below.

2.1 Pre-construction Stage

Pre-construction phase of any construction industry comprises of design of a building and contract document. It includes specifications of materials, design drawings, work schedules, and legal papers binding the stakeholders of a project. In the construction industry, design and contract issues are perceived to impact the project's overall success significantly. The design aspect includes the design of building, plumbing fixtures, electrical provisions. The decision regarding design affects the production of construction and demolition waste. The failure of designers to interpret and implement measures to reduce waste generation results in about 33% waste generation. Most of the construction debris comprises of electrical and mechanical fixtures. This is attributed to errors in design resulting in reworking and change of fixtures. Many designers have a lack of understanding with respect to waste management and often consider it as an issue specific to construction site and not design process. However, early considerations by designers can substantially reduce waste generation.

Contract document is observed to be another source of Construction and Demolition Waste. In a study conducted on contractual clauses of various manuals of construction projects, it was found that imposition of a contract clause to develop coordination among stakeholders in the pre-construction stage leads to less generation of site waste. It was also established that regular revision, elimination of useless clauses, and alteration of contract clause pertaining to inspection, workmanship, and quality help in minimizing waste generation.

Two significant impacts of errors in contract documents are deterioration of buildings and collapse of structures (Dosumu 2018). This may happen due to wrong design calculations, omission of important design parameters, forgetting items in bill of quantities, wrong cost estimate, incomplete knowledge and inexperience of engineer and contractor, incompetent structural engineer as well as dimensional errors in calculations. Hence, in order to ensure development of a quality project with minimal waste generation, it is mandatory to evaluate and consider the relation between pre-construction phase and management of waste.

2.2 Construction Stage

There were four elements that were found to be associated with waste generation during the construction stage are material handling, management of site, supervision, human resource, and storage of materials. Material handling concerns itself with material flow on the site of the project. On the other hand, general instructions for on-site waste management techniques need to be coordinated and monitored in the context of site management and supervision. However, waste management problems brought on by workers at the project site are among the factors relating to human resources. Storage of materials concentrates on the issues related to the storage of building materials.

The connection between trash creation and site management strategies has been emphasized in many literatures. Proper site management ensures that the construction mistakes are reduced, building materials are stored properly, good quality materials are used which ensures desired strength of structure and right quantity of materials are ordered to minimize wastage. Experienced contractors and site engineers can also ensure that materials can be reused and recycled at the end of their useful life and even at the construction stage recycled and reclaimed materials like brick, concrete, etc. can be used for temporary structures.

During the construction of roads and highways, the main cause of waste is the shortage of construction equipment. A substantial portion of the waste generated at the facility is caused by ordering, shipping, and product procurement problems that do not meet standards. The mapping method was used by Hasmori et al. (2020) to identify the management techniques that promote waste minimization during project building. Off-site utilization of products, enhancing workforce education, waste segregation, and appropriate material handling were found to be beneficial.

2.3 Post-construction Stage

Waste generation in the post-construction phase is influenced by residual-materials-related factors. According to Fadiya et al. (2014), residual trash makes for 30% of the project's overall cost. Given the staggering amounts of money spent on residual waste, it is vital to investigate the sources of waste and identify appropriate mitigation strategies in order to limit the associated risks.

The primary cause of physical waste is residuals. The incidence of material waste from the construction of public building projects was examined by Meghani et al. (2011). The results showed the amount of building material that gets wasted like sand- 13.95%, bricks- 10.29%, cement- 4.16%, and reinforcement steel- 4.15%. The difference between the anticipated and actual material consumption was used to calculate waste.

Sources of C&DW are mostly leftover materials on the site of construction, waste from disasters, waste produced due to errors in operation, and procurement of materials. Other indirect sources of generation of C&DW are budgeting errors in management of construction waste, lack of proper cooperation among workers, communication and supervision, lack of consciousness on part of contractor, and usage of poor-quality materials (Islam et al. 2019). The leading reasons for generation of waste in construction sites are improper use of construction materials, traditional construction methods, unskilled workers.

The C&DW can be categorized to be produced from majorly eight sources which are the various procedures of any construction activity like design, procurement procedure, operation, site management and supervision, residual wastes, handling and storage of materials, and external sources like.

3 Impacts of Construction and Demolition Waste

The production of C&DW has several harmful effects, such as increasing the amount of land used for waste landfilling, endangering public health, harming the environment, wasting natural resources, and placing a financial strain on businesses and governments. The impacts of C&DWs are covered below.

3.1 Impact on Environment

3.1.1 Air Pollution

Fast urbanization and growth in many nations lead to the regular presence of volatile soil, sand, and construction and demolition dust in the air. Furthermore, the transportation of C&DWs contaminates the air, causing severe haze and blurs in urban areas. In 2015, the percentage of air pollution due to C&DW for the Chinese cities of Beijing, Shenzhen, and Guangzhou was 14%, 12%, and 10%, respectively. The incineration of C&DWs results in several issues, including the production of toxic compounds such as dioxins from plastic waste, exhausted gases from uncontrolled combustion that cause severe air pollution and acid rain, and the possibility of heavy metals and other contaminants in the bottom residue, or fly ash (Duan et al. 2019).

3.1.2 Soil and Water Pollution

Because emerging towns often lack leachate accumulation systems, hazardous pollutants such as landfill leachate may contaminate nearby surface water, groundwater, and soil during periods of rain. For example, a C&DW landfill site in Shenzhen, China, is situated next to a source of drinking water. As a result, humans as well as other flora and wildlife may be impacted by this contaminated soil and water (Duan et al. 2019).

3.1.3 Toxic Effects

Since most building materials are poisonous or hazardous elements, inert, such as heavy metals and brominated flame retardants (BFRs), may be present in C&DW or cause pollution. Moreover, the biodegradation of gypsum drywall in C&DW landfills may emit hydrogen sulfide (H_2S) (Yang et al. 2006).

3.2 Impact on Natural Resources, Social Life and Economics

3.2.1 Loss of Natural Resources

The land will be lost if the C&DW amount grows steadily, as more landfills will be required for disposal. Earth lands are harmed because of the removal of raw materials for construction activities (Aslam et al. 2020). Construction operations use about 32% of the world's resources, including up to 40% of energy and 12% of water. About 40% of the raw materials extracted from the ground are used in construction, and 25% are made entirely of wood.

3.2.2 Societal Impacts

People need to be made aware of the harmful environmental impacts of C&DW. This means that there will be a chance to educate and teach people on the construction and redesign of environmentally friendly structures (Aslam et al. 2020).

3.2.3 Economic Impacts

There are expenses related to landfills as well as the procurement, accumulation, and sorting of C&DW. On the other hand, after C&DW is recycled and reused, there are financial gains. Job opportunities can also be created in waste management sector (Aslam et al. 2020). Recycled aggregate obtained through demolition of buildings can be utilized as an aggregate which can lead to economic benefits of around $35 million and it will also reduce greenhouse gas emissions. As per the codal standards, around 20–25% of coarse aggregates and fine aggregates can be replaced with recycled aggregates. Similarly, recycling of aggregates can also reduce the overall cost of the project as well as reduce the environmental damage due to landfill emissions (Ghisellini et al. 2018).

4 Strategies for Implementation of Circular Economy in Construction and Demolition Waste Management Process

The circular economy concept is applied as a management tool in the later stages (building, operation, and end of life) or as a preventive strategy in the earlier stage (design phase).

4.1 Design Phase

If the C&DW management is implemented during the first step of design, it would help reduce the quantity of C&DW generated. Because the design choices made at this stage directly impact the stage that follows, a well-thought-out design approach could contribute to waste reduction, increased recycling rates, and material reuse throughout the project. In design stage, the choice of materials used for construction purposes should be planned in such a way that sustainable materials should be promoted, wherever feasible, and the use of cement-based and concrete structures could be reduced which may reduce the load on the environment. Use of steel and timber frames can be promoted which will lead to reduction in waste generated during both construction and demolition phase of the building. Individual components of the structure can be designed by reusing materials like bricks, timber, recycled aggregates, furniture, etc. Moreover, during the design stage only the choice of materials for different components of the structure should be such that they can be easily deconstructed and utilized for different projects even at the end of the life of structure. Numerous studies have highlighted the importance of implementing a circular economy throughout the project's early design phase (Akanbi et al. 2018). The buildings in the circular economy concept are not made for a single use, but rather for a lifecycle.

4.2 Construction Phase

The quantity of waste produced during construction phase is significantly influenced by the construction process. Using steel or timber frames for building or off-site manufacturing could save waste production by 90 percent. The basic goals of the circular economy implementation during the construction phase are to minimize the waste produced on-site, and the usage of reused and recycled materials should be promoted (Groh and Dubik 2018). Adams et al. (2017) suggested off-site building and purchasing recyclable and reuse materials in light of this.

4.3 Operation Phase

To increase a building's lifespan, its design should minimize maintenance, making repairs easier and allowing for future upgrades to accommodate changing needs (Groh and Dubik 2018). To extend the life of the building, Zimmann et al. (2016) also recommend that low-energy and inexpensive sensor technology be used on a regular basis for building management. There are various sensor technologies which can be used to monitor the performance of the building, like, temperature sensors, chemical sensors, pressure sensors, humidity sensors, air quality and water quality sensors, as well as motion sensors. Temperature sensors are effectively utilized in fire alarms, lighting controls and HVAC systems to maintain and regulate the temperature of the building. Air quality sensors detect the harmful pollutants present in the air and are used in air purifiers. Pressure sensors are used to measure the load on critical elements of the structure as well as in bridges, tunnels, foundations, etc. Water quality sensors measure the pH level, suspended solids, cations and anions present in water. Humidity sensors monitor the humidity level in hospitals and museums.

4.4 End-of-Life Phase

The stage of the building lifecycle that produced the greatest C&DW was found to be the end-of-life stage (Ruiz et al. 2020). Circular economy discourages structure demolition. The building's components could be reused in a redesign, or the stakeholders could choose to enlarge or reduce the structure.

5 Waste Management Hierarchy: Construction and Demolition Waste

Divergent perspectives on managing C&DW have created incoherent and incompatible waste management ideologies (Lu and Yuan 2011). As a result, recycle, reuse, or reduce initiatives have been initiated. Effective waste prevention techniques should be used for handling all kinds of waste. The implementation of a waste management hierarchy is necessary to accomplish this goal. This hierarchy dictates that created waste should be retrieved before it is disposed of in landfills, depending on whether it may be reduced, reused, or recycled (Pickin et al. 2018).

The Waste Management Hierarchy (WMH) has five stages: reduce, reuse, recycle, recover and disposal as represented in Fig. 2. Among these, the three main steps, i.e. reuse, reduce, recovery in the Waste Management Hierarchy structure, have undergone extensive research over the past few years. Nevertheless, recovery is classified as a general waste management technique, and disposal is not a very effective option. The C&DW management hierarchy prioritizes strategies based on the most desirable outcome (avoidance of waste generation) and the least desirable outcome (disposal).

Some scholarly perspectives on C&DW management strategies suggest that the three R techniques are the major components of waste minimization technique among the

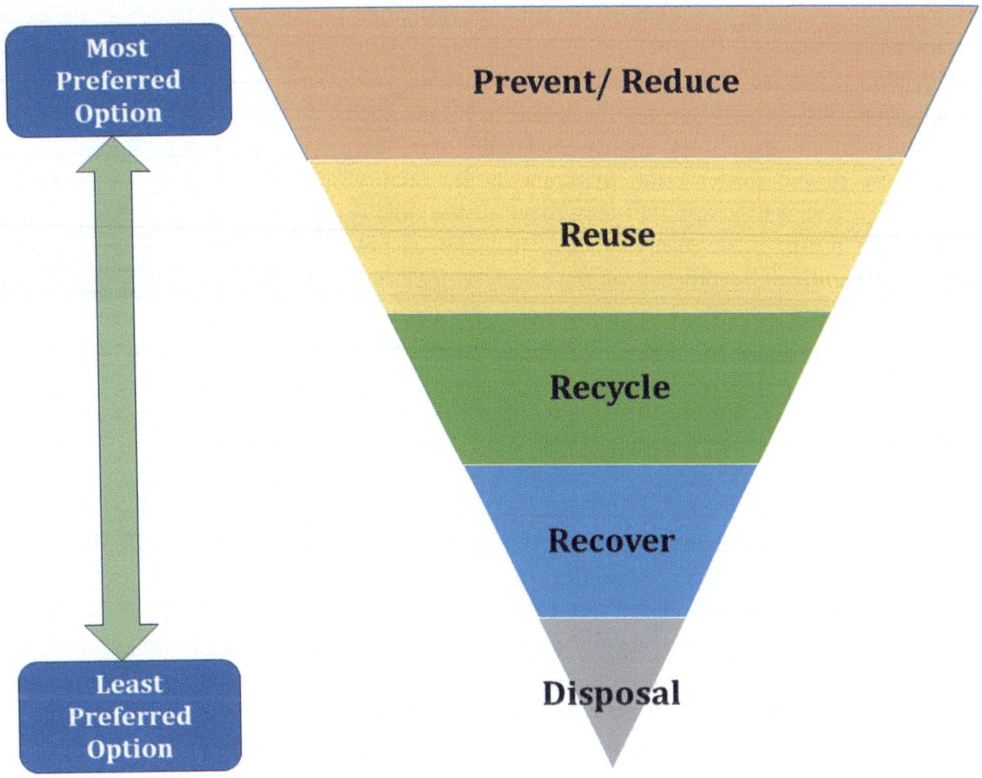

Fig. 2 Waste management hierarchy

five phases (Huang et al. 2018). In the past, C&DW management study has found that the reduce, reuse, and recycle management hierarchy worked well.

5.1 Reduce

Reduction has the least environmental effect and is the best C&DW management measure. Thus, it is given priority over other techniques when creating C&DW management plans. It is crucial to reduce waste at the generation end. However, waste generation will inevitably be reduced completely; hence, the next step should immediately be to think of ways to reuse the wasted material. Finally, recycling the materials is important if this phase is also impossible. Disposal is the last regime in waste management when the productive value of the material is perceived to be completely lost. A few advantages of reducing site-generated waste are the reduction of excessive resource exhaustion, the economic benefits of this reduction in resource extraction, the transportation of waste to landfills, and environmental benefits like cutting CO_2 emissions (Pickin et al. 2018). The most environment-friendly and economical way to decrease the development of construction site waste is to use materials of standard sizes and quantity, minimize offcuts, and minimize rework due to subpar workmanship and workers' mistakes (Ding et al. 2016). One of the main obstacles to implementing waste reduction is the communication gap, lack of shared understanding among stakeholders, and lack of cooperation in the construction industry. For effective management of C&DW, focusing on the reduction strategy to minimize waste is imperative. Implementing a waste reduction strategy in the building industry is necessary owing to the explosive global expansion of C&DW generation (Esa et al. 2017).

5.2 Reuse

'Reuse' refers to using building materials repeatedly for their intended or other purpose until it reaches its end of life (Huang et al. 2018). The majority of demolished C&DW can be recycled. Reducing and reusing are the best alternatives for environmental protection, natural resource conservation, and economic benefits. To sustain the environment for future generations, reusing building waste and maximizing its reuse potential helps reduce greenhouse gas emissions (Park and Tucker 2017). Environmental and construction sustainability can be ascertained by recovering and reusing salvaged building materials produced from existing buildings. This process involves careful dismantling of buildings to preserve materials that can be reused. It not only helps in economic advantages but a significant reduction in environmental impacts. Use of circular economy concept in construction industry requires salvaging and reusing the products in new construction projects.

Some of the key features of reusing construction and demolition waste are: careful dismantling of the buildings using skilled labour, precise tools and techniques to remove the reusable materials, identification of materials to be salvaged based on their compatibility, quality, and condition, then followed by proper preservation to maintain the quality; they are then tested to assess that they meet all the quality standards for reuse, an inventory of these materials is prepared to streamline the process of reuse, these materials can be used in new designs if the designers integrate innovative ways ensuring safe, aesthetic and quality structures (Bragança et al. 2025)

To address the issues related to global climate, an adaptive reuse process needs to be incorporated to transform and reuse the existing buildings. It's a strategy for preserving ancient structures of cultural significance, requiring skills to transform the old buildings into new buildings. This can significantly improve building durability and functionality, develop cultural identity, build social inclusion of such buildings, reduce waste production and complete life costs, improve economic potential and aesthetics, and help in the regeneration of urban areas. Adaptive reuse can involve operations like thermal insulation, window replacement, using HVAC for retrofitting, using photovoltaic panels (Bragança et al. 2025).

Reusing involves recycling all kinds of building materials, salvaging them from demolition, renovation, and construction sites, and selling/saving them for future use or use in the current project. However, extra caution is required when handling harmful C&D materials, like latex paint, adhesives, and chemical solvents. Furthermore, demolition operations require prior consideration of the age of the structures, which is an influential decision-making element for recycling C&DW (Akinade et al. 2015). For example, older structures use asbestos, which is prohibited in new construction. Reuse of construction and demolition waste (C&DW) can be effectively done by sorting and collecting C&DW by hiring skilled workers, offering incentives for recycling it, utilizing standard technology, materials, and design, and, lastly, creating a recycled products market.

5.3 Recycle

C&DW recycling refers to disassembling materials utilized in construction to develop fresh materials; nevertheless, several barriers to C&DW recycling include inadequate technology, immature markets for recycled goods, and inexperienced administration of the practice. Recycling plays an important role in targeting circular economy, environmental concerns, and sustainable construction. It helps to reduce environmental impacts and conserve natural resources by processing waste into valuable construction materials. The key steps involved in recycling process are waste collection and its transportation to recycling facilities, where the waste gets sorted into defined categories. Sorted waste is then subjected to different processes to make them suitable for reuse. Woods are shredded or chipped to be used as raw material in production of engineering wood, it can also be used as fuel. The larger concrete, asphalt, and masonry pieces are crushed for easy processing of materials further. Metal pieces are magnetically separated and used as per their shape and size configuration or they are sent to industries for preparing new structural elements.

Recycling of C&DW can be either on the site or away from the site depending upon the facilities and capabilities of the project. Construction sites render certain materials like wood, concrete, roofing materials, metal, asphalt, corrugated cardboard, and plasterboard as waste that can be recycled. Building material waste generation, when reduced, can reduce the extraction of new raw materials, subsequently reducing their long-distance transportation, hence reducing CO_2 emission. It reduces the need for landfilling, hence reducing landfill capacity, energy saving, and the economic and environmental benefits gained out of them. Businesses connected to Recycling can generate job possibilities. Quality-assured recycled materials have greater marketing potential. They are used in landscapes, foundations, road construction, sports fields, and noise-absorbing walls. Government involvement is necessary to ensure successful recycling outcomes (Fatemi and Imaninasab 2016).

Since concrete releases Carbon Dioxide into the atmosphere and is resource intensive, i.e. natural resources get exhausted. Ideally, recycled fillers or fibers can be used in place of cement and aggregate. Industrial byproducts and wastes from waste streams can be reduced using recycled materials. Many research suggests that concrete can be produced from recycled materials like plastic, glass, ceramics, bricks, metals, agricultural waste, etc. This reduces the quantity of waste dumped in landfills.

The construction industry is responsible for 39% of the greenhouse emissions, which can be reduced by reducing energy consumption through recycling construction waste. This can help mitigate the risks posed by hazardous substances through segregation and on-site recycling. Landfilling wastes can be diverted to recycling to save space and reduce environmental impacts (Bragança et al. 2025).

Many innovative techniques are nowadays incorporated to revolutionize the construction industry. Data Management Systems are used to monitor waste and maintain a record of the progress in recycling operations. It is proving to be very helpful to make the process of recycling more effective. Recycling rates have improved due to using better technologies for categorizing waste, hence making the process time effective. Usage of portable recycling equipment helps the construction industries to reduce the transportation of waste and rather recycling the waste on the site itself.

5.3.1 Aggregates Replacement

The construction and demolition waste can be recycled and utilized as partial replacement for aggregates in concrete. One of the major components of C&D waste which can be used as replacement for aggregates is glass. Glass helps to improve the tensile and flexural strength of the concrete and when it combines with cement, a pozzolanic material is formed which reduces the greenhouse gas (CO_2 and NO_2) emissions from concrete (Ismail and Al-Hashmi 2008). Glass can be used in the form of fine powder, fine glass aggregate and coarse glass aggregate in concrete. Glass has also got high thermal conductivity; hence it provides thermal stability to the buildings.

Plastic waste in the form of PET, i.e. Polyethylene terephthalate can be utilized as an aggregate replacement in concrete. It helps improve ductility and reduce shrinkage fractures caused due to variations in moisture content of concrete (Choi et al. 2005). In addition to this, it also makes the concrete light in weight which can be utilized to lessen the dead weight of structures while also reducing the slump, density, elastic modulus, workability, and tensile strength. It works well with corrosion-resistant and lightweight concrete.

Tiles, marbles, and ceramics are used as aggregate to improve the quality of concrete. Coarse granules of ceramics increase the compressive strength by 10–20%. However, it experiences a reduction in the specific weight without affecting the ability of concrete to absorb water (Tennich et al. 2015). Ceramic properties also vary due to manufacturing processes and other factors. It is pertinent to test such concrete before using because their efficacy may be impacted. Burnt bricks can be used to replace sand. A practicable and affordable means for manufacturing concrete is using clay bricks. However, it may negatively impact the reinforced concrete by corroding steel.

Wasted rubber and tires have little potential for recycling. However, when added to concrete, it reduces the material's rigidity and increases fire resistance and flexural strength. Investigation on control samples made of concrete and experimental samples made of rubber and concrete suggests that control samples showed fractures and splitting due to brittleness on excessive deformation. However, experimented samples showed deflection but no collapse (Yilmaz and Degirmenci 2009). Additionally, compressive strength was observed to improve and resistance to freeze–thaw increased by combining silica powder with the rubber and cement paste (Gesoglu et al. 2014). Though rubber is an excellent additive, further studies are needed to understand its impact on strength and durability.

Recycling is also very crucial in the agricultural sector as almost one-fourth of its output gets wasted. Almond and coconut shells find effective utilization in production of lightweight concrete. The use of coconut and almond shells in concrete as partial replacement for aggregate improves the tensile and flexural strength of concrete, but reduces the compressive strength. However, this reduction in compressive strength can be compensated by decreasing the water-cement ratio.

5.3.2 Supplementary Cementitious Materials (SCMs)

The concept of recycled concrete being used as a construction material in concrete evolved after World War II. From the previous studies, it has been observed that the use of C&D waste in concrete leads to shrinkage issues, which can be compensated by adding fly ash to the mix design. When the recycled aggregates obtained from C&D waste are used in concrete, it may lead to reduction in compressive strength which can be compensated by usage of cement made from clay brick powder (Letelier et al. 2017). Also, special care should be taken that the replacement of fine aggregates and coarse aggregates with recycled aggregates should not be more than 25%, otherwise it can significantly impact on the lifespan, durability and strength of structure.

Slag is a by-product that is generated during production of metal. It has been observed that coarse aggregates when replaced with slag in concrete leads to high shear modulus and chemical stability in both acidic and alkaline solutions. Slag with high-performance concrete increases the water absorption capacity, density, workability, as well as tensile, compressive and flexural strength of concrete. Hence, the strength of concrete increases with addition of slag (Harrison et al. 2020).

Synthesis of silica metal generates a by-product, i.e. silica fume, which is 'super pozzolanic' in nature and can enhance Portland cement's characteristics. Adding silica fluid by replacing 10–15% of cement improves its strength during the initial drying phase. However, some studies suggest the negative impact of silica fume on concrete's durability. Hence, Silica fumes should be used in accordance with design specifications. In another study conducted by Zhang et al. (2021), it was observed that coal fly ash when used with concrete, showed improved mechanical characteristics, durability, and impact resistance. Varying amounts of nano-silica, ranging from 1 to 5% of binder weight, were incorporated into concrete, which exhibited improved freeze–thaw resistance, superior mechanical properties, compression strength improved by 15.5%, split tensile strength improved by 19%, and flexural (bending) strength improved by 27.3%. Even nano-silica in basal fiber-modified recycled aggregate concrete can help reduce microcrack formation and its spread (Zheng et al. 2021). Nano-silica fills the micro cracks and also facilitates the hydration of cement. Even geo-polymerization of mortar can benefit from the addition of nano-silica. However, its excess usage may badly influence the mechanical properties (Zhang et al. 2021).

Rice husk ash is also one of the most desirable waste materials which can be used in concrete, as burning of rice husk increases its pozzolanic qualities. Adding rice husk to concrete containing micro-silica causes the cement to hydrate, which decreases the cement's porosity (Huang et al. 2017). Moreover, improvements are also noted in the water absorption, resistance against chloride attack, and compressive strength due to addition of rice husk ash in concrete. Nations with limited manufacturing facilities can also effectively utilize rice husk ash to improve the properties of mortar. Rice husk ash is generally used as cementitious material in concrete, but it should be used efficiently and in limited proportions to get the desired results.

Power stations using coal produce fly ash as the by-product. 40–60% replacement of cement with fly ash increases the characteristic compression strength of concrete. Class F fly ash with good pozzolanic quality has low permeability for chloride, good durability, and strong mechanical qualities. A study found that in the early curing stages, 5–10% of glass powder in a cement mix has better compressive strength than fly ash, but after 90 days, fly ash yields concrete with better water absorption and higher compressive strength (Schwarz et al. 2008). It is important to use silica in concrete to reduce Alkali-Silica reaction which otherwise has negative impacts on properties and strength characteristics of concrete.

Pozzolanic characteristics were also discovered in the ceramic tile during early investigations into its chemical properties. According to a study, it is possible to achieve almost 90% of the strength of regular concrete by substituting clay brick waste for other ingredients in cement (Pacheco-Torgal and Jalali 2011). The replacement cement also improves efficiency and decreases the permeability of the concrete. As an alternative, silica fume and tile powder can be combined to create concrete with characteristics similar to the controlled sample.

6 Benefits of Construction and Demolition Waste Management

Recycling is employed in construction since it protects the natural resources from being misused and squandered. This way, the three key foundations of sustainability, i.e. the economy, the environment, and society, will get the benefits of such good practices. The advantages of using recycled building materials are discussed as follows:

6.1 Environmental Benefits

Reducing, recycling, and reusing building waste to be landfilled increases the useful life of products for future uses. The pollution gets exacerbated at the waste dump sites due to the excessive use of chemical additives in building materials. This hazardous waste gets leached through groundwater intrusion and enters streams and natural waterways. CO_2 emissions also get reduced due to reduced waste transportation to landfill sites.

6.2 Economic Benefits

On the one hand, the elimination of landfills results in the loss of jobs for industry participants. On the other hand, these losses can be offset by creating new job opportunities in the recycling field. Certain modifications are done in recycled materials to improve the properties of recycled products so that they maintains the physical standards and fulfils its purpose in the building. Reused materials, however, retain their original form. Certain skill sets are necessary to exhibit reuse and recycling, hence there is a huge scope of job creation. This will not only help boost the country's economy but also lessen the detrimental effects of construction waste on the environment.

6.3 Societal Benefits

Land development is the need of the hour for the rising population growth. Recycling in the construction sector helps reduce the amount of land turning into landfills and the availability of high-quality land to meet the housing demand. This also controls the dumping of waste into landfills which wind up into natural streams. Such situations have the potential to jeopardize health of community members. Even the smell of degrading waste can spread to the neighbouring communities through strong winds. Hence, rendering the place unsuitable to live.

7 Construction and Demolition Waste Management Tools

7.1 Information Technology Tools

C&DWM can be aided by information technology (IT) tools that enhance processes and technologies throughout a project's lifespan. These instruments can also aid in waste reduction and environmental law compliance.

7.1.1 Building Information Modelling (BIM)

The most important aspect of C&DW management is the prediction of the amount of waste for effective waste management. Building Information Model (BIM) is used to simulate building designs in advance before construction. Material Volume is considered during the planning and design phase of the project using BIM. According to Cheng and Ma (2013), building information modeling (BIM) is used in waste estimation and planning for the same. It can help extract material and quantity information that helps the project participants to develop technology and procedures during design and planning phase, and necessary information about construction and demolition to effectively manage the waste generated in the process.

7.1.2 Radio Frequency Identification (RFID)

Radio transmissions of various frequencies are used as a technology for communication in Radio Frequency Identification (RFID) to identify a specific target without direct contact or line of sight, in real time in challenging environments. RFID is frequently used on construction sites to reduce human error, and it is seamlessly integrated with systems for project management to facilitate easier and faster tracking and control of building materials. RFID technology is being used in many technical projects for serving several objectives, including identification of tools, control over people, entry and exit of things, vehicles and management of warehouses. Its primary benefits are its quickness, accuracy, dependable transmission of data, inspection, and high level of control (Bandeira et al. 2019).

7.1.3 Geographic Information System (GIS)

GIS helps ease down finding relationships, trends, and patterns in tabular or written data that would be hidden (Li et al. 2020). GIS has been used to identify unlawful C&DW dumping sites, environmental impact assessment for the construction sites, and determine whether landfill sites are suitable (Seror and Portnov 2018). GIS integration improves waste management decision-making, efficiency, and sustainability by leveraging geographical data to develop educated solutions. Remote sensing and Geographic information systems (GIS) have been successfully applied in twenty-first century for selection of landfill sites. An analysis hierarchical process (AHP) was also utilized to discover the ideal location for solid waste disposal. GIS tools can be used to undertake a variety of spatial analyses related to C&DW management, including mapping unlawful dumping sites, simulating environmental dangers, and determining ideal regions for landfills and recycling plants.

7.1.4 Global Positioning System (GPS)

GPS is capable of offering accurate location. It can be installed in carriage trucks to track C&DW movement in real time (Li et al. 2020). The on-site material and equipment management system has reduced construction waste through the use of GPS coupled with GIS. A GPS tracking device can be used to track and verify that dump trucks are not illegally disposing of their cargo. GPS can be utilized for real-time tracking, optimization of routes, geo-fencing as well as for generating detailed reports using analytics.

GPS tracking enables trash management organizations to locate routes, reduce usage of fuel, and obtain operational effectiveness, resulting in minimizing expenditures and a more environmentally friendly waste management process. Supervisors can use live monitoring to monitor the location and activities of waste collection vehicles in real time, resulting in better resource management and a quicker reaction to operational difficulties or emergencies. Waste management firms can use GPS data to optimize collection routes, reduce fuel usage, and lower the environmental effect of their operations.

7.1.5 Big Data Technology

Big Data Technology helps to examine and store huge amount of data. It is increasingly being used for analysis and storage of C&DW for some years now. Lu et al. utilized waste production rate as a major performance metric to understand the performance of C&DW (Lu et al. 2015). This investigation utilized a Big Data collection on Hong Kong's building waste management to maximize data utilization. Big Data analysis was used in another study to measure the performance of building waste management using data from both public and private sectors. The utilization of this technology in managing C&DW through the processing and storing of substantial amounts of data during the design stage, procurement phase, as well as construction, and demolition has also been the subject of another study (Ram et al. 2019).

7.1.6 Image Analysis and Image Recognition Technology (IRT)

Image analysis has been used to locate and estimate the landfills in Guangming, Shenzhen, China, that are susceptible to landslides, as well as to evaluate the distribution size of C&DW for recycling aggregate (Wang et al. 2019). When renovating or demolishing existing structures, Image Recognition Technology (IRT) assists in determining the gross floor area using Google Earth. For example, Yu et al. (2019) used a recently developed technique on a hybrid trilogy based on two indicators, gross floor space and waste generation rate, to anticipate the output of demolition trash during large-scale urban regeneration in Shenzhen, China. In this investigation, Google Earth and IRT were used to calculate total floor area.

7.1.7 Blockchain Technology

Blockchain technology helps to establish a separate environment for data transactions, where trusted third parties are not needed for transaction execution, storage, or control. Blockchains can enhance records containing information about construction and demolition activities on the site, including logbooks, carried out activities, material quantities used, historical data records about waste, and reporting with transparency, traceability, immutability, reliability, and a high degree of trust (Wang et al. 2017). Blockchain has the potential to transform the current process of gathering, storing, copying, and tracking waste data at each trash movement. This helps evaluate waste management's effectiveness and fulfil compliance requirements related to waste.

7.1.8 Barcode Technology

In many different domains, including assembly checking, labour distribution, records administration, control and tracking of inventory, stock taking, billing, receiving, transporting, warehouse picking, remittance processing, and barcode technology facilitates the automatic collection and identification of data. In the 1980s, it was first used in the building sector. It has been used to reduce construction waste on-site by creating a strategy that combines barcode technologies with an incentive reward system (Li et al. 2020). Lu et al. (2011) conducted a comparison of RFID, and Barcode technology in order to determine the benefits of RFID technology for building project management.

7.2 Waste Management Approaches

7.2.1 Circular Economy

In addition to turning wastes into resources, the idea of a circular economy strongly emphasizes promoting the most effective resource usage, material recycling, and energy efficiency. Reduce, reuse, and recycle compose the fundamental idea of a circular economy (Huang et al. 2018). By setting up a closed loop of resources for economic activity, encouraging good production, and boosting industrial bionomics, circular economy (CE) seeks to improve harmony between the ecosystem and the economic system. Pomponi and Moncaster's (2017) study took into account a number of circular economy research dimensions, such as sociological, political, technological, environmental, and behavioural.

7.2.2 Lean Construction

According to some experts, practically all waste can be eliminated by applying lean principles centered on the production process that creates value through authentic processing. The lean construction approach aims to design a construction and demolition production system that maximizes abundant value creation by minimizing material waste and labour work while utilizing lean principles and related tools and techniques. It boosts worker productivity, reduces reworks and faults, enhances coordination and communication, improves performance of project, and ultimately supports sustainable development in the building industry (Ahmed et al. 2020).

7.2.3 Zero-Waste Approach

Because of extensive waste streams, stakeholders were forced to choose environmentally harmful and ineffective waste management landfills. Due to insufficient landfills in urban areas, alternate waste management solutions have become important. Zero waste has emerged as an alternative for all the waste concerns implemented by various cities in recent decades, including Adelaide, San Francisco, and Vancouver (Zaman and Lehmann 2011). Policymakers have also embraced this concept since it encourages consumption and production on sustainability lines, maximizes recycling and resource recovery, and keeps waste out of landfills and incinerators. Several studies have successfully claimed that zero-waste goals can be achieved using waste recovery methods like incineration, but this zero-waste theory must be expanded to implement widespread application and forbid incineration and landfills. According to a study by Zaman and Lehmann (2011), a city with zero-waste should achieve a 100% recycling rate and recover its entire resources from waste.

7.2.4 Green Rating System

The sustainability of construction-related operations is evaluated using Green Rating Systems such as LEED (Leadership in Energy and Environmental Design), Green Star, and BREEAM (Building Research Establishment Environmental Assessment Method). These have been widely used to encourage environment-friendly construction practices.

Green Star is an international rating system widely used in South Africa and Australia to evaluate and assess the suitability and sustainability of various construction projects. This rating system is applicable to all residential, commercial and industrial buildings where it determines energy usage, indoor air quality, building materials used, water usage, emissions, transportation and land use. The major intention of Green Star Rating systems is to help construction agencies select sustainable construction materials and reduce energy usage.

LEED is a green rating system which was introduced by the US in 1996 and this certification is completed with third party verification or on-site verification. LEED provides certification to all types of structures, which includes newly constructed buildings and already existing buildings. Certified, silver, gold and platinum are four certification stages in LEED rating system. The major objective of LEED rating system is to provide safe, healthy and sustainable environment to all the inhabitants of certified buildings throughout the life of structure. It also ensures that there is effective utilization of all the resources like water, energy, materials, etc.

BREEM was founded in 1990 and is one of the oldest green building certification programs. The major objective of BREEM rating system is to escalate the efficiency and performance of buildings while also making the buildings more sustainable. The performance of the buildings is assess based on several parameters like sustainable materials, land use, pollution, water, health, transportation and environment.

7.2.5 Site Waste Management Plan (SWMP)

The SWMP tool is obtaining great demand these days to help the construction stakeholders foresee and officially record the kind and amount of C&DW so as to take the right decisions at management level. The entire lifespan of a building project, ranging from planning stage to demolition, is the focus of this plan (Esa et al. 2017). Many countries implement SWMP by law for construction activities. For example, the Site Waste Management Plan rule, a statutory framework in the United Kingdom, mandates the creation of SWMP for projects costing more than £300,000 before beginning the construction stage. Another illustration of this is the implementation of Site Waste Management Plan, 2003 for the construction sector in Hong Kong, which received bad reviews from certain C&DW practitioners due to their belief that it would reduce productivity. The SWMP is also a prerequisite for planning clearance for some building projects. A standard SWMP should include strategies for pre-construction activities and a detailed description of suggested techniques for managing waste during and after construction. Assuring appropriate waste sorting, collection, and auditing, setting waste diversion targets, boosting profitability, and guaranteeing waste reduction, and recycling, and effectiveness of waste management are carried out sufficiently and correctly are the key goals of developing a SWMP (Ajayi et al. 2015).

7.3 Waste Management Technologies

7.3.1 Prefabrication

Prefabrication is a manufacturing method where a product is made by combining multiple materials together at a specialized facility. Furthermore, it is beneficial to use prefabricated and modularized building materials to get rid of common problems like waste, subpar results, low productivity, excessive variability, and inadequate safety. Modular building materials are basically the prefabricated building materials which are produced at the workshop and are finally assembled at the construction site based on the requirement. Few examples of modular building materials include steel and concrete used to create wall panels and floor decks, glass in wired, tempered or laminated form which may be prepared at workshop and used at site for different requirements, steel frames, etc. It can cut waste production from concrete and timber formwork by approximately 55% and 80%, respectively. Furthermore, it can save waste by up to 85% when compared to traditional construction processes (Tam et al. 2007). When concrete components are produced in a factory or pre-casting yard and then transported to the building sites, prefabrication can dramatically reduce the generation of trash on the construction sites. Wall panels, stairways, roof trusses, and other two-dimensional flat surfaces may be considered by prefabrication building components.

7.3.2 Modularization

Though the phrases "modular" and "prefabrication" are sometimes used interchangeably, modular construction is a three-dimensional (3D) prefabricated construction. The distinction between individual components and complete modules should be taken into account when comparing prefabrication and modularization processes. Because modularization reduces the number of components that employees must touch, it is a cost-saving strategy that enhances worker safety. Additionally, modularization, which is used as the disassembly design, allows for rapid and simple modifications to the construction structure, which lowers construction waste. Modularization combined with additive manufacturing and 3D printing technologies may also cut down on construction waste and shorten project timelines (Mignacca et al. 2020). Prefabrication and modularization reduce project timeframes by about 65%.

8 Construction and Demolition Waste Management Rules, 2016

Realizing the issues of C&DW, the Ministry of Environment and Forest and Climate Change (MoEF&CC) produced the "Construction and Demolition Waste Management Rules" on 29th March 2016. These regulations, which include the duties of various stakeholders such as waste generators, local governments, Urban Development Departments, State Pollution Control Boards, etc., are extensive. The responsibilities of various stakeholders in construction and demolition waste management are listed as follows (Ministry of Housing and Urban Affairs 2018).

8.1 Waste Generator

(a) Appropriately gathering and storing waste, making sure it doesn't overflow or combine with municipal solid waste.
(b) Place waste in the places specified by the local government.
(c) Get approval and submit a waste management plan before beginning any construction or demolition activity.
(d) Pay the appropriate fees for waste gathering and disposal as the local government advises.

8.2 Utility Service Providers

(a) Create a thorough waste management strategy.
(b) Gather and safely store waste while preventing pollution or disturbance in the area.
(c) Make arrangements for disposal with the urban local authority, covering all applicable charges.

8.3 Local Authority

(a) Principal figure in charge of waste management. May hire a third party, but still in charge of the supervision and result. Ought to complete a feasibility analysis prior to approving the management strategy.
(b) Prepare bylaws for handling C&D waste and establishing the appropriate penalties.
(c) In cooperation with state authorities, identify potential locations for intermediate waste collecting stations and processing facilities.
(d) Examine, accept, and collect the appropriate payments for the generators' waste management plans.
(e) Make plans for processing, transportation, and collection under a private contract.
(f) Create a database for C&D waste creation by connecting waste management permits and keep an eye on its compliance.
(g) Maintain ongoing Information, Education and Communication (IEC) initiatives for all parties involved.
(h) Incentivize to encourage using recycled materials, such as preferential purchasing clauses in contracts with municipalities.

8.4 State Pollution Control Board

(a) Observe how the local government is putting the rules into practice.
(b) Authorize C&D waste processing facilities in accordance with standards, and keep an eye on environmental adherence.
(c) Create the annual reports of Central Pollution Control Board (CPCB).

8.5 State Government

(a) Create a document outlining C&D waste management policy.
(b) When needed, assist cities in identifying land for trash management.
(c) Enable all state entities to purchase recycled materials at a favourable rate.

8.6 Central Pollution Control Board

(a) Create C&D waste management guidelines.
(b) Prepare the central government's yearly compliance report by analysing the data that State Pollution Control Boards (SPCBs) have collected.

8.7 Bureau of Indian Standards/ Indian Road Congress

(a) Establish specifications for use of recycled materials appropriately from C&D debris in road building and construction activities.

8.8 Central Government

(a) Facilitation of compliance by Ministry of Rural Development and Ministry of Housing and Urban Affairs.
(b) MoEF&CC evaluation of the implementation.

9 Conclusions

Inefficient C&DW management in India is even now related to several issues which scholars have found difficult to identify. To protect natural habitats and preserve valuable resources, managing Construction and Demolition Waste effectively and sustainably is imperative. Two key factors need to be kept in mind to achieve this: first, it should integrate the successful elements of C&DW management like tools, sustainability concepts, stakeholder attitudes, life cycle of projects. Next, the waste management hierarchy of reduce, reuse, and recycle should be adhered.

Many overlapping concepts of C&DW are there, yet for efficient management, both contribution of waste management and 3R principles need to be paid attention. Future research should focus on evaluating the impacts of waste management hierarchy elements-reduce, reuse, recycle and recover on overall C&DW management effectiveness. In addition to this, new tools and technologies, materials and waste management processes should be examined. For example, prefabrication and circular economy strategies exemplify innovative IT solutions for this field. However, efficient waste management requires exploring various practical solutions for real-world scenarios.

Many studies have concentrated on the C&DW management stakeholders. Still, advanced research is needed to determine the effects of C&DW stakeholders' attitudes on the diversion rate through various techniques, including rewards, training, penalties and fines, etc. While C&DW management is prioritized over the entire life cycle of C&DW generation project, the minimization of waste should only be considered during construction.

This study aids in the recognition and classification of aspects related to efficient management of C&DW, yet a metric for evaluating C&DW efficacy ought to exist. C&DW management efficacy has not yet been well benchmarked, even though some researchers have assessed the efficacy and fruitfulness of C&DW management by concentrating on certain indexes like Life Cycle Assessment (LCA) from a sustainability point of view, waste generation rate, etc. Moreover, there is a pressing need to create a more thorough mechanism to evaluate the performance of C&DW management.

References

Adams K, Osmani M, Thorpe A, Thornback J (2017) Circular economy in construction: current awareness, challenges and enablers. Waste Res Manage 170:1–11. https://doi.org/10.1680/jwarm.16.00011

Ahmed S, Hossain MM, Haq I (2020) Implementation of lean construction in the construction industry in Bangladesh: awareness, benefits and challenges. Int J Build Pathol Adapt 39:368–406. https://doi.org/10.1108/IJBPA-04-2019-0037

Ajayi SO, Oyedele LO, Bilal M, Akinade OO, Alaka HA, Owolabi HA, Kadiri KO (2015) Waste Effectiveness of the construction industry: understanding the impediments and requisites for improvements. Resour Conserv Recycl 102:101–112. https://doi.org/10.1016/j.resconrec.2015.06.001

Akanbi LA, Oyedele LO, Akinade OO, Ajayi AO, Delgado MD, Bilal M, Bello SA (2018) Salvaging building materials in a circular economy: a BIM based whole-life performance estimator. Resour Conserv Recycl 129:175–186. https://doi.org/10.1016/j.resconrec.2017.10.026

Akinade OO, Oyedele LO, Bilal M, Ajayi SO, Owolabi HA, Alaka HA, Bello SA (2015) Waste minimisation through deconstruction: a BIM based deconstructability assessment score (BIM-DAS). Resour Conserv Recycl 105:167–176. https://doi.org/10.1016/j.resconrec.2015.10.018

Aslam MS, Huang B, Cui L (2020) Review of construction and demolition waste management in China and USA. J Environ Manage 264:110445. https://doi.org/10.1016/j.jenvman.2020.110445

Bandeira SR, Maciel JBS, de Oliveira JCS, Sanches AE (2019) Construction and demolition waste management practices at construction sites. Int J Adv Eng Res Sci 6:35–45. https://doi.org/10.22161/ijaers.610.7

Bragança L, Griffiths P, Askar R, Salles A, Ungureanu V, Tsikaloudaki K, Bajare D, Zsembinszki G, Cvetkovska M (eds) (2025) Circular economy design and management in the built environment. Springer Nature Switzerland. https://doi.org/10.1007/978-3-031-73490-8

Cheng JCP, Ma LYH (2013) A BIM-based system for demolition and renovation waste estimation and planning. Waste Manage 33:1539–1551. https://doi.org/10.1016/j.wasman.2013.01.001

Choi Y-W, Moon D-J, Chung J-S, Cho S-K (2005) Effects of waste PET bottles aggregate on the properties of concrete. Cem Concr Res 35:776–781. https://doi.org/10.1016/j.cemconres.2004.05.014

Ding Z, Yi G, Tam VWY, Huang T (2016) A system dynamics-based environmental performance simulation of construction waste reduction management in China. Waste Manage 51:130–141. https://doi.org/10.1016/j.wasman.2016.03.001

Dosumu OS (2018) Assessment of the likelihood of risk occurrence on tendering and procurement of construction projects. J Constr Bus Manage 2:20–32. https://doi.org/10.15641/jcbm.2.1.95

Duan H, Miller TR, Liu G, Tam VWY (2019) Construction debris becomes growing concern of growing cities. Waste Manage 83:1–5. https://doi.org/10.1016/j.wasman.2018.10.044

Esa MR, Halog A, Rigamonti L (2017) Developing strategies for managing construction and demolition wastes in Malaysia based on the concept of circular economy. J Mater Cycles Waste Manage 19:1144–1154. https://doi.org/10.1007/s10163-016-0516-x

Fadiya O, Georgakis P, Chiniyo E (2014) Quantitative analysis of the sources of construction waste. J Constr Eng 2014:651060. https://doi.org/10.1155/2014/651060

Fatemi SS, Imaninasab R (2016) Performance evaluation of recycled asphalt mixtures by construction and demolition waste materials. Constr Build Mater 120:450–456. https://doi.org/10.1016/j.conbuildmat.2016.05.117

Gesoglu M, Güneyisi E, Khoshnaw G, Ipek S (2014) Abrasion and freezing–thawing resistance of pervious concretes containing waste rubbers. Constr Build Mater 73:19–24. https://doi.org/10.1016/j.conbuildmat.2014.09.047

Ghisellini P, Ripa M, Ulgiati S (2018) Exploring environmental and economic costs and benefits of a circular economy approach to the construction and demolition sector. A literature review. J Cleaner Prod 178:618–643. https://doi.org/10.1016/j.jclepro.2017.11.207

Giannarakis G, Garefalakis A, Lemonakis C, Konteos G (2017) The drivers of social responsible stock index: the case of Dow Jones sustainability index world. Corp Ownersh Control 14:173–180. https://doi.org/10.22495/cocv14i2c1p3

Groh J, Dubik P (2018) BIM as a tool to implement circular economy into construction projects' life-cycle. Master's thesis, The Faculty of Engineering and Science, Aalborg University, Aalborg, UK. https://projekter.aau.dk/projekter/files/294311131/BIM_as_a_tool_to_implement_circular_economy_into_construction_projects__life_cycle.pdf

Harrison E, Berenjian A, Seifan M (2020) Recycling of waste glass as aggregate in cement-based materials. Environ Sci Ecotechnology 4:100064. https://doi.org/10.1016/j.ese.2020.100064

Hasmori MF, Zin AFM, Nagapan S, Deraman R, Abas N, Yunus R, Klufallah M (2020) The on-site waste minimization practices for construction waste. IOP Conf Ser: Mater Sci Eng 713(012038):1–11. https://doi.org/10.1088/1757-899X/713/1/012038

Huang H, Gao X, Wang H, Ye H (2017) Influence of rice husk ash on strength and permeability of ultra-high performance concrete. Constr Build Mater 149:621–628. https://doi.org/10.1016/j.conbuildmat.2017.05.155

Huang B, Wang X, Kua H, Geng Y, Bleischwitz R, Ren J (2018) Construction and demolition waste management in China through the 3R principle. Resour Conserv Recycl 129:36–44. https://doi.org/10.1016/j.resconrec.2017.09.029

Islam R, Nazifa TH, Yuniarto A, Uddin ASMS, Salmiati S, Shahid S (2019) An empirical study of construction and demolition waste generation and implication of recycling. Waste Manage 95:10–21. https://doi.org/10.1016/j.wasman.2019.05.049

Ismail ZZ, Al-Hashmi EA (2008) Reuse of waste iron as a partial replacement of sand in concrete. Waste Manage 28:2048–2053. https://doi.org/10.1016/j.wasman.2007.07.009

Kabirifar K, Mojtahedi M, Wang C, Tam VWY (2020) Construction and demolition waste management contributing factors coupled with reduce, reuse, and recycle strategies for effective waste management: a review. J Clean Prod 263:121265. https://doi.org/10.1016/j.jclepro.2020.121265

Letelier V, Tarela E, Moriconi G (2017) Mechanical properties of concretes with recycled aggregates and waste brick powder as cement replacement. Procedia Eng 171:627–632. https://doi.org/10.1016/j.proeng.2017.01.396

Li CZ, Zhao Y, Xiao B, Yu B, Tam VWY, Chen Z, Ya Y (2020) Research trend of the application of information technologies in construction and demolition waste management. J Clean Prod 263:121458. https://doi.org/10.1016/j.jclepro.2020.121458

Lu W, Yuan H (2011) A framework for understanding waste management studies in construction. Waste Manage 31:1252–1260. https://doi.org/10.1016/j.wasman.2011.01.018

Lu W, Huang GQ, Li H (2011) Scenarios for applying RFID technology in construction project management. Autom Constr 20:101–106. https://doi.org/10.1016/j.autcon.2010.09.007

Lu W, Chen X, Peng Y, Shen L (2015) Benchmarking construction waste management performance using big data. Resour Conserv Recycl 105:49–58. https://doi.org/10.1016/j.resconrec.2015.10.013

Meghani MD, Vyas CM, Bhavsar JJ, Hingu RJ (2011) A study on basic materials waste in building industry: main causes and prevention. In: National conference on recent trends in engineering & technology. B.V.M. Engineering College, V.V. Nagar, Gujarat, India

Mignacca B, Locatelli G, Velenturf A (2020) Modularisation as enabler of circular economy in energy infrastructure. Energy Policy 139:111371. https://doi.org/10.1016/j.enpol.2020.111371

Ministry of Housing and Urban Affairs (2018) Strategy for promoting processing of construction and demolition (C&D) waste and utilisation of recycled products. https://www.niti.gov.in/sites/default/files/2019-03/CDW_Strategy_Draft%20Final_011118.pdf

Narcis N, Ray I, Hosein G (2019) Construction and demolition waste management actions and potential benefits: a perspective from Trinidad and Tobago. Buildings 9:150. https://doi.org/10.3390/buildings9060150

Pacheco-Torgal F, Jalali S (2011) Compressive strength and durability properties of ceramic wastes based concrete. Mater Struct 44:155–167. https://doi.org/10.1617/s11527-010-9616-6

Park J, Tucker R (2017) Overcoming barriers to the reuse of construction waste material in Australia: a review of the literature. Int J Constr Manag 17:228–237. https://doi.org/10.1080/15623599.2016.1192248

Pavlides T, Vardopoulos I, Papamichael I, Voukkali I, Stylianos M, Zorpas AA (2023) Environmental sustainability assessment of excavation, construction, and demolition waste conditions and practices across Greece and Cyprus. IOP Conf Ser: Earth Environ Sci 1196:012037. https://doi.org/10.1088/1755-1315/1196/1/012037

Pickin J, Randell P, Trinh J, Grant B (2018) National Waste Report 2018. Blue Environment

Pomponi F, Moncaster A (2017) Circular economy for the built environment: a research framework. J Clean Prod 143:710–718. https://doi.org/10.1016/j.jclepro.2016.12.055

Ram J, Afridi NK, Khan KA (2019) Adoption of big data analytics in construction: development of a conceptual model. Built Environ Proj Asset Manage 9:564–579. https://doi.org/10.1108/BEPAM-05-2018-0077

Rayhan DSA, Bhuiyan IU (2023) Review of construction and demolition waste management tools and frameworks with the classification, causes, and impacts of the waste. Waste Disposal Sustain Energy 6:95–121. https://doi.org/10.1007/s42768-023-00166-y

Río Merino MD, Gracia PI, Azevedo ISW (2009) Sustainable construction: construction and demolition waste reconsidered. Waste Manage Res 28:118–129. https://doi.org/10.1177/0734242X09103841

Ruiz LPL, Ramon XR, Domingo SG (2020) The circular economy in the construction and demolition waste sector—a review and an integrative model approach. J Clean Prod 248:119238. https://doi.org/10.1016/j.jclepro.2019.119238

Schwarz N, Cam H, Neithalath N (2008) Influence of a fine glass powder on the durability characteristics of concrete and its comparison to fly ash. Cement Concr Compos 30:486–496. https://doi.org/10.1016/j.cemconcomp.2008.02.001

Seror N, Portnov BA (2018) Identifying areas under potential risk of illegal construction and demolition waste dumping using GIS tools. Waste Manage 75:22–29. https://doi.org/10.1016/j.wasman.2018.01.027

Tam VW, Tam CM, Zeng SX, Ng WCY (2007) Towards adoption of prefabrication in construction. Build Environ 42:3642–3654. https://doi.org/10.1016/j.buildenv.2006.10.003

Tennich M, Kallel A, Ouezdou MB (2015) Incorporation of fillers from marble and tile wastes in the composition of self-compacting concretes. Constr Build Mater 91:65–70. https://doi.org/10.1016/j.conbuildmat.2015.04.052

Velenturf APM, Purnell P (2021) Principles for a sustainable circular economy. Sustain Prod Consumption 27:1437–1457. https://doi.org/10.1016/j.spc.2021.02.018

Wang H, Zhang J, Lin H (2019) Satellite-based analysis of landfill landslide: the case of the 2015 Shenzhen landslide. Int J Geotech Eng 16:293–300. https://doi.org/10.1080/19386362.2019.1610605

Wang J, Wu P, Wang X, Shou W (2017) The outlook of blockchain technology for construction engineering management. Front Eng Manage 4:67–75. https://doi.org/10.15302/J-FEM-2017006

Yang K, Xu Q, Townsend TG, Chadik P, Bitton G, Booth M (2006) Hydrogen sulfide generation in simulated construction and demolition debris landfills: impact of waste composition. J Air Waste

Manag Assoc 56:1130–1138. https://doi.org/10.1080/10473289.2006.10464544

Yilmaz A, Degirmenci N (2009) Possibility of using waste tire rubber and fly ash with portland cement as construction materials. Waste Manage 29:1541–1546. https://doi.org/10.1016/j.wasman.2008.11.002

Yu B, Wang J, Li J, Zhang J, Lai Y, Xu X (2019) Prediction of large-scale demolition waste generation during urban renewal: a hybrid trilogy method. Waste Manage 89:1–9. https://doi.org/10.1016/j.wasman.2019.03.063

Zaman AU, Lehmann S (2011) Challenges and opportunities in transforming a city into a "zero waste city." Challenges 2:73–93. https://doi.org/10.3390/challe2040073

Zhang P, Wang K, Wang J, Guo J, Ling Y (2021) Macroscopic and microscopic analyses on mechanical performance of metakaolin/fly ash based geopolymer mortar. J Clean Prod 294:126193. https://doi.org/10.1016/j.clepro.2021.126193

Zhang P, Sha D, Li Q, Zhao S, Ling Y (2021) Effect of nano silica particles on impact resistance and durability of concrete containing coal fly ash. Nanomaterials 11:1296. https://doi.org/10.3390/nano11051296

Zheng Y, Zhuo J, Zhang P (2021) A review on durability of nano-SiO_2 and basalt fiber modified recycled aggregate concrete. Constr Build Mater 304:124659. https://doi.org/10.1016/j.conbuildmat.2021.124659

Zimmann R, Brien HO, Hargrave J, Morrell M (2016) Circular economy in the built environment. https://www.arup.com/perspectves/publications/research/section/circular-economy-in-the-built-environment

Towards a Circular Economy: Achieving *Zero Waste* Through Integrated Farming System (IFS)

Lumbini Kalita and Bhabesh Gogoi

Abstract

The future of the world hinges on economic principles that acknowledge the necessity for individuals to integrate themselves within a cyclical ecological system capable of perpetually regenerating waste materials. To establish such a regenerative system, it is crucial to prioritize the valorization of waste or the conversion of waste into valuable resources. This approach focuses on minimizing resource input, reducing liberated waste and energy leakage, and shifting away from linear material flows towards recycling processes while avoiding its disposal. Agricultural activities contribute significantly to biomass waste, including animal manure, green residues, and non-marketable products. Research on integrated crop and livestock systems has demonstrated their high productivity and environmental sustainability. Integrated Farming Systems (IFS) have transformed the conventional farming practices across livestock, aquaculture, horticulture, and agro-industry. IFS aligns with the principles of Circular Economy (CE), striving for sustainable development, environmental quality, economic prosperity, and social equity to secure a better future for current and future generations. Adopting IFS presents an opportunity to unlock the potential of each product through recycling, yielding dual benefits. This system facilitates the reuse of waste or by-products from one component as inputs for other compatible components, optimizing farm space utilization for increased production. Moreover, the continuous flow of money within the system alleviates financial burdens on farmers. Importantly, IFS directly contributes to reducing natural resource consumption by repurposing materials that would otherwise become waste. Additionally, IFS prolongs the useful life of materials/products, enhancing their potential for valorization.

Keywords

Integrated farming system · Biomass · Valorization · Waste-to-wealth · Circular economy

L. Kalita (✉) · B. Gogoi
Department of Soil Science, Assam Agricultural University, Jorhat, India, Assam
e-mail: lumbini_k27@yahoo.com

B. Gogoi
e-mail: bhabesh.gogoi@aau.ac.in

Advanced Centre for Integrated Farming Systems Research, (AICRP on IFS under ICAR-IIFSR), Assam Agricultural University, Jorhat, India

1 Introduction

In the pursuit of a sustainable future, the traditional linear economy founded on the '*take-make-dispose*' model stands as a formidable barrier. The relentless consumption of finite resources and heedless generation of waste have led to a global crisis, marked by climate change and ecosystem degradation. Globally, agricultural waste (agro-waste) makes up a considerable percentage of total waste production, and its improper management can result in environmental degradation and significant health hazards. This needs to be replaced by the concept of 'resource recovery and recycle' to ensure sustainable growth. Thus, there emerges the concept of Circular Economy (CE). The Ellen MacArthur Foundation (Capalbo and Bauen 2019) defines CE as 'an economic system of closed loops in which raw materials, components and products keep their quality and value for the longest possible and systems are fueled by renewable energy sources' (Kumar et al. 2018), further refined the concept of CE as an 'economic model aimed at the efficient use of resources through waste minimization, long-term value retention, reduction of primary resources, and closed loops of products, product parts, and materials within the boundaries of environmental protection and socioeconomic benefits'. CE

A. Mandpe et al. (eds.), *Sustainable Waste Management*, Advances in Science, Technology & Innovation,
https://doi.org/10.1007/978-3-031-85685-3_13

represents a promising strategy for saving relevant resources and reducing the negative environmental impacts caused by agricultural activities' while improving economic performance (Morseletto 2020). The concept of circular economy describes the use of waste from one industry as raw material to another one and is based on the Reduce, Reuse, Refuse, Recycle, And Recover (5R) principle of the sustainable development. The challenge of transitioning from a linear to a CE model is still ongoing and requires the development and application of new knowledge, leading to innovative, technological, and sustainable processes, products, and services (Abad-Segura et al. 2020). CE in reference to agriculture sector can be defined as the set of activities designed to not only ensure economic, environmental and social sustainability in agriculture through practices that pursue the efficient and effective use of resources in all phases of the value chain but also guarantee the regeneration of and biodiversity in agro-ecosystems and the surrounding ecosystems.

Integrated Farming System (IFS), with its holistic approach to agriculture, embodies this ethos with unparalleled efficacy. IFS is an interrelated complex matrix of soil, water, plant, animal and environment and their interaction with each other enabling the system to be more viable and profitable. Research suggests a synergistic relationship between crops and livestock within an IFS. This system emphasizes resource recycling to maximize utilization of available resources. Crop residues serve as a source of feed for livestock, while livestock waste and by-products are returned to the agro-ecosystem in the form of manures or soil amendments. These by-products, rich in nutrients, act as natural soil fertilizer, thereby enhancing agricultural productivity and reducing dependence on chemical fertilizers. By seamlessly integrating diverse agricultural practices within a closed-loop system, IFS offers a pathway to achieving zero waste while advancing the Sustainable Development Goals (SDGs) laid out by the United Nations. IFS highlights a sincere commitment to sustainability, deep rooted in the principles of conservation and efficiency. Unlike conventional monoculture farming, which often depletes soil fertility and relies heavily on external inputs, IFS harnesses the power of symbiotic relationships among crops, livestock, and natural resources. This synergy not only maximizes resource utilization but also minimizes waste generation, creating a regenerative cycle that sustains both ecosystems and livelihoods with regard to agro-waste generation. By diversifying agricultural outputs and optimizing resource allocation, IFS enhances food security and promotes access to nutritious produce for communities worldwide. Moreover, its emphasis on agro-ecological practices fosters resilience in the face of climate variability, ensuring that vulnerable populations are not left behind in the fight against hunger and malnutrition and thus fulfil SDG 2(Zero Hunger). IFS embodies the spirit of SDG 12(Responsible Consumption and Production) by prioritizing efficiency, minimizing waste at every stage of the agricultural process, it sets a precedent for sustainable consumption patterns that resonate throughout the supply chain. IFS emerges as a potent ally in the fight against climate change (SDG 13) through practices such as agroforestry, conservation tillage, and rotational grazing. By preserving natural habitats, minimizing chemical inputs, and fostering habitat diversity, IFS safeguards the intricate web of life that sustains our planet's ecosystems. (SDG 15).

IFS is based on the concept that **'there is no waste'** and **'waste is only a misplaced resource'** which means waste from one component becomes an input for another part of the system. This approach is not only a reliable way of obtaining fairly high productivity with substantial fertilizer economy but also deriving maximum compatibility and replenishment of organic matter by way of effective recycling of organic residues/wastes obtained through integration of various land-based enterprises (Agrifarming n.d.). It focuses on 5Rs namely "Reduce, Reuse, Refuse, Recycle, And Recover." The continuous flow of money from one or other components within the system also minimizes the financial burden on the farmers.

2 Scope of Circular Economy (CE) in Agriculture

In response to escalating environmental concerns and the imperative to address resource scarcity, the concept of circular economy has emerged as an important concept across various sectors, including agriculture. The main objective of a circular economy system is to couple economic growth with sustainable resource use (Cullen 2018; Pauliuk 2018) through the implementation of three key principles (Ellen MacArthur Foundation 2013) namely:

(1) Regenerate natural systems through the control of finite stocks and balancing of renewable resource flows;
(2) Optimize resource yields by keeping materials within biological and technological cycles for as long as possible and
(3) Design-out waste and pollution from production and consumption (Baishya et al. 2014).

These core principles of CE in agricultural system be accommodated in a way that can be elucidated as follows:

- **Reduce**: Reduction entails minimizing the consumption of resources, energy, and inputs to mitigate environmental impacts. In agriculture, this principle encompasses practices such as precision farming, wherein technologies like sensors and drones optimize resource utilization by precisely targeting inputs like water and fertilizers based on real-time data (Baishya et al. 2014). IFS entails the principle of reduction wherein consumption of resources is minimised by using outputs of another enterprise as input for one. As for instance, crop residues can be used for animal feed, while livestock and livestock by-product production and processing can enhance agricultural productivity by intensifying nutrients that improve soil fertility, reducing the use of chemical fertilizers. Adoption of agro-ecological approaches that emphasize natural processes and biodiversity conservation helps in reducing the dependency on synthetic inputs and mitigating ecological degradation (Altieri and Nicholls 2020) and thus optimize resource use efficiency.

- **Reuse**: IFS consist of using by-products and organic wastes of one enterprise as inputs for another enterprise. This is one example of reuse in the agricultural context by incorporation of IFS. One way to close nutrient loops and lessen dependency on synthetic substitutes is through the composting of agricultural wastes. This waste may include crop stalks, animal litter, etc. that can be used to create organic fertilizers within the IFS setup. By utilizing the synergies between various crops, circular practices like crop rotation and intercropping can enable the efficient use of land and resources (Gadde et al. 2009). In IFS, from animal components, animal excreta can be reused as a fertilizer, feed and fuel. Excreta can be dried, composted, or liquid-composted for the production of biogas and energy for household use (e. g. cooking, lightning) or for rural industries (e.g. powering mills and water pumps).
- **Refuse**: Embracing the practice of reducing waste generation can lead to significant benefits. By opting out of purchasing chemical fertilizers, pesticides, and other external inputs and instead incorporating Integrated Farming System (IFS) practices, we minimize reliance on these external resources. This approach not only supports a healthier farming system but also reduces the waste typically associated with these inputs.
- **Recycle**: The process of valuing agricultural waste streams and by-products to produce biogas, biofuels, or bio-based materials is known as recycling. Furthermore, the utilization of closed-loop systems, like aquaponics or livestock integration, reduces nutrient losses and maximizes resource utilization, which boosts the productivity and resilience of agricultural systems (Bhat et al. 2018). By harnessing bioenergy technologies such as anaerobic digestion and biomass gasification, agricultural residues can be converted into renewable energy sources, reducing dependency on fossil fuels and mitigating greenhouse gas emissions (Gobardhan-Intiative n.d.). In IFS, crops and livestock interact to create a synergy. Crop residues can be used for animal feed, while livestock and livestock by-product production and processing can enhance agricultural productivity by intensifying nutrients that improve soil fertility, reducing the use of chemical fertilizers. Furthermore, the utilization of bio-based materials derived from agricultural residues fosters the development of a bio-economy, promoting sustainable consumption and production patterns.
- **Repurpose**: Repurposing may also be called 'upcycling'. The concept of repurposing in IFS can be seen for instance under management of Crop and Livestock. Different components of the system can be interlinked to maximize resource use. Grazing of livestock can help manage crop residues, and their manure can be repurposed to fertilize crop fields. Additionally, water from aquaculture systems can be used for irrigating crops, benefiting from the nutrients present in the water.

3 From Linear to Closed-Loop Circular Systems in Agriculture

Agriculture has traditionally been following a linear trajectory, with resources being extracted, processed, and then disposed off after use. This linear model of economy is characterized by a linear flow of materials; inputs, such as water, fertilizers, and pesticides, are applied to agricultural production systems, yielding crops and livestock products as outputs. This strategy is fundamentally unsustainable, resulting in waste production, environmental deterioration, and resource depletion (Gill et al. 2009).

The linear 'take-make-dispose' model (Fig. 1) presents serious obstacles to resilience and sustainability. The long-term sustainability of food production systems is jeopardized, and environmental pressures are exacerbated as a result of its contribution to soil degradation, water pollution, biodiversity loss, and greenhouse gas emissions. Furthermore, the linear model increases vulnerability to resource scarcity and price volatility by maintaining a reliance on limited resources (Manjunath and Itnal 2003). There is increasing momentum in farming to switch from linear to closed-loop systems (Fig. 2) in response to these difficulties.

By closing the loop of material flows within agricultural systems, closed-loop systems, also referred to as circular or regenerative agriculture aim to reduce waste, maximize resource utilization, and improve ecosystem services (Giller et al. 2021). Rethinking conventional farming methods and adopting concepts like reduce, reuse, refuse, recover and recycle are part of this paradigm shift. There are many benefits in transitioning to a Circular bio-economy in agriculture by incorporation of IFS practices. This includes:

- **Environmental benefits**: Reducing dependence on chemical inputs, reduced resource utilization, and

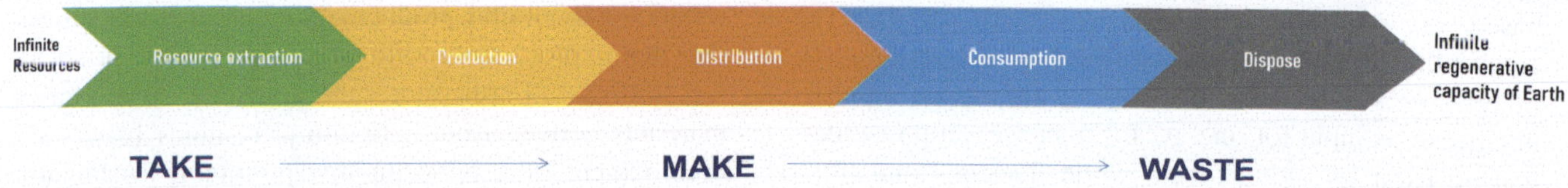

Fig. 1 Linear model of economy (*Source* www.bilinvestmentinsights.com)

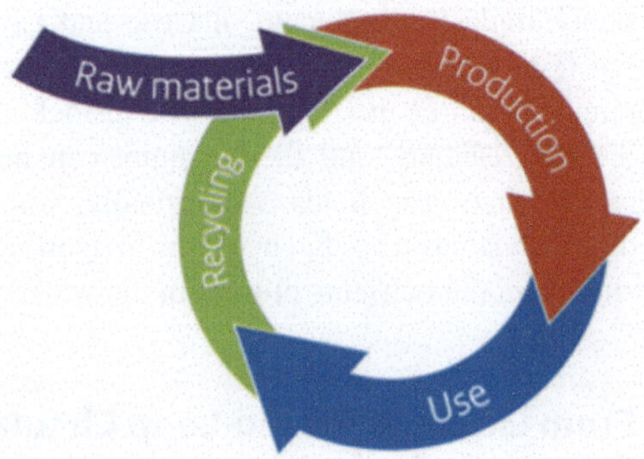

Fig. 2 Closed-loop system (circular system) (*Source* www.Googleimages.com)

reduced waste generation lead to a more environmentally friendly agricultural system. Improved soil health with the incorporation of CE model can result in higher crop yields and greater resilience to insect pests and diseases within the IFS. Various IFS techniques like cover crops and composting using various agricultural waste enhance the structure, fertility, and general health of the soil.

- **Resource efficiency**: Throughout the agricultural cycle, more particularly IFS by incorporating CE principles optimize resource use, minimizing waste and maximizing the value extracted from inputs.
- **Economic opportunities**: Farmers and rural communities may be able to take advantage of new business models and financial prospects brought about by innovations in CE technologies and practices and incorporate it by taking in the practice of IFS.

4 Concept and Components of IFS

Integrated Farming System (IFS) is defined as a biologically integrated farming system that integrates natural resources and regulation mechanisms into farming activities to achieve maximum replacement of off-farm inputs, secures sustainable production of high-quality food and other products through ecologically preferred technologies, sustains farm income, eliminates or reduces sources of present environment pollutions generated by agriculture and sustains the multiple functions of agriculture. IFS endeavours to minimize inefficiencies within farm subsystems by repurposing wastes or by-products from each subsystem as inputs for others. This strategy enhances overall productivity while reducing production costs across various subsystems (Ellen MacArthur Foundation 2013). In light of increasing population density, limited land availability, and the imperative to address agricultural waste and its potential valorization, IFS emerges as a viable solution. Integration of two or more appropriate combinations of each farm according to the availability of resources can help to sustain and satisfy the necessities of the farmer as well as tackle waste management on the farm. Through the integration of diverse farm components (Fig. 3), including Aquaculture, Field crops, Vegetables, Mushrooms, Horticultural crops, Forest tree crops, Composting, livestock (cow, buffalo, sheep, pig, rabbit, etc.), Poultry units, Goatery, Apiaries, biogas, agroforestry, etc. IFS optimizes resource utilization and fosters sustainable agricultural practices. IFS combine them in an expanded symbiotic or synergistic system, so that the wastes of one process become the input for other processes, with or without treatment to provide the means of production, such as energy, fertilizer, and feed for optimum productivity at minimum costs. IFS has great promise for improving the profitability by reducing the cost of production and/or enhancement of productivity by adopting sustainable management. IFS has the ability to reduce cost of production through recycling wastes as by-products of one enterprise become input to other enterprises (Manjunath and Itnal 2003; Ravisankar et al. 2007) and thus minimizing the need for external inputs.

In agriculture, while growing crops is the main focus, relying solely on crops for income often isn't enough to support farm families year-round. To ensure a steady income, farmers can integrate other activities alongside growing crops. By carefully choosing and combining different activities, such as raising animals or growing other types of plants, farmers can make more money, especially if they have small plots of land. This integrated approach not only helps financially but also allows farmers to use leftovers and waste effectively. Various grains like rice, wheat, and maize, as well as oilseeds like peanuts and sesame seed are considered the crop component of IFS. Farmers may also grow vegetables, fruits, root crops, sugarcane, and trees like moringa and teak under IFS. Keeping animals like cows, goats, chickens, and pigs is also part of IFS. This helps to stabilize the farm system because animals can eat leftover crops and in turn their manure can be used to

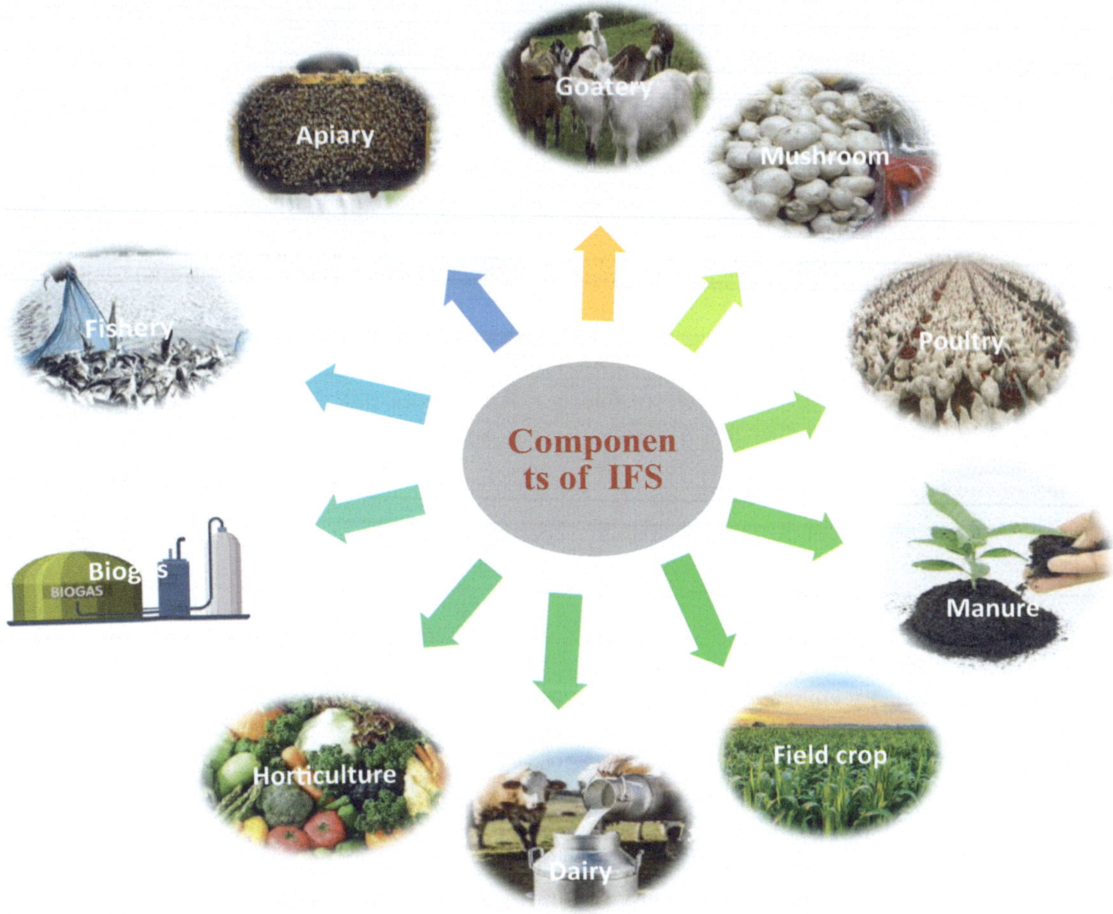

Fig. 3 Different components of an integrated farming system (IFS) unit (*Source* www.googleimages.com)

fertilize the soil, which enhances crop growth. IFS incorporates pig tractor systems, which involve confining pigs in crop fields prior to planting, allowing them to dig and uproot crop residues, effectively ploughing the field and preparing it for the next crop. Cattle or other livestock are allowed to graze over crops on farms that contain both cropland and pasture within IFS. Construction of animal houses over a pond is often done in IFS so that animal waste fall directly into the pond on which fish feed. Agroforestry, another aspect of IFS, goes beyond just growing trees for wood. It involves growing trees that produce other useful things like gum, oils, and medicines. Certain tree species are planted along borders to enrich the soil and provide additional benefits. Other activities like mushroom cultivation and various crafts such as making baskets or candles can also be part of the farm. These activities provide extra income for the family. By combining different activities like beekeeping, fish farming, and raising livestock, farmers can maximize their land use without sacrificing crop production. It's important for all these activities to be linked together in a coordinated way for the farming system to be successful, holistic and sustainable.

5 Waste Valorization in IFS

According to a conservative estimate, around 600–700 Mt of agricultural waste is available in the country every year, but most of it is lost due to dumping of this waste by the side of the road, in streams, and in the ocean and also to burning practices in the field which results in a loss of 0.824 Mt of NPK from the soil (Cullen 2018) which can otherwise be composted or directly incorporated back into the soil thereby preventing breeding of pests and insects and contributing to reduction in global warming potential and also most importantly acting as a nutrient source for growing crops. The choice of selecting a particular waste stream for recycling and upcycling depends on accessibility, affordability and output/yield. IFS with its peculiar and effective approach offers a unique way to agricultural production by promoting resource efficiency and minimizing waste to maximum extent possible. IFS utilizes and manages waste streams, transforming them into valuable resources. This is achieved by taking into consideration different ways of waste valorization. This may be stated under the subheads as follows:

5.1 Resource Recycling

IFS embraces a closed-loop system where waste from one component becomes a resource for another. Farm families around the world and more specifically in India raise cattle (cows or buffaloes) and goats. However, there's a missed opportunity. Traditionally, people burn animal manure (dung) as fuel after drying it into cakes. Recycling this manure on farms as compost can significantly reduce the dependence on expensive chemical fertilizers. Similarly, plant waste like leaves and weeds can be composted using worms and returned to the soil. By embracing a more integrated approach, farmers can combine crops, vegetables, livestock, and even fish farming. This 'closed-loop system' reduces reliance on external inputs, improves soil health in the long run, and boosts profitability. Waste from one component becomes a resource for another, creating a sustainable and cost-effective agricultural system. This approach offers significant environmental and economic benefits. Waste materials like crop residues, animal manure, and food scraps generated by one component are repurposed as inputs for another in the IFS unit. These processes transform organic matter into nutrient-rich vermicompost and compost which thereby reduces reliance on synthetic fertilizers and promotes soil health, leading to increased crop yields and improved farm resilience (Arancon et al. 2006). It directly affects the reduction of natural resource usage through recycling/reuse of the waste materials. For example, fertile silty materials generated from the de-siltation of fish ponds in an IFS system could be used for the cultivation of horticultural crops. (Arumugam et al. 2017) De-siltation was carried out after every two years while maintaining a suitable depth, besides silt was found to be very fertile containing 0.495% organic carbon, 0.0293% N, 0.025% P and 0.0175% K.

5.2 Composting

Composting is a natural bio-process that uses microbial activity to break down organic waste from various sources, including agricultural, industrial, and animal waste. It is a fundamental waste valorization technique in IFS. Organic waste materials like crop residues, animal manure, and food scraps can be composted to create nutrient-rich soil amendments. This effort not only reduces waste volume but also provides a valuable organic fertilizer, promoting soil health and reducing reliance on synthetic fertilizers and also reduces environmental pollution. Composting is proven to reduce crop residual waste by about 5%, resulting in a stable product called 'compost'. Co-composting rice straw with animal waste is likely to yield better results than composting alone. IFS, in its novel approach, often integrates livestock with crop production, creating a closed-loop system where manure from animals becomes compost for growing crops, minimizing need for external input.

Apart from this, the process of turning organic debris into worm castings (Vermicompost and Vermiwash) is also a great, natural and efficient method for converting organic waste materials into a nutrient-rich fertilizer using earthworms. The term 'vermicompost' describes using of earthworms to compost biodegradable organic materials. (Arumugam et al. 2017) According to an estimate, 0.1 kg of earthworms can consume 0.1 kg of waste every day. The 'castings', i.e. the worms' excretions are a rich source of nitrate as well as readily available forms of Phosphorus, Potassium, Calcium, and Magnesium, all of which increase soil fertility (Basso and Antle 2021). The growth of bacteria and actinomycetes is also facilitated by the presence of earthworms in the soil. This process of vermicomposting indeed offers a sustainable solution for waste management and adds value to waste material, particularly within the framework of IFS. Vermicompost contains 5 times the available N, 7 times the available P, and 1.5 times more Ca than found in good topsoil. (Pretty et al. 2018). A very high B:C ratio of 7.86 was obtained with application of vermicompost in an IFS unit in a study (Arumugam et al. 2017) which may be attributable to the availability of raw materials from IFS unit itself and selling out of out extra earthworm from the unit for extra profitability.

Farmyard manure (FYM) constitutes a recycled organic amendment produced through the controlled decomposition of plant residues, animal excreta, and bedding materials. FYM preparation offers a two-fold benefit for farm management. Firstly, it serves as a sustainable waste management strategy by diverting organic materials from waste streams. Secondly, the resulting decomposed product acts as a nutrient-rich soil amendment, promoting robust plant growth. This practice fosters a circular nutrient economy within the farm by minimizing reliance on external chemical fertilizers and transforming farm wastes into a valuable resource. Liquid manure is another product which is produced by recycling in an IFS unit. Output of cow dung and urine mixture produces this liquid manure. One litre of liquid manure is found to contain 6.6 kg Nitrogen, 3.2 kg Phosphorus and 5.0 kg Potassium (Arumugam et al. 2017) (Table 1).

Table 1 Nutrient content of compost/amendments produced in an IFS unit

Compost/amendment	N (%)	P (%)	K (%)
Compost	0.5–1.5	0.2–0.8	0.5–1
Vermicompost	1.5–2	1–1.2	0.8–1
Vermiwash	0.05–0.1	0.02–0.03	0.03–0.05
FYM	0.8–1.5	0.4–0.8	0.5–1
Liquid manure	0.2–0.8	0.1–0.4	0.2–0.3

5.3 Transforming Waste into Biochar

Biochar production in an IFS unit involves utilizing agricultural waste to create biochar, which is then incorporated back into the farming system to improve soil health and fertility. Various types of agricultural residues such as crop residues, wood chips, animal manure, or even invasive plant species can serve as feedstock for biochar production in an IFS unit. The choice of feedstock depends on its availability, cost-effectiveness, and environmental sustainability. The selected feedstock then undergoes pyrolysis that thermally decomposes the organic materials into biochar, along with by-products such as bio-oil and gas. The solid carbon-rich residue left behind after pyrolysis, is then collected. This forms the biochar.

Biochar is a stable form of carbon that can persist in soils for hundreds to thousands of years, making it a valuable tool for carbon sequestration and soil improvement. It can be used as an excellent adsorbent, catalyst, soil conditioner, in fuel cells, supercapacitors and hydrogen storage. The biochar is proven to help improve soil structure, water retention, nutrient retention, and microbial activity in the soil. Moreover, by utilizing agricultural residues for biochar production, farmers can effectively manage waste while simultaneously producing a valuable soil amendment. Biochar has an excellent growing market too. This growing biochar market, projected to reach 484 million USD by 2030 (according to reports), presents an additional income opportunity to the farmers and holds immense prospects for moving towards a successful circular economy.

Biochar is primarily made up of Carbon, Nitrogen, and Hydrogen, with smaller amounts of minerals like Potassium (K), Calcium (Ca), Sodium (Na), and Magnesium (Mg) (Wako 2021). Higher pyrolysis temperatures (300–800 °C) typically lead to increased carbon content in biochar, while nitrogen and hydrogen content decreases. However, Biochar often has a high CEC, and thus attract and hold positively charged cations (nutrient ions) like Calcium, Magnesium (Mg), and Potassium. This slow release can make nutrients more available to plants over time on a farm.

5.4 Biogas Production

An effective waste valorization strategy in IFS involves converting organic waste materials into bioenergy sources. Anaerobic digestion is a promising technique where organic matter is broken down by microbes in an oxygen-free environment, producing biogas (methane) as a by-product. Cowdung and other biomass produced in an IFS unit can be utilized for biogas production through this anaerobic digestion process. By this process, the bulk waste is reduced in volume. This biogas can be used for various purposes, including generating electricity, heating greenhouses, or fueling farm machinery. Additionally, the digestate, a leftover residue (biogas slurry) from the process, can be used as a nutrient-rich fertilizer, further reducing waste and creating a valuable resource. Moreover, by replacing fossil fuels with biogas, we can potentially curb air pollution and climate change in several ways. Biogas combustion generally produces fewer pollutants, leading to cleaner air, especially in areas with significant reliance on solid fuels for cooking and heating in comparison to the use of fossil fuels like firewood or coal. Studies suggest biogas from various feedstocks can reduce greenhouse gas emissions by 50–90% compared to fossil fuels (Fig. 4).

6 Zero-Waste IFS Case Studies

There's no one-size-fits-all approach to IFS. The ideal model depends on factors like location, resources, and suitability. Each IFS model fosters unique interactions between its components. The ultimate goal is to create a sustainable system that's not only technologically feasible and economically profitable, but also environmentally friendly and accepted by the community.

To understand the benefits and variations of IFS, let's explore some real-world case studies. These success stories showcase how different models implement zero-waste strategies, creating a more sustainable agricultural approach (Fig. 5).

IFS is a complex web of interacting elements like crops, livestock, soil, labour, and machinery. Farmers manage these elements based on their resources and circumstances to get the most output and profit while also considering long-term sustainability. The Apatani tribe residing in the hills of Arunachal Pradesh, India is an example of a unique and sustainable farming system. The Apatani community, where rice is a primary food source, employs a unique settled agriculture system that combines paddy cultivation with fish and finger millet production. This integrated approach enhances economic returns while ensuring judicious land use (Tayo 2017). The Apatani tribe is particularly

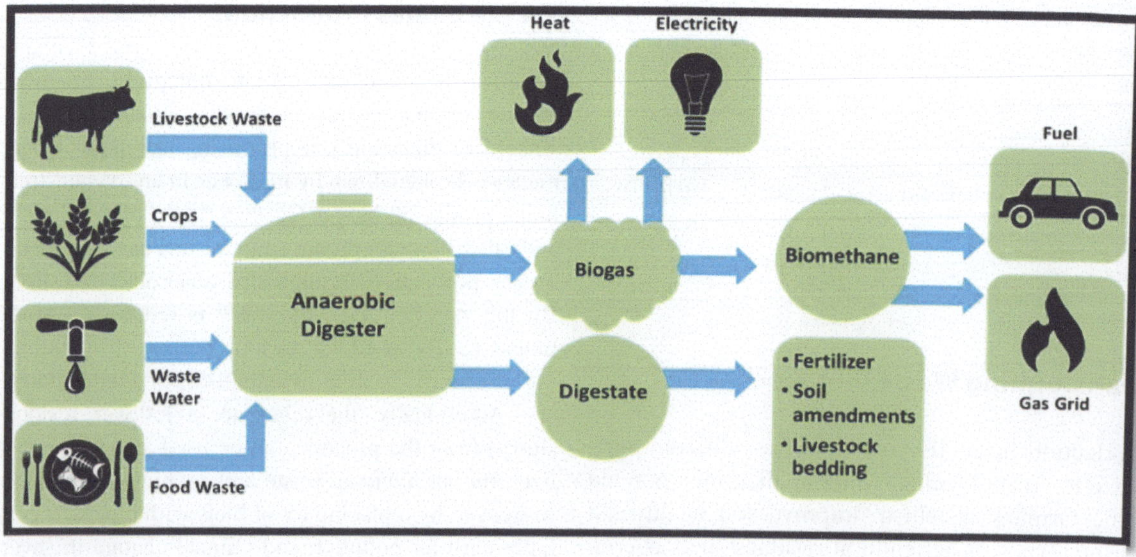

Fig. 4 Production process of biogas from agricultural waste (*Source* www.insightsonindia.com)

celebrated for its innovative land use pattern and natural resource management strategies, setting a benchmark for sustainable practices. They combine rice and fish farming for better yields and resource utilization. This rice-fish farming system is known for its efficient resource utilization and minimal waste generation. Outlining the strategies for achieving it are as follows:

(a) **Reduced need for external inputs**: Fish in the paddy fields feed on natural sources like algae and insects, minimizing the need for additional fish feed or pesticides and thus reduces additional waste generation and balances the economy.
(b) **Improved rice yields**: Fish movement and waste produced act as natural fertilizers thus boosting rice growth by 10–15% on an average thus minimizing minimizing external inputs.
(c) **Water management**: The paddy field system itself conserves water and reduces the need for additional water application.
(d) **Natural pest control**: Fish helps to control weeds and insects in the paddy field, reducing reliance on pesticides for their control and helps boost ecosystem and harbour biodiversity.
(e) **Increased biodiversity**: Fish contribute to a richer ecosystem in the paddy field, attracting beneficial organisms for further pest control.
(f) **Improved water quality**: By consumption of organic matter, fish helps to keep the paddy field water clean and aerated.

Fig. 5 Apatani rice-fish farming (*Source* www.agrifarming.in)

(g) **Additional income source**: Farmers can harvest the fish for additional income and as a protein source for farm family consumption.

This efficient use of resources and minimal generation of external waste makes fish farming in paddy fields practice a sustainable agricultural practice aligned with the concept of CE incorporating two major components in an IFS unit.

Apart from this, researchers at Assam Agricultural University, Jorhat, Assam have successfully implemented a highly productive IFS model on a 1-hectare rain-fed farm. This system combined crops (field and horticultural), cattle, fishery, and apiculture. The land was strategically divided to accommodate different crop planting cycles. Fruit and fodder crops were planted around the edge of a pond. This integrated approach achieved a net return of ₹1,45,284 ($2409.75) and a benefit–cost ratio of 1.94. It also included activities like producing and processing liquid manure, biogas, and vermicompost. The report suggests that this model, with appropriate adjustments based on individual resources, is well-suited for small and marginal farmers (Arumugam et al. 2017). The processed bioresources produced (*i.e.* value of inputs generated/ recycled) within a 1 hectare IFS was found to contribute 29.5%, whereas value of inputs purchased and farm labour engaged shared 34.7% and 35.9%, respectively to the total input cost of the system (Fig. 6). Subsequently, adoption of zero-waste IFS model (following the principle of circular economy) by the farmers holds promises to satisfy all needs at the grass-root levels as the system proves to be socially acceptable, economically viable and environmentally safe in the long term. The valorization methods in IFS that aim at resource recovery from waste thus consider all value domains (environmental, economic, social, or technical).

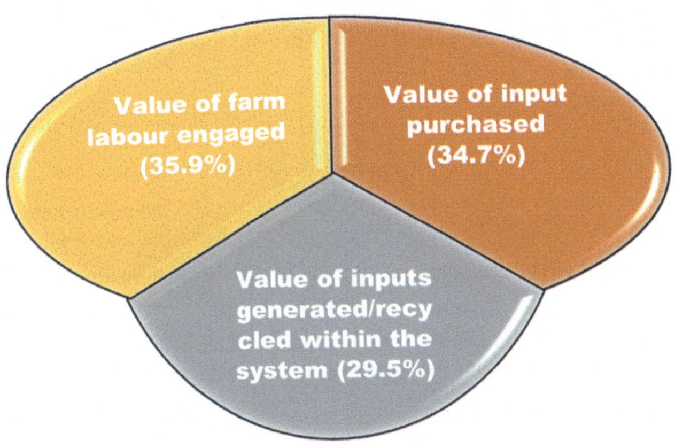

Fig. 6 Total inputs' cost and percent share of inputs in IFS unit (*Source* Baishya et al. 2014)

7 Barriers to Adopting Zero-Waste IFS Practices in Agriculture

(1) **Economic Constraints**: Implementing zero-waste practices through IFS often requires upfront investments initially while setting up the enterprises and interlinking them.Many farmers, particularly small holders or those operating on tight profit margins, may find it difficult to justify these costs without immediate returns.

(2) **Technological Limitations**: Some zero-waste solutions, such as advanced recycling or waste-to-energy technologies, may not yet be sufficiently developed or affordable for widespread adoption in agriculture. Additionally, certain regions may lack access to necessary technologies due to infrastructure limitations. All of these factors may hinder the adoption of advanced practices in IFS and thus act as a bottleneck in implementing the zero-waste policy using IFS in agriculture.

(3) **Policy Challenges**: Existing agricultural policies and regulations may not incentivize or support the adoption of zero-waste practices at the moment. Without supportive policies, farmers may struggle to navigate legal requirements, obtain necessary permits, or access financial incentives for implementing sustainable practices which may prove to be a big barrier.

8 Potential Economic Advantages of Transitioning Towards a CE Through IFS

(1) **Resource Efficiency**: By adopting CE principles, such as reducing waste and reusing resources as in a IFS unit, farmers can optimize the use of inputs like water, energy, and nutrients. This resource use efficiency can lead to cost saving and increased productivity over the long term.

(2) **Value-Added Products**: CE practices in a IFS unit can enable farmers to diversify their income streams by creating value-added products from agricultural waste. For example, converting crop residues into biofuels, animal feed, or organic fertilizers can generate additional revenue streams for the farmers.

(3) **Regulatory Compliance and Risk Mitigation**: Transitioning towards a circular economy in agriculture by moving towards IFS can help farmers comply with increasingly stringent environmental regulations. By reducing waste and minimizing pollution, farmers can mitigate the risk of fines or penalties associated with non-compliance and also contribute to healthy environmental practices.

(4) **Long-Term Sustainability**: Embracing CE principles in agriculture can contribute to the long-term sustainability of the sector by preserving natural resources, reducing greenhouse gas emissions, and enhancing ecosystem resilience. This not only benefits farmers but also society as a whole by ensuring food security and environmental stewardship.

9 Government Policies and Incentives in Promoting CE Through IFS

Government policies and incentives play a crucial role in promoting CE practices within the agricultural sector in India. The government has already laid out some policies in line with it (Table 2) and can implement more policies on the following frameline:

1. **Regulatory Frameworks**: Governments can establish regulatory frameworks that mandate or encourage the adoption of CE practices in agriculture in a more stringent manner. This can include setting standards for waste management, resource efficiency, and sustainable farming practices. For example, regulations may require farmers to properly manage agricultural waste through composting or recycling or to minimize the use of synthetic inputs through organic farming practices and repurposing waste as plant nutrient sources.
2. **Financial Incentives**: Governments can provide financial incentives to farmers to encourage the adoption of circular economy practices like IFS and others. This can take the form of grants, subsidies, tax incentives, or low-interest loans for investments in sustainable infrastructure, technology, and training. Financial support can help offset the upfront costs associated with transitioning to circular farming systems and make sustainable practices more economically viable for farmers.
3. **Research and Innovation Funding**: Governments can allocate funding for research and innovation in farming systems to promote circularity in agriculture, supporting the development of new technologies, processes, and practices that promote resource efficiency, waste reduction, and sustainable farming. This can include research grants, academic-industry partnerships, and innovation hubs focused on sustainable agriculture. By investing in research and innovation, governments can help drive technological advancements and knowledge dissemination in the agricultural sector.
4. **Capacity Building and Technical Assistance**: Governments can provide capacity building and technical assistance to farmers to help them adopt CE practices. This can include training programs, extension services, and knowledge-sharing platforms that educate farmers about sustainable farming techniques, waste management strategies, and alternative production systems. They may also be made aware of practices like IFS with enterprises suited to their particular area. By empowering farmers with the necessary skills and knowledge, governments can facilitate the transition to circular agriculture and overcome barriers to adoption.
5. **Market Incentives and Certification Programs**: Governments can create market incentives and certification programs that reward farmers for adopting CE practices and producing sustainable agricultural products
6. **Tax Benefits and low-interest loans**: Potential tax breaks for farmers who implement circular practices within their IFS may be promoted thereby encouraging resource recycling and waste reduction. Additionally, access to loans with favourable interest rates to help farmers invest in infrastructure and equipment needed for IFS practices will also encourage more and more farmers to take up IFS and promote circularity in the agricultural economy.
7. **Cross-Sectoral Collaboration**: Governments can foster collaboration between different stakeholders, including

Table 2 Different government policies and incentives for promotion of CE

Sl. No.	Scheme/Policy/Initiative	Objective
i	Rashtriya Krishi Vikas Yojana (RKVY)	Offer financial assistance to set up and upgrade IFS models including livestock units, biogas plants, composting facilities, and more
ii	Krishi Vigyan Kendras (KVKs)	To demonstrate hands-on practices to framers for implementing IFS practices
iii	National Mission on Sustainable Agriculture (NMSA)	• To explicitly promote IFS models under its Rain-fed Area Development program, recognizing their potential for resource efficiency and sustainability
iv	• Sub-Mission on Agricultural Mechanization (SMAM)	• To offer subsidies for farm machinery that can support IFS practices, such as machinery for efficient composting or better management of livestock waste
v	• National Resource Efficiency Policy (NREP)	• To focus on sustainable production and consumption in agriculture, aligning well with the circular principles of IFS

farmers, industry, academia, and civil society, to promote CE practices in agriculture. This can involve multi-stakeholder platforms, public–private partnerships, and collaborative initiatives that facilitate knowledge exchange, innovation diffusion, and policy dialogue. By bringing together diverse perspectives and expertise, governments can leverage collective action to address complex sustainability challenges and drive systemic change in the agricultural sector.

Overall, government policies and incentives play a critical role in creating an enabling environment for the adoption of CE practices in agriculture, driving innovation, investment, and behaviour change towards more sustainable and resilient food systems.

10 Future Prospects and the Way Forward

IFS hold immense promise for creating a circular and sustainable agricultural future. To achieve this vision, further research is crucial to optimize these systems for minimal waste generation and maximum resource efficiency. There is upfront call for research needs aimed at optimizing IFS for Zero-Waste Agriculture.

Research should be conducted to develop comprehensive models that integrate agronomic, ecological, and economic factors to optimize IFS for minimal waste generation and maximum resource efficiency. This would involve assessing. Investigating innovative technologies for valorizing agricultural waste streams into value-added products is another stream that needs to be harnessed. This includes exploring bioconversion processes like anaerobic digestion, fermentation, etc. to produce biogas, biofuels, biobased chemicals, and biopolymers from crop residues, animal manure, and food waste. Carbon Sequestration and Climate Resilience should be more highlighted to promote the adoption of IFS. Exploring strategies to enhance carbon sequestration and climate resilience within IFS, such as agroforestry, conservation agriculture, and soil carbon management practices are to be considered. Research is needed to quantify the carbon sequestration potential of different IFS configurations and assess their long-term impacts on soil health and climate change mitigation.

Integration of precision agriculture technologies, such as remote sensing, drones, and IoT sensors, to optimize resource use efficiency, reduce input waste, and enhance decision-making in IFS. These technologies enable real-time monitoring of soil health, crop growth, and environmental conditions, facilitating targeted interventions and adaptive management practices. Moreover, biotechnological tools can also be utilized to engineer microbial communities for improved soil health, nutrient cycling, and waste decomposition within IFS. This may include development of genetically engineered crops with enhanced nutrient use efficiency, pest and disease resistance, and stress tolerance to reduce reliance on external inputs and minimize waste in agricultural production. Integration of renewable energy technologies, such as solar photovoltaics, wind turbines, and biomass gasification, to meet the energy demands of IFS and enhance energy self-sufficiency. Renewable energy sources can power on-farm operations, such as irrigation, machinery, and processing facilities, while also enabling the production of bioenergy from agricultural residues and biomass.

By addressing these research needs and embracing future technologies and innovations, agriculture can transition towards a more circular and sustainable future, where resources are utilized efficiently, waste is minimized, and ecosystems are restored and resilient.

11 Conclusion

The transition towards a circular economy in agriculture is no longer a distant ideal, but a necessity for a healthier planet and a more secure future. Integrated Farming Systems (IFS), with their focus on zero-waste strategies, offer a powerful solution. By minimizing waste and maximizing resource use, IFS promotes environmental sustainability by reducing pollution and conserving natural resources. Economically, zero-waste IFS empowers farmers. Reliance on expensive external inputs like chemical fertilizers diminishes with the IFS approach, leading to increased profitability and resilience in the face of fluctuating market prices. The creation of valuable by-products from IFS like compost generates new income streams. Additionally, Zero-waste IFS contributes directly to food security. By optimizing resource utilization and promoting soil health, these systems enhance agricultural productivity, ensuring a stable and sustainable food supply for generations to come.

While IFS offer a promising path towards zero-waste agriculture, significant challenges remain. Economic constraints, technological limitations, and a lack of supportive policies can hinder the widespread adoption of these practices. However, the potential economic advantages are substantial. By embracing resource efficiency, creating value-added products, and achieving regulatory compliance, farmers can transition towards a more sustainable and profitable future. Government intervention is crucial for overcoming these challenges and unlocking the full potential of circular economy in agriculture. By addressing these challenges and harnessing the power of IFS, we can create a future where agriculture is not only productive but also environmentally responsible, contributing to long-term sustainability and a more secure food system for generations to come. As we move forward, embracing and scaling up these zero-waste models is crucial for building a more sustainable and resilient agricultural sector, one that nourishes both the environment and its inhabitants and also cares for a sustainable economy in the long run.

References

Abad-Segura E, Batlles de la Fuente A, Gonz´ alez-Zamar MD, Belmonte-Ureña LJ (2020) Effects of circular economy policies on the environment and sustainable growth: worldwide research. Sustainability 12(14):5792.https://doi.org/10.3390/su12145792

Altieri MA, Nicholls CI (2020) Agroecology and the reconstruction of a post-COVID-19 agriculture. J Rural Stud 79:23–28

Arancon NQ, Edwards CA, Lee S, Byrne R (2006) Effects of humic acids from vermicomposts on plant growth. Eur J Soil Biol 42:S65–S69

Arumugam K, Seenivasagan R, Kasimani R, Sharma N, Babalola O (2017) Enhancing the post consumer waste management through vermicomposting along with bioinoculumn. Int J Eng Trends Technol 44:179–182

Baishya A, Hazarika J, Gogoi B, Borah M, Bora AS, Rajbongshi A, Bordoloi D, Hazarika JP (2014) Integrated farming system-the answer to livelihood security?

Basso B, Antle JM (2021) Adaptation and mitigation strategies in agriculture: an analysis of greenhouse gas emissions and food security under climate change. Eur Rev Agric Econ 48(2):231–263

Bhat SA, Singh S, Singh J, Kumar S, Vig AP (2018) Bioremediation and detoxification of industrial wastes by earthworms: vermicompost as powerful crop nutrient in sustainable agriculture. Bioresour Technol 252:172–179

Capalbo SM, Bauen A (2019) Bioenergy and the circular economy: strategic opportunities and research needs. J Clean Prod 214:962–975

Cullen J (2018) Toward a circular materials economy. Indian J Hill Farming 30(1)

Ellen MacArthur Foundation (2013) Towards the circular economy: opportunities for the consumers goods sector. EMF (Accessed 30 June 2020). www.ellenmacarthurfoundation.org/publications

Gadde B, Bonnet S, Menke C, Garivait S (2009) Air pollutant emissions from rice straw open field burning in India, Thailand and the Philippines. Environ Pollut 157(5):1554–1558

Gill MS, Singh JP, Gangwa K (2009) Integrated farming system and agriculture sustainability. Indian J Agron 54(2):128–139

Giller KE, Andersson JA, Corbeels M, Kirkegaard J, Mortensen D, Erenstein O (2021) Beyond conservation agriculture. Front Sustain Food Syst 5:623831

Godfray HCJ, Beddington JR, Crute IR, Haddad L, Lawrence D, Muir JF, Pretty J, Robinson S, Thomas SM, Toulmin C (2010) Food security: the challenge of feeding 9 billion people. Science 327(5967):812–818

Heller MC, Keoleian GA, Willett WC (2013) Toward a life cycle-based, diet-level framework for food environmental impact and nutritional quality assessment: a critical review. Environ Sci Technol 47(22):12632–12647

https://www.agrifarming.in/rice-fish-farming

https://www.bilinvestmentinsights.com/the-circular-economy-more-than-just-recycling/

https://www.insightsonindia.com/2023/08/01/gobardhan-initiative/

Kumar S, Bhatt BP, Dey A, Shivani S, Kumar U, Idris MD, Mishra JS, Kumar S (2018). Integrated farming system in India: Current status, scope and future prospects in changing agricultural scenario. Indian J Agric Sci 88(11):1661–1675

Manjunath BL, Itnal CJ (2003) Integrated farming system in enhancing the productivity of marginal rice (Oryza sativa) holdings in Goa. Indian J Agron 48(1):1–3

Möller K, Müller T (2012) Effects of anaerobic digestion on digestate nutrient availability and crop growth: a review. Eng Life Sci 12(3):242–257

Morseletto P (2020) Targets for a circular economy. Resour Conserv Recycl 153:104553

Pauliuk S (2018) Critical appraisal of the circular economy standard BS 8001: 2017 and a dashboard of quantitative system indicators for its implementation in organizations. Resour Conserv Recycl 129:81–92

Pretty J, Benton TG, Bharucha ZP, Dicks LV, Flora CB, Godfray HC, Goulson D, Hartley S, Lampkin N, Morris C, Pierzynski G, Vara Prasad PV, Reganold J, Rockström J, Smith P, Thorne P, Wratten S, Sutherland WJ (2018) Global assessment of agricultural system redesign for sustainable intensification. Nat Sustain 1(8):441–446

Ravisankar N, Pramanik SC, Rai RB, Nawaz S, Biswas TK, Bibi N (2007) Study on integrated farming system in hilly upland areas of Bay Islands. Indian J Agron 52(1):7–10

Stegmann P, Londo M, Junginger M (2020) The circular bioeconomy: its elements and role in European bioeconomy clusters. Resour Conserv Recycl: X 6:100029

Tayo T (2017) Integrated paddy fish and finger millets cultivation by Apatani tribes in the Eastern Himalayan region- Arunachal Pradesh.

Wako RE (2021) Preparation and characterization of vermicompost made from different sources of materials. Open J Plant Sci 6(1):42–48

Zhang H, Voroney RP, Price GW (2015) Effects of temperature and processing conditions on biochar chemical properties and their influence on soil C and N transformations. Soil Biol Biochem 83:19–28

The Potential Utility of Terrestrial Weed Biomass in Bioenergy Production

Ashita Pawaiya, Manish Chaudhary, and Surindra Suthar

Abstract

The quest for renewable energy resources prompted researchers to find alternative feedstock for bioenergy programs without diverting the food crops for these purposes. The invasive terrestrial weeds pose a significant challenge to policymakers and farmers as chemical and mechanical tools are frequently utilized for their management, which seems costly and non-eco-friendly. The utility of weed biomass in renewable bioresource generation could be a sustainable and cost-effective approach. Terrestrial weeds such as *Parthenium hysterophorus* and *Lantana camara* have been significant problems in different land use systems worldwide. The high productivity, availability of green foliage and woody substances, and adaptability to the local environment proposed them as valuable sources of lignocellulosic biomass. Recent investigations suggested that such weeds could be effectively utilized for developing bioenergy sources (bioethanol, biomethane, biohydrogen, etc.), and woody substances could be a suitable feedstock for pyrolysis products. This chapter summarised the potential of *Parthenium* and *L. camara* biomass in bioenergy production under a waste-to-wealth approach system. Also discusses innovative, cost-effective and sustainable strategies for managing and mitigating weed biomass in the context of bioenergy production.

Keywords

Waste valorisation · Bioenergy · Sustainable development · Circular economy

1 Introduction

In recent years, the quest for renewable and sustainable energy has intensified due to the depletion of conventional energy resources. Most countries rely on traditional energy sources, but the rapid increase in urbanization, industrialization, and depletion of fossil fuels leads to a tremendous energy crisis. Waste biomass-based energy generation is a notable solution for economic and sustainable energy production. Bioenergy and bioethanol production using sugar crops and grain crops are widely practised worldwide (Swain and Mohanty 2018). However, using food crops as feedstock for bioenergy generation could lead to a global food crisis. Researchers are discovering alternative biomass sources, such as weeds, crop residues, municipal waste materials, etc., to address global challenges for bioenergy generation. (Adak et al. 2018a, b, c, d; Kataki and Kataki 2022).

Terrestrial weeds are invasive or unwanted plants that are non-native to the ecosystem and grow along with food crops, posing a risk to food security, biodiversity, and economic development. (Ramya et al. 2023). The high lignocellulosic content of plant weeds makes them a potential cost-effective feedstock for bioenergy generation. Using biomass in biofuel production could help control weed biomass. Recent studies have shown the potential of weeds as feedstock for biofuel generation and the resources needed to develop fertilizer (Bashar et al. 2021). This chapter will discuss the potential of *Parthenium* and *L. camara* biomass for producing bioenergy (ethanol, biogas, and biochar) within a waste-to-wealth system. It will also emphasize the need for pre-treatments to enhance the yield of biofuel generated.

A. Pawaiya · M. Chaudhary · S. Suthar (✉)
School of Environment and Natural Resources, Doon University, Dehradun, Uttarakhand, India
e-mail: suthariitd@gmail.com

A. Mandpe et al. (eds.), *Sustainable Waste Management*, Advances in Science, Technology & Innovation,
https://doi.org/10.1007/978-3-031-85685-3_14

2 Terrestrial Weeds: Issues and Opportunities

2.1 *Parthenium*

Parthenium, a highly prolific and noxious annual weed from the family Asteraceae, originated in northeast Mexico (Rai and Suthar 2020). It has a significant global impact in Asia, Africa, and around 30 countries worldwide (Adkins and Shabbir 2014). In India, it was first introduced as a stray plant on rubbish heaps in Maharashtra in 1955. Since then, it has become highly prevalent in India, impacting land use, crops, animals and humans (Adak et al. 2018d). It is considered one of the World's most devastating weeds, causing various health issues for animals and humans. Also responsible for biodiversity loss, they compete with the crops for land and nutrients due to their extensive growth and colonizing potential over various habitats and climatic conditions. They are semi-woody annual weeds, have high seed productivity, and can produce 15,000–100,000 per plant. They adapt highly to environmental conditions and grow in various habitats (Saini et al. 2014). They can easily survive at optimum temperatures between 12 and 27 °C, so they are usually seen throughout the year at different life cycle stages and can show three generations per year (Kushwaha and Maurya 2012). They are present in natural and agroecosystem environments, impacting agriculture, biodiversity, and humans. It causes allergies to humans and animals. Its vigorous growth affects the vegetation by replacing the native species and sometimes alters the habitat properties, resulting in biodiversity loss. It significantly impacts the agricultural ecosystem and reduces crop productivity, leading to economic loss to the nation. It competes with crops for land, nutrients and sunlight and indirectly inhibits crop growth by releasing allelochemicals in the surrounding soil, which hinders the growth and germination of crops (Saini et al. 2014).

2.2 *Lantana Camara*

L. camara, commonly known as red sage, belongs to the Verbenaceae family. It is a shrub growing to 0.5–2 m tall, having broadly ovate leaves with strong odour. It originated in subtropical and tropical America as an ornamental plant. It has established over 60 countries, including India, with around 650 varieties. (Negi et al. 2019). In India, it was first introduced as an ornamental plant in West Bengal in 1807 and successfully propagated throughout the Indian Sub-continent. It is a highly tolerant plant that can survive in extensive habitats such as wetlands, rainforest edges, railway tracks, etc. (Dogra et al. 2009). *L. camara's* rapid invasion of natural and agricultural fields negatively impacts biodiversity and crop productivity, disrupting the life cycle of native species. However, studies have highlighted the therapeutic values of *L. camara*. The alkaloids present in its leaves help in lowering blood pressure and promote intestinal movements (Chatterjee 2015). Additionally, it has been utilized as a treatment for cancer, asthma, chicken pox, etc. (Sharma et al. 2007).

3 Terrestrial Weed Management and Its Utility in Solving Environmental Issues

Managing terrestrial weeds is problematic due to their high proliferation rates and adaptability to the environment. Various methods have been used conventionally to manage such weeds: physical, chemical and biological. The physical/mechanical method includes hoeing, ploughing, uprooting, and burning weeds. This method has some limitations, as hoeing and uprooting can lead to the dispersion of seeds in the soil. In ploughing, weed nutrients can be incorporated into the soil, affecting crop productivity. Burning could lead to a decrease in soil quality and pollution of the environment. *Parthenium* and *L. camara* are controlled in chemical methods using herbicides such as 2,4-D, atrazine, triclopyr, picloram, etc. Such chemicals can effectively prevent weed biomass but are economically expensive and harmful to the environment. Another method, biological control, is an environment-friendly and long-term method in which insects (bio-control agents) such as *Zygogramma bicolorata* (leaf beetle), *Carmenta ithacae* (stem boring moth), *Prospodium tuberculatum*, *Puccinia lantanae*, attacks on weeds plant parts and damages their stages of life cycle, leads to metabolic disruptions and mortality (Saini et al. 2014).

The mentioned issues, such as eradication, control, and impact on environments, can be resolved using weeds in different applications. The associated benefits of *Parthenium* and *L. camara,* such as medicinal effects, industrial products, etc., could be utilized on a large scale. The high lignocellulosic content makes them qualified as a potential source for biofuel production (bioethanol, biogas, biochar). They can be utilized to treat the dye from wastewater. The enzymes extracted from the weed plant can degrade the aromatic dye rings, resulting in dye removal (Banu et al. 2023; Nzilu et al. 2024). The high macro and micronutrient availability in the weeds makes them potentially utilized as compost and vermicompost (Rai et al. 2021; Rawat and Suthar 2014; Suthar and Sharma 2013). Lignocellulosic content makes it a sustainable source for pulp and paper-making (Pande 1970). Out of all such applications, the production of bioenergy could be the best

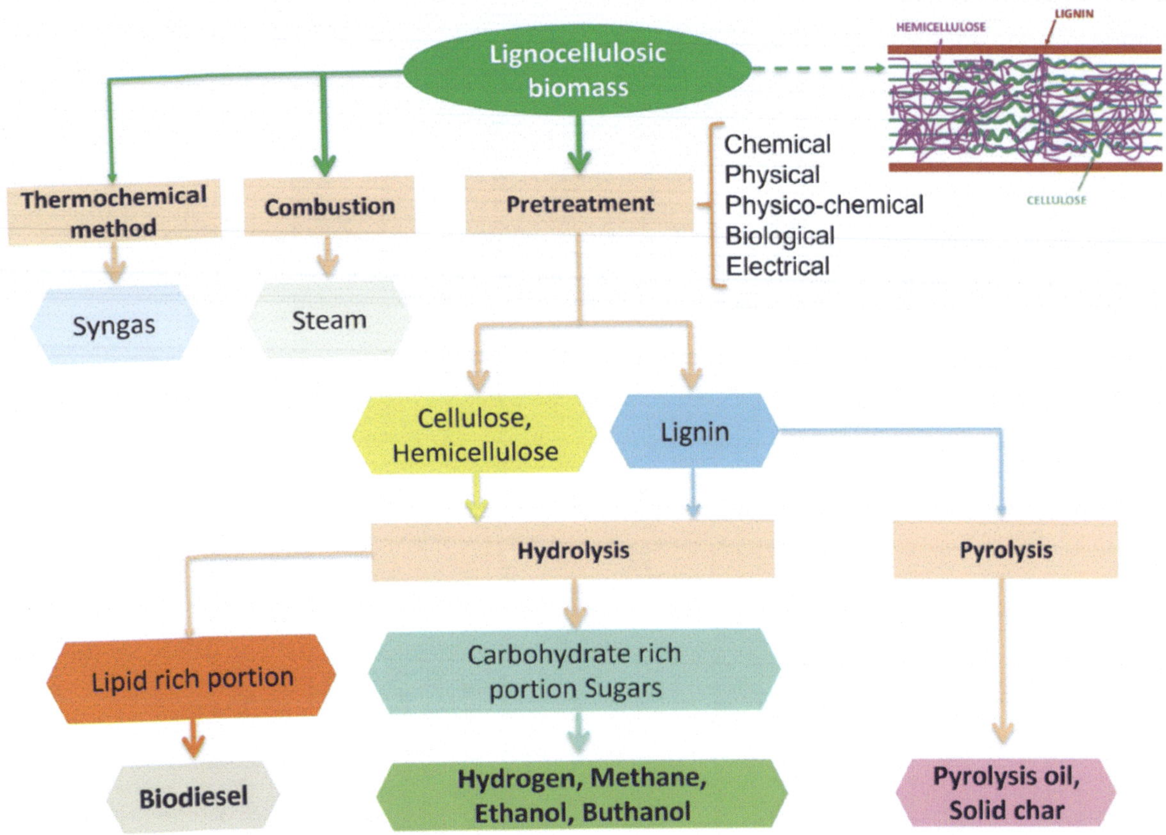

Fig. 1 Possible utilities of lignocellulosic biomass in bioenergy and industrial chemical production. *Source* (Blasi et al. 2023)

sustainable approach to weed management and also contribute to renewable energy production from waste biomass (Fig. 1).

4 Weeds Biomass Application in Bioenergy Production

The high cellulose, hemicellulose, lignin, starch and lipid content of the weeds makes them a potential feedstock for bioenergy production. Table 1 illustrates the effect of different pretreatment methods on organic matter solubilization. *Parthenium* and *L. camara* contain a significant amount of cellulose (35–45%), hemicellulose (22–26%), and lipid content (20–23%), indicating the potential biomass for renewable energy operations (Adak et al. 2018a, b, c, d). Lignocellulose is a complex substance, and its breakdown is required to make it suitable for enzymatic hydrolysis and further biofuel production. However, pretreatment of lignocellulosic content is needed for enhanced hydrolysis and bioenergy production (Barua and Kalamdhad 2017; Toquero and Bolado 2014). Studies have revealed that selecting a suitable pretreatment method is mandatory for harvesting biofuel with an optimal range (Sarto et al. 2019; Toquero and Bolado 2014).

Pre-treatment is crucial for transforming lignocellulosic biomass into easily degradable monomers, significantly boosting biofuel yield. Such terrestrial weeds are comprised of complex lignocellulosic structures. Pre-treatment facilitates the breakdown of crystalline cellulose, hemicellulose, and lignin compounds into simpler structures, thereby improving the accessibility to digestable cellulose and hemicellulose for biofuel production (Wang et al. 2021). Additionally, it increases the surface area of biomass, facilitating chances of chemical reactions and microbial activities and enhancing the yield and properties of ethanol, biogas, and biochar produced. Pre-treatment is classified into physical, physiochemical, chemical and biological categories (Fig. 2). Physical method treatment includes the size reduction of compounds to 1 mm, improving the feedstock's surface area and enhancing the biomass's digestibility of cellulose and hemicellulose (Karunanithy and Muthukumarappan 2011). Physiochemical pre-treatments involve a steam explosion, CO_2 explosion and Ammonia fibre explosion (AFEX). In steam explosion, biomass treated with high-pressure steam leads to holocellulose

Table 1 Pretreatment methods followed in *Parthenium* and *Lantana camara* to enhance biofuel production

Substrate	Pretreatment	Biofuel	References
Thermal pretreatment			
Lantana camara	Autoclave at 110 °C for 80 min	Methane production of 3656 mL	(Saha et al. 2023)
Chemical pretreatment			
Parthenium hysterophorus	1% w/v, NaOH at room temperature for 30 min	Reducing sugar yield 513.1 ± 41.0 mg/g dry substrate	(Pandiyan et al. 2014)
Parthenium hysterophorus	Ultrasonic-assisted 0.5–2.5% KOH at 20–50 °C for 10–15 min	Reducing sugar yield 128.15 ± 0.06 mg/g	(Kumar et al. 2023a, b, c)
Parthenium hysterophorus	Organo-solvent treatment with 2% H_2SO_4 (as catalyst) at 180 °C for 60 min	Total reducing sugar yield of 24.82 ± 1.2 g	(Bharti et al. 2022)
Parthenium hysterophorus	Sodium hydroxide, 1.0% w/v	Reducing sugar yield of 110.36 ± 0.05 mg/g substrate	(Kumar et al. 2023a, b, c)
Parthenium hysterophorus	Ultrasound-assisted deep eutectic solvent	148.54 mg/g biomass	(Nargotra et al. 2022)
Lantana camara	1% (v/v) H_2SO_4 assisted autoclaving	Reducing sugar yield 132.7 mg/g	(Kumar et al. 2020)
Physiochemical pretreatment			
Parthenium hysterophorus	Autoclaving for 30 min with 1% v/v, H_2SO_4	Total fermentable sugar released 39.77 g/100 g raw biomass	(Singh et al. 2014)
Lantana camara	Enzymatic pretreatment with Laccase	Bioethanol production 5.14% (v/v)	(Kuila and Banerjee 2014)
Lantana camara	3.0%, v/v H_2SO_4, 120 °C for 45 min combined pretreatment of sodium sulphite (5.0% w/v) and sodium chlorite (3.0% w/v)	Reducing sugars 187.14 mg/g 87.2% delignification	(Kuhad et al. 2010)
Biological pretreatment			
Lantana camara	Laccase and Chlorite pretreatment followed by SSF	Maximum bioethanol production 6.01% v/v Chlorite pretreatment improves saccharification up to 86–92%	(Gupta et al. 2011)
Parthenium hysterophorus	*Trametes hirsuta* ITCC136 Lignin degrading micromycete fungus *Myrothecium roridum* LG7	Delignification of substrate higher lignin recovery (1.92-fold); 485.64 mg/gds sugar upon enzymatic hydrolysis	(Rana et al. 2013)

degradation and lignin transformation. CO_2 explosion helps in the degradation of lignin and enhances the digestibility of substrate using CO_2, improving the hydrolysis rate of cellulose and hemicellulose. AFEX is mainly used to treat fibrous biomass, disturb the fibres and transform cellulose and lignin using anhydrous ammonia at high pressure and 60–120 °C temperature (Adak et al. 2018a). Chemical treatment involves acid, alkali, oxidizing agents, etc., to degrade lignocellulose. The acid treatment is used to enhance the enzymatic hydrolysis of biomass to fermentable sugars (glucose, xylose, etc.). Alkali pretreatment is mainly used to eliminate lignin from biomass, increase the accessibility of cellulase enzymes and enhance biomass's porosity (Asgher et al. 2013). Organic solvents such as methanol, acetone, etc., help hydrolyze hemicellulose into monosaccharides, acetic acid, etc. Partial hydrolysis of cellulose takes place, and lignin is hydrolyzed into low-weight fragments that dissolve in the solvents (Adak et al. 2018a). Biological pre-treatments are facilitated by microorganisms (actinomycetes, white rot fungi and bacteria) that produce enzymes that help degrade lignocellulosic structure (Ramos et al. 2022). Studies indicate that chemically pre-treated *Parthenium* biomass produces maximum reducing sugars in the range 110.36 ± 0.05 mg/g–24.82 ± 1.2 g substrate (Bharti et al. 2022; Mittal et al. 2023). A study demonstrated that autoclaving the *Parthenium* biomass with 1% v/v, H_2SO_4, releases 39.77 g/100 g raw biomass fermentable sugar (Singh et al. 2014). Similarly, *L. camara* autoclaved with 1% v/v, H_2SO_4 produces a reducing sugar yield of 132.7 mg/g biomass (Kumar et al. 2020). A combined pretreatment of *L. camara* by 3.0%, v/v H_2SO_4, 120 °C for 45 min, sodium sulphite (5.0% w/v) and sodium chlorite (3.0% w/v) produces a reducing sugar yield of up to 187.14 mg/g. Additionally, this shows 87.2% lignin removal was observed (Kuhad et al. 2010). The study reported the maximum bioethanol production of up to 6.01% v/v when pre-treated by Laccase, followed by enhancing saccharification to 86–92% with chlorite treatment (Gupta et al. 2011). Likewise, *Parthenium* undergoes biological pretreatment via *Trametes hirsuta* ITCC136, enhancing lignin recovery by 9.12-fold, producing 485.64 mg/gds sugar during enzymatic hydrolysis (Rana et al. 2013). Thus, pretreatment is essential for optimizing the conversion of biomass into biofuel. This approach significantly improves the yield and quality of biofuel, providing sustainable weed management and bioenergy generation.

4.1 Ethanol

Studies have revealed that lignocellulosic biomass can serve as a viable feedstock for bioethanol production due to its high lignocellulosic content, which can hydrolyze to produce fermentable monomeric sugars, leading to ethanol formation (Dimos et al. 2019). Figure 3 shows the step-by-step procedure for ethanol production using different lignocellulosic materials. The lignocellulose must be pre-treated to transform into ethanol. Pre-treatment is essential to convert holocellulose and lignin into monomer structures to make them available to enzymes that can convert them into fermentable sugars. Post–pre-treatment, the lignocellulose is hydrolyzed by carbohydrate degrading enzymes such as cellulase, etc., to fermentable sugar in the saccharification process. Such sugars are then fermented to

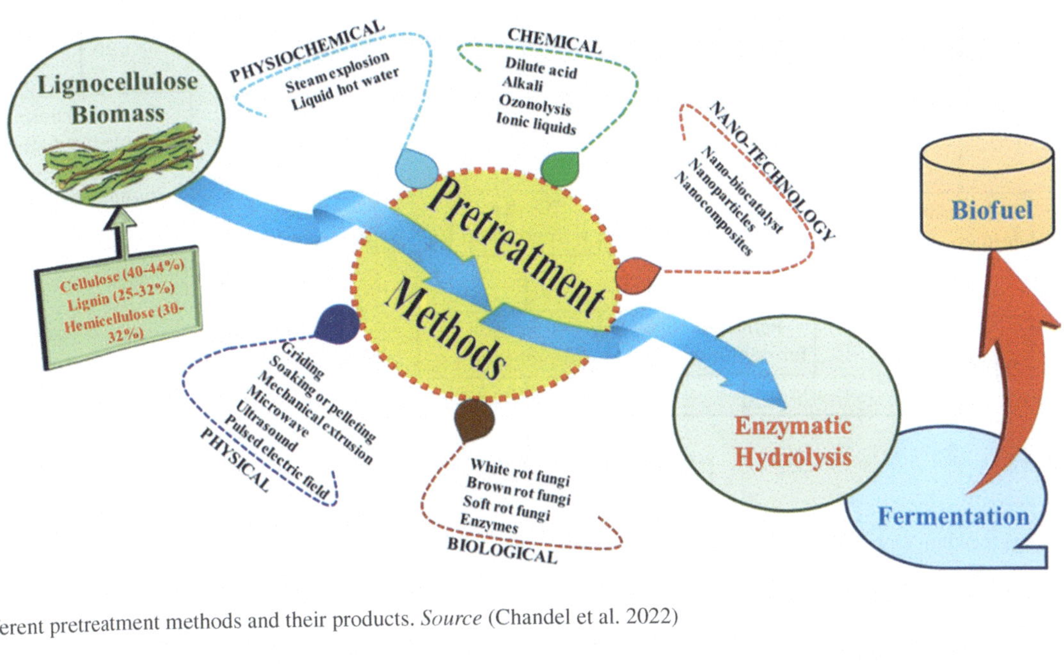

Fig. 2 Different pretreatment methods and their products. *Source* (Chandel et al. 2022)

ethanol using fermenting microbes via different technologies. Various cost-effective strategies can be employed during hydrolysate fermentation to enhance the high ethanol yield, including separate hydrolysis and fermentation (SHF) and simultaneous saccharification and fermentation (SSF) (Toquero and Bolado 2014). This process involves the fermentation of lignocellulose hydrolysate using thermotolerant microbes such as *Kluyveromyces marxianus* for enzymatic hydrolysis (Neves et al. 2007). Consolidated bioprocessing (CBP) and simultaneous saccharification and co-fermentation (SSCF) are potential cost-effective strategies for fermenting cellulose to ethanol in a single reactor. This involves mono or co-culture microorganisms such as *Clostridium thermocellum*, *S. cerevisiae*, etc. for fermentation (Adak et al. 2018b). Following fermentation, the product undergoes distillation to obtain the desired bioethanol yield. Recent studies have demonstrated, *Parthenium* is an efficient biomass for bioethanol production. Nargotra et al. (2022) reported that *Parthenium* could yield up to 148.54 mg of ethanol per gram of biomass using the CBP technique. Additionally, *Parthenium* biomass treated with ultrasound-assisted deep eutectic solvent (DES) followed by SHF of sugar yielded 110.20 mg/g biomass ethanol. It has also been claimed that Fungal-based fermentation could produce ethanol using *Parthenium* feedstock. In a study, fungal FPase enzymes were generated under solid-state fermentation conditions, which were enzymatically hydrolyzed to 0.37 ± 0.01 g/g reducing sugars within 96 h, maintaining enzyme and substrate levels of 30 FPU/100 ml and 3% (w/v), respectively. The hydrolysate was fermented using *Saccharomyces cerevisiae,* yielding 0.45 g ethanol per gram biomass (Kumar and Aggarwal 2024). Furthermore, maximum yield of 157.57–219.67 μL/mL ethanol was reported by enzymatic and acid hydrolysis of *Parthenium* leaves and stem, followed by fermentation using *S. cerevisiae* (Gupta et al. 2017). Similarly, *L. camara* has been recognized as a potential feedstock for bioethanol. Pasha et al. (2007) reported the fermentation of acidic hydrolysate of *Lantana* using *S. cerevisiae*, indicating 83.7% fermentation efficiency results in the 0.431 ± 0.018 g ethanol/g sugar yield. The acid and enzymatic pretreatment of *Lantana* holocellulose and lignin content show 80.0% saccharification. Hydrolysate fermentation produces 5.16 – 17.7 g/L ethanol, yielding 0.32 – 0.48 g/g (Kuhad et al. 2010). A study reveals the production of a maximum bioethanol concentration of 6.01% (v/v) using a mutant strain of *S. cerevisiae* via the SSF process (Kuila and Banerjee 2014). Additionally, a theoretical bioethanol yield of 0.01 g/g was obtained from the acid-treated hydrolysate of the *L. camara* stem, contributing the maximum 196 mg/g pre-treated biomass total sugar (Kumar et al. 2020). Such invasive weeds can be potentially utilized as a cost-effective and sustainable feedstock for bioethanol production, contributing to renewable energy sources and effective weed management.

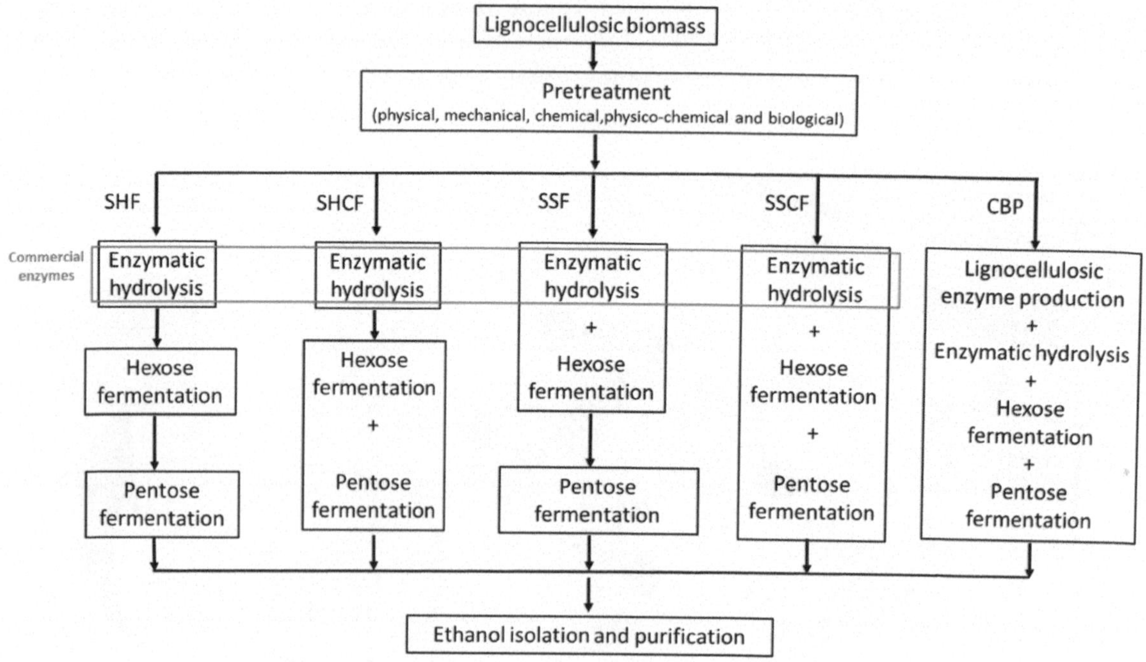

Fig. 3 Step-by-step procedure for ethanol production. *Source* (Beluhan et al. 2023)

4.2 Biogas

Anaerobic digestion (AD) is an effective solution for managing excessive weed growth in agricultural fields. It is a biological process that transforms organic biomass into biogas, which is a cost-effective approach due to limited capital investment per unit of production. In addition, nutrient-rich by-products can be used as fertilizer (Sharma et al. 2023). The AD comprises four stages: hydrolysis, acidogenesis, acetogenesis, and methanogenesis. Different microorganisms facilitate each stage. During hydrolysis, the hydrolytic microbes produce enzymes that convert lignocellulose into monomeric glucose and xylose. The hydrolyzed products then permeate into the cell membrane of acidogenic bacteria, which ferment them into volatile fatty acids and alcohol by acidogenesis (Anukam et al. 2019). In the subsequent stage, volatile fatty acids are converted into acetic acid, hydrogen and carbon dioxide by acetogenic bacteria during acetogenesis.

Finally, during methanogenesis, methanogens transform them into methane and carbon dioxide resulting in biogas production (Kothari et al. 2014). The high availability of organic matter per hectare in both weeds can be considered for biogas production. A study reported the cumulative methane production of 4165 mL CH_4 g$-^{1}$ VS and 5026 mL CH_4 g^{-1} VS from the anaerobic co-digestion of *Parthenium* and *L. camara* with cow dung and food waste, respectively (Saha et al. 2020). Another study examined the impact of electro-hydrolysis pretreatment of *Parthenium* biomass, which enhances biogas production by boosting the efficiency of VFA and sCOD, where a maximum cumulative methane production of 4242 ± 17 mL methane per gram VS was achieved (Saha et al. 2021). The aggregate outcome of 572.5 mL biogas was reported when AD of *Parthenium* with goat manure (3:1) was conducted (Hailu and Chimdessa 2020). Research suggests the potential utilization of *Parthenium* as feedstock for biogas generation, with 60–70% methane when anaerobically co-digested with cattle manure. However, its lignocellulosic content may hinder production potential which can be overcome by pre-treatments before AD (Yadav et al. 2019). A batch scale study was conducted with pre-treated *L. camara* in co-digestion with cow dung at different concentrations, maintaining the constant total solids and 10–15% (V/V) seeding, suggesting the total methane content of 13.18 L and 21.09 L, respectively (Saini et al. 2003). Another study reported that partially decomposed *L. camara* can produce biogas to 339.6 L (Verma et al. 2020).

4.3 Pyrolysis

Biochar, a carbon-based product derived by pyrolysis without oxygen, is one sustainable alternative (Fig. 4). It is produced at high temperatures in anaerobic conditions. It can be utilized as soil nutrients to improve soil properties and act as a slow-releasing fertilizer due to its large surface area and nutrient retention capacities (Sharma et al. 2023). Thus, biochar can enhance crop productivity by managing weed biomass sustainably. Few research studies have suggested the positive impact of *Parthenium* biochar on crops such as *Zea mays* and rice (Kumar et al. 2013; Qurat-ul-Ain et al. 2016). A study demonstrated the effect of temperature on the properties of *Parthenium* pyrolyzed products. A study shows the pyrolysis of *Parthenium* biomass at 300, 450 and 600 °C temperatures, revealed that biochar and bio-oil

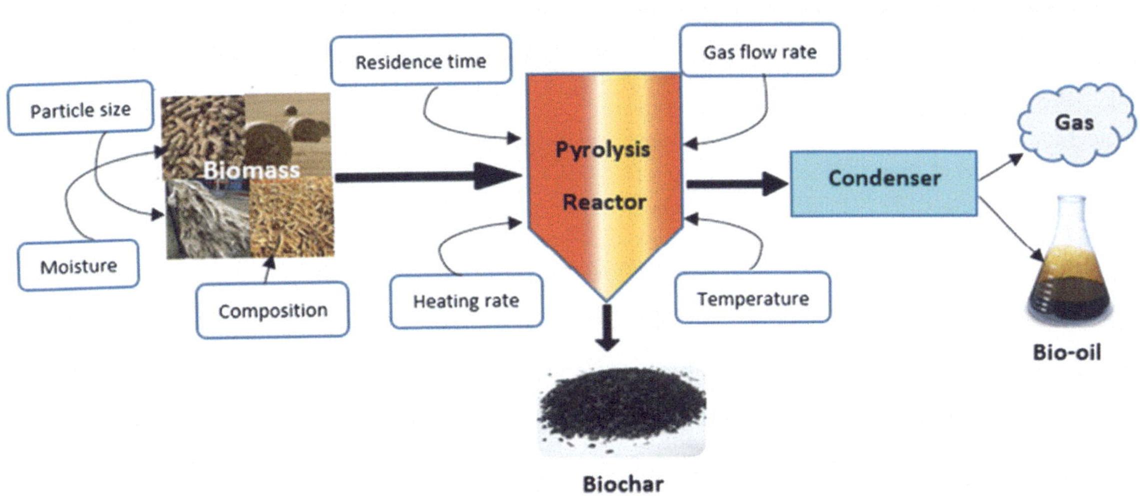

Fig. 4 Typical concept of pyrolysis process. *Source* (Sakhiya et al. 2020)

produced at 450 °C have higher quality and more excellent stability, making them suitable as soil additives without phytotoxic chemicals in the soil (Qurat-ul-Ain et al. 2021) Another study reported, production of bio-oil in the range 23.11–32.30% when *Parthenium* sp. was pyrolyzed at 400–600 °C. Also, 26.94% biochar was produced during the process, followed by syngas with a maximum exergy of 11.6 MJ kg − 1 at 600 °C found using the Friedman method (Ramesh and Somasundaram 2022). Similarly, (Tahir et al. 2020) reported that 34.66% of biochar is formed through the nano-catalytic gasification of *Parthenium* employed by Co and Ni as nanocatalysts. Apart from this, it can also be utilized as a natural product for removing contaminants from wastewater. Recent studies demonstrated the role of *Parthenium* biochar as a tool for textile wastewater treatment by eliminating methylene blue dye with an efficiency of 99.99%. The adsorption of dye on the biochar was investigated by the Freundlich and Temkin isotherm models and the pseudo-second-order model (Teweldebrihan et al. 2024). 89% chromium removal using *Parthenium* biochar from the tannery wastewater with an adsorptive capacity of 24.8 mg/g (Bedada et al. 2020) Likewise, the removal of cadmium from surface and groundwater was reported by the *Parthenium* biochar prepared at 500 and 700 °C, resulting in cadmium removal by 97–99% with the removal capacity of 129.2 mg g − 1 and 59.3 mg $g-^{1}$, respectively (Kumar et al. 2023a, b, c). *L. camara* is another invasive weed that affects the native vegetation and disturbs the food chain. It can be a considered feedstock for pyrolysis due to its high carbon and volatile matter percentage and low moisture content. A thermogravimetric analysis at varied temperatures (20–900 °C) using pyrolysis, resulted in the activation energy of the biochar in the range 149.39–169.98 kJ/mol, suggesting the suitability of *L. camara* as a potential biomass for pyrolysis (Sahoo et al. 2021). Research findings demonstrated, the utilization of modified *L. camara* biochar in the elimination of pharmaceutical compounds (ibuprofen), with a maximum adsorption capacity of 106.4 mg g^{-1} (Ganesan et al. 2021), removal of methylene blue dye (Kundu et al. 2024). Additionally, Chen et al. (2022) reported that *L. camara* biochar effectively removes heavy metals like vanadium from wastewater with a maximum adsorption capacity of 77.38 mg $g-^{1}$. Negi et al. (2022) reported utilization of *L. camara* biochar in constructed wetlands to remove high nutrient load in wastewater leads to 90.7% lead removal. Also, it serves as an efficient bio-absorbent for contamination in wastewater. Parihar et al. (2022) propose to design a low-cost wastewater treatment system by removing nitrate, ammonium nitrogen, phosphate, sulphate and COD from wastewater in a constructed wetland setup as possible. Thus, biochar can be seen as a sustainable and cost-effective product of weed biomass for controlling excessive weed production along with enhancing soil fertility and wastewater treatment.

5 Conclusion

The chapter emphasizes the potential of terrestrial weeds, specifically *Parthenium* and *L. camara*, as a sustainable resource for bioenergy generation. The main findings highlight biomass as an alternative to food crops in addressing the food crisis by transforming them into bioethanol, biogas and biochar after appropriate pre-treatments. The high lignocellulosic content makes them a valuable feedstock for bioenergy production, with effective weed management, contributing to global energy security in an eco-friendly and low-cost manner. It also supports the global transition to renewable energy sources, emphasizing the waste-to-wealth approach. The implications of these findings align with a global aim to reduce dependency on fossil fuels, mitigate environmental impacts and promote economic sustainability by transforming waste weed biomass into valuable energy feedstock. Moreover, it could be helpful to policymakers, industries and researchers in developing a plan of action that integrates weed management with bioenergy generation, encouraging to shift to a more sustainable energy future.

Acknowledgements Authors are thankful to the contributors of these books for their cooperation and patience in the compilation of this task. Authors are also thankful to Springer team, for their constant support in the publication of this work.

References

Adak A, Singh S, Lavanya AK, Sharma A, Nain L (2018a) Sustainable production of biofuels from weedy biomass and other unconventional lignocellulose wastes. In: Singh OV, Chandel AK (Eds.), Sustainable biotechnology- enzymatic resources of renewable energy, pp 83–116. Springer International Publishing. https://doi.org/10.1007/978-3-319-95480-6_4

Adak A, Singh S, Lavanya AK, Sharma A, Nain L (2018b) Sustainable production of biofuels from weedy biomass and other unconventional lignocellulose wastes. In: Sustainable biotechnology- enzymatic resources of renewable energy, pp 83–116. Springer International Publishing. https://doi.org/10.1007/978-3-319-95480-6_4

Adak A, Singh S, Lavanya AK, Sharma A, Nain L (2018c) Sustainable production of biofuels from weedy biomass and other unconventional lignocellulose wastes BT–Sustainable biotechnology- enzymatic resources of renewable energy. Singh OV, Chandel AK (Eds), pp 83–116. Springer International Publishing. https://doi.org/10.1007/978-3-319-95480-6_4

Adak A, Singh S, Lavanya AK, Sharma A, Nain L (2018d) Sustainable production of biofuels from weedy biomass and other unconventional lignocellulose wastes BT–Sustainable biotechnology- enzymatic resources of renewable energy. Singh OV, Chandel AK (Eds), pp 83–116. Springer International Publishing. https://doi.org/10.1007/978-3-319-95480-6_4

Adkins S, Shabbir A (2014) Biology, ecology and management of the invasive parthenium weed (Parthenium hysterophorus L.). Pest Manag Sci 70(7):1023–1029. https://doi.org/10.1002/ps.3708

Anukam A, Mohammadi A, Naqvi M, Granström K (2019) A review of the chemistry of anaerobic digestion: methods of accelerating and optimizing process efficiency. Processes 7(8):1–19. MDPI AG. https://doi.org/10.3390/PR7080504

Asgher M, Ahmad Z, Iqbal HMN (2013) Alkali and enzymatic delignification of sugarcane bagasse to expose cellulose polymers for saccharification and bio-ethanol production. Ind Crops Prod 44:488–495. https://doi.org/10.1016/J.INDCROP.2012.10.005

Banu R, Bhagavanth Reddy G, Ayodhya D, Ramakrishna D, Kotu GM (2023) Biogenic Pd-nanoparticles from Lantana trifolia seeds extract: Synthesis, characterization, and catalytic reduction of textile dyes. Results Chem 5:100737. https://doi.org/10.1016/J.RECHEM.2022.100737

Barua VB, Kalamdhad AS (2017) Effect of various types of thermal pretreatment techniques on the hydrolysis, compositional analysis and characterization of water hyacinth. Biores Technol 227:147–154. https://doi.org/10.1016/j.biortech.2016.12.036

Bashar HMK, Juraimi AS, Ahmad-Hamdani MS, Uddin MK, Asib N, Anwar MdP, Rahaman F (2021) A mystic weed, parthenium hysterophorus: threats, potentials and management. Agronomy 11(8). https://doi.org/10.3390/agronomy11081514

Bedada D, Angassa K, Tiruneh A, Kloos H, Fito J (2020) Chromium removal from tannery wastewater through activated carbon produced from Parthenium hysterophorus weed. Energy, Ecol Environ 5(3):184–195. https://doi.org/10.1007/s40974-020-00160-8

Beluhan S, Mihajlovski K, Šantek B, Ivančić Šantek M (2023) The production of bioethanol from lignocellulosic biomass: pretreatment methods, fermentation, and downstream processing. Energies 16(19). Multidisciplinary Digital Publishing Institute (MDPI). https://doi.org/10.3390/en16197003

Bharti AK, Kumar A, Dutt D (2022) Organosolv pre-treatment and enzymatic hydrolysis of Parthenium hysterophorus for bioethanol production. Biomass Convers Biorefinery. https://doi.org/10.1007/s13399-022-03432-x

Blasi A, Verardi, A, Lopresto CG, Siciliano S, Sangiorgio P (2023) Lignocellulosic agricultural waste valorization to obtain valuable products: an overview. Recycling 8(4):61. https://doi.org/10.3390/RECYCLING8040061

Chandel H, Kumar P, Chandel AK, Verma ML (2022) Biotechnological advances in biomass pretreatment for bio-renewable production through nanotechnological intervention. Biomass Convers Biorefinery 14(3):2959–2981. https://doi.org/10.1007/S13399-022-02746-0

Chatterjee R (2015) Impact of Lantana camara in the Indian society. Int J Environ 4(2):348–354. https://doi.org/10.3126/ije.v4i2.12663

Chen K, Ma D, Yu H, Zhang S, Seyler BC, Chai Z, Peng S (2022) Biosorption of V(V) onto Lantana camara biochar modified by H3PO4: characteristics, mechanism, and regenerative capacity. Chemosphere 291:132721. https://doi.org/10.1016/J.CHEMOSPHERE.2021.132721

Dimos K, Paschos T, Louloudi A, Kalogiannis KG, Lappas AA, Papayannakos N, Kekos D, Mamma D (2019) Effect of various pretreatment methods on bioethanol production from cotton stalks. Fermentation 5(1). https://doi.org/10.3390/fermentation5010005

Dogra K, Kohli R, Sood S (2009) An assessment and impact of three invasive species in the Shivalik hills of Himachal Pradesh, India. Intv J Biodivers Conserv 1:4–10

Ganesan S, Eswaran M, Chokkiah B, Dhanusuraman R, Lingassamy AP, Ponnusamy VK, Kumar G, Pugazhendhi A (2021) Facile and low-cost production of Lantana camara stalk-derived porous carbon nanostructures with excellent supercapacitance and adsorption performance. Int J Energy Res 45(12):17440–17449. https://doi.org/10.1002/er.5730

Gupta G, Gour VS, Sharma P, Kothari SL (2017) Acid and enzymatic hydrolysis mediated bioethanol production from biomass of a noxious weed-Parthenium hysterophorus L. Environ Prog Sustain Energy 36(1):294–296. https://doi.org/10.1002/ep.12495

Gupta R, Khasa YP, Kuhad RC (2011) Evaluation of pretreatment methods in improving the enzymatic saccharification of cellulosic materials. Carbohyd Polym 84(3):1103–1109. https://doi.org/10.1016/J.CARBPOL.2010.12.074

Hailu G, Chimdessa M (2020) Anaerobic digestion of parthenium weed with goat manure to generate biogas energy. Int J Energy Power Eng 9(3):35. https://doi.org/10.11648/j.ijepe.20200903.12

Karunanithy C, Muthukumarappan K (2011) Optimization of switchgrass and extruder parameters for enzymatic hydrolysis using response surface methodology. Ind Crops Prod 33(1):188–199. https://doi.org/10.1016/J.INDCROP.2010.10.008

Kataki R, Kataki MD (2022) Chapter 20 - Weeds as a renewable bioresource: prospects for bioconversion to biofuels and biomaterials through a cascade of approaches, (Gurunathan B, Sahadevan R, Z. A. B. T.-B. and Zakaria B (Eds) pp 437–461). Elsevier. https://doi.org/10.1016/B978-0-323-85269-2.00021-6

Kothari R, Pandey AK, Kumar S, Tyagi VV, Tyagi SK (2014) Different aspects of dry anaerobic digestion for bio-energy: an overview. Renew Sustain Energy Rev 39:174–195. https://doi.org/10.1016/J.RSER.2014.07.011

Kuhad RC, Gupta R, Khasa YP, Singh A (2010) Bioethanol production from Lantana camara (red sage): pretreatment, saccharification and fermentation. Biores Technol 101(21):8348–8354. https://doi.org/10.1016/J.BIORTECH.2010.06.043

Kuila A, Banerjee R (2014) Simultaneous saccharification and fermentation of enzyme pretreated Lantana camara using S. cerevisiae. Bioprocess Biosyst Eng 37(10):1963–1969. https://doi.org/10.1007/s00449-014-1172-y

Kumar A, Singh S, Rajulapati V, Goyal A (2020) Evaluation of pretreatment methods for Lantana camara stem for enhanced enzymatic saccharification. 3 Biotech 10(2):37. https://doi.org/10.1007/s13205-019-2029-5

Kumar N, Aggarwal NK (2024) Process development for fungal cellulase production and enzymatic saccharification of Parthenium hysterophorus for bioethanol production. Waste Manag Bull 2(1):1–8. https://doi.org/10.1016/J.WMB.2023.11.004

Kumar N, Mittal M, Yadav A, Saini DK, Aggarwal NK (2023) Statistical optimization of enzymatic saccharification of sodium hydroxide pretreated parthenium hysterophorus biomass using response surface methodology. J Wood Chem Technol 43(1):1–12. https://doi.org/10.1080/02773813.2022.2145312

Kumar N, Saharan V, Yadav A, Aggarwal NK (2023) Ultrasound-assisted alkaline pretreatment of Parthenium hysterophorus for fermentable sugar production using a response surface approach. Sustain Chem Clim Action 2:100027. https://doi.org/10.1016/J.SCCA.2023.100027

Kumar S, Masto RE, Kaushik G (2023) Thermally processed biochar: preparation, characterisation and their application for cadmium removal from surface and groundwater. Int J Environ Anal Chem 103(18):6539–6558. https://doi.org/10.1080/03067319.2021.1958324

Kumar S, Masto RE, Ram LC, Sarkar P, George J, Selvi VA (2013) Biochar preparation from Parthenium hysterophorus and its potential use in soil application. Ecol Eng 55:67–72. https://doi.org/10.1016/J.ECOLENG.2013.02.011

Kundu D, Sharma P, Bhattacharya S, Gupta K, Sengupta S, Shang J (2024) Study of methylene blue dye removal using biochar derived from leaf and stem of Lantana camara L. Carbon Research 3(1):22. https://doi.org/10.1007/s44246-024-00108-1

Kushwaha VB, Maurya S (2012) Biological utilities of Parthenium hysterophorus. J Appl Nat Sci 4(1):137–143. www.ansfoundation.org

Nargotra P, Sharma V, Sharma S, Kapoor N, Bajaj BK (2022) Development of consolidated bioprocess for biofuel-ethanol production from ultrasound-assisted deep eutectic solvent pretreated Parthenium hysterophorus biomass. Biomass Convers Biorefinery 12(12):5767–5782. https://doi.org/10.1007/s13399-020-01017-0

Negi GCS, Sharma S, Vishvakarma SCR, Samant SS, Maikhuri RK, Prasad RC, Palni LMS (2019) Ecology and Use of Lantana camara in India. Bot Rev 85(2):109–130. https://doi.org/10.1007/s12229-019-09209-8

Negi A, Joshi SK, Bhandari NS (2022) Estimation of sorption–desorption characteristics of biosorbent of Lantana camara leaves for removal of Pb (II) ions from wastewater. Environ Monitor Assessment, 195(1):42. https://doi.org/10.1007/s10661-022-10629-0

Neves MA, Shimizu N, Das Neves MA, Kimura T, Nakajima M (2007) State of the art and future trends of bioethanol production, dynamic biochemistry, process biotechnology and molecular biology State of the Art and Future Trends of Bioethanol Production. https://www.researchgate.net/publication/285028340

Nzilu D, Madivoli E, Makhanu D, Wanakai S, Kirui G, Mwangi V, Kareru P (2024) Synthesis and Characterization of *Parthenium hysterophorus*-Mediated ZnO nanoparticles for methylene blue dye degradation. J Chem 2024:1088430. https://doi.org/10.1155/2024/1088430

Pande P (1970) Evaluation of Parthenium for pulp and paper making. https://www.researchgate.net/publication/292047124

Pandiyan K, Tiwari R, Rana S, Arora A, Singh S, Saxena AK, Nain L (2014) Comparative efficiency of different pretreatment methods on enzymatic digestibility of Parthenium sp. World J Microbiol Biotechnol 30(1):55–64. https://doi.org/10.1007/s11274-013-1422-1

Parihar P, Chand N, Suthar S (2022) Treatment of high nutrient-loaded wastewater in a constructed floating wetland with different configurations: role of lantana biochar addition. Sustainability (Switzerland) 14(23). https://doi.org/10.3390/su142316049

Pasha C, Nagavalli M, Venkateswar Rao L (2007) Lantana camara for fuel ethanol production using thermotolerant yeast. Lett Appl Microbiol 44(6):666–672. https://doi.org/10.1111/j.1472-765X.2007.02116.x

Qurat-ul-Ain F-e-B, Shafiq M (2016) Management of the Parthenium hysterophorus through biochar formation and its application to rice-wheat cultivation in Pakistan. Agr Ecosyst Environ 235:265–276. https://doi.org/10.1016/J.AGEE.2016.10.027

Qurat-ul-Ain, Shafiq M, Capareda SC, Firdaus-e-Bareen (2021) Effect of different temperatures on the properties of pyrolysis products of Parthenium hysterophorus. J Saudi Chem Soc 25(3):101197. https://doi.org/10.1016/J.JSCS.2021.101197

Rai R, Singh RK, Suthar S (2021) Production of compost with biopesticide property from toxic weed Lantana: quantification of alkaloids in compost and bacterial pathogen suppression. J Hazard Mater 401:123332. https://doi.org/10.1016/J.JHAZMAT.2020.123332

Rai R, Suthar S (2020) Composting of toxic weed Parthenium hysterophorus: nutrient changes, the fate of faecal coliforms, and biopesticide property assessment. Biores Technol 311:123523. https://doi.org/10.1016/J.BIORTECH.2020.123523

Ramesh N, Somasundaram M (2022) Thermochemical conversion of Parthenium hysterophorus biomass for bio-oil synthesis: kinetics and techno-economic analysis. Biomass Convers Biorefinery 12(8):3145–3159. https://doi.org/10.1007/s13399-020-00790-2

Ramos A, Monteiro E, Rouboa A (2022) Biomass pre-treatment techniques for the production of biofuels using thermal conversion methods—a review. Energy Convers Manag 270:116271. https://doi.org/10.1016/J.ENCONMAN.2022.116271

Ramya R, Adur Alaknanda J, Raajasubramanian D, Srinivasan S, Narendra K, Manjushree M (2023) Weed—an alternate energy source, pp 165–193. https://doi.org/10.1007/978-981-99-3002-9_10

Rana S, Tiwari R, Arora A, Singh S, Kaushik R, Kumar Saxena A, Dutta SC, Nain L (2013) Prospecting Parthenium sp. pretreated with Trametes hirsuta, as a potential bioethanol feedstock. Biocatal Agric Biotechnol 2(2):152–158. https://doi.org/10.1016/J.BCAB.2013.02.002

Rawat I, Suthar S (2014) Composting of tropical toxic weed lantana camera L. Biomass and its suitability for agronomic applications. Compos Sci Util 22(3). https://doi.org/10.1080/1065657X.2014.895455

Saha B, Barua VB, Khwairakpam M, Haq I, Kalamdhad AS, Varjani S (2023) Thermal pretreatment of Lantana camara for improved biogas production: process parameter studies for energy evaluation. Environ Res 216:114661. https://doi.org/10.1016/J.ENVRES.2022.114661

Saha B, Kalamdhad AS, Khwairakpam M (2021) Efficiency of electrohydrolysis pretreatment on terrestrial weed (Parthenium hysterophorus) to cut down the hydrolysis stage during the anaerobic digestion process and continuous reactor study. Energy Rep 7:3547–3555. https://doi.org/10.1016/J.EGYR.2021.06.023

Saha B, Mohammed Yunus P, Khwairakpam M, Kalamdhad AS (2020) Biochemical methane potential trial of terrestrial weeds: evolution of mono digestion and co-digestion on biogas production. Mater Sci Energy Technol 3:748–755. https://doi.org/10.1016/J.MSET.2020.09.003

Sahoo A, Kumar S, Kumar J, Bhaskar T (2021) A detailed assessment of pyrolysis kinetics of invasive lignocellulosic biomasses (Prosopis juliflora and Lantana camara) by thermogravimetric analysis. Biores Technol 319:124060. https://doi.org/10.1016/J.BIORTECH.2020.124060

Saini A, Aggarwal NK, Sharma A, Kaur M, Yadav A (2014) Utility Potential of *Parthenium hysterophorus* for Its Strategic Management. Adv Agric 2014:381859. https://doi.org/10.1155/2014/381859

Saini R, Kanwar SS, Sharma OP, Gupta MK (2003) Biomethanation of lantana weed and biotransformation of its toxins. World J Microbiol Biotechnol 19(2):209–213. https://doi.org/10.1023/A:1023250530977

Sakhiya AK, Anand A, Kaushal P (2020) Production, activation, and applications of biochar in recent times. Biochar 2(3):253–285. Springer Science and Business Media B.V. https://doi.org/10.1007/s42773-020-00047-1

Sarto S, Hildayati R, Syaichurrozi I (2019) Effect of chemical pretreatment using sulfuric acid on biogas production from water hyacinth and kinetics. Renew Energy 132:335–350. https://doi.org/10.1016/j.renene.2018.07.121

Sharma A, Swami S, Pawaiya A, Mishra A, Chaudhary M, Bhatt N, Khan AA, Mutiyar PK, Suthar S (2023) Developing slow-release fertilizer based on pyrolysis of domestic biogas digestate cake: a circular economy approach for rural farmers. Biomass Bioenergy, pp 178. https://doi.org/10.1016/j.biombioe.2023.106966

Sharma OP, Sharma S, Pattabhi V, Mahato SB, Sharma PD (2007) A review of the hepatotoxic plant lantana camara. Crit Rev Toxicol 37(4):313–352. https://doi.org/10.1080/10408440601177863

Singh S, Khanna S, Moholkar VS, Goyal A (2014) Screening and optimization of pretreatments for Parthenium hysterophorus as feedstock for alcoholic biofuels. Appl Energy 129:195–206. https://doi.org/10.1016/J.APENERGY.2014.05.008

Suthar S, Sharma P (2013) Vermicomposting of toxic weed–Lantana camara biomass: chemical and microbial properties changes and assessment of toxicity of end product using seed bioassay. Ecotoxicol Environ Saf, pp 95. https://doi.org/10.1016/j.ecoenv.2013.05.034

Swain M, Mohanty S (2018) Bioethanol production from corn and wheat: food, fuel, and future, pp 45–59. https://doi.org/10.1016/C2017-0-00234-3

Tahir N, Tahir MN, Alam M, Yi W, Zhang Q (2020) Exploring the prospective of weeds (Cannabis sativa L., Parthenium hysterophorus L.) for biofuel production through nanocatalytic (Co, Ni) gasification. Biotechnol Biofuels 13(1):148. https://doi.org/10.1186/s13068-020-01785-x

Teweldebrihan MD, Gnaro MA, Dinka MO (2024) The application of magnetite biochar composite derived from parthenium hysterophorus for the adsorption of methylene blue from aqueous solution. Front Environ Sci 12. https://doi.org/10.3389/fenvs.2024.1375437

Toquero C, Bolado S (2014) Effect of four pretreatments on enzymatic hydrolysis and ethanol fermentation of wheat straw. influence of inhibitors and washing. Biores Technol 157:68–76. https://doi.org/10.1016/j.biortech.2014.01.090

Verma, ML, Saini R, Sharma S, Rani V, Jana A K (2020) Suitability of the lantana weed as a substrate for biogas production. In Srivastava N, Srivastava M, Mishra PK, Gupta VK (Eds.), Substrate analysis for effective biofuels production pp. 51–78. Springer Singapore. https://doi.org/10.1007/978-981-32-9607-7_3

Wang F, Ouyang D, Zhou Z, Page SJ, Liu D, Zhao X (2021) Lignocellulosic biomass as sustainable feedstock and materials for power generation and energy storage. J Energy Chem 57:247–280. https://doi.org/10.1016/J.JECHEM.2020.08.060

Yadav M, Paritosh K, Pareek N, Vivekanand V (2019) Coupled treatment of lignocellulosic agricultural residues for augmented biomethanation. J Clean Prod 213:75–88. https://doi.org/10.1016/J.JCLEPRO.2018.12.142

Critical Evaluation of the Mechanical Performance of One-Part Geopolymer Mortar for Sustainable Industrial Waste Solutions

Sajan Sharma and Shailja Bawa

Abstract

Geopolymer mortar can serve as a new-age construction material due to its outstanding performance characteristics and reduced carbon footprints. Derived from conscious industrial by-products such as fly ash, rice husk ash, and red mud, geopolymer mortar has remarkable qualities and offers a viable green substitute for conventional cement-based mortar. The traditionally used two-part geopolymers can pose scalability limitations for large-scale engineering applications due to challenges associated with handling and storing corrosive alkaline activator solutions during transportation and prefabrication. Considering this limitation, a one-part geopolymer mortar is formulated. This behaviour results from pre-incorporated alkaline binding agents being activated. These materials are mostly silicate and aluminate, derived from recycled wastes that are rich in silica and alumina. By reducing carbon emissions, this innovative method complies with global environmental goals while improving the performance and management of one-part geopolymer mortar. This paper has been formulated on the basis of significant recent findings by researchers. It aims to highlight the distinctive mechanical performance of one-part geopolymer mortar. It focuses on how feasible it is as an environmentally friendly industrial waste management solution within the framework of the building sector. The study meticulously synthesizes recent information about the unique mechanical characteristics of one-part geopolymer mortar and highlights its possible uses in environmentally friendly construction methods. The material is analysed as a resilient and eco-friendly replacement for repair and building applications, showcasing numerous advantages over traditional cement-based mortar. This study signifies a paradigm shift towards the application of one-part geopolymer mortar in construction projects, providing valuable insights for researchers, engineers, and industry professionals looking for an easy, resilient, and eco-friendly alternative with superior mechanical performance.

Keywords

Alkali-activated · One-part geopolymer mortar · Waste recycling · Binders · Mechanical performance · Environmentally friendly solutions

1 Introduction and Background

As society progresses more and urbanization increases, the demand for construction increases more, leading further to the construction of more buildings. In response to the increasing public consciousness about the potential risks associated with climate change, the construction industry has implemented substantial precautions to mitigate greenhouse gas (GHG) emissions. The construction industry is said to be generating up to half of all greenhouse gases emitted in the environment. On the other hand, out of all construction materials, Ordinary Portland Cement (OPC) is the first and dominant choice among engineers.

During the production process of Ordinary Portland Cement, significant ecological imbalance happens due to the large amount of carbon dioxide emission and the high energy consumption (Zhao et al. 2023). According to researchers, about 5% of the world's carbon dioxide (CO_2)

S. Sharma (✉) · S. Bawa
Department of Civil Engineering, Dr. B. R. Ambedkar National Institute of Technology, Jalandhar, Punjab, India
e-mail: sajans.ce.22@nitj.ac.in

S. Bawa
e-mail: bawas@nitj.ac.in

A. Mandpe et al. (eds.), *Sustainable Waste Management*, Advances in Science, Technology & Innovation,
https://doi.org/10.1007/978-3-031-85685-3_15

emissions are due to the production process of Ordinary Portland Cement (Zhang et al. 2021). Reports state that up to 0.6 to 0.82 kg of carbon dioxide is anticipated to be released into the atmosphere during the manufacturing of 1 kg of OPC (Srinivasa et al. 2023). About 7% to 8% of greenhouse gases (GHGs) released into the atmosphere are due to OPC production (Srinivasa et al. 2023).

To minimize these greenhouse gas emissions in the environment and optimize resource utilization in cement production, researchers tried substituting cement partially or completely. This replacement is done with additional cementitious materials, which were earlier treated as waste in the society, such as slags, silica fume, ashes, calcined clays, and other natural pozzolans. Introduced by Davidovits (Davidovits 1991) in the 1970s, geopolymers are the most potential substitute for traditional cement in the construction industry among all the other choices. Geopolymers are recognized for their superior qualities, such as high strength, durability, and reduced carbon emission and are said to be the most promising alternative to traditional cement (Deng and Zheng 2023).

Geopolymers are inorganic polymers exhibiting three-dimensional network structures of polymerized aluminate (AlO_4) and silicate (SiO_4) units derived from precursor materials. Industrial by-products containing reactive silicon dioxide (SiO_2) and aluminium oxide (Al_2O_3) serve as viable precursors for geopolymers. The aluminosilicate material reacts with the activator solution, forming geopolymeric aluminosilicate gels. However, the efficacy of geopolymers and binders is contingent upon numerous parameters, comprising the molar ratio of SiO_2/Al_2O_3, the activation solution, and the curing process (Duxson et al. 2007). In contrast to hydraulic binders, geopolymers and binders are governed by the chemical makeup of the reactants (Mahfoud et al. 2023).

Depending on the amount of calcium exists in the unprocessed binders, these materials can be divided into two categories: low-calcium structures (e.g. metakaolin, fly ash type F), termed geopolymers, predominantly yielding sodium aluminosilicate hydrate (NASH) gels featuring a multifaceted structure comprising AlO_4 and SiO_4 tetrahedral structure linked by shared oxygen atoms (Jia et al. 2018). Conversely, high-calcium systems (e.g. slag) are designated as reactive materials, producing gels like calcium-aluminium–silicate-hydrate (CASH) structure with dense matrix (Ameri et al. 2019). The coexistence of these alkaline activated gel types has been extensively researched by number researchers. Increased calcium content causes the transition from NASH structure to CASH structure, likely to increase shrinkage and also improve the mechanical performance of the matrix (Duxson et al. 2007; Marvila et al. 2021). These gels are resistant to sulphate and chloride, and they also have a high-density and durability (Dong et al. 2020).

1.1 Comparison of Two-Part and One-Part Geopolymer Mortar

As a sustainable substitute for conventional cement-based mortars, geopolymer mortars are becoming more and more popular, offering environmental benefits, better durability, and improved mechanical properties. Furthermore geopolymer mortar can be described as one-part and two-part, each type having its own mixing method and material composition. These types of geopolymer mortar are extensively studied by the researcher, comparing the pros and cons of geopolymers to understand their best applications in construction works.

Concentrated alkaline solution is mixed with solid precursors which have high amount of SiO_2 and AlO_2 likely recycled wastes of ash and slag, combining with fine aggregates to produce two-part geopolymer mortar as shown in Fig. 1. This method is mostly accepted by researchers, for its simple mixing ability to produce geopolymer mortars with high mechanical strength and durability. However, the use of concentrated alkali solutions produced by mixing hydrates and silicates raises challenges of safety, storage, and handling, which can be major concerns for large-scale production of geopolymer mortar (Ghazy et al. n.d.). Despite this safety and handling risks, the two-part geopolymer method remains the favourite in both the research and industrial sectors due to its reliability and familiarity (Segura et al. 2022).

On the other hand, the one-part geopolymer mortar highlights the process by pre-mixing in the dry form of alkali activators with the solid precursor and then adding water, like traditional cement mortar. This process of producing geopolymer mortar eliminates the need for handling corrosive concentrated alkali solutions, as illustrated in Fig. 1. This method offers a number of benefits, including high safety, easier handling, and a reduced environmental footprint (Segura et al. 2022). Additionally, one-part geopolymer process are particularly suitable for cast-in-situ applications, where on-site mixing and placement are required. According to research, the one-part method shows slightly lower mechanical strength compared to the two-part geopolymer method due to variations in reaction kinetics and gel formation (Srinivasa et al. 2023; Ren 2021). Despite this drawback, latest advancements in one-part geopolymer formulations and processing techniques have successfully countered these challenges. As a result of this performance, there is a growing interest in the adoption of a one-part geopolymer process in both the academic and industrial sectors (Mahfoud et al. 2023).

There exist several critical elements that significantly impact the performance and attributes of both one and two-part geopolymer mortars. Performance of geopolymer majorly depends on the choice of precursor materials,

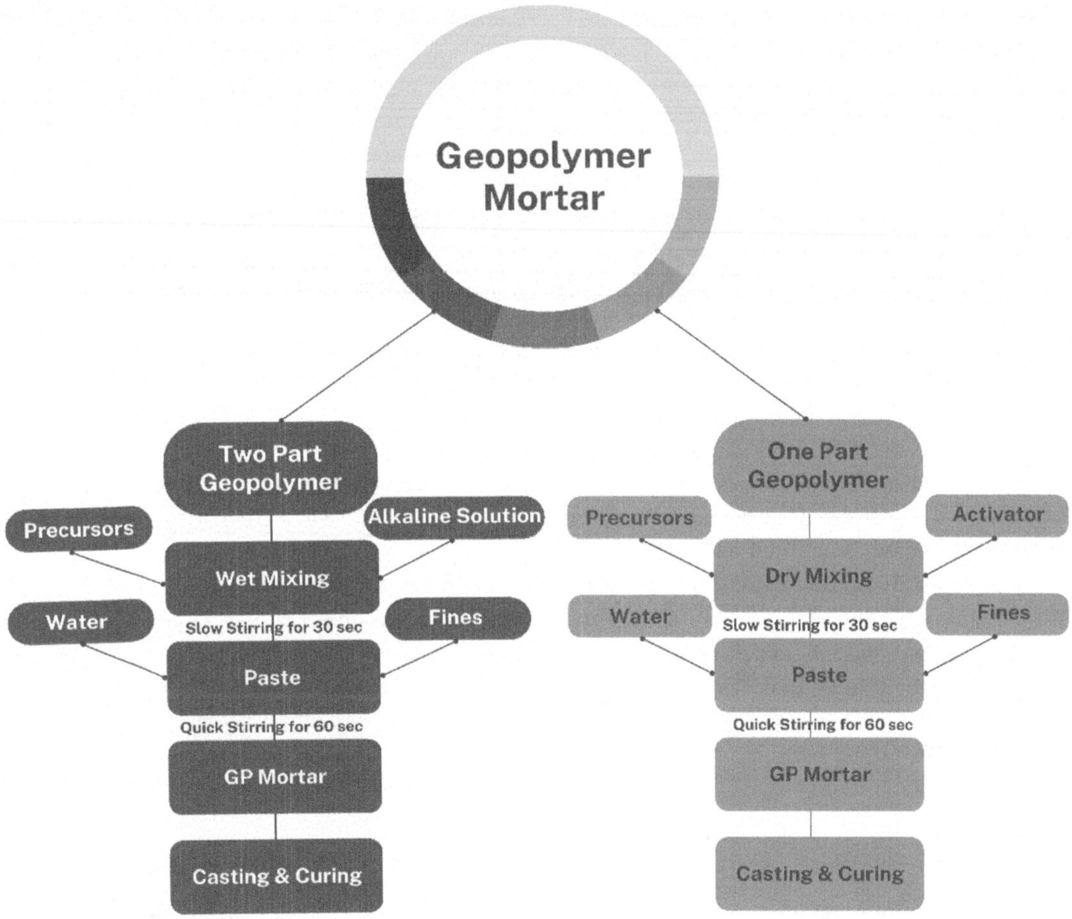

Fig. 1 Manufacturing process of two-part geopolymer and one-part geopolymer mortar

their chemical composition and reactivity. It also varies on the alkali activator type and concentration, the curing temperature, and whether or not superplasticizers or other supplemental additives are used. Researchers are continuously working to optimize these necessary parameters to improve the mechanical strength, durability, and workability of geopolymer mortars, all while reducing production costs and ecological impact.

2 Geopolymer Chemistry and Material Composition

2.1 Overview of the Polymerization Process

Understanding of polymerization process is key component before understanding the synthesis process of one-part geopolymer mortar. Geopolymerization, pioneered by Joseph Davidovits in 1978 (Davidovits 1991), has created "geopolymers", or three-dimensional alumina-silicate materials that are amorphous to semi-crystalline in structure. This unique polymerization process, rooted in geochemistry, involves amalgamation between the different alumina-silicate oxides, typically those in the group of aluminium cation (Al^{3+}) (IV–V fold) in reacting with silicates, in the presence of extremely alkaline conditions (Xu and Van Deventer 2000). The subsequent polymeric Si–O–Al–O structural bonds are the necessary and most required bonds of geopolymers, offering a rich properties matrix mass for various construction application (Xu and Van Deventer 2000).

$$\begin{aligned} & n(Si_2O_5,\ Al_2O_2) + 2nSiO_2 \\ & + 4nH_2O + NaOH \text{ or } (KOH) \\ & \rightarrow Na^+, K^+ + n(OH)^{3-}Si - O - Al^- - O - Si - (OH)_3 \end{aligned}$$

$$(Si - Al\ \text{material}) \,|\, (\text{geopolymer precursor})$$

The structure of geopolymers is majorly based on the sialate (silicon-oxo-aluminate) structure, where silicate and aluminate tetrahedra are connected, consequently connecting with all O_2 atoms. Positive ions like sodium ion (Na^+), potassium cation (K^+), and calcium ions (Ca^{++}) coexist with Al^{3+} ions in the structural cavities preserving charged neutrality (Davidovits 1991). This poly(sialate) further takes form of ring and chain polymers, with Si^{4+} and Al^{3+} ions in four fold combination with O_2 (Davidovits 1991; Komnitsas and Zaharaki 2007). This chemical structure, $Mn(-(SiO_2)z-AlO_2)n \cdot wH_2O$, describes the degree of polycondensation (n) and the existance of monovalent cations (M) like K^+ or Na^+ in this reaction (Komnitsas and Zaharaki Nov. 2007; Xu and Van Deventer n.d.). As shown in Fig. 2, the diverse form of polysialates shown chain and sheets type three-dimensional network interconnected with SiO_4 and AlO_4 tetrahedra, which explain the structural complexity of geopolymers (Xu and Van Deventer 2000; Komnitsas and Zaharaki 2007).

The geopolymer setting and hardening process involves number of different steps, which majorly includes dissolution, orientation, and reprecipitation via polycondensation (Xu and Van Deventer n.d.). It seems that in order to initiate the dissolving of alumina and silica and catalyse the condensation reaction, alkali metal hydroxides and salts are required. The quantity of aluminium in the initial components may have an impact on how well it bonds to other aluminium in the final matrix (Davidovits 1991).

Different source materials are available containing high amount of Al-Si, such as furnace slag, fly ash, pozzolan, kaolinite, and meta kaolinite, have been used to determine their potential in polymerization process (Zhang et al. 2021; Srinivasa et al. 2023; Dong et al. 2020). The intrique nature of geopolymerization of different minerals and waste materials necessitates the determination of precise reaction parameters, including mass ratios among source materials, solid/liquid ratio, alkalinity/water ratio, and alkaline solution/binder ratio (Zhao et al. 2023; Zhang et al. 2021; Adesanya et al. 2018). Research studies have shown that these parameters greatly impact the mechanical properties of geopolymers, highlighting the importance of optimization (Ren 2021; Komnitsas and Zaharaki 2007; Geraldo et al. 2022).

2.2 Characteristics of Activator and Precursor in One-Part Geopolymer Mortar

The characterization of materials used as activators and precursors in one-part geopolymer matrices is a multi-faceted endeavour specially for having a comprehensive understanding and optimizing the performance of these sustainable, eco-friendly construction materials. Hydrate and silicate are most commonly used activators, with a

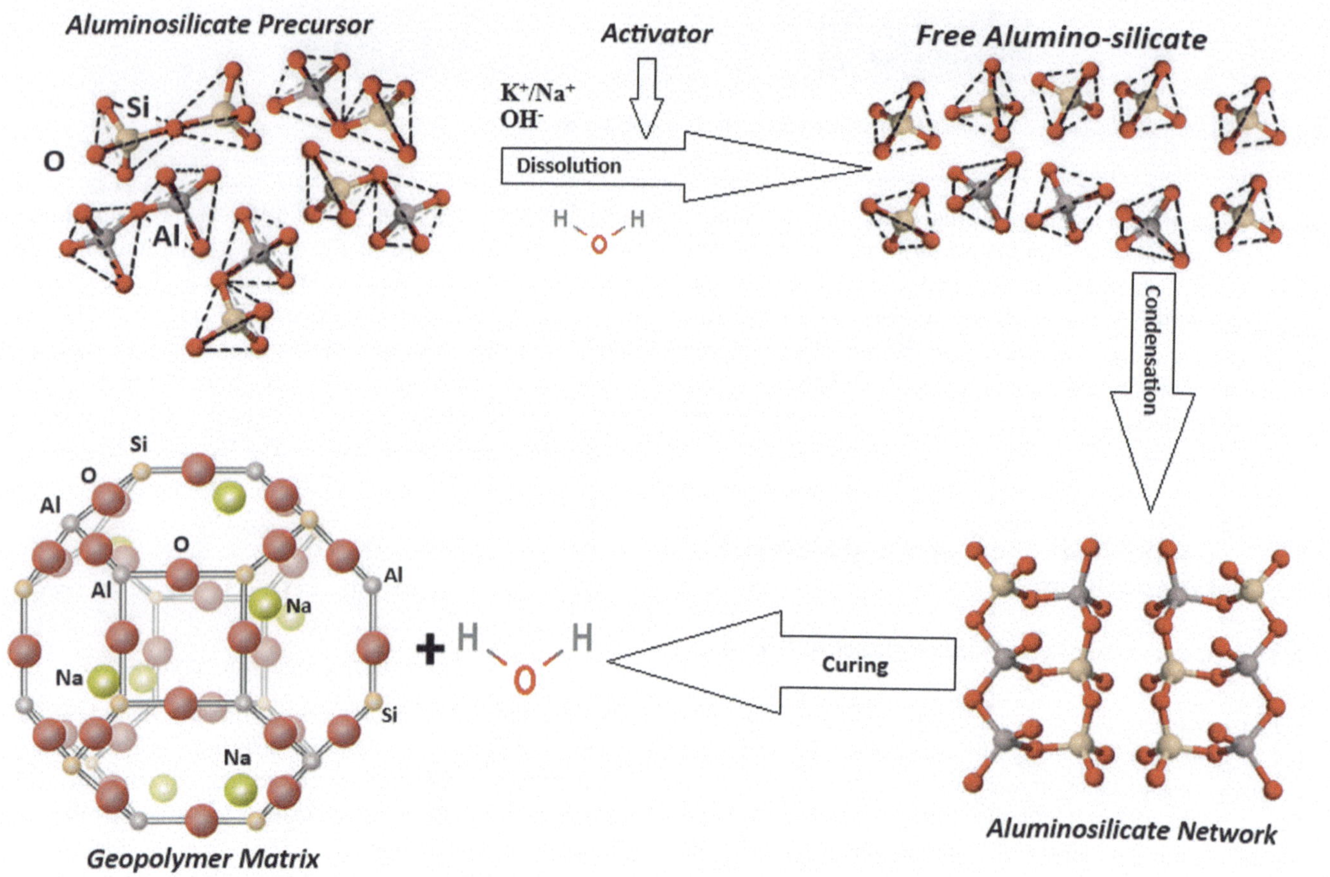

Fig. 2 Geopolymerization process

combination of commercial-grade sodium hydrate (NaOH) powder and instant sodium silicate (Na_2SiO_3) powder being prevalent. Interestingly, anhydrous Na_2SiO_3 powder with SiO_2/Na_2O mass ratio of 1.0 shows superior performance (Zhao et al. 2023). Both instant and anhydrous Na_2SiO_3 powders are commercially available easily. Standardized NaOH and Na_2SiO_3 solutions are essential for optimum sample preparation, maintaining a total mass ratio of SiO_2/Na_2O to 1.0 for uniform consistency (Zhao et al. 2023). The precursor fly ash (FA) has ratio of SiO_2/Al_2O_3 equal to 2.3, and Ground Granulated Blast-Furnace Slag (GGBS) has ratio of SiO_2/Al_2O_3 equal to 2.1, as shown in Table 1.

Researcher compares one-part and two-part geopolymer matrix using FA and metakaolin (MK) at a 1:1 mass ratio as precursors (Zhang et al. 2021). Two different solid activators are used for geopolymerization, anhydrous Na_2SiO_3 and a blend of hydrous Na_2SiO_3 with NaOH. In one-part systems, activators were directly mixed with the precursor and fine aggregate, further water is added to start geopolymerization process, while in two-part systems, alkaline

Table 1 Properties of different precursors

References	Binder	SiO_2	Al_2O_3	Fe_2O_3	CaO	MgO	LOI
Zhao et al. 2023	FA	56.74	24.58	6.550	4.870	1.200	2.340
Zhao et al. 2023	GGBS	33.21	15.76	0.710	37.14	8.510	03.00
Zhang et al. 2021	FA	47.05	36.21	4.300	0.870	0.020	4.830
Zhang et al. 2021	MK	53.37	42.21	0.310	0.160	0.000	1.890
Deng and Zheng 2023	LZT	77.27	5.390	2.260	8.560	1.260	0.000
Deng and Zheng 2023	S	34.50	17.70	1.030	34.00	6.010	0.000
Srinivasa et al. 2023	FA	60.65	28.62	3.190	1.700	1.840	4.520
Srinivasa et al. 2023	GGBS	32.45	16.72	1.220	33.75	9.610	0.400
Vu 2020	FA	58.15	22.29	6.950	5.390	1.280	1.090
Vu 2020	S	33.89	14.14	0.450	40.46	6.840	−0.11
Adesanya et al. 2018	GGBS	32.33	9.580	1.230	38.50	10.24	−1.30
Adesanya et al. 2018	PS	0.700	1.000	0.100	95.10	0.800	50.00
Abdel-Gawwad et al. 2019	GGBS	37.81	13.92	0.290	32.73	9.510	0.000
Abdel-Gawwad et al. 2019	CoW	17.31	1.460	1.000	38.50	9.200	30.86
Abdel-Gawwad et al. 2019	LBS	46.10	1.540	4.160	8.080	0.900	13.78
Shen 2023	FA	45.85	35.98	4.960	6.900	0.810	0.000
Shen 2023	EFNS	50.98	4.660	10.37	1.190	30.70	0.000
Shen 2023	S	32.65	16.15	8.380	38.74	0.530	0.000
Shen 2023	C	22.20	4.710	3.580	65.48	2.100	0.000
Mahfoud et al. 2023	FA	51.00	26.10	5.800	2.300	1.600	3.800
Mahfoud et al. 2023	SD	39.62	9.640	5.120	10.57	0.880	27.63
Mobili et al. 2020	K	51.00	34.00	0.000	0.000	0.000	0.000
Mobili et al. 2020	BA	09.00	1.700	0.000	46.10	3.600	0.000
Geraldo et al. 2022	MK	49.36	41.07	2.080	0.070	0.820	2.450
Geraldo et al. 2022	RHA	93.19	1.270	0.050	0.580	0.330	2.080
Shoaei et al. 2024	Z	66.20	14.20	0.140	4.600	0.500	9.900
Chen et al. 2024	FA	56.83	30.17	3.120	3.300	1.150	2.260
Chen et al. 2024	GGBS	32.50	15.02	0.890	40.67	6.210	1.560
Ren 2021	GGBS	31.86	16.53	0.430	39.81	6.890	0.0200
Ouyang et al. 2020	FA	50.20	30.10	12.70	3.240	0.440	0.000
Ouyang et al. 2020	GGBS	34.80	16.30	0.000	38.50	0.000	0.000
Segura et al. 2022	MK	53.00	43.80	0.430	0.020	0.030	0.460
Segura et al. 2022	GGBS	32.22	9.580	1.230	38.51	10.24	0.000

Fly ash (FA), Ground Granulated Blast-Furnace Slag (GGBS), Metakaolin (MK), lead zinc tailing (LZT), slag (S), paper sludge (PS), concrete waste (CoW), lead-bearing sludge (LBS), electric furnace nickel slag (EFNS), cement (C), raw sediments (SD), kaoline (K), biomass Ash (BA), rice husk ash (RHA), zeolite (Z) solutions were formed before mixing it with precursor and fine aggregate (Zhang et al. 2021). In this study, a specific experimental procedure for characterizing precursor synthesis is used. Initially NaOH is mixed with low-grade zinc tailings (LZTs) and then thermally activated in a muffle furnace (Deng and Zheng 2023). After controlled heating and cooling, the activated material is mixed with slag to form the precursor mixture. Further adding water for geopolymerization maintaining water/solid ratio of 0.35 with solid facilitates paste formation, cast into moulds, and curing (Deng and Zheng 2023). This meticulous procedure provides valuable insights into material characteristics, especially the influence of activated LZTs in the matrix for further advance research in one-part geopolymer matrices.

In a study locally brought granular anhydrous sodium metasilicate powder is used as solid activator for one-part geopolymer. The hydration modulus of GGBS, calculated as the ratio of ($CaO+MgO+Al_2O_3$) to SiO_2, is 1.85 (Srinivasa et al. 2023). The ratio of Al_2O_3/SiO_2 is 0.51, indicating high activity slag as it exceeds 0.25. The alkalinity coefficient of matrix, analysed from ($CaO+MgO$) by ($SiO_2+Al_2O_3$), is 0.88>1.0 (Srinivasa et al. 2023). This type of activator was crucially used in initiating the polymerization process within the matrix. The silicate module of SiO_2/Na_2O in the sodium metasilicate powder was determined to be 0.97 (Srinivasa et al. 2023). Class F FA with a SiO_2/Al_2O_3 ratio of 2.6 and ground GGBS with a SiO_2/Al_2O_3 ratio of 2.39 are the binder materials used in this research (Vu 2020). Sodium metasilicate in anhydrous form and a carbonate mineral that releases hydroxide anions (OH^-) when hydrated are illustrations of solid activators. The GGBS analysed contains inherently reduced species like iron, sulphate, and manganese, which undergo mass gain during heating or oxidising processes. Consequently, slag often exhibits a near-zero or slightly negative loss on ignition (LOI) of −0.11 (Vu 2020). The geopolymer powder ratio employed was maintained at 2.5.

In another study GGBS, concrete waste (CoW), lead-bearing sludge (LBS), and NaOH is used based on specific properties and sources. GGBFS has a specific gravity of 3.10, with an amorphous structure and distinctive akermanite peaks (Abdel-Gawwad et al. 2019). CoW from the construction demolition displays a light grey hue and specific gravity of 3.15, with crystalline peaks associated with dolomite, quartz, and calcium carbonate. LBS is yellow with a specific gravity of 1.90, with an amorphous structure and crystalline peaks attributed to calcium carbonate (Abdel-Gawwad et al. 2019). NaOH taken is 99.99% pure, ensuring consistency in chemical reactions. Chemical treatments involving CoW interaction with solid NaOH produce chemically treated CoW (CTC), which is crucial for one-part AAC formulation. Pastes undergo meticulous drying and grinding for consistent fineness. GGBFS, CTC, and LBS particle size distributions ensure uniformity and optimal performance. Formulation of one-part AAC involves dry mixing GGBFS with CTC, adjusting Na_2O content to 2 wt % to mitigate environmental concerns (Abdel-Gawwad et al. 2019).

Another study investigates the properties of FA, GGBS, electric furnace nickel slag, and Ordinary Portland Cement (OPC) for one-part geopolymer matrices (Shen 2023). Notably, secondary FA had an average particle size of 62.52 μm and a water demand of 80%, while slag exhibited an average particle size of 15.6 μm. The range of particle size of nickel slag is from 3 to 4 mm. Anhydrous sodium silicate (Na_2O-$nSiO_2$) and anhydrous sodium carbonate (Na_2CO_3) were used as solid activators in this research for polymerization process (Shen 2023). In this research, primary alkaline activator used for geopolymerization is calcium hydroxide ($Ca(OH)_2$) having purity of 90%. To keep achieve good strength and workability additional additives like gypsum powder, micro silica, and the superplasticizer were incorporated (Kogbara et al. 2024). The mechanical properties and flowability got improved using superplasticizer with reduced ratio of the water-to-cementitious material, for 3D printing (Kogbara et al. 2024). In this study, four different binders as the source of aluminosilicate for one-part geopolymer matrices, with class F FA as the control binder and activated sediment as the other are incorporated (Mahfoud et al. 2023). The alkaline activator comprised an activation powder synthesized from a commercial sodium silicate and NaOH blend. To increase polymerization reaction calcined powders is grinded to finer particles. Alkaline activator powder used in this study is in the amount of 25% wt of NaOH and 75% wt of Na_2SiO_3, was pivotal in initiating geopolymerization (Mahfoud et al. 2023).

It is important to characterize materials in one-part geopolymer matrices, when solid activators powders are used in dry form. It is been studied that sodium metasilicate emerges as a prominent solid activator powder. Studies report use of calcium oxide (CaO) powder shows more heat generation during hydration in comparison with $Ca(OH)_2$, underscoring the significance of activator choice. Preference to the $Ca(OH)_2$ over NaOH as an activator is given due to corrosive and hygroscopic nature of solid NaOH (Ghazy et al. n.d.). Investigations on geopolymer matrix have explored the synergistic effects of $Ca(OH)_2$ with sodium carbonate (Na_2CO_3), revealing in-depth insights. Comparative analysis shows that one-part geopolymers activated by $Ca(OH)_2$ and Na_2CO_3 results in remarkable increase in the compressive strength compared to those matrix activated by NaOH alone (Ghazy et al. n.d.).

This investigation focuses on characterizing materials in a one-part geopolymer matrix to evaluate alkali-activated binders. Metakaolin (MK), rice husk ash (RHA), and sodium hydroxide (NaOH) serve as precursor materials, with natural river quartz sand for mortar formulation (Geraldo et al. 2022; Abdollahnejad et al. 2015). MK majorly contains 50–55% SiO_2 and 40 to 45% Al_2O_3, with minor impurities like TiO_2, while RHA has a high SiO_2 content (>90%), confirming its suitability for geopolymer synthesis (Geraldo et al. 2022). This research emphasizes sodium hydroxide, sodium metasilicate, FA (Murtaza Zaidi, Sharma, and Student 2021) and GGBS (Ren 2021). Sodium hydroxide, crucially utilized in pellet form, is closely linked to its purity for cost-effectiveness. Sodium metasilicate, an activator, boasts versatility in various industries due to its bonding properties. GGBS's density, fineness, and chemical composition are pivotal in geopolymer formulations.

Temperature dynamics during mixing reveal distinctions between one-part and two-part geopolymer binders, shedding light on thermodynamic intricacies in geopolymerization reactions (Ren 2021; Ali Murtaza Zaidi, Sharma, and Student 2021). This study uses FA with SiO_2 50.2% and Al_2O_3 30.1% and GGBS with SiO_2 34.8% and Al_2O_3 16.3% oxide compositions and loss of ignition (LOI) 1.22 and 0.55, respectively. These precursors are enhanced with fast-dissolving sodium silicate and sodium metasilicate pentahydrate, as well as high-purity sodium carbonate (99%) and calcium hydroxide (95%) (Ouyang et al. 2020). Activator selection plays a pivotal role in modulating geopolymer properties, with variations in sodium silicate and sodium carbonate explored for their effects on geopolymerization kinetics (Mobili et al. 2020). Mortar synthesis involves systematic exploration of activator compositions and ratios, with meticulous control over binder content and density (Guo et al. 2022). Fly ash, and GGBFS account for 70% and 30% of the binder content, respectively (Ouyang et al. 2020).

2.3 Relation Between Water/Binder, Sand/Binder, Alkaline/Binder Ratio

One-part geopolymer mortars' rheological, mechanical, and microstructural performance is greatly influenced by the complex interactions between the water/binder (W/B), sand/binder (S/B), and alkaline/binder (A/B) ratios as shown in Fig. 3. One of the main factors affecting workability and compressive strength is the W/B ratio. Increased water content makes the mixture more workable by making it more fluid, which is necessary for handling and putting the mortar with ease. On the other hand, too much water might increase porosity, which will reduce compressive strength. In a study (Zhao et al. 2023), a range of S/B and A/B ratios, a fixed W/B ratio of 0.45 is employed to preserve a balance between mechanical performance and

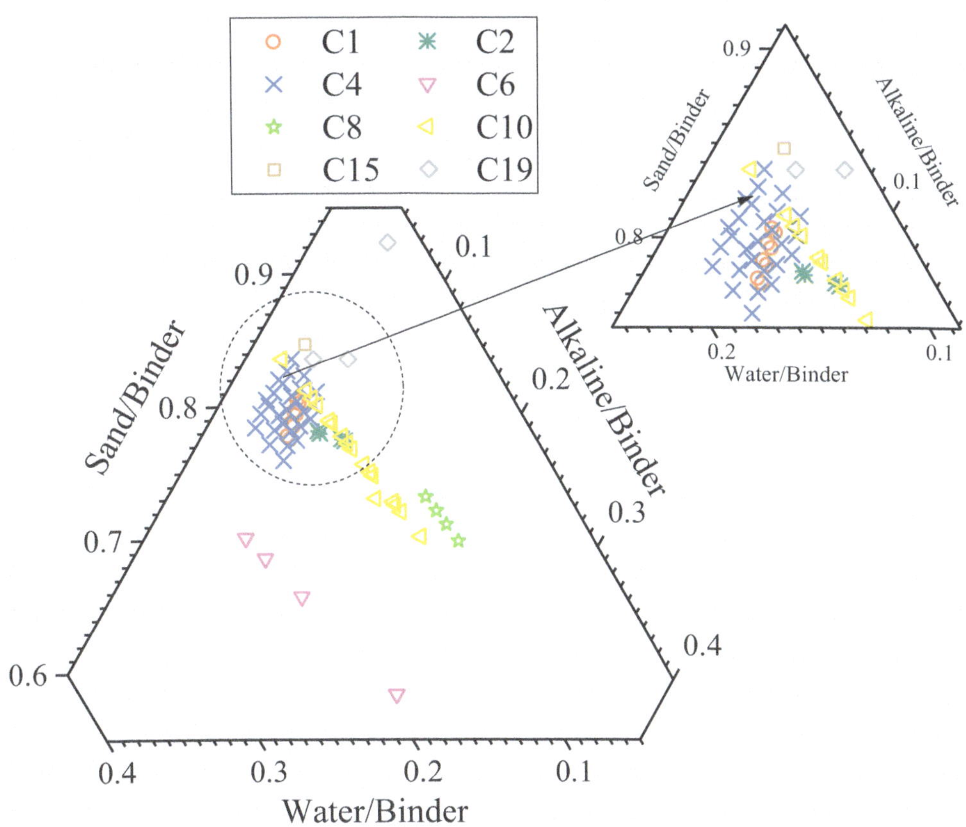

Fig. 3 Relation between water/binder, sand/binder, and alkaline/binder ratio. (C1 (Zhao et al. 2023), C2 (Zhang et al. 2021), C4 (Srinivasa et al. 2023), C6 (Adesanya et al. 2018), C8 (Shen 2023), C10 (Mahfoud et al. 2023), C15 (Shoaei et al. 2024), C19 (Ouyang et al. 2020))

workability. This study suggests a more consistent way of determining other factors without adding more water content variation to the mix. However, a different researcher (Zhang et al. 2021) conducted with W/B ratios varying from 0.53 to 0.57 in an attempt to examine the ideal W/B ratio that would maximize compressive strength without disturbing the workability of the geopolymer matrix. Comparably, a range of W/B ratios from 0.38 to 0.54 shows an in-depth analysis of how varying the water content affects the geopolymer matrix's fresh and hardened properties (Srinivasa et al. 2023).

The density, workability, and mechanical properties of the geopolymer matrix are also greatly impacted by the S/B ratio. However, a higher sand quantity in a geopolymer matrix usually results in a denser and stronger mortar, but on the other hand, it can also make the mixture less workable. In a study, the use of sand percentage in a geopolymer matrix is adjusted using the S/B ratios, varying from 2.12 to 2.50, to balance workability and density (Zhao et al. 2023). However, greater S/B ratios, varying from 2.99 to 3.21, indicate that compressive strength and density are prioritized (Zhang et al. 2021). Utilizing a low S/B ratio of 0.94, likely due to the high-density materials used, such as slag and pre-treated paper sludge (Adesanya et al. 2018). Without requiring a high sand percentage, these recovered waste materials significantly increase the mortar's overall density and strength. In a different investigation, the researcher standardized the mix using a S/B ratio of 3.00 in order to assess the impact of varying A/B ratios methodically (Mahfoud et al. 2023).

The consistent S/B ratio ensures that variations in mechanical properties can be attributed to changes in the A/B ratio rather than fluctuations in sand content. An essential factor in the geopolymerization process, the A/B ratio greatly influences the mortar's mechanical properties as well as the activation of aluminosilicate binders. In general, the chemical processes involved in the creation of geopolymers are enhanced by higher alkali concentrations, resulting in a more strong binder matrix. However, this can also impact setting times, compressive strength and flexural strength. Zhao et al. (Zhao et al. 2023) used the A/B ratio ranges from 0.15 to 0.16, with slight variations, to optimize binder activation and mechanical strength. On the other hand, Zhang et al. (Zhang et al. 2021) investigated greater A/B ratios of 0.29 to 0.40, suggesting attempts to improve reaction kinetics and strength by adding more alkali content. In another study, a thorough analysis of fine-tuning activation energy for various mix proportions to produce ideal mechanical properties, with A/B ratios ranging from 0.09 to 0.19 (Srinivasa et al. 2023). Shen et al. (Shen 2023) employed A/B ratios varying from 0.46 to 0.59, showing the substantial influence on the durability and mechanical properties of increased alkali content on the geopolymer matrix.

From a research perspective, the multi-objective optimization seen in Srinivasa et al. (Srinivasa et al. 2023) involves adjusting all three ratios (W/B, S/B, A/B) to achieve desired properties. To deeply understand the interrelationship between these ratios and how they affect the final characteristics of the geopolymer matrix, rigorous testing and analysis are required for such optimization. Higher W/B ratios often lead to increased workability, whereas higher S/B and A/B ratios show negative effects, decreasing workability and, at the same time, increasing mechanical properties. Lower W/B ratios lessen the porosity of geopolymer mass, and increase the A/B ratio helps in the activation of binders, which as a result, improves the compressive and flexural strength. The compressive strength of geopolymer mortar is also greatly affected by the S/B ratio; up to an optimal point, more solid content improves the density and mechanical properties of geopolymer mortar. The microstructure of the geopolymer matrix is also influenced by the A/B ratio, as it determines the degree of polymerization. The matrix with a higher alkali concentration is usually denser and has a more interconnected bond network, which, in fact, improves the mechanical properties of the matrix. By understanding these intricate relationships, researchers can tailor mix proportions to achieve the desired performance of one-part geopolymer mortars, considering factors such as workability, strength, and durability depending on the application of mortar.

2.4 Methodology

The fluidity of fresh RGM mixture (Zhao et al. 2023), one-part and two-part geopolymer mixture (Zhang et al. 2021) is measured using a jolting table according to Chinese standard GB/T 2419–2005 (GBT2419–2005. n.d.), similar to ASTM C1437-15 (ASTM C1437–15. n.d.). The procedure involved placing a test cone on the table, filling it with fresh mortar, lifting the cone to allow even flow, and operating the table for 25 jolts (Zhao et al. 2023). The average diameter of the spread mixture recorded. The diameter of the OPGM (Srinivasa et al. 2023) flow in its fresh state is measured following ASTM C1437 (ASTM C1437–15. n.d.) guidelines. Samples of OENG (Shen 2023) prepared according to GB/T 17,671 (GBT17671–1999. n.d.) had their fluidity tested under lab conditions (20 $\pm$ 2) °C and $\geq$ 50% relative humidity, per GB/T 2419–2005 (GBT2419–2005, n.d.).

The compressive and flexural strengths of various mortar mixes are assessed using standardized methods and testing machines. According to Chinese standard JGJ/T70-2009 (JGJ, T70–2009. n.d.), cube axial compression tests are conducted (Zhao et al. 2023). Twelve specimen groups underwent flexural and compression tests with a UTM5205 machine (Zhang et al. 2021), following GB/T 17,671–1999 (GBT17671–1999. n.d.) and ASTM C348 (ASTM C0348–21. n.d.) standards, with loading rates of 50 N/s for bending and 2400 N/s for compression (Adesanya et al. 2018). OPGM compressive strength (Srinivasa et al. 2023) is evaluated using 70.6 mm cubes per IS 4031-part 6 (IS4031. n.d.), and flexural strength using 40 × 40 × 160 mm moulds per ASTM C348-97 (ASTM C0348–21. n.d.).

Unconfined compressive strength (UCS) and flexural strength tests conducted on four samples at 7, 28, and 50 days with a Zwick machine, applying 2.4 kN/s for UCS and 0.05 kN/s for flexural strength (Adesanya et al. 2018). GB/T 17,671–1999 (GBT17671–1999. n.d.) is used to determine compressive and flexural strengths of OENG mortars with 40 × 40 × 160 mm specimens (Shen 2023), using a DYE-300 machine at 2400 ± 200 N/s. UCS and flexural strength tests on 4 × 4 × 16 cm specimens at 7 and 28 days are performed according to NF EN 196–1 (NF En196–1. n.d.) with a mechanical press/5500R—150 kN (Instron) (Mahfoud et al. 2023). Compressive strength of 50 mm cubes (Shoaei et al. 2024) is measured at 3, 7, 14, and 28 days per ASTM C109 (ASTM C0109. n.d.) with an ELE International machine at 0.9 kN/s (Ouyang et al. 2020). The 28-day flexural strength of 160 × 40 × 40 mm beams is evaluated by ASTM C348 (ASTM C0348–21. n.d.) three-point bending tests using an ELE International machine at 0.04 kN/s. Compressive strength is also determined using a digital compression machine at 2.4 kN/s (Ouyang et al. 2020).

3 Results and Discussion

3.1 Fluidity

The fluidity or workability property of one-part geopolymer mortar with rubber waste is greatly affected by the alkaline activator type and percentage of recycled rubber waste used in the mix. It is observed that solid activators reduce the fluidity of rubberized geopolymer mortar (RGM) by 10% to 30% as compared to alkali solutions due to greater friction from rapid polymerization. Adding solid NaOH pallets as compared to NaOH solution displays lower fluidity (Zhao et al. 2023). Further, adding higher rubber content also decreases RGM fluidity, which is attributed to rougher surfaces and entrapped air. However, the impact of rubber is less pronounced, hinting at improved rubber interface performance with NaOH incorporation.

With a water/binder ratio of 0.45, sand/binder ratio of 2.12 to 2.5, and fluidity between 110 and 190 (Zhao et al. 2023). Flow tests were conducted on various mortar groups to assess the impact of activator concentration, modulus, and type on the workability of one-part geopolymer mortar. For instance, with a water/binder ratio of 0.55, a sand/binder ratio of 2.99, and an alkaline/binder ratio of 0.29, the fluidity decreased by 27% (Zhang et al. 2021). Comparing one-part to conventional two-part geopolymer mortars, the former exhibits slightly better fluidity under similar formulations. This superiority is attributed to the direct addition of water into the precursor and solid activator mixture, unlike the two-part process, where a higher viscosity activator solution is poured into the precursor (Zhang et al. 2021).

The incomplete dissolution of solid activators in one-part mortars leads to a lower accurate concentration of Na_2SiO_3, resulting in a less viscous matrix. Furthermore, one-part mortars activated by anhydrous Na_2SiO_3 having less fluidity than hydrous Na_2SiO_3, mostly because granular anhydrous Na_2SiO_3 dissolves more slowly (Zhang et al. 2021). The water-to-geopolymer solid (W/GS) ratio influences the fluidity. Increasing the W/GS ratio enhances flowability, with a maximum value of 240 mm observed at a ratio of 0.49 (Srinivasa et al. 2023). Activator dosage variation from 8% to 16% leads to approximately 14% to 25% flow improvements (Srinivasa et al. 2023). Higher activator content can hinder flow due to excess anhydrous Na_2SiO_3, reducing the dissolution of raw materials. Increasing GGBS content reduces flowability by about 8% to 23% due to the larger specific surface area and irregular shape of the slag particles, which hinder both flow and reaction kinetics (Srinivasa et al. 2023).

The fluidity of one-part geopolymer mortar is minimally affected by alkali dosage and silicate modulus. Optimal fluidity was achieved in electric furnace nickel slag-based geopolymer mortar with a silicate modulus of 1.2 and an alkali dosage of 6.5% (Shen 2023). Both very low and high silicate moduli raise viscosity, thereby decreasing fluidity. A low alkali dosage leads to poor fluidity due to inadequate particle dispersion, whereas increasing the Na_2O concentration enhances fluidity by improving particle dispersion (Shen 2023). At higher alkali dosages, increased polymerization marginally decreases fluidity. These findings underscore the delicate balance between alkali dosage, silicate modulus, and fluidity in one-part geopolymer mortar formulations. Fluidity values range from 181 to 202 mm, reflecting variations in mix proportions of water/binder 0.27, sand/binder 2, and alkaline/binder ratios 0.5 to 0.59 (Shen 2023) as shown in Fig. 4.

3.2 Compressive Strength

The activator type and the addition of rubber significantly influence the compressive strength of one-part geopolymer mortar (RGM). In RGM, the compressive strength of

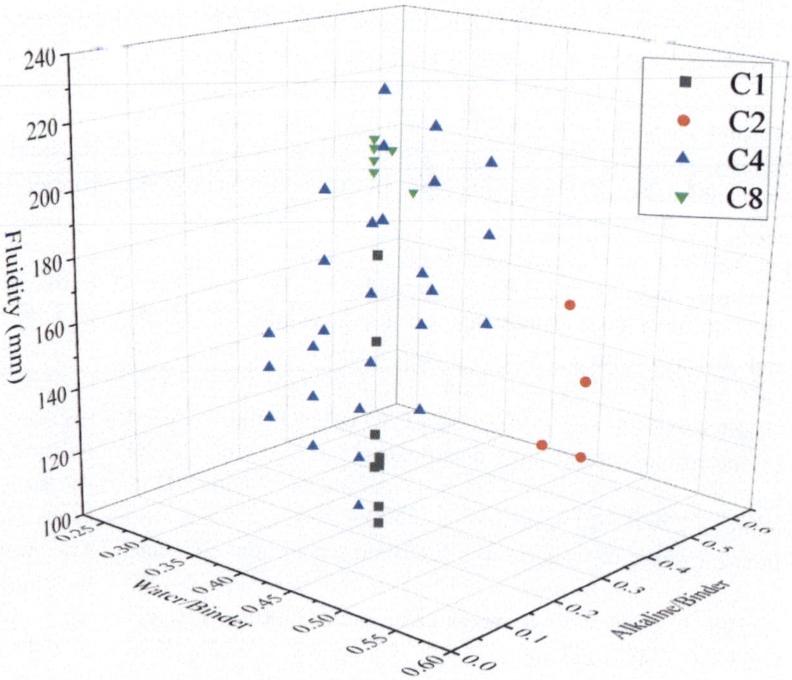

Fig. 4 Fluidity of one-part geopolymer mortar (C1 (Zhao et al. 2023), C2 (Zhang et al. 2021), C4 (Srinivasa et al. 2023), C8 (Shen 2023))

one-part rubberized geopolymer mortar (ORGM) tends to be lower than that of two-part rubberized geopolymer mortar (TRGM), primarily attributed to differences in activator dissolution. The compressive strength of ORGM-anhydrous Na_2SiO_3 (NS) powder decreases by less than 15%. In comparison, ORGM-instant Na_2SiO_3 (NH) powder experiences a reduction exceeding 65%, mainly due to incomplete dissolution of solid activators and premature reaction with precursors (Zhao et al. 2023). Additionally, adding rubber leads to decreased compressive strength, with ORGM-NS experiencing declines of 21%, 35%, and 46% at 5%, 10%, and 15% rubber content, respectively (Zhao et al. 2023). The influence of rubber content on ORGM-NH compressive strength is weaker, possibly due to NaOH pre-treatment of rubber improving interface performance.

The curing age also affects compressive strength, with TRGM exhibiting typical early-strength behaviour, reaching 85% of 28-day strength within the first 7 days, while ORGM shows slower strength development, resembling ordinary cement mortar, attributed to differences in activator dissolution and polymerization (Zhao et al. 2023). Higher activator concentration generally increases strength, similar to trends observed in two-part geopolymer specimens. Lower activator modulus enhances alkalinity, intensifying the polymerization reaction between activator and precursor, resulting in higher strengths (Zhang et al. 2021). One-part geopolymer mortar (OPGM) specimens using anhydrous activators exhibit higher strengths than those using hydrous activators due to geopolymer gel bonding and microstructure density differences. However, increased activator modulus diminishes the strength disparity between one and two-part OPGM (Zhang et al. 2021). The highest compressive strength of OPGM reaches 49.2 MPa, only 10% lower than the corresponding two-part geopolymer. Strength development from the 7-day to 28-day curing age is more significant in OPGM due to gradual activator dissolution and prolonged geopolymerization reaction, especially when using hydrous activators (Zhang et al. 2021).

Various factors influence compressive strength, such as water-to-geopolymer solid (W/GS) ratio, activator dosage, and slag content. Higher W/GS ratios decrease strength due to hindered geopolymerization, with significant drops observed (Srinivasa et al. 2023). Conversely, increasing slag content enhances compressive strength regardless of activator dosage and W/GS ratio (Srinivasa et al. 2023). The impact of activator dosage on strength development is evident, with approximately doubling activator dosage raised from 8% to 12%. However, excessive activator content beyond 12% leads to marginal decreases in strength due to matrix degradation caused by activator leaching and efflorescence (Srinivasa et al. 2023). Increasing the quantity of dry activator in blended geopolymer cement affects mechanical strength at different curing ages. With a water/binder ratio of 0.31, sand/binder ratio of 0.94, and alkaline/binder ratio of 0.09 to 0.36, compressive strength ranges between 18 and 26 MPa after 7 days of curing and ranging between 34 and 43 MPa after 28 days of curing (Adesanya et al. 2018).

The compressive mortars increased with curing age, exhibiting rapid development in early stages followed by

Fig. 5 Compressive strength **a** at 7 days **b** at 28 days. Compressive strength at 7 days. (C1 (Zhao et al. 2023), C2 (Zhang et al. 2021), C4 (Srinivasa et al. 2023), C6 (Adesanya et al. 2018), C8 (Shen 2023), C10 (Mahfoud et al. 2023), C19 (Ouyang et al. 2020)). Compressive strength at 28 days. (C1 [1], C2 (Zhang et al. 2021), C4 (Srinivasa et al. 2023), C6 (Adesanya et al. 2018), C8 (Shen 2023), C10 (Mahfoud et al. 2023), C15 (Shoaei et al. 2024), c19 (Ouyang et al. 2020))

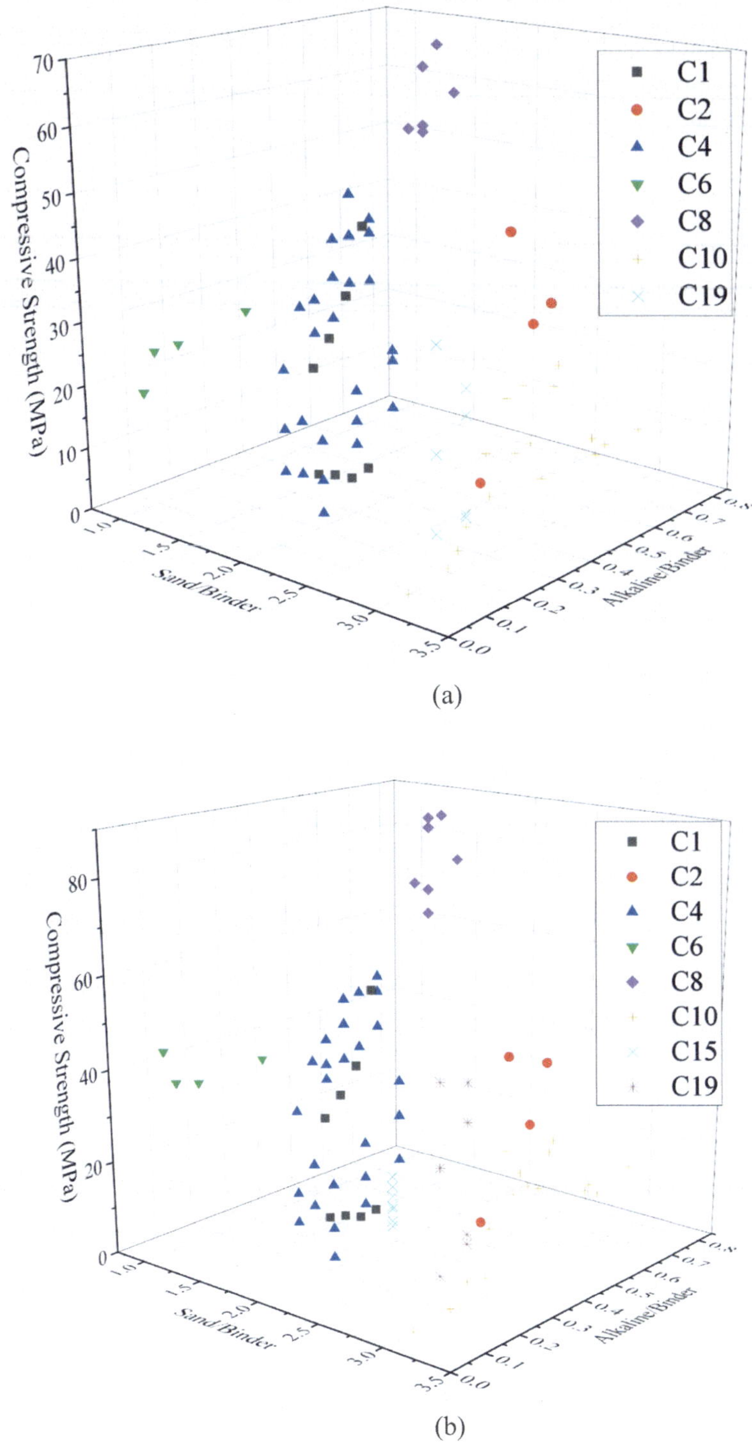

slower progress, consistent with previous research. With rising silicate modulus, strength initially increased. The rise in strength is correlated with increased polymerization due to a decreased pH with a higher silicate modulus, while excessive SiO_2 inhibits further dissolution, reducing strength (Shen 2023). Similarly, compressive strengths increased initially with alkali dosage up to 6.0%, then decreased at higher dosages. Excessive alkali dosage led to rapid pre-reaction, coating unreacted particles, hindering further reactions, and impeding strength enhancement ranging between 58 and 70 MPa at 7 days, as shown in Fig. 5a and 70–90 MPa at 28 days, as shown in Fig. 5b (Shen 2023).

The compressive strength of one-part geopolymer mortar is influenced by the SiO_2/Al_2O_3 molar ratio, with optimal ratios of 3.5 (Mahfoud et al. 2023). The addition of sediment enhances strength by improving source reactivity and promoting the formation of gels like C-A-S-H alongside N-A-S-H, filling voids and enhancing mechanical properties. Heating the fly ash and micronized dredged sediments further boosts reactivity, facilitating polymerization with a maximum compressive strength of 25 MPa (Mahfoud et al. 2023). Mortars incorporating quick-dissolving sodium silicate experienced slower strength growth than those using sodium metasilicate pentahydrate. The modulus ratio of the solid activator, influenced by the high SiO_2 content of quick-dissolving sodium silicate, affects strength development (Ouyang et al. 2020). Mortars with sodium metasilicate pentahydrate alone achieved acceptable strength, with notable performance observed when combined with sodium carbonate, nearly matching the strength of alkali solution-based mixes. With a water/binder ratio of 0.09 to 0.43, sand/binder ratio of 3, alkaline/binder ratio of 0.16/0.24 compressive strength ranging between 8 and 35 MPa after 7 days and 9–48 MPa after 28 days (Ouyang et al. 2020) as shown in Fig. 5.

3.2.1 Flexural Strength

The flexural strength of one-part geopolymer mortar is influenced by various factors, mainly depending on the concentration of the activator. However, the modulus of activators plays a crucial role; lower moduli result in more substantial alkalinity, enhancing the polymerization reaction and thus increasing strength. Comparisons between different types of sodium silicate reveal that anhydrous activators yield higher strengths due to denser microstructure and more robust bonding performance (Zhang et al. 2021). Despite lower flexural strengths compared to two-part geopolymers, one-part specimens exhibit notable performance, especially when optimizing activator combinations (Zhang et al. 2021). Additionally, the gradual dissolution of solid activators leads to prolonged geopolymerization, resulting in increased strength over time. Anhydrous sodium silicate may pose challenges in further dissolution, affecting long-term strength development (Zhang et al. 2021).

With a water/binder ratio of 0.53 to 0.57 and an alkaline/binder ratio of 0.29 to 0.47, flexural strength ranges from 2.7 to 9 at 7 days and 3 to 9.7 at 28 days (Zhang et al. 2021). A discernible decrease in flexural strength corresponds to an escalation in the water-to-geopolymer solids (W/GS) ratio, which impedes the geopolymerization kinetics, thereby diminishing strength characteristics (Srinivasa et al. 2023). Excessive activator dosage exceeding 12% may yield a marginal decrement strength. The sand/binder ratio is 2.10 to 2.61, the alkaline/binder ratio is 0.09 to 0.19, and flexural strength is 1.8 to 4.8 at 7 days and 2.1 to 9.2 at 28 days (Srinivasa et al. 2023). Flexural strength development follows a similar trend across samples, with minor deviations. All samples demonstrate comparable flexural strength with a sand/binder ratio of 0.94, alkaline/binder ratio of 0.12 to 0.36, and flexural strength of 6.7 to 8 at 7 days, as shown in Fig. 6a and 8.6 to 11 as shown in Fig. 6b (Komnitsas and Zaharaki 2007).

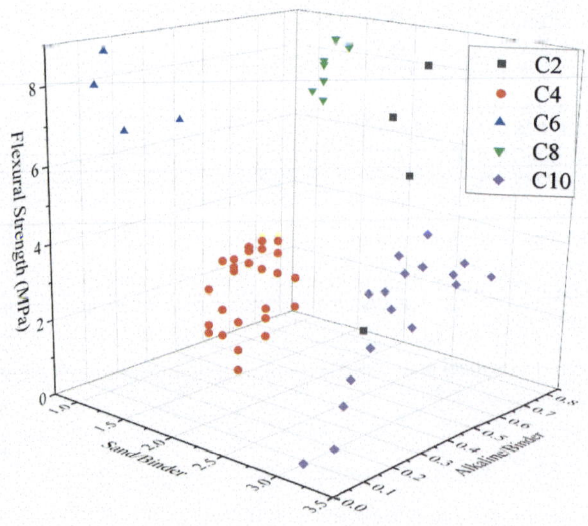

(a)

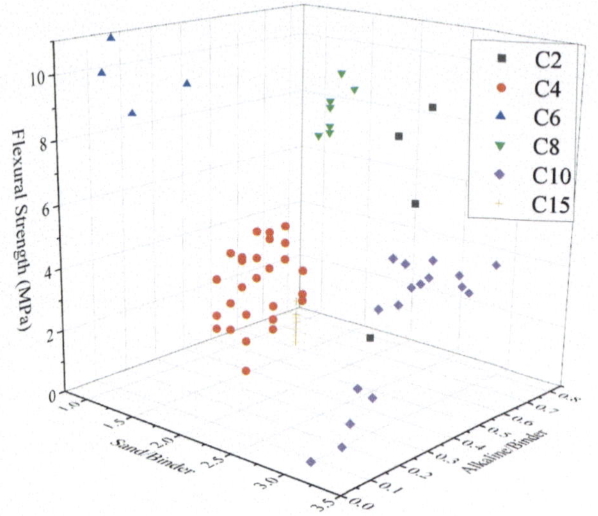

(b)

Fig. 6 Flexural strength **a** at 7 days **b** at 28 days. Flexural strength at 7 days (C2 (Zhang et al. 2021), C4 (Srinivasa et al. 2023), C6 (Adesanya et al. 2018), C8 (Shen 2023), C10). Flexural strength at 28 days (C2 (Zhang et al. 2021), C4 (Srinivasa et al. 2023), C6 (Adesanya et al. 2018), C8 (Shen 2023), C10 (Mahfoud et al. 2023), C15 (Shoaei et al. 2024))

Silicate modulus and alkali dosage influence the flexural strength of one-part electric furnace nickel slag geopolymer (OENG) mortars. This phenomenon is attributed to the interplay between silicate modulus, pH levels, and polymerization reactions, wherein an optimal balance promotes strength enhancement while deviations lead to diminished mechanical properties (Shen 2023). Deviations from this

curing exhibited a porous structure with unreacted particles, as shown in Fig. 7 (C15), indicating incomplete geopolymerization and lower mechanical strength (Shoaei et al. 2024). Heat curing resulted in a more compact microstructure and reduced unreacted particles, enhancing ITZ and mechanical strength. Increased zeolite content led to a more porous structure with unreacted particles, reducing strength. Heat curing slightly improved material reactivity, but high zeolite content still resulted in a loosely packed microstructure and lower strength (Shoaei et al. 2024).

Morphological observations of alkali-activated slag (AAS) pastes and mortars showed crystals, as shown in Fig. 7 (C18), on the surface of one-part AAS paste specimens attributed to hydrated sodium carbonate (Ren 2021). Efflorescence, dependent on pore connectivity, was absent in normal AAS pastes. One-part specimens had more interconnected pores and fewer C-A-S-H gels, allowing more free alkalis to migrate to the surface (Ren 2021). Normal AAS binders were more densified due to various hydration products, resulting in a more compact matrix (Ren 2021).

SEM analysis of alkali-activated metakaolin and GGBFS revealed amorphous particles with rough edges and some needle-like particulates, indicating unreacted metakaolin (Segura et al. 2022). The matrix in both one-part and two-part GGBFS mortars showed micro-cracks, pores, and particulates linked to unreacted slag grains, as shown in Fig. 7 (C21). The formation of C-A-S-H and N-A-S-H gels was also observed (Segura et al. 2022). SEM images of mixes cured for 90 days showed a dense microstructure with minimal micro-cracks (Dong et al. 2020). High Na_2O content led to efflorescence, resulting in a rough microstructure and shrinkage cracking. Mixes with microsilica had dense and smooth matrices, as shown in Fig. 7 (C23), though gel development was incomplete in submerged specimens (Dong et al. 2020). Coarse metasilicate mixes had less smooth matrix regions and lower compressive strength. Mixes with 100% GGBS and 3% Na_2O showed significant ettringite formation, reducing matrix development and compressive strength (Dong et al. 2020).

3.4 Cost Analysis

In this study, the cost analysis of one-part geopolymer mortar, based on the local price of the material near Punjab, India, reveals significant variations in cost depending on the quality and variety of the incorporated materials. The primary binders considered include FA, GGBS, MK, and Z. Each binder is used in varying amounts across different mixtures, affecting the total cost of one-part geopolymer mortar.

For instance, a study (Zhao et al. 2023) combines FA and GGBS with an additional activator, sodium silicate powder. The cost of Na_2SiO_3 varies based on its quantity, with additional chemicals like barium chloride ($BaCl_2$) or sodium hydroxide (NaOH) influencing the overall price, as shown in Fig. 8. The cost ranges from approximately ₹4220–4717. This variation is mainly due to the differing amounts of fine

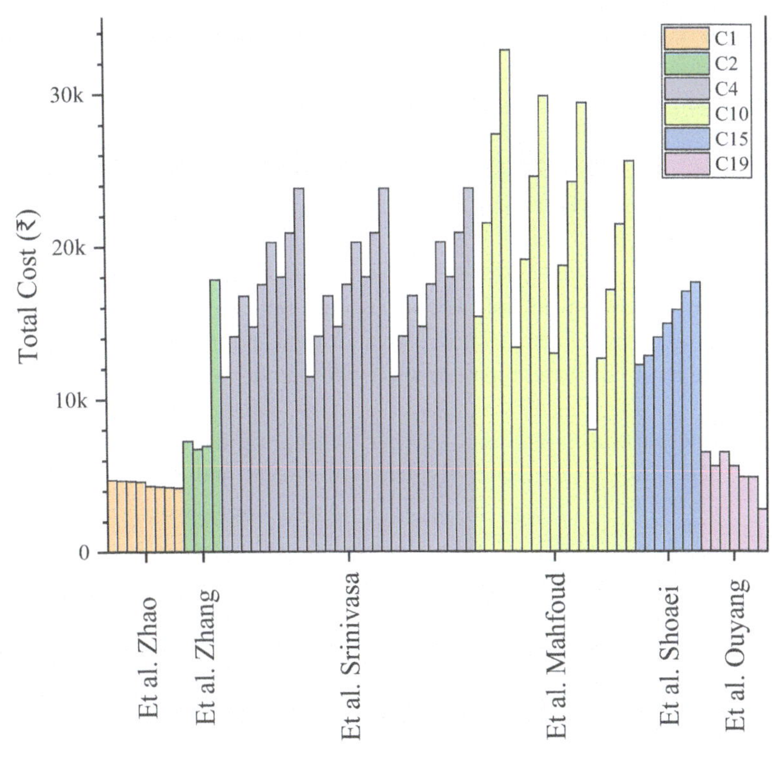

Fig. 8 Cost analysis of geopolymer mortar. (C1 (Zhao et al. 2023), C2 (Zhang et al. 2021), C4 (Srinivasa et al. 2023), C10 (Mahfoud et al. 2023), C15 (Shoaei et al. 2024), C19 (Ouyang et al. 2020))

aggregates and activators. In another study (Zhang et al. 2021), a geopolymer mix uses FA and MK, with significant quantities of anhydrous/hydrous Na_2SiO_3, resulting in costs ranging from ₹6720–17,804. The considerable variation is attributed to the different states (anhydrous or hydrous) of sodium silicate and its proportion in the mix.

On the other hand, in another study, researcher (Srinivasa et al. 2023) predominantly used FA and GGBS, with large quantities of anhydrous sodium metasilicate powder as an activator ranging from ₹11,464 to ₹23,798. The substantial increase in cost is because of the higher amounts of anhydrous sodium metasilicate powder used, significantly raising the price. In this study (Mahfoud et al. 2023), the mix relies mainly on FA, with some variations, including GGBS. The primary cost driver is again the large quantity of Na_2SiO_3, especially in anhydrous form, pushing the total cost from ₹7952–29,402.

In another study (Shoaei et al. 2024), the researcher uses GGBS and zeolite with a constant amount of anhydrous Na_2SiO_3. This mix shows a more consistent cost range, from ₹12,210–17,610, reflecting the steady use of materials and activators. Enhancing mechanical properties and cost efficiency can be achieved through a comprehensive analysis of the mechanical, durability, microstructure, and replacement percentages of various waste materials as precursors. However, the results are highly contingent on the specific mix design, emphasizing the importance of factors such as the Si/Al ratio, water/binder, sand/binder, and alkaline/binder ratio. The absence of established standards or guidelines for geopolymer development exacerbates existing uncertainties. Therefore, further research is imperative to determine the optimal combination of design parameters, facilitating a systematic approach to consistently achieve the desired strength using different precursors.

4 Conclusion

Based on the extensive research done by researchers, these are the following technical conclusions can be drawn regarding the alkaline activator, precursor type and mechanical properties of the one-part geopolymer matrix:

(1) One-part mix exhibited longer setting times compared to two-part geopolymer, with initial and final setting times being 30% to 50% longer, respectively. Solid anhydrous sodium silicate (Na_2SiO_3) proves to be a superior activator compared to the blend of hydrous sodium silicate and sodium hydrate for MK and FA mixtures. This superiority arises from its ability to promote a more extensive geopolymerization reaction at equivalent concentrations and activator moduli.

(2) The meticulous characterization of activators and precursors in one-part geopolymer matrices is crucial for optimizing the performance of these eco-friendly materials. These values include the SiO_2/Na_2O ratio of 1 in the activator and the SiO_2/Al_2O_3 ratio ranging between 2.4 and 4.5, indicating a precursor's efficacy.

(3) A one-part geopolymer's compressive strength was generally lower than its two-part counterpart. However, one-part geopolymer activated by anhydrous Na_2SiO_3 powder showed improved compressive strength compared to other activators.

(4) Similar to compressive strength, the flexural strength of the one-part mix showed a decline as compared to the two-part geopolymer. The addition of rubber also adversely affected flexural strength, creating weak zones where cracks extended along Interfacial Transition Zones (ITZs).

(5) Multi-objective optimization revealed an optimum mix for one-part geopolymer mortar with specific parameters such as S/B, water-to-geopolymer solid ratio, and activator content.

(6) SEM analysis reveals that the microstructural integrity and compressive strength of alkaline activated matrix are greatly affected by the degree of geopolymerization, activator composition, and precursors, with optimized formulations yielding denser matrices and enhanced mechanical properties.

(7) The primary factors influencing the cost of geopolymer mortar are the types and quantities of binders and activators used. The extensive use of anhydrous sodium metasilicate significantly elevates the price, while combinations of FA and GGBS offer a more economical solution. Careful selection and optimization of these components are essential for cost-effective one-part geopolymer mortar production.

The research provides actionable insights for advancing the development of one-part geopolymer mortar systems, highlighting critical factors such as activator selection and optimization of recycled waste content. However, there remains a need for deeper exploration into the structural performance and long-term durability of these formulations. Moreover, while the utilization of waste materials in geopolymers holds significant promise for environmental sustainability and engineering applications, comprehensive research is required to fully elucidate their performance characteristics.

References

Abdel-Gawwad HA, Rashad AM, Heikal M (2019) Sustainable utilization of pretreated concrete waste in the production of one-part alkali-activated cement. J Clean Prod 232:318–328. https://doi.org/10.1016/j.jclepro.2019.05.356

Abdollahnejad Z, Pacheco-Torgal F, Aguiar JB, Jesus C (2015) Durability performance of fly ash based one-part geopolymer mortars. In Key Engineering Materials, Trans Tech Publications Ltd, 2015, pp 113–120. https://doi.org/10.4028/www.scientific.net/KEM.634.113
Adesanya E, Ohenoja K, Luukkonen T, Kinnunen P, Illikainen M (2018) One-part geopolymer cement from slag and pretreated paper sludge. J Clean Prod 185:168–175. https://doi.org/10.1016/j.jclepro.2018.03.007
Ali Murtaza Zaidi S, Sharma P, Student M (2021) Preparation of geopolymer cement with dry activators by one part system. JETIR. [Online]. Available: www.jetir.org
Ameri F, Shoaei P, Zareei SA, Behforouz B (2019) Geopolymers vs. alkali-activated materials (AAMs): a comparative study on durability, microstructure, and resistance to elevated temperatures of lightweight mortars. Constr Build Mater 222:49–63. https://doi.org/10.1016/j.conbuildmat.2019.06.079
ASTM C0109. [Online]. Available: https://www.astm.org/c0109_c0109m-20.html
ASTM C0348-21. [Online]. Available: https://www.astm.org/c0348-21.html
ASTM C1437-15. https://www.astm.org/c1437-15.html
Chen K et al (2024) Performance characteristics of micro fiber-reinforced ambient cured one-part geopolymer mortar for repairing. Constr Build Mater 415:135086. https://doi.org/10.1016/j.conbuildmat.2024.135086
Davidovits J (1991) Geopolymers. J Therm Anal 37(8):1633–1656. https://doi.org/10.1007/BF01912193
Deng P, Zheng Z (2023) Mechanical properties of one-part geopolymer masonry mortar using alkali-fused lead–zinc tailings. Constr Build Mater 369. https://doi.org/10.1016/j.conbuildmat.2023.130522
Dong M, Elchalakani M, Karrech A (2020) Development of high strength one-part geopolymer mortar using sodium metasilicate. Constr Build Mater 236:117611. https://doi.org/10.1016/j.conbuildmat.2019.117611
Duxson P, Mallicoat SW, Lukey GC, Kriven WM, van Deventer JSJ (2007) The effect of alkali and Si/Al ratio on the development of mechanical properties of metakaolin-based geopolymers. Colloids Surf A 292(1):8–20. https://doi.org/10.1016/j.colsurfa.2006.05.044
GBT17671-1999. [Online]. Available: https://www.chinesestandard.net/PDF.aspx/GBT17671-1999
GBT2419-2005. [Online]. Available: https://www.chinesestandard.net/PDF/English.aspx/GBT2419-2005
Geraldo RH, Gonçalves JP, Camarini G (2022) Production process of an eco-friendly one-part alkali-activated binder. Mat Res 25. https://doi.org/10.1590/1980-5373-MR-2021-0433
Ghazy MF, Elaty MAA, Mostafa SM (n.d.) Properties of one-part versus two-part geopolymers composites-a review. Am J Eng Res [Online]. Available: www.ajer.org
Guo G, Lv C, Liu J, Wang L (2022) Properties of fiber-reinforced one-part geopolymers: a review. Polymers 14(16). https://doi.org/10.3390/polym14163333
IS 4031. [Online]. Available: https://ia600306.us.archive.org/35/items/gov.in.is.4031.6.1988/is.4031.6.1988.pdf
JGJ/T70-2009. https://www.chinesestandard.net/PDF/English.aspx/JGJT70-2009
Jia Z, Yang Y, Yang L, Zhang Y, Sun Z (2018) Hydration products, internal relative humidity and drying shrinkage of alkali activated slag mortar with expansion agents. Constr Build Mater 158:198–207. https://doi.org/10.1016/j.conbuildmat.2017.09.162
Kogbara RB, Al-Zubi A, Mortada Y, Hammoud A, Masad EA, Khraisheh MK (2024) Lime-activated one-part geopolymer mortars from construction, demolition and industrial wastes. Results Eng 21. https://doi.org/10.1016/j.rineng.2023.101739
Komnitsas K, Zaharaki D (2007) Geopolymerisation: a review and prospects for the minerals industry. Miner Eng 20(14):1261–1277. https://doi.org/10.1016/j.mineng.2007.07.011
Mahfoud E, Ndiaye K, Maherzi W, Aggoun S, Benzerzour M, Abriak N-E (2023) Mechanical properties and shrinkage performance of one-part-geopolymer based on fly ash and micronized dredged sediments. Dev Built Environ 16:100253. https://doi.org/10.1016/j.dibe.2023.100253
Marvila MT, Azevedo ARG, Delaqua GCG, Mendes BC, Pedroti LG, Vieira CMF (2021) Performance of geopolymer tiles in high temperature and saturation conditions. Constr Build Mater 286:122994. https://doi.org/10.1016/j.conbuildmat.2021.122994
Mobili A, Tittarelli F, Rahier H (2020) One-part alkali-activated pastes and mortars prepared with metakaolin and biomass ash. App Sci (Switzerland) 10(16). https://doi.org/10.3390/app10165610
NF EN 196-1. [Online]. Available: https://www.boutique.afnor.org/en-gb/standard/nf-en-1961/methods-of-testing-cement-part-1-determination-of-strength/fa184622/57803
Ouyang S, Chen W, Zhang Z, Li X, Zhu W (2020) Experimental study of one-part geopolymer using different alkali sources. In: Journal of physics: conference series, institute of physics publishing. https://doi.org/10.1088/1742-6596/1605/1/012155
Ren J et al (2021) Experimental comparisons between one-part and normal (two-part) alkali-activated slag binders. Constr Build Mater 309. https://doi.org/10.1016/j.conbuildmat.2021.125177
Segura IP et al (2022) Comparison of one-part and two-part alkali-activated metakaolin and blast furnace slag. J Sustain Metall 8(4):1816–1830. https://doi.org/10.1007/s40831-022-00606-9
Shen Y et al (2023) Effects of silicate modulus and alkali dosage on the performance of one-part electric furnace nickel slag-based geopolymer repair materials. Case Stud Constr Mater 19. https://doi.org/10.1016/j.cscm.2023.e02224
Shoaei P, Momenzadeh A, Hosseini H, Rajaei S, Ameri F, Pilehvar S (2024) One-part slag/zeolite geopolymer mortars under ambient and heat curing conditions. Case Stud Constr Mater 20:e02677. https://doi.org/10.1016/j.cscm.2023.e02677
Srinivasa AS, Yaragal SC, Swaminathan K, Rakesh Kumar Reddy R (2023) Multi-objective optimization of one-part geopolymer mortars adopting response surface method. Case Stud Constr Mater 409. https://doi.org/10.1016/j.conbuildmat.2023.133772
Vu TH et al (2020) Assessing carbonation in one-part fly ash/slag geopolymer mortar: change in pore characteristics using the state-of-the-art technique neutron tomography. Cem Concr Compos 114. https://doi.org/10.1016/j.cemconcomp.2020.103759
Xu H, Van Deventer JSJ (2000) The geopolymerisation of alumino-silicate minerals. [Online]. Available: www.elsevier.nlrlocaterijminpro
Xu H, Van Deventer JSJ (n.d.) Geopolymerisation of multiple minerals/ [Online]. Available: www.elsevier.com/locate/mineng
Zhang HY, Liu JC, Wu B (2021) Mechanical properties and reaction mechanism of one-part geopolymer mortars. Constr Build Mater 273. https://doi.org/10.1016/j.conbuildmat.2020.121973
Zhao J, Xie J, Wu J, Zhao C, Zhang B (2023) Workability, compressive strength, and microstructures of one-part rubberized geopolymer mortar. J Build Eng 68. https://doi.org/10.1016/j.jobe.2023.106088

Winning Sustainability Through Waste Management: Potential Waste-to-Energy (WtE) Techniques and Sustainable Strategies Analysis

Harsh Vishwakarma, Vipul Keerti Sharma, and Kiran Billore

Abstract

Amidst the vital pillars of human survival—food, shelter, and clothing—lies another cornerstone: Energy, essential for our very existence. With the global population steadily increasing and socioeconomic progress advancing rapidly, the energy demand is soaring. In today's world, the idea of living without energy is virtually inconceivable. This pressing need must be met through the utilization of safe renewable energy sources, as opposed to non-renewable alternatives that contribute to global warming and climate change, to ensure the well-being of our planet. To meet this energy demand sustainably, various types of waste, including municipal, industrial, and agricultural waste, can be tapped as sources of cleaner and safer energy. Effective management and treatment of waste through waste-to-energy (WtE) techniques have the potential to contribute significantly to achieving many sustainability goals outlined in initiatives like the United Nations Sustainable Development Goals and India's Sustainable Vision 2047, thereby supporting overall sustainability efforts. The escalating global population correlates with a swift increase in waste generation. Projections suggest that solid waste generation could reach 2.2 billion tons by 2025 and 9.5 billion tons by 2050 worldwide, according to the World Bank. To harness energy from waste, a multitude of techniques and possibilities for advancement are available. This chapter involves a detailed analysis of WtE techniques, encompassing thermal conversion methods like incineration, pyrolysis, gasification, and torrefaction; biological conversion techniques involving bioagents like bacteria and fungi in processes such as anaerobic digestion and ethanol fermentation; landfilling approaches; and mechanical biological treatment. These techniques yield various forms of energy, including heat, char, oil, syngas, biogas, and landfill gas. The chapter also addresses strategies for sustainable development and waste management, current challenges, principles of the circular economy, and sustainability economics.

Keywords

Bioeconomy · Bioenergy · Climate · Microbial · Renewable · Sustainability

1 Introduction

Garbage disposal thus becomes a growing global problem as the world's population germinates, and people develop infrastructure and economize. Waste generation is on the increase, and therefore it is necessary to adopt measures that reduce the effects of waste on the environment and, at the same time, ensure that energy is reclaimed as much as possible. The global forecasts report that in the year 2050, the total bulk of the solid waste will be at 3.4 billion tons; this is based on projections of 2016, where total solid waste production was at 2.01 billion tons, mainly as a result of the world's growing urbanization coupled with population aging. The general impacts of poor waste management on the environment, according to the United Nations Environment Program (UNEP 2024), include greenhouse gas emissions, soil and water pollution, and disease vector breeding (UNEP 2024). While the IEA also notes there is very large potential for WtE, particularly incineration with energy recovery and anaerobic digestion

H. Vishwakarma (✉) · K. Billore
Department of Biotechnology, Government Holkar Science College, Indore, India
e-mail: vishwakarmah68@gmail.com

V. K. Sharma
Department of Zoology, Government Holkar Science College, Indore, India

A. Mandpe et al. (eds.), *Sustainable Waste Management*, Advances in Science, Technology & Innovation,
https://doi.org/10.1007/978-3-031-85685-3_16

to accommodate a circular economy transition and mitigate greenhouse gas emissions arising from waste management, and demands that WtE be accompanied by proper source management to avoid incremental pollution. Further, it is identified that municipal solid waste landfills are among the third biggest sources of human-induced methane emissions in the United States, contributing to about 15.1% of MSW emissions in 2019 (US EPA 2024). According to the WHO of the world, if waste is not disposed of properly, it results in pollution of water, soil, and air, which is dangerous to the health of workers as well as members of the community (WHO 2022).

This chapter seeks to give an overview of different types of waste with reference to landfill waste, municipal waste, agricultural residues, power plant waste, and other wastes and also look at how WtE technologies and processes deal with these wastes. In this chapter, we attempt to improve the understanding of capacity building in waste management for a circular economy and sustainable development by comparing characteristics, challenges, and opportunities of various waste types with WtE practices and recent advancements. The mission uses complex interactions between environmental, social, and economic contexts to demonstrate how everyone's joint efforts can result in the efficient management of wastes and the utilization of these wastes to help in the protection of the environment.

In this respect, the contributions of this chapter are, therefore, manifold. It not only describes the relationship between waste management and sustainability but also gives an understanding of certain WtE technologies that can be implemented in the disposal of waste into energy sources. Thus, recognizing the environmental, social, and economic aspects of waste management discussed in this chapter reinforces the notion that strategies for efficient waste management should involve stakeholders in one coordinated approach. Additionally, it stresses the significance of such policy directives that will enable the shift toward more practice of the circular economy while at the same time enhancing sustainable practices.

The organization of this chapter has been arranged in such a way that the authors take the consumers through a smooth sequence of thoughts. This is followed by a detailed discussion on several WtE technologies comprising thermal processes like direct combustion and gasification together with the bio process of digestion. The strengths and weaknesses of each method will also be compared and contrasted concerning sustainable development. Further sections shall elaborate on India's efforts toward sustainability, encompassing two of their missions, which include "Carbon-Free India by 2070" and "India Vision 2047," both of which stand as examples of national efforts to attain sustainable development through efficient ways of waste management. In a broader and final perspective, this chapter is intended to present a comprehensive idea to reduce waste, recycle it, and in the process, make major strides toward environmental solutions and increased energy production as a way of addressing increasing acute scarcity and environmental pollution around the globe.

2 Methodology

This chapter was designed following a plan- and structure-driven approach to identify and evaluate the state of the art in waste-to-energy (WtE) technology and its applications for sustainable waste management. There were three stages in this methodological approach. First, a literature review was carried out involving both published articles from reputable journals, namely Springer Nature, Scopus, and Google Scholar, and technical papers and policy documents. Such terms and phrases as "waste-to-energy," "renewable energy," "sustainable waste management," and "circular economy" formed the basis for the search, restricting the material to only those published in the past decade. Second, the collected literature was sorted based on themes in order to explore WtE technologies, including incineration, pyrolysis, gasification, and anaerobic digestion. The four categories were further examined with regards to technological progress, sustainability effects on the three pillars, and strengths and weaknesses identified. Third, the chapter included cases and examples of the successful application of identified findings and general tendencies of WtE technologies development. As a result, the generative framework was built by combining environmental science, engineering, and economics disciplines. The use of figures and tables, for instance, was incorporated in the material since they make it easier to understand concepts that involve a lot of data and procedures. This approach gave a comprehensive and stringent analysis of WtE technologies and sustainable practices.

3 India and Sustainability: Carbon-Free India by 2070, and Vision 2047: Sustainable India

India is a nation that is constantly working toward the goal of becoming a developed country. With a vast and diverse population, the country is putting in a lot of effort to excel in all aspects of development. One of the key focus areas for India is sustainable development. The country has adopted various strategies and plans to achieve this, including "India Vision 2047: Sustainable India" and "Carbon-free India 2070." These initiatives are aimed at ensuring that India's growth is sustainable and that the country moves toward a greener and more environmentally friendly future.

The "India Vision 2047: Sustainable India" initiative is a long-term plan that aims to make India a sustainable and inclusive economy by the year 2047, which is the 100th year since India's independence. The plan focuses on several key areas, including clean energy, water conservation, waste management, and sustainable agriculture. For instance, it is apparent in the Swachh Bharat Mission (Urban) 2.0 is part of this effort, which includes the scientific treatment of waste, the minimization of landfill waste, and the support of a circular economy by 2047. The objective is to create a sustainable and resilient India that can meet the needs of its citizens while also preserving the environment for future generations.

Another ambitious initiative taken by India is "Carbon-free India 2070." This plan aims to make India a carbon–neutral country by the year 2070, which means that the country's net carbon emissions will be zero (Gupta and Gupta 2022). This is a significant step toward combating climate change and ensuring a sustainable future for the planet. The initiative has defined some detailed targets like reaching 50% of cumulative installed energy capacity from non-fossil fuel based resources by the year 2030 which was announced by the Indian government at COP 26. The plan focuses on reducing carbon emissions by renewable energy sources, promoting energy efficiency, and encouraging sustainable transportation.

The various initiatives undertaken by India in recent times are a testament to the country's unwavering dedication toward sustainable development. By adopting a holistic approach toward creating a greener and cleaner future, India is striving to emerge as a global leader in sustainable development. These well-thought-out plans and strategies not only demonstrate the country's commitment to sustainable development but also set a remarkable example for other nations to emulate.

4 Strategies in Sustainable Development and Waste Management

Waste management plays a crucial role in achieving sustainable development, a concept encompassing environmental protection and conservation, economic prosperity and financial well-being, and social well-being and community welfare (Roy et al. 2023; United Nations 2022). Sustainable waste management strategies directly contribute to several Sustainable Development Goals (SDGs) established by the United Nations in the year 2015. Among the 17 SDGs, sustainable waste management can help fulfill three major goals: SDG 6 (clean water and sanitation) by minimizing pollution from landfills, SDG 12 (responsible consumption and production) by promoting resource recovery and reducing waste generation, and SDG 13 (climate action) by changing the path of waste from landfills, which produce a potent greenhouse gas, methane, high levels of it is responsible for causing various adverse health effects along with warming the globe.

"Strategy" in the field of waste management implies a range of measures aimed at mitigating the environmental impacts caused by these processes. These measures include reducing energy, managing and treatment of waste, reducing water consumption (including water footprint), diminishing the overall environmental footprint, and promoting to influence social perceptions and behaviors. The use of sustainable resources and the adoption of environmental management systems like ISO 14001, EMAS, or Ecolabel are some strategies that can be implemented. At the same time, they relate to a variety of activities that deal with prevention, minimization, reuse, restoration, as well as remanufacturing, taking into consideration their impact on society (Zorpas 2022).

It is often known that inadequate waste management leads to economic losses as well as threats to human health and the well-being of the planet (Kinnaman 2006; Van den Bergh 2008; Somplak et al. 2019; Zorpas 2020). Therefore, effective waste management strategies take into account the three pillars of sustainability: environment, economy, and society.

Economically: Efficiently planned waste management systems can be financially beneficial. Efforts in reducing waste can enhance businesses to reduce costs on services and products and, save money by using essential materials. Here, recycling and waste-to-energy (WtE) approaches and technologies would play an important role. They can also generate income by selling recyclable materials and producing various forms of energy that can be used internally or commercially. Furthermore, effective waste management can lead to the creation of environmentally friendly jobs in waste collection, processing, and waste-to-energy facilities (Tsui and Wong 2019; Elsheekh et al. 2021).

Socially: Sustainable waste management safeguards public health by minimizing pollution and preventing the spread of diseases linked to improper waste disposal. In developing countries, the medical waste disposal system frequently faces deficiencies and shortcomings, exacerbating the transmission of infections and pathogens. Garbage collectors are regularly exposed to health hazards due to improper sorting and recycling practices concerning hazardous waste (Elsheekh et al. 2021). Community participation in waste reduction endeavors nurtures environmental consciousness and societal accountability. Moreover, a clean environment enhances citizens' quality of life.

Environmentally: Recycling constitutes the heart of a "recyclable circular economy" and the pinnacle of its pyramid. Reducing waste at source is the highest priority, followed by reuse, then recycling, and finally waste-to-energy (WtE). By following this paradigm, the ecological impact of waste is minimized and multiplied natural resources are conserved, while the problem of landfill pollution is significantly attenuated. Sustainable waste management can potentially provide clean energy using WtE. It is widely recognized as a competitive and affordable solution to current environmental concerns and global challenges, including climate change and accelerated global warming. WtE is at the forefront of modern waste management and could reduce the need for fossil fuels. The energy extracted from landfill gas collected from geo-exchange-covered sites can be used to produce heat and power (Zhou et al. 2017).

The waste management hierarchy offers direction on what approaches should take priority. For instance, waste prevention is at the top of the hierarchy, followed by reuse, recycling, and WtE, among other approaches, which should be landfilling (Tsui and Wong 2019). Landfilling does not offer long-term sustainability and should only be implemented as a last resort. Although WtE is a critical waste management approach, the most cost-effective technique should be selected. Various techniques come under WtE, such as incineration, gasification, and pyrolysis, all with their pros and cons. Incineration, for example, has a high amount of energy recovery, but its exhaust stream calls for purification technologies. Pyrolysis involves heating solid waste materials in the absence of oxygen to break down large molecules into smaller ones. It is a sustainable alternative for disposing of polymer waste. Gasification accepts various waste streams but may require additional clean-up of the syngas (Beyene et al. 2018).

The main element is to bring it into a comprehensive waste management plan, and this is the secret! The first step in cultivating recycling and in creating waste reduction should prioritize WtE projects. Firstly, government policies and regulations offer significant opportunities for stimulating sustainable waste management operations. The regulations determine the role of reducing waste, and recycling as well as compel various plants to keep environmental concerns such as emissions containing at the top of the agenda.

Through the systematized approach that seeks to prioritize waste prevention, promote resource recoveries, and sensible use of WtE, we can achieve a sustainable future that will elevate waste to the resource instead of burden.

5 Current Trends and Scenario in Sustainability and Waste-to-Energy Techniques

5.1 Current Trends and Scenario in Sustainability

The concept of sustainability has grown much into a vital agenda on the global stage in an era when environmental problems are on the rise, resources are increasingly limited, and societal inequalities persist. It is a multifaceted concept that encompasses three interdependent pillars: Environmental, society-related, and/or economic (Milutinović et al. 2014; Roy et al. 2023). The environmental pillar aims at ecological protection and conservation of all natural resources, ecosystems, and biodiversity. The social pillar encompasses racial equality, respect for human dignity, and the development of residents and communities alike. The economic dimension looks to induce economic development and growth, enhance the prudent use of resources, and develop strategies for the fair distribution of wealth. (Milutinović et al. 2014; Khan and Kabir 2020; Hariram et al. 2023).

The commitment to sustainability is transcending greater heights following increasing world problems like global warming, poor resource management, and environmental deterioration. The main reason for climate change is when carbon dioxide and other gas emissions from human activity enter the atmosphere (Friedrich and Trois 2011). This causes severe consequences that affect the ecosystems, biodiversity, and human settlement. The unsustainable mining and consumption of resources have threatened the complete exhaustion of the finite reserves which does not allow the future generations to be prosperous. The environment includes air pollution, water pollution, destruction of trees, and destruction of habitats, which brings about serious adverse health, food supply, and ecosystem benefits consequences.

Sustainability as a concept has been in the spotlight and realization, in the past few years when countries and organizations have been replacing unsustainable practices. Sustainable development has been embraced across various sectors. Governments, businesses, and communities have put many natural conservation steps including in management and decision-making (Bonnedahl et al. 2022).

In the energy sector enormous moves toward conditional energy production have been recorded, for example, by solar, wind, and hydroelectric power. According to the data brought in from the International Renewable Energy

Agency (IRENA), 29% of world electricity was generated from renewable sources in 2020 and the trend that is expected for future years is a growing one (IRENA, 2021). The Paris Agreement, as well as the SDGs of the UN, has brought up even more of the ambition to go green and handle the energy transition from high to low carbon.

The agricultural industry saw many sustainable farming approaches emerging including organic farming, agroforestry, and precision agriculture. These steps are considered for the minimal use of the inorganic manufacturing of fertilizers and pesticides which thereby also protects the soil's health and water use (Muhie 2022). For example, the Food and Agriculture Organization (FAO) particularly advertises sustainable farming as an effective strategy for achieving both food safety as well as environmental conservation (FAO 2024).

Industrial actions like environmental efficiency, circular economy, and industrial symbiosis implementation in the manufacturing industries have quickly gained acceptance (Barros et al. 2021). While these policies are focused on eliminating wastage, increasing resource efficiency, and promoting the going back of materials, they also put the public where they can directly participate in the whole recycling process. Programs like the Ellen MacArthur Foundation's Circular Economy continue to take firm strides in shifting focus toward a more eco-sustainable industrial domain.

Greenhouse gas emissions reduction and air pollution abatement in road transport have been the focus of innovation such as in UVs, fuel biofuel, and efficient public transportation systems. Over the years cities across the world have put in place ways to enhance active (pedestrianism and cycling) means of transport and sustainable urban planning (Friedrich and Trois 2011).

Consequently, the implementation of sustainable practices is extremely complicated to accept due to many factors that are obstacles to their allocation to many people. Economic problems and increased challenges which are part of the decision to have both faster economic growth and improvement of environmental conditions can stand in the way (Cantarero 2020). Uncertainty in awareness, inadequate level of technology, and improper policy framework all indirectly create a hindrance (Bachmann et al. 2022). Another layer of complexity that comes along with the three pillars of sustainability is the possibility of conflicts occurring which would then need a keen understanding of priorities and working toward balancing each pillar (Cantarero 2020).

Sustainment of waste management is very important in the process of achieving sustainability objectives. Improper waste management results in undesirable outcomes such as the pollution of the environment, the depletion of resources as well as the development of social problems. Waste greatly pollutes the air, water, and soil by discharging contaminants which in turn endanger human health and natural environments (Elsheekh et al. 2021). Resource exhaustion happens when the valuables are thrown away and the possibility to recover them and reuse them is lost. Additionally, disordered waste treatment is a serious threat to the spread of diseases which may get worse and will naturally exacerbate the already existing social inequities.

The circular economy concept is getting more traction as an appropriate solution for these problems, recycling waste as a precious material instead of being a load. This paradigm shift is based on a profound view of the well-known "take, make, and dispose" model, and establishes a "waste prevention, reuse, recycling, and recovery principles" grid (Lisbona et al. 2023).

A winsome method in the conservation of sustainable waste management is the use of waste-to-energy (WtE) techniques. These technologies enable energy production from non-recyclable materials which results in a lower quantity of emissions to landfills as well as greenhouse gas emissions while recycling these types of waste is not possible (Ram et al. 2021). Different methods for recovering energy from waste are available including incineration, gasification, anaerobic digestion, and landfill gas harvesting, each possessing advantages and limitations.

As mankind is on the way to a sustainable future, it is vitally important not only to set up WtE techniques but also other sustainable solutions to address today's sustainability challenges successfully. Encompassing a comprehensive and interdisciplinary approach to sustainability we can carve out the path for a more robust, sea-standing, and economically thriving world that is friendly for the current generation as well as the future.

5.2 Current Trends and Scenario in Waste-to-Energy Techniques

Given the increasing attention to waste-to-energy (WtE) technologies, WtE techniques have developed as a promising way to combine the task of waste management, intending to harness energy. These technologies are the ones that through a mechanism of converting non-recyclable wastes into different energy sources that include, electric, fuel, or heat. At present, breakthroughs in WtE technology, widespread adoption, and exertion to combat environmental contamination and community health issues are the main trends in this field (Lisbona et al. 2023).

The world is facing major sustainable development issues and efficiently managing waste has become extremely crucial and dominant, so novel solutions are required to address the problem. This is because the world is growing rapidly with urbanization and the manufacturing of goods and the population, the global generation of waste has greatly increased, which amounts to serious environmental and public health problems in many countries (Khan and Kabir 2020). Energy capable of clean energy sources is rising in its demand as the threats of climate change and the exhaustion of limited conventional fossil fuels are beginning to materialize. The IEA disclosed the basis upon which renewable energy sources should bring over fifty percent of the total energy mix if the climate change goal is to be met by 2030. In addition to being the key to solving both the waste management and renewable energy dilemmas, the WtE processes thus, emerge as a dependable technology that everyone can readily depend on (IEA 2022).

The types of waste-to-energy techniques span a wide variety of technologies including combustion, pyrolysis, gasification, or plasma this emanates useful forms of energy such as heat, steam, electricity, or mollas which can be used for various purposes. The methods of recycling and harnessing the power of waste not only put some pressure off waste management but also add up to the amount of generated renewable energy, which is a good thing because it replenishes the expensive and exhaustible fossil fuels (Moya et al. 2017; Lisbona et al. 2023).

Below are the dominant waste-to-energy (WtE) techniques currently employed, each with its own advantages and considerations:

Advanced Thermal Treatments

The waste-to-energy (WtE) sector continues to evolve in these processes, such as gasification, and pyrolysis, which are preferred nowadays. They are using these already efficient technologies to combat emissions because burning with these technologies emits less harmful emissions than traditional burning methods. Conversion of wastes into crucial materials like synthetic gas (syngas) or bionic fluid (bio-oils) using gasification and pyrolysis, would be suitable for energy generation usage or chemical feedstocks production (Nandhini et al. 2022).

Carbon Capture and Storage will also be Integrated with the Bio-oil Production Process

Moreover, the integration of Waste-to-Energy (WtE) processes with Carbon Capture and Storage (CCS) systems is also planned. This option will promote an increase in the degree of environmental friendliness by achieving a balancing of the carbon footprint of the WtE processes via storing and capturing the CO_2, thus, contributing to the reduction of carbon emissions (Wienchol et al. 2020).

Waste-to-Fuel and Biological Treatments

In light of the pre-emption of sustainability in the WtE sector, a new focus is arising for waste-to-fuel conversion that is turning some of the abundant biomass waste into biodiesel, ethanol, and other biofuels, through promising technologies like anaerobic digestion, gasification, and hydrothermal liquefaction. Biological treatments undoubtedly are the core of the whole thing, and anaerobic digestion in particular is proven to serve the purpose of processing the organic waste into biogas effectively, which can be, in turn, utilized as electricity, heat, or an eco-friendly fuel source for transport. Some development also occurs in the direction of the simultaneous digestion of various waste streams to manage waste and increase biogas production (Ram et al. 2021; Jain et al. 2022).

Distributed-Module Systems

The WtE decentralized moving toward distribution and quarterly based capacities in which small, decentralized plants can provide individual waste streams and energy needs of local communities. Such systems will maintain budget efforts while reducing transportation costs (Woima Corporation 2024).

Waste-to-Chemicals

The WtE sector is now more and more oriented on chemical recovery technologies that strive to yield valuable chemicals and materials from waste materials processing. Shown below are examples of such interactions, where hydrogen as well as synthesis gas and basic chemicals are manufactured from virgin waste of plastics and biomass (Mishra et al. 2023).

6 Need of Sustainability and Waste-to-Energy Techniques

6.1 Need for Sustainability

The environmental footprint of waste generation requires an urgent sustainability remedy to eliminate the maximum landfilling and incineration. The World Bank project, predicts global waste generation to grow by 70% until 2050 (Kaza et al. 2018), leading to more environmental effects on waste disposal as a result. Methane produced by landfills is an important contributing factor to the greenhouse effect (Friedrich and Trois 2011); soil and groundwater can be polluted if the landfills are not treated properly. An incinerating method enables to reduction of quite a lot of waste and at the same time produces a lot of harmful pollutants such as dioxins, furans, and particles in the air and poses health risks. In the EU Waste Framework Directive (2008/98/EC) prevention, reuse, recycling, and energy recovery, all are given a high priority which minimizes the level of waste disposal (European Union 2008).

Encouraging a circular economy by changing waste into valuable raw materials to relieve environmental issues is another crucial challenge. The circular economy stresses retaining materials in use for as much time as necessary while taking the most out of them and recovering as well as regenerating the products and materials after their full life cycle. It is implied by the World Economic Forum that the earnings of a circular economy in the amount of the sum of $4.5 trillion are supposed to be gained in 2030 (World Economic Forum 2023). Waste-to-energy technologies could not only recover valuable materials but also the energy from the waste streams. This would in turn pose no threat to the circular economy, which was once the accepted approach.

Meeting the soaring demand for power while diminishing dependence on fossil fuels is another sustainability goal that waste-to-energy technologies can help achieve. The IEA in its reports suggests that global energy consumption stands to grow more than 25% at the same time expected to reach 2040 (IEA 2018). Renewable energy sources including solar and wind power are the key forces, but solid waste-based power can also be used to diversify the energy mix with good performance. According to the World Energy Council, waste-to-energy could generate as much as 10% of the overall electricity share and help to secure the energy supply (Ram et al. 2021).

Concerning reducing greenhouse gas output and joining global reduction in climate change rates, it also means much for sustainable efforts. The sector of waste which refers to the waste management sector accounts for about 4 and 6% percent of the global greenhouse gas emissions through landfilling and incineration processes (Kaza et al. 2018). Using waste-to-energy technologies would eliminate emissions from landfills and eventually push out any fossil fuels as well as energy generators as an alternative source of power. Waste-to-energy could avoid millions of tons of CO_2 emissions annually by 2050.

Production of economic opportunities and green employment jobs with enormous potential by waste management and energy recovery is one of the major sustainability needs that has been growing in these years. The size of the world-scale waste management market in 2032 will be about 3498.2 billion dollars according to Allied Market Research, the organization which studies and analyses changes in the market (Allied Market Research 2023). Making and operating waste-to-energy plants generate employment that will help the construction sector or maintenance forces, and support industries like waste collection and transportation. The adoption of a circular economy is expected to create millions of new jobs, benefiting both local and overseas workers.

Yet, on the other hand, expanding public awareness and providing support toward behavioral change toward environment-friendly waste management practices is one more sustainability challenge. Proper waste management not only involves public involvement and engagement in recycling, source separation, and proper disposal but also goes a step further in engaging the community on awareness about the dangers of waste pollution. Education, campaigns, and information could bring about sustainable waste-handling principles and also show good reasons why waste-to-energy technologies ought to be applied in wider use. Actions such as the implementation of UNEP's Waste Wise Cities program and initiatives like the Swachh Bharat Abhiyan (Clean India Mission) in India can be used to sensitize people and provide them with tools to improve sustainable waste management in urban areas.

6.2 Need for Waste-to-Energy Techniques

The presence of mounting pressure to combat waste management problems as well as the implementation of sustainable waste management strategies has made waste-to-energy (WtE) technique vital. Modern cities alongside their economic progress become the generators of substantial solid waste and commonly used approaches like dumping overflow waste into landfills are less environmentally friendly day by day. WtE sets into motion these technologies that are quite capable of helping the world get rid of waste that ends up at landfills, recovering other precious resources, and energy at the same time. This not only limits but also means that security and independence on fossil resources are strengthened as a result. More importantly, WtE supports the recovering economy concepts in which as many assets are extracted as possible from the waste for reuse before disposal.

Environmental benefits aside, the assimilation of WtE technologies into the world's order also proposes lucrative economic prospects. Employment creation among the waste management and energy sector, low landfill costs, and revenue harvest from energy production suggest that WtE is a more than profitable venture. The recent technological advancements in the issue such as the development of sophisticated conversion techniques like gasification and pyrolysis have expanded the scope of waste types that can be effectively mined for energy recovery purposes. The local and national regulations and the policy drivers which

Table 1 Waste types to be treated, along with key points explaining their characteristics and features

Waste type	Key points
Municipal solid waste (MSW)	• Composition: Organic, paper, plastics, glass, metals, etc. • Segregation and pretreatment: Shredding, drying, removing non-combustibles • Challenges: Heterogeneous nature, non-combustibles, environmental concerns • Opportunities: Sustainable waste management, energy generation, resource recovery
Industrial waste	• Sectors: Pulp and paper, food processing, textile, chemical • Characteristics: High moisture content, varying calorific values, hazardous components • Pretreatment and sorting: Shredding, drying, pelletizing, and removing hazardous materials • Co-processing in cement kilns or other industrial processes • Environmental considerations: Emission control, hazardous waste treatment
Agricultural waste	• Types: Crop residues, animal manure, agro-industrial waste • Energy recovery processes: Anaerobic digestion, gasification, direct combustion • Challenges: Collection and transportation logistics • Opportunities: Decentralized or localized WtE facilities, integrated waste management
Power plant waste	• Waste streams: Fly ash, bottom ash, flue gas desulfurization waste • Potential: Energy recovery, resource utilization (construction materials) • Co-processing with other waste streams • Environmental considerations: Emission control, heavy metal treatment

include waste management regulations; renewable energy targets and potential carbon pricing schemes, more so, also encourage the adoption of the WtE technologies. Of great importance too is the awareness campaigns, and acceptance, at large, of the benefits of WtE, and international cooperation to share knowledge. They are among the drivers of the high rate of adoption of these waste management strategies that are environmentally sustainable.

7 Wastes to Be Treated: Municipal, Industrial, Agricultural, and Power Plants

The volume of trash thrown across the planet persistently reaches critical points led by causes like population growth, urbanization, and economic development rate. It is forecasted by the World Bank that global waste generation will increase by around 70% from 2016 figures of 2 billion tons and will reach 3.4 billion tons by 2050 (Kaza et al. 2018).

MSW is considered one of the main feedstocks for waste-to-energy (WtE) plants in a way where the World Bank estimated 2.01 billion tons were generated in the year 2016 (Kaza et al. 2018). This number will likely get to 3.4 billion metric tons toward 20% years from now, with many of these represented from those regions fast growing in urban development like some parts of Asia and Sub-Saharan Africa. The world's agriculture generates close to 1 billion tons of waste every year and that majorly takes place from through crop residues and animal waste (Karić et al. 2022). The amount of waste exists to a very large extend that complies with region and agricultural practices.

Generally, coal-fired power plants at their worst find themselves responsible for fly ash and bottom ash in third place among the major contributors to the overall quantity of waste generated. In 2019, the American Coal Ash Association (ACAA) released a report that revealed plants in the US produced significant quantities of coal ash that exceeded 129 million tons (USEPA 2024). Thus, it brings to attention the type of waste that exists with this kind of energy (Table 1).

8 Review of Present Waste-to-Energy Techniques

8.1 Thermal Conversion Technologies

Thermal conversion technologies encompass the use of heat to work on different waste streams (Nandhini et al. 2022); the outcome is the processing of the waste into char, oil, and syngas. Such systems will be working well in the recovery of energy from waste with the immediate solution of the pollution issues linked with landfilling or direct incineration.

Incineration

Incineration has been a most favored waste treatment technology used in about 50% of MSW facilities across the world and involves the heating of the waste materials up to about 750–1100 °C. This process is mainly applied in connection with the treatment of MSW, industrial waste as well as certain quantities of dangerous waste. The heat produced from burning at the incineration site, can either be utilized to generate electricity, generate steam, or serve as district heating, depending on the need.

The incineration process begins with the acceptance and sorting of waste materials every day. Inert and non-combustible materials like metals are skimmed away, and the waste that is left is shredded and fed into the incinerator. Combustion occurs in a carefully controlled, and extremely meticulous environment with pollutant controls that efficiently reduce emissions while concurrently, in severe situations, generating power.

Another great benefit of incineration is the huge decline in the volume of waste, more than 90%, freeing up the space, time, and money that had to be allocated to curing landfills. Also, the recoverable energy is used to power a generator to make electricity or heat that is involved in the energy production from renewable resources. Nevertheless, incineration is confronted with many problems, including the emission of hazardous air pollutants which may happen if not adequately controlled it could lead to the release of dioxins, furans, etc.

To overcome this issue, modern incinerators are designed with an advanced air pollution control system that includes catalyst reduction (SCR) for cutting down nitrogen oxide (NOx), scrubbers to remove acid gases, and fabric filters or electrostatic precipitators for suspendable particles. These units, which are commonly used in incinerators, are designed to enable the emissions from the incinerator process to achieve very high compliance with the environmental rules.

Additionally, the ash entailed during incinerating, made of accumulated concentrated heavy metals and so on, must be addressed well by convenient means. These products are often tough and with minimal metal, they can be disposed of in the designated landfills for hazardous waste or recycled for some uses like when driving from one place to another with functional chemicals after they are treated properly (Ram et al. 2021).

Incineration has been around as the most popular method of waste management and blending in areas where the availability of land is limited, or where the waste-to-energy (WtE) initiatives are being favored. The pollution of air, however, even when the emissions of greenhouse gases and toxic persistent organic pollutants increase constantly there have been stronger and more strict laws introduced (Friedrich and Trois 2011). Consequently, the sector becomes more finely tuned as the efforts are focused on a high-efficiency rate of energy, low emissions, and thorough improvement in alternative thermal treatment technologies.

Pyrolysis

Pyrolysis is a thermochemical process whose mechanism happens in a non-oxygen or oxygen-deficient environment. This technology employs the thermal treatment of organic materials, like biomass or plastics, for an extended period at average temperatures from 300 to 1300 °C. This process involves complicated chemical reactions that strip the organic molecules of their complex structures and create products like solids, liquids, and gases.

The essential product in pyrolysis is the oily substance and bio-oil, the viscous dark-colored liquid is known to be used to manufacture various goods, from transportation fuels to chemicals, to feedstock in refineries (Nandhini et al. 2022). Besides the gaseous products, pyrolysis produces a solid residue characterized as char or biochar that can be used as fuel or soil amendment agents. The gaseous phase, called syngas, basically is a mixture of carbon monoxide, hydrogen, and other hydrocarbons which have the potential to generate thermal and electrical energy or chemical synthesis once chemically treated.

The fact that the pyrolysis process is extremely universal, that it can handle a wide variety of feedstock including biomass, plastics, tires, and MSW is one of the key advantages of co-pyrolysis (Mishra et al. 2023). The scaling-up and adjustment of the technology is easy to achieve and explains its use in waste management and resource recovery in a sustainable fashion. Moreover, pyrolysis can be used in the production of high-performance materials with low environmental impact because the process occurs in an oxygen-deficient space, which is responsible for the drastic reduction of the harmful emissions formation.

The development of new technologies in the field of pyrolysis nowadays is aimed at making the products of restoration more effective and the process itself more efficient. An example can be found in catalytic pyrolysis which uses the catalyzers so that a concentrate is formed. They are meant to cause reactions that get closer to the targeted product. Furthermore, the combination of pyrolysis with other processes such as gasification or burning is considered a manageable and more efficient system that can produce a lot more applications.

One of the new areas pyrolysis is used is waste-to-energy (WtE), and parts of the hazardous waste are turned into fuels or other chemical products. This method can not only deal with waste management-related issues but also contribute to the circular economy by reusing input coming from the flow of waste.

Nevertheless, there are technical difficulties linked to the pyrolysis process, such as the technique of temperature control and working with possibly dangerous byproducts. Furthermore, these assets can be equipped with emergency power systems, backup generators, or other redundancies to ensure continuous operations.

Pyrolysis technology provides the possibility of environment environment-friendly approach toward waste management and resource recovery thus offering a chance for byproduct production with almost zero impact on nature.

Gasification

Gasification is a thermal conversion process that is commonly used to produce energy from coal, biomass, waste, and even municipal garbage. The process occurs at high temperatures, about 800–1200 °C in an oxygen-deficient environment. The goal of the procedure is to get the solid feedstock turned into a syngas—the gaseous mixture made up primarily of carbon monoxide, hydrogen, and methane.

By dryness, pyrolysis as well as oxidation of the process are gasification operational stages. Primarily during the evaporation stage, water is removed from the biomass in the subsequent pyrolysis stage, the dried residue undergoes thermal decomposition in the absence of oxygen forming charcoal, tars, and gaseous stream. Ultimately, the last stage in the unit operation of the autothermal reactor is that of limited oxygen reacting with char and volatile products resulting in the production of syngas.

It is by this produced syngas that we can take advantage of their different applications. It can be modified directly for gas turbine (GT) engines or power systems generation. Besides, syngas can be still used as a raw material to produce multiple chemicals through Fischer–Tropsch processes such as hydrogen, methanol, or diesel/gasoline substitutes.

By distinguishing its flexibility to gasify a wide range of feedstocks like waste, biomass, and fossil fuels, gasification is one of its main benefits it. As such, this will ensure the right direction for both sustainable waste management as well as renewable energy innovation. But the gasification is also noted to be more environmentally friendly than direct combustion as the particulates and nitrogen oxides also, will be emitted if remediated.

Thanks to the most recent advances in research and development, the focus of the scientists has been air gasification quality and reliability, as well as the use of modernized gas cleaning to get rid of gas contaminants. One of the strategies that have been investigated to further improve the CCGT output and, reduce GHG (Greenhouse Gas) emissions is the integration of gasification with combined cycle power plants or carbon capture utilization and storage (CCUS) (Wienchol et al. 2020).

The future of sustainable energy generation and waste solutions is rife with opportunities; however, significant technical and economic challenges yet remain when it comes to effective gas cleaning, byproduct/tar containment, and profitability—major impediments for smaller plants. In any event, gasification technology still goes forward keeping in mind optimization, versatility, and environmental impact, which are considered as a part of the green economy type.

Torrefaction

Biomass torrefaction is a kind of mild pyrolysis process involving biomass heating with inert gases in a temperature varying from 200 to 300 °C. The decomposing biomass encounters partial decomposition that results in the evaporation of moisture and several of the volatile compounds in this process. The process involves various steps of drying, heating, and cooling until a solid, dry, and brittle result is known as torrefied biomass or biocoal.

Torrefied biomass is superior in the sense that it has a higher energy density and grindability. It has improved upon it, and now it is also hydrophobic (Tran et al. 2013; Nandhini et al. 2022). The second generation facilitates its use for diverse purposes, for instance, it can be co-fired with coal in power plants, undergo gasification, and be burned in boilers specifically developed for biomass.

With the novel torrefaction technology being in the spotlight, problematic aspects of the process parameters like temperature, residence time, and feedstock composition are being addressed to improve the output and quality of the torrefied product. That said, research is also devoted to the development of continuous torrefaction systems and the integration of torrefaction with other biomass conversion methods so that output is higher and viable.

8.2 Biological Conversion Technologies

Biological finite technologies utilize micro-living forms (bacteria, algae, fungi, etc.) and the pathways of their metabolisms to convert organic waste materials into useful stuff such as biogas, biofuel, and biochemicals. They provide an ecologically and economically sustainable methodology of waste utilization which at the same time source renewable sources of energy. Biological resource conversion technologies that are widely used include anaerobic digestion and ethanol fermentation.

Anaerobic Digestion

One of the advanced methods of the treatment of solid waste, which is anaerobic digestion, is a bioprocess in which the decomposition of organic matter is performed by a consortium of microorganisms without the presence of oxygen. It constantly takes place both in the wild nature and various surrounding areas, for example, wetlands, landfills, and the bellows of ruminant animals. Nevertheless, the emission is also an inevitable consequence of our lives and waste handling will result in landfills producing methane gas which can be captured in anaerobic digester bioreactor for further use.

The anaerobic digestion process typically occurs in four sequential stages, each facilitated by different groups of microorganisms: The anaerobic digestion process typically occurs in four sequential stages, each facilitated by different groups of microorganisms (Ram et al. 2021).

Hydrolysis: Aerobic microorganisms such as complex organic polymers which include carbohydrates, proteins, and fats are converting these into simpler molecules, that is, monomers and oligomers, through hydrolysis. Mediating this process are enzymes, the extraction of which is affected by bacteria.

Acidogenesis: These molecules generate volatile organic acids like acetic acid, propionic acid, butyric acid, and carbon dioxide, besides hydrogen and alcohol as final products. During this stage, cellulose is converted by acidogenic bacteria.

Acetogenesis: In this step, acetogenic bacteria and other VFAs turn acetate, hydrogen, and carbon dioxide from the anaerobic degradation stage.

Methanogenesis: The last stage represents the bacteria Methanogenic transformation of acetic acid, carbon dioxide, and hydrogen into methane and carbon dioxide. As for the remedy of the energy issue, biogas made up of methane, the primary constituent, is used as a fuel for multiple applications such as electricity generation, as well as heating and transport.

The overall chemical reaction for anaerobic digestion can be represented as:

$$\text{Organic material} + H_2O \rightarrow CH_4 + CO_2 + \text{Digestate} \tag{1}$$

Gas normally produced while bacteria undergo anaerobic decomposition represents a mixture of 50–70% methane and 30–50% carbon dioxide with minor amounts of other gases such as hydrogen sulfide and ammonia present. The digestate, a sort of byproduct during the digestion process that might be solid, or even semi-solid, is incredibly rich in nutrients that can be used as a replacement for commercial fertilizers or a soil conditioner.

The latest achievements in anaerobic digestion have been achieved to make it more efficient, intensifying biogas production, and enlarging the spectrum of feedstock that can be processed (Bharathiraja et al. 2018). These include:

Pretreatment Methods: Pretreatments comprising physical, chemical, and biological methods are applied to enhance the degradability of feedstocks, lignocellulosic biomass, which is a common form of feedstock, which is usually comprised of polymers that are not broken down by microbes.

Reactor Designs: With the introduction of high-performance reactor systems such as continuously stirred tank reactors (CSTRs), up-flow anaerobic sludge blankets (UASBs), and dry fermentation systems, the process of biogas formation is being improved by a parallel increase of process stability.

Co-digestion Strategies: Co-digestion entails the concurrent digestion of varied waste streams including sewage sludge, food, and agricultural residues among others to gain a better nutrient constitution, high biogas yield, and more stable operation.

Microbial Community Optimization: A modern approach under microbial molecular technologies is applied to analyze and improve the anaerobic digestion microbiota community making it faster and more stable.

Biogas Upgrading and Utilization: Biogas upgrading technology such as membrane separation, water scrubbing, and amine scrubbing is being developed to clean-up impurities, and increase methane content, which in turn allows biogas to be useful in different sectors for example either as a fuel for heavy-duty transportation or as an injection to natural gas grids.

Anaerobic digestion is a significant part of eco-friendly waste disposal procedures offering the potential to convert organic waste into renewable energy and valuable byproducts along with the positive influence on the atmosphere (like emission-reducing) and decreasing energy costs (because of fuel source independence).

Ethanol Fermentation

Ethanol fermentation is a bioprocess that is based on yeast strains, which are responsible for transforming simple sugars into ethanol and carbon dioxide in the microorganism's presence. This process is of special note since it provides a way of alternative renewable biofuel production, encompassing various biomass feedstocks, such as agricultural residues, energy plants, and solid waste.

The ethanol fermentation process typically involves the following steps (Gavahian et al. 2019):

Pretreatment: Feedstock selection leads to numerous pretreatment options, such as physical, chemical, or biological procedures, which prepare these complex carbohydrates from them and make fermentable sugars like glucose or xylose.

Hydrolysis: The next step, enzymes which are also called endolytic enzymes (e.g., cellulases, amylases) are used to decompose the pretreated biomass into simple sugars which can be fermented by the microorganism.

Fermentation: In this phase, which could be the most vital, the fermentable sugars are changed to two ethanol and carbon dioxide by the activity of yeast as well as other fermentative microorganisms. S. cerevisiae, one of the most frequently used yeast strains for ethanol fermentation, is considered the most common.

The overall chemical reaction for ethanol fermentation can be represented as:

$$C_6H_{12}O_6(\text{Glucose}) \rightarrow 2C_2H_5OH(\text{Ethanol}) + 2CO_2(\text{Carbon dioxide}) \quad (2)$$

The yeast cells ferment the glucose by respiration, in turn, they produce ethanol and carbon dioxide. This process takes place in a dark, anaerobic environment (oxygen-deficient conditions).

Distillation and Dehydration: During the fermenting stage, the produced liquid (the first known form of beer) consists of ethanol, water, and some other byproducts. Engaging the expertise of the distillation process, the ethanol is recovered from water and other materials through it, leading to the purification of ethanol solution in high concentration. Further degassing modes are useful such as the manner of molecular sieves well as membrane-based techniques to remove the remaining water and have hydrate-free ethanol for a biofuel.

Developments in recent years in ethanol technology have been carried out in search of enhancing its productivity and economics and enlarging the variety of sources that could be used for more effective fermentation. These include:

Development of Genetically Modified Microorganisms: Researchers choose a few of the yeast strains and other microorganisms that can transform many sugars (e.g., xylose, arabinose) that are normally found in the lignocellulosic biomass, and therefore, increase the production of ethanol.

Consolidated Bioprocessing (CBP): CBP stands for consolidated bioprocessing, which utilizes enzyme production, biomass hydrolysis, and fermentation in a singular step. The integration of these features into a single step may drive down the total cost and complexity of the process.

Integrated Biorefinery Concepts: Ethanol production being integrated with other biomass conversion processes like anaerobic digestion and pyrolysis enables biorefineries to produce different products (e.g. ethanol, biogas, chemicals) from the same feedstock, through resource utilization and viability as well as improved economics.

Process Intensification: Methods including for instance SSF and SSCF are now utilized to enhance process efficacy by minimizing overall processing time through.

Ethanol fermentation is undoubtedly significant in the joining force of renewable transportation fuels to make the transition to a more low carbon, sustainable economy a reality. Nevertheless, the advances made in this area include improving feedstock availability among others, the treatment costs, and the process economy subject to ongoing research and development efforts (Fig. 1).

8.3 Landfilling

Landfill Gas Generation

The organic matter decomposition process leads to the production of gas mixed with different organic compounds at landfill sites that is what landfill gas is. When the microorganisms that break down organic matter do so in the presence of oxygen (aerobic degradation), the organic carbon in the material they are digesting tends to be converted to

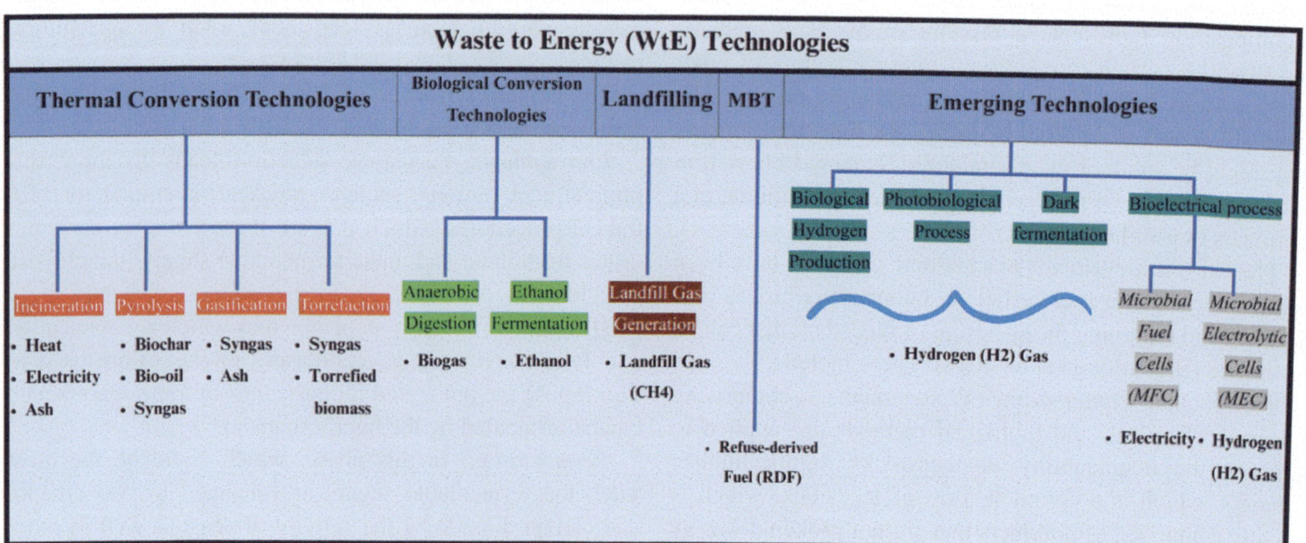

Fig. 1 Waste-to-energy (WtE) technologies and their products

carbon dioxide (CO_2), a gas with a composition, unlike the methane that is produced in as anaerobic environment (i.e., the absence of oxygen) such as that found in a landfill.

The formation of landfill gas is distinct from other bacterial processes in that, it involves multiple steps and particular microorganisms initiate each stage of the process. Firstly, aerobic bacteria use up a combination of oxygen and organic matter to conduct their respiration, thereby reducing the amount of oxygen available. In the end, oxygen depletion carb hat or anaerobic bacteria condition mainly depends on the process of decomposition and thus, the production of landfill gas.

The chemical makeup of landfill gas can vary substantially at times because of parameters like the age of the landfill, the type of waste generated at that landfill, and the environmental circumstances in the landfill. In most cases, the gas composition of landfills is about 50% methane and another 50% carbon dioxide, and there are, of course, trace amounts of other gases.

Landfill gas can be collected and utilized as a renewable energy source through various methods, including:

Extraction Wells: Vertical or horizontal wells are drilled to the landfill from where flash gas is extracted flowing through the network of gas pipes and finally becoming collected at a central gathering point.

Horizontal Collectors: A pipe network is embedded into the daily uphill waste layers at the beginning of the landfill construction. Gas collection occurs simultaneously.

Membrane Covers: This is done by spreading water-permeable membranes or tarps on top of the landfill. After that, a network of pipes is installed to capture the gas that is being released from the remains of the waste.

The collected landfill gas can be utilized in various ways (Nanda and Berruti 2020):

Electricity Generation: Landfill gas is easily burnable, thereby being able to drive fuel that powers machines or turbines to generate electricity, which could be supplied to the local grid or used on-site.

Direct Use: In the place of landfill gas, we can also use boilers for direct heat purposes and industrial processes.

Upgrading to Biomethane: Landfill gas can go through a purification process to expel undesirable compounds and the methane content will increase by doing this resulting in biomethane that can be used in pipeline systems or vehicles.

Landfill gas collection and utilization not only provide us with a green energy source but also helps us reduce the anxiety about greenhouse gas emissions as it captures methane which has a higher warming potential than carbon dioxide.

Although landfill gas production is not an unlimited process, since the speed of decomposition of organic matter gradually declines, there are still methods we can use to effectively reduce landfill gas emissions. Consequently, gas collection and utilization from landfills might stay just to be a transitional solution, meanwhile, full attention should be paid to the development of more environmentally friendly waste management practices including waste reduction, material recycling, and usage of alternative energy, etc.

8.4 Mechanical Biological Treatment (MBT)

Mechanical biological treatment (MBT) is an integrated type of MSW treatment that combines mechanical processes and biological ones to treat and stabilize MSW with a view to its further disposal or treatment. The major achievements of MBT involve recovering recyclable materials, decreasing the amount and weight of waste, winding out the organic fraction that is good for the land, and recovering energy (Beyene et al. 2018).

The MBT process typically involves the following steps:

Mechanical Processing: Sorting and shredding incoming waste into plastic, non-ferrous materials, and organics fractions.

Biological Treatment: The organic fraction is run through biological treatment either through aerobic composting or anaerobic digestion to break the biological components causing biodegradations.

Refinement: Alternatively, once the organic fraction is stabilized by the above-mentioned process, it can go through drying or screening processes, which turn it into refuse-derived fuel (RDF) (Beyene et al. 2018).

MBT installations could very well be a goal result of various methods that have been aimed at the maximum recovery of materials, production of high-quality refuse-derived fuel, or minimizing the volume of wastes for landfills. The exact layout of these facilities together with the chosen processing methods may vary regionally based on laws, waste complexity, and desired end products (Beyene et al. 2018).

MSW treatment by MBT brings about closed-cell landfill area performance, lessened greenhouse gas emissions, and a chance for energy recovery through stabilized organic fraction or refuse-derived fuel utilization. Although this technology has a high investment and operational cost, and the produced outputs may differ from ware involved in the recycling process and the waste composition, MBT can produce valuable materials.

8.5 Emerging Technologies

The realm of waste-to-energy continues to develop and innovative technologies, as researchers attempt to implement new and improved processes that can reuse the resources found in the waste streams. These emerging technologies based on biotechnology, nanotechnology, and materials science are innovative areas in sustainable waste management and energy production, which show better performance than the conventional way. These frontline and top-notch technologies range from biological courses like microbial fuel cells and dark fermentation to the photobiological processes which are mostly dependent on the photosynthetic bacteria. Also, the hybridization of the established ones, such as anaerobic digestion and gasification, with these developing technologies opens a path for enriched efficiency and resource usage. The technologies develop alongside and eventually will be able to change the seamless sector of waste-to-energy even giving a chance to a circular and sustainable economy.

Biological Hydrogen Production

A technology that is relatively new is biological hydrogen production, which is based on using different microorganisms to transform organic waste into hydrogen, relying on a renewable and clean energy source. Circulating-current electrolysis is a clean source to produce hydrogen without the use of fossil fuels or energy-consuming processes.

There are several routes (of biological pathways) of hydrogen production, depending on the kind of the microorganisms and the metabolisms involved. The most prominent methods include (Akhlaghi and Darzi 2020):

Dark Fermentation: This procedure consists of different organisms that are capable of fermenting organic substances, like Clostridium or Enterobacter ones, in order to produce hydrogen gas, carbon dioxide, and organic acids (e.g., acetic acid, butyric acid) under anaerobic conditions.

$$\text{Chemical equation}: C_6H_{12}O_6 + 2H_2O \rightarrow 2CH_3COOH + 2CO_2 + 4H_2 \quad (3)$$

Photo-Fermentation: The bacteria that makes use of light energy to work like that of *Rhodospirulum* and *Rhodobacter* fixes organic acids (produced in dark fermentation) into H_2 and CO_2.

$$\text{Chemical equation}: C_3H_5OH + 2H_2O + \text{Light energy} - 4H_2 + 2CO_2 \quad (4)$$

Biophotolysis: In this procedure, microalgae and cyanobacteria utilize sunlight and water to break water down, releasing hydrogen gas and oxygen as the ultimate products.

$$\text{Chemical equation}: \text{Water molecules input}: H_2O + \text{Sunlight} \Rightarrow H_2 + O_2 \quad (5)$$

The organic materials for biohydrogen production can range from solid wastes like municipal waste, organic residues from agriculture, spoiled food, and wastewater. Unsaturated approaches like hydrolysis and fermentation are also used in this process to simplify the complex organic matter for the microorganisms so that they can be digested easily into simple compounds.

Recent technological developments in biological hydrogen production include the creation of genetically modified microorganisms that generate hydrogen in higher amounts, the optimization of reactors for better efficiency, and the integration of multiple biological processes (e.g., dark fermentation and photo fermentation) together (Akhlaghi and Darzi 2020).

Among the important gains of biological hydrogen production is the capacity for the process to be carbon–neutral or even carbon-negative operations, by the input feedstock are waste organics that function as carbon sinks. Moreover, when compared to other processes, the process produces residual wastes with minimum impact and the remaining biomass can be used for other purposes like biofertilizers, or as feed for animals.

Photobiological Process

The photobiological processes focus on using photosynthetic microorganisms typically microalgae and cyanobacteria that transforms organic and carbon dioxide industrial waste into money-making husk such as biofuel, biomass, and chemicals. They employ the very capacity of these organisms to absorb light energy, carbon dioxide, and nutrients to utilize them in photomethylation to form organic compounds in course of process of photosynthesis.

The overall chemical equation for photosynthesis can be represented as:

$$CO_2 + H_2O + \text{Light energy} \rightarrow (CH_2O) + O_2. \quad (6)$$

Along with these (CH_2O) denotes the biological allies that are physically present in the system.

Microorganisms in reactor tanks who consume of waste streams such as wastewater and agricultural residues serve as nutrient bases in the process photobiological use. The microbes will instead utilize these nutrients alongside carbon dioxide and light energy to produce biomass which can be transformed (for example, biofuels such as biodiesel or bioethanol) or turned into other yieldful products.

In the sphere of photosynthetic biotechnology, there have been great improvements that involve genetic manipulation of microalgae strains to optimize their productivity and the development of cultivation systems, like photobioreactors, to allow the maximal utilization of light and nutrients.

Hence, photobiological processes become those eco-friendly technologies which simultaneously manage the polluted organic waste streams and produce valuable products. Such technologies are believed to give a boost to the circular bioeconomy development. Additionally, this is a process and catches CO_2 that helps in reducing the greenhouse gas.

Dark Fermentation

During dark fermentation, occur microorganisms that are anaerobic, and in the conditions that their growth occurs in, selective termination products are produced: Hydrogen, carbon dioxide, and organic acids. Biosolids process is valuable and desirable technology used for the production biohydrogen based on various targeted waste streams mentioned above (Rasheed et al. 2021). These include agricultural residues, food waste, and municipal solid waste.

The overall chemical equation for dark fermentation can be represented as:

$$C_6H_{12}O_6 + 2H_2O \rightarrow 2CH_3COOH + 2CO_2 + 4H_2 \quad (7)$$

When glucose ($C_6H_{12}O_6$) is run through a process of fermentation by carbon dioxide (CO_2), hydrogen gas (H_2), and acetic acid (CH_3COOH) by lactic acid bacteria, such as *Clostridium* and the *Enterobacters*, the main fermentative bacteria.

The first organizing process in dark hydrolysis is that certain complex carbohydrates are actually broken down into simpler sugars under the hydrolysis process. Sugars such as glucose are then broken down by the bacteria giving rise to hydrogen gas, carbon dioxide, and organic compounds like butyric and acetic acids as byproducts.

Modern dark fermentation technology has advanced exponentially thanks to the development of effective pretreatment techniques to increase the rate of organic waste digestion, investigation of reactor designs for better results out of the processes, and co-culture systems that are created by the mixing of different strains of the bacteria.

Dark fermentation stands out among other biohydrogen production methods due to its ability to produce biohydrogen from a variety of waste streams, including both renewable energy generation and waste management. Nevertheless, the advancement of this area still faces the issue of reaching higher yields, decreasing the generation of other products, and, perhaps the greatest challenge, developing an effective link to other processes for efficient utilization of the remaining organic acids and the leftover biomass.

Bioelectrical Process

Microbial Fuel Cells (MFC)

Microbial fuel cells (MFCs) are bioelectrochemical devices that draw energy from the process by which some of the microbes can transfer electrons during their metabolism as well as the bacteria reduce protons and carbon dioxide while they transmit electrons to oxidize and thereby, they acquire the energy (Vishwanathan 2021). In the MFC electrogenic microorganisms or electrochemically active microorganisms (EAM) use in situ organic matter which is oxidized for the generation of electrons. These electrons transport to the external circuit producing an electrical current (Rasheed et al. 2021).

The overall reaction in an MFC can be represented as:

$$\text{Anode: Organic matter} + \text{Oxidized mediator} \rightarrow CO_2 + \text{Reduced mediator} + \text{Electrons.}$$

$$\text{Cathode: Oxidized mediator} + \text{Electrons} + O_2 \rightarrow \text{Reduced mediator} + H_2O.$$

The microbes at the anode perform oxidation of the organic matter where the flow of electrons occur from the anode to the anode electrode. These electrons are the carriers of the current, and they flow externally through a circuit (Vishwanathan 2021). They move toward water-forming the cathode where they combine oxygen and protons to create water.

Addition of new MFC features amid the recent advancements includes the produce of more effective electrode materials, testing of alternative membranes or membraneless designs, and integration of MFCs with other waste treatment processes for the improved performance and recovery of resources.

MFCs provide several benefits, including the release of renewable electricity from waste streams, the possibility of wastewater treatment and the fact that it is possible to have them functioning at ambient temperatures. While these challenges exist in scaling up the MFC systems as well as in increasing the power density and addressing cathode performance and membrane fouling issues, great opportunities also lie ahead for clean energy generation from MFCs.

Microbial Electrolytic Cells (MEC)

The technology of microbial electrolytic cells (MECs) is closely related to that of membrane-based fuel cells which operate in reverse. They use an external power source, usually electricity from renewable sources to drive the production of hydrogen gas or methane liked products from organic waste.

In the MEC, there exist some hydrogen- or methane-producing microorganisms at the cathode, where they get electrons, protons, and CO_2 to produce hydrogen and methane (the products are determined by the working conditions).

The overall reaction in an MEC for hydrogen production can be represented as:

$$\text{Anode: } 2H_2O \rightarrow O_2 + 4H + + 4e - .$$

Cathode: $4H^+ + 4e^- + CH_2O$

$\rightarrow CH_4 + 2H_2O$ (for methane production).

$2H^+ + 2e^- \rightarrow H_2$ gases (for hydrogen production)

In the cathode of the fuel cell, at the same time, the water is oxidized into protons, electrons, and oxygen gas. In the external circuit, it is electrons that are the ones flowing through this circuit, to the cathode—where the electrons combine with protons and carbon dioxide—to either produce methane or hydrogen gas, by the operating conditions (Kadier et al. 2020).

Sheer MEC development recently aims at the deployment of state-of-the-art cathodes, the testing of new membranes and membrane-less concepts, and the incorporation of MECs in other wastewater treatments to elevate the efficiency of the systems.

MECs carry on their suction via producing worthwhile products like hydrogen or methane but also tend to take the waste management. Nevertheless, some difficulties need to be resolved in energy efficiency principal optimization, a recreation of operating conditions, and the problem of cathode performance and membrane fouling imposition.

9 Circular Economy and Sustainability Economics

The shift to a circular economy is speedily gaining prominence as a sustainable and cost-effective approach for handling the waste management crisis and the creation of new, green businesses, generating jobs. The circular economy principles offer a paradigm change out of the originally dominant "take-make-dispose" linear model to the restorative and rejuvenating pattern.

The pursuit of new business models that leverage material waste recovery is at the core of the circular economy's broader development. Companies are exploring different ways to extract benefits from the waste streams, which can be either for the reuse of materials in recycling or to produce energy through gasification, combustion, and other WtE technological means. Granted, these circular business models not only provide revenue but also minimize excessive use of resources and contribute to waste reduction.

Correspondingly, the circular economy will point to the new employment creation within the sustainable waste management area. Industries sustaining jobs for sorting, collection, recycling, remanufacturing, and WtE, at the same time as additionally contributing to the local economy and ecology, are the ones that provide employment opportunities (Tsui and Wong 2019; Elsheekh et al. 2021).

The government acts as one of the key participants in enacting a circular economy by developing policy arrangements and market incentives oriented toward sustainable waste management (Bonnedahl et al. 2022). As part of it, the policy will incorporate measures like extended producer responsibility (EPR), landfill taxes, and subsidies for the recycling and WtE projects (Mayanti and Helo 2023). Failure to put such mechanisms in place would lead to an uneven business landscape favoring those who engage in circular economic systems over others thus paying a huge price for the present generation and those to come after.

In addition, digital technology integration is of paramount importance for evaluating flows and establishing circular business models of the economy. Leapfrog developments including the Internet of Things (IoT), the blockchain, and artificial intelligence (AI) allow materials to be followed and traced, which certainly helped with such tasks as the better organizing of waste and the higher rate of resource recovery.

Despite this fact, it should be kept in mind that the transition to the circular economy needs a system-based approach that implies the participation of all main stakeholders—businesses, governments, customers, and civil society organizations. Efficient and educative communications as well as a campaign that is an eye-catcher are the basis of a valorization of responsible consumption and also the disposal of waste (Wijkman et al. 2019).

Through a circular economy framework, societies are reborn into opportunities for creating jobs and mitigating challenging circumstances like waste management, zero resources, and environmental destruction. This approach touches upon all three necessary dimensions of sustainability, hence contributing to sustainable development as well as sustaining economies that are resilient and will still be able to thrive in the future.

10 Sustainability Impacts on Environmental Health and Climate Change

The WtE alternative technology is not only very capable of averting the problem of poor air quality and smog in a country, but it is also a competent contributor to the global fight against climate change. Highly, the conventional approaches to waste management, i.e., landfilling and incineration have posed severe threats to environmental health and global environments as well.

Often, air quality and environmental health are among the key problems related to routine waste management. Along with the landfills, methane which is a powerful greenhouse gas, the emission of incineration also consists of the particulate matter, dioxins, and furans, which are known to worsen allergies, asthma, or other respiratory problems. Diversion from landfills and developments of modern emission control systems by WtE technologies are vital to help in the reduction of air pollution that puts public health at risk.

Water pollution is another critical environmental health issue that is being highlighted by the WtE technics. In addition, leachate from landfills is a possible contaminant of groundwater and river/water bodies that may even put drinking water supply and aquatic ecosystems at risk. WtE facilities, provided they are designed and operated following proper standards, will be much less involved in the generation of leachate and will use state-of-the-art wastewater treatment techniques, hence water resources being safeguarded, and human health being further protected.

In particular, WtE methods contribute to two processes that are important in solving the climate change problem—the first is that such methods reduce emissions of greenhouse gases. The landfills are mainly responsible for methane emissions, although incineration and carbon dioxide generation from the use of fossil fuel to produce energy are obligations as well (Friedrich and Trois 2011). The technology of WtE can strike the balance between recovering energy from sustainable sources and the combustion of organic streams, while it also efficiently reduces fossil fuel consumption and carbon gas emissions (Lisbona et al. 2023).

The latter may also be addressed with the circular economy approach as it leads to the recovery of resources and reduction of landfill disposal. These methods facilitate a decrease in the burden on the planet's environment through the extraction of valuable materials and energy from waste flows while at the same time adding to resource conservation and resource management that are sustainable.

As a result, in addition to replacing landfills, WtE facilities should be restrained in terms of emissions control measures and following environmental regulations to avoid competing characteristics of bad air quality, polluted water resources, and possibly human health. High-end means of technology such as gasification and pyrolysis allow options for cleaner energy retrieval processes that release less emissions than that of conventional incineration is there.

Apart from the ecological benefits, WtE methods can generate economic benefits by creating employment, revenue streams from energy sales, and saving on costs as a result of offensive waste disposal. These financial benefits may not only lead to attracting waste management practices. It encourages and even makes it less costly to use WtE.

11 Sustainability Challenges and Opportunities in WtE Innovation

The mainstreaming of waste-to-energy (WtE) techniques is hindered by the fact that they are identified as facing some challenges that need to be addressed for successful implementation and wide adoption. The public's view and approval are some of the obstacles that such WtE facilities need to overcome, as a lot of them may be opposed due to claims about emissions of gases, odors, and health risks. Communication that is accurate and with an environment that is accountable for pollution are the major determinants of earning trust from the public. On top of that being environment responsible are necessary. Beyond that, it is hard to have a proper waste feedstock flow with the right calorific number and kind composition as well as regions where there is good waste recycling or seasonal variation in waste generation.

Though the implementations of WtE come with some hurdles, the outlook for the technology is bright under a scenario of consistent innovation and technology improvements. New technologies like advanced plasma gasification, hydrothermal carbonization, and microbial fuel cells also give opportunities to have cleaner and faster energy recovery from different types of waste. There are more factors such as integrating WtE with other approaches of waste management like material recovery facilities (MRF), recycling programs, and circular economy models to build integrated and sustainable waste management systems that further recover resources and reduce negative impacts on the environment (Tisi et al. 2023).

The integration of innovative technologies like adding carbon capture systems in WtE plants can be a viable way carbon capture makes it possible to use this carbon for other processes (Wienchol et al. 2020), for example, enhanced oil recovery and creation of value-added products. Furthermore, the emergence of waste-to-chemicals and biofuel processes is carving the way for direct utilization of waste streams as feedstock to produce high-valued chemicals, biofuels, and other products. In this sense, it could constitute a source of revenue and a further realization of sustainability objectives (Jain et al. 2022).

Implementation of waste-to-energy capabilities gives hope that contributes to sustainability. Firstly, they can drive toward a renewable energy production strategy by either making electricity from energy contained in the waste or generating other useful forms of energy while using fewer fossil fuels as an alternative resource. Besides that, organic waste dumps that reduce landfill methane and decrease fossil fuel combustion can enable facilities to decrease greenhouse emissions and sustain climate change mitigation attempts. Additionally, waste treatment facilities can extract beneficial materials and products out of the waste flow that are consistent with the principles of the so-called circular economy, which supports sustainable utilization of resources.

Unlike other power plants, WtE is localized in nature; therefore, it maintains diversity in job creation by creating opportunities in various sectors (construction, running, maintenance, and other supporting industries like waste collection and transportation). With the ongoing need for the planet to be respected and the world struggling with the issues of sustainable waste management and energy security, WtE techniques will gain more ground and therefore be a key determinant. Alternative use of energy is a key conclusion, but its victimized implementation involves continuous research and development in raising the conversion efficiency, engaging in reducing emissions as well as developing the ability to treat the wide types of waste streams.

12 Discussion

This chapter analyzes the concept of waste management and sustainability, with a focus on waste-to-energy (WtE) systems. Due to the increase in the world population, energy demand is also expected to increase in the future decades, as found in the analysis above. According to the World Bank, there might be 3.4 billion tons of solid waste by 2050, and so the disposal of waste also materializes into new paths through research and development of effective waste management and new renewable energy sources. The authors opine that integrated sustainable systems for proper disposal, collection, treatment, and recycling of municipal, industrial, and agricultural waste can significantly go a long way in meeting the sustainability bar laid down in various global and India-specific initiatives such as the United Nations Sustainable Development Goals for Sustainable Development Sustainable Development for All by 2030, among others, and the Prime Minister's vision of Making India A Sustainable Nation.

The chapter is also a detailed discussion of different WtE technologies, the thermal conversion technologies including incineration and gasification, and the biological conversion processes that entail the digestion of wastes through the anaerobic digestion process. Both methods are then analyzed based on their ability to turn waste into useful energy forms, including heat, biogas, and syngas, as well as their effects on the environment. The discussion also points out that as an integrated part of incineration, energy recovery rate is quite promising but requires the application of high technologies in controlling emissions that degrade the air. On the other hand, anaerobic digestion is a relatively sustainable option; however, it has likely issues with feedstock fluctuations and issues with scalability. Overall, the chapter explains the techniques described above to highlight the need for proper identification of WtE methods in the context of sustainable waste management.

Also, the chapter describes how India regulates sustainable development through concepts like "Carbon-Free India 2070" and "India Vision 2047." These initiatives are aimed at very noble goals to transform India into a green economy with increased focus on clean energy and efficient waste management. This is very well demonstrated by the Swachh Bharat Mission, which targets scientific waste treatment and limited use of dump sites while promoting a circular economy. Through synchronization of domestic policies in connection with international sustainability standards, India aims at placing itself at the vanguard among those countries whose performances in the area of sustainable development will serve as an example for emulation by other countries.

Summing up this chapter, there is a set of recommendations for the further development of waste management as a way to achieve sustainable development. Through the EFARR approach, societies can turn disadvantages of waste challenges into advantages of using waste as a resource for energy generation. The authors also call for multi-sectoral policy and regulatory frameworks to support the high-level policies of waste reduction and resource recovery to be formulated and implemented to support and promote a sustainable future where waste is beyond being seen as garbage but as business opportunities for development. This multifaceted approach not only addresses pressing environmental challenges but also promotes economic development and social well-being, thereby reinforcing the interconnectedness of sustainability's three pillars: Here are some of the pillars: Ecological health, economic sustainability, and social justice.

13 Conclusion

Sustainability with the switch from a linear economy to a circular one necessitates a systematic and multifaceted approach to tackle the waste management obstacles via the latest waste-to-energy technologies and strategic plans. Rising global waste production stipulates urgent solutions that don't only divert waste from landfills but also recover waste and recycle these resources into energy.

Waste-to-energy (WtE) processes such as incineration, pyrolysis, gasification, anaerobic digestion, and modern methods like microbial fuel cells and dark fermentation, are more perspectives for using waste for energy production and waste treatments. These technologies can accomplish a range of tasks, such as reducing greenhouse gas emissions, minimizing pollution, practicing resource recovery, and reaching the fullest potential of the circular economy.

Nonetheless, it is appreciating that innovative WtE ways are possible if we take over challenges such as the acceptance of people, supply of feedstock, and technical obstacles. Non-stop renovations, system improvements, and investigations into the innovations that yield increased conversion efficiency, reduced emissions, and more treatable waste streams.

Involving WtE operations with other waste management strategies like material recovery facilities (MRFs), recycling programs and circular economy models can help set up sustainable waste management systems that help to recover valuable resources and shrink environmental impacts.

Moreover, sustainable waste management strategies must consider the three pillars of sustainability: Environmental sustainability, economic development, and social health. The devising of the right policies, market stimulation, educational campaigns, and collaborations with stakeholders constitute the fundamental tools for steering this economy toward a sustainable and circular economic system.

Through following waste-to-energy and sustainable strategies, society will be able to work toward the supply of renewable power, control climate change, generate numerous jobs, and enable sustainable construction. This is why all of us must at all costs, consciously battle waste management and resource recovery for effective waste minimization, which leads to an environmentally friendly society with sustainable development and economy.

References

Akhlaghi N, Darzi GN (2020) A comprehensive review on biological hydrogen production. Int J Hydrogen Energy 45(43):22492–22512. https://doi.org/10.1016/j.ijhydene.2020.06.182

Allied market research (2023) Waste management market expected to reach $3498.2 billion by 2032 [Press release]. Allied Market Research, https://www.alliedmarketresearch.com/, All right reserved 2024. Retrieved 26 April 2024, from https://www.alliedmarketresearch.com/press-release/waste-management-market.html

Bachmann N, Tripathi S, Manuel B, Jodlbauer H (2022) The contribution of data-driven technologies in achieving the sustainable development goals. Sustainability 14(5):2497. https://doi.org/10.3390/su14052497

Barros MV, Salvador R, Prado G, De Francisco AC, Piekarski CM (2021) Circular economy as a driver to sustainable businesses. Clean Environ Syst 2:100006. https://doi.org/10.1016/j.cesys.2020.100006

Basic information about landfill gas|US EPA (2024) US EPA. https://www.epa.gov/lmop/basic-information-about-landfill-gas

Beyene HD, Werkneh AA, Ambaye TG (2018) Current updates on waste to energy (WtE) technologies: a review. Renew Energy Focus 24:1–11. https://doi.org/10.1016/j.ref.2017.11.001

Bharathiraja B, Sudharsana T, Jayamuthunagai J, Praveenkumar R, Chozhavendhan S, Iyyappan J (2018) RETRACTED: biogas production–a review on composition, fuel properties, feed stock and principles of anaerobic digestion. Renew Sustain Energy Rev 90:570–582. https://doi.org/10.1016/j.rser.2018.03.093

Bonnedahl KJ, Heikkurinen P, Paavola J (2022) Strongly sustainable development goals: overcoming distances constraining responsible action. Environ Sci Policy 129:150–158. https://doi.org/10.1016/j.envsci.2022.01.004

Cantarero MMV (2020) Of renewable energy, energy democracy, and sustainable development: a roadmap to accelerate the energy transition in developing countries. Energy Res Soc Sci 70:101716. https://doi.org/10.1016/j.erss.2020.101716

Elsheekh KM, Kamel RR, El-Sherif DM, Shalaby AM (2021) Achieving sustainable development goals from the perspective of solid waste management plans. J Eng Appl Sci 68(1). https://doi.org/10.1186/s44147-021-00009-9

European union (2008) Directive–2008/98–EN–waste framework directive–EUR-LEX. https://eur-lex.europa.eu/legal-content/EN/TXT/?uri=celex%3A32008L0098

FAO (2024) Sustainable food and agriculture. Food and Agriculture Organization of the United Nations. https://www.fao.org/sustainability/en/

Friedrich E, Trois C (2011) Quantification of greenhouse gas emissions from waste management processes for municipalities—a comparative review focusing on Africa. Waste Manage 31(7):1585–1596. https://doi.org/10.1016/j.wasman.2011.02.028

Gavahian M, Munekata PES, Eş I, Lorenzo JM, Khaneghah AM, Barba FJ (2019) Emerging techniques in bioethanol production: from distillation to waste valorization. Green Chem 21(6):1171–1185. https://doi.org/10.1039/c8gc02698j

Gupta V, Gupta N (2022) India's target to be a carbo-free country: "Life" (Lifestyle for the environment). www.dmejournals.com. https://doi.org/10.53361/dmejm.v3i01.10

Hariram NP, Mekha KB, Suganthan V, Sudhakar K (2023) Sustainalism: an integrated socio-economic-environmental model to address sustainable development and sustainability. Sustainability 15(13):10682. https://doi.org/10.3390/su151310682

IEA (2018) World energy outlook 2018—analysis–IEA. In: IEA. https://www.iea.org/reports/world-energy-outlook-2018

IEA (2022) Renewable energy market update – May 2022 IEA Paris. https://www.iea.org/reports/renewable-energy-market-update-may-2022, Licence: CC BY 4.0

Jain A, Sarsaiya S, Awasthi MK, Singh R, Rajput R, Mishra U, Jia Q, Shi J (2022) Bioenergy and bio-products from bio-waste and its associated modern circular economy: current research trends, challenges, and future outlooks. Fuel 307:121859. https://doi.org/10.1016/j.fuel.2021.121859

Kadier A, Al-Shorgani NKN, Jadhav DA, Sonawane JM, Mathuriya AS, Kalil MS, Hasan HA, Alabbosh KF (2020) Microbial electrolysis cell (MEC). In: Bioelectrosynthesis (eds) Wang A, Liu W, Zhang B, Cai W, pp 95–128. https://doi.org/10.1002/9783527343829.ch4

Karić N, Maia AS, Teodorović A, Atanasova N, Langergraber G, Crini G, Ribeiro AR, Đolić M (2022) Bio-waste valorisation: Agricultural wastes as biosorbents for removal of (in)organic pollutants in wastewater treatment. Chem Eng J Adv 9:100239. https://doi.org/10.1016/j.ceja.2021.100239

Kaza S, Yao L, Bhada-Tata P, Van Woerden F, World bank (2018) What a waste 2.0: a global snapshot of solid waste management to 2050. In: Washington, DC: World Bank eBooks. https://doi.org/10.1596/978-1-4648-1329-0

Khan I, Kabir SZ (2020) Waste-to-energy generation technologies and the developing economies: a multi-criteria analysis for sustainability assessment. Renew Energy 150:320–333. https://doi.org/10.1016/j.renene.2019.12.132

Kinnaman TC (2006c) Policy watch: examining the justification for residential recycling. ˜the œJournal of Economic Perspectives/˜the œJournal of Economic Perspectives 20(4):219–232. https://doi.org/10.1257/jep.20.4.219

Lisbona P, Pascual S, Pérez VH (2023) Waste to energy: trends and perspectives. Chem Eng J Adv 14:100494. https://doi.org/10.1016/j.ceja.2023.100494

Mandpe A, Kumari S, Kumar S (2020) Composting: a sustainable route for processing of biodegradable waste in India. In: Springer eBooks, pp 39–60. https://doi.org/10.1007/978-3-030-36283-6_3

Mayanti B, Helo P (2023) Circular economy through waste reverse logistics under extended producer responsibility in Finland. Waste Manage Res 42(1):59–73. https://doi.org/10.1177/0734242x231168801

Milutinović B, Stefanović G, Dassisti M, Marković D, Vučković G (2014) Multi-criteria analysis as a tool for sustainability assessment of a waste management model. Energy 74:190–201. https://doi.org/10.1016/j.energy.2014.05.056

Mishra KB, Siwal SS, Nayaka SC, Guan Z, Thakur VK (2023) Waste-to-chemicals: green solutions for bioeconomy markets. Sci Total Environ 887:164006. https://doi.org/10.1016/j.scitotenv.2023.164006

Moya D, Aldás C, López G, Kaparaju P (2017) Municipal solid waste as a valuable renewable energy resource: a worldwide opportunity of energy recovery by using waste-to-energy technologies. Energy Procedia 134:286–295. https://doi.org/10.1016/j.egypro.2017.09.618

Muhie SH (2022) Novel approaches and practices to sustainable agriculture. J Agric Food Res 10:100446. https://doi.org/10.1016/j.jafr.2022.100446

Nanda S, Berruti F (2020) Municipal solid waste management and landfilling technologies: a review. Environ Chem Lett 19(2):1433–1456. https://doi.org/10.1007/s10311-020-01100-y

Nandhini R, Berslin D, Sivaprakash B, Rajamohan N, Vo DN (2022) Thermochemical conversion of municipal solid waste into energy and hydrogen: a review. Environ Chem Lett 20(3):1645–1669. https://doi.org/10.1007/s10311-022-01410-3

Ram C, Kumar A, Rani P (2021b) Municipal solid waste management: a review of waste to energy (WtE) approaches. Bioresources 16(2):4275–4320. https://doi.org/10.15376/biores.16.2.ram

Rasheed T, Anwar MT, Ahmad N, Sher F, Khan SU, Ahmad A, Khan R, Wazeer I (2021) Valorisation and emerging perspective of biomass based waste-to-energy technologies and their socio-environmental impact: a review. J Environ Manage 287:112257. https://doi.org/10.1016/j.jenvman.2021.112257

Roy S, Rautela R, Kumar S (2023) Towards a sustainable future: nexus between the sustainable development goals and waste management in the built environment. J Clean Prod 415:137865. https://doi.org/10.1016/j.jclepro.2023.137865

Sharma S, Basu S, Shetti NP, Kamali M, Walvekar P, Aminabhavi TM (2020) Waste-to-energy nexus: a sustainable development. Environ Pollut 267:115501. https://doi.org/10.1016/j.envpol.2020.115501

Šomplák R, Kůdela J, Šmejkalová V, Nevrlý V, Pavlas M, Hrabec D (2019) Pricing and advertising strategies in conceptual waste management planning. J Clean Prod 239:118068. https://doi.org/10.1016/j.jclepro.2019.118068

The international renewable energy agency (IRENA). World adds record new renewable energy capacity in 2020 (2021) https://www.irena.org/News/pressreleases/2021/Apr/World-Adds-Record-New-Renewable-Energy-Capacity-in-2020. (2021, April 5). World adds record new renewable energy capacity in 2020 [Press release]. https://www.irena.org/News/pressreleases/2021/Apr/World-Adds-Record-New-Renewable-Energy-Capacity-in-2020

Tisi YSaB, Matos FA, Carneiro MLN (2023) Development of waste-to-energy through integrated sustainable waste management: the case of ABREN WtERT Brazil towards changing status quo in Brazil. Waste Dispos Sustain Energy/Waste Dispos Sustain Energy 5(3):295–308. https://doi.org/10.1007/s42768-022-00127-x

Tran K, Luo X, Seisenbaeva GA, Jirjis R (2013) Stump torrefaction for bioenergy application. Appl Energy 112:539–546. https://doi.org/10.1016/j.apenergy.2012.12.053

Tsui T, Wong JWC (2019) A critical review: emerging bioeconomy and waste-to-energy technologies for sustainable municipal solid waste management. Waste Dispos Sustain Energy/waste Dispos Sustain Energy 1(3):151–167. https://doi.org/10.1007/s42768-019-00013-z

UN environment programme (2024) Solid waste management. UNEP–UN environment programme. Retrieved 30 April 2024, from https://www.unep.org/explore-topics/resource-efficiency/what-we-do/cities/solid-waste-management

United Nations (2022) Global sustainable development report 2023: times of crisis, times of change: science for accelerating transformations to sustainable development. United Nations, New York. https://desapublications.un.org/publications/global-sustainable-development-report-2023

United States Environmental protection agency (2024) Coal ash basics|US EPA. US EPA. https://www.epa.gov/coalash/coal-ash-basics

Van Den Bergh JC (2008) Environmental regulation of households: an empirical review of economic and psychological factors. Ecol Econ 66(4):559–574. https://doi.org/10.1016/j.ecolecon.2008.04.007

Vishwanathan A (2021) Microbial fuel cells: a comprehensive review for beginners. 3 Biotech 11(5). https://doi.org/10.1007/s13205-021-02802-y

Wienchol P, Szlęk A, Ditaranto M (2020) Waste-to-energy technology integrated with carbon capture—challenges and opportunities. Energy 198:117352. https://doi.org/10.1016/j.energy.2020.117352

Wijkman A, Climate KIC, Osdoba T, Heineman S, Acharya D, ARUP (2019) Circular economy in cities requires a systems approach. In: Background paper for an OECD/EC workshop on 5 July 2019 within the workshop series "managing environmental and energy transitions for regions and cities." OECD/EC. https://www.oecd.org/cfe/regionaldevelopment/Wijkman-2019-Circular-Economy-Cities-Requires-Systems-Approach.pdf

Woima Corporation (2024) Waste-to-energy|modular power plant–Woima corporation. Woima corporation. https://woimacorporation.com/technical-solution/

World economic forum (2023) Making the $4.5 trillion circular economy opportunity a reality. https://www.weforum.org/impact/helping-the-circular-economy-become-a-reality/#:~:text=A%20circular%20economy%20is%20a,additional%20economic%20output%20by%202030.

World health organization (2022) Compendium of WHO and other UN guidance on health and environment, 2022 update. In: World health organization 2022. Retrieved 30 April 2024, from https://www.who.int/tools/compendium-on-health-and-environment

Zhou Z, Tang Y, Chi Y, Ni M, Buekens A (2017) Waste-to-energy: a review of life cycle assessment and its extension methods. Waste Manage Res 36(1):3–16. https://doi.org/10.1177/0734242x17730137

Zorpas AA (2020) Strategy development in the framework of waste management. Sci Total Environ 716:137088. https://doi.org/10.1016/j.scitotenv.2020.137088

Zorpas AA (2022) The role of end of waste criteria in the framework of circular economy strategy. Glob NEST Int Conf Environ Sci and Technol. https://doi.org/10.30955/gnc201900980

Unleashing Waste-to-Energy Technologies

Shweta Kalia and Neha Chandra

Abstract

In today's scenario, increasing population, economic development, and urbanization have intensified pressure on traditional energy sources and exacerbated environmental concerns. In response, there is an emerging need for sustainable solution for energy. Municipal solid waste (MSW) management emerges as a significant opportunity as it offers a dual benefit of minimizing the environmental issues related to waste disposal and generation renewable energy, with an emphasis on circular economy. In many developing nations, the MSW management poses a problem due to lack of energy availability and traditional practices of waste disposal that case many environmental and socio-economic problems. As a result, researchers are looking into various waste-to-energy conversion strategies to produce clean energy in safe and eco-friendly ways. Several technologies have been developed to utilize energy from MSW, such as incineration, pyrolysis, gasification, ethanol fermentation, landfill recovery, and bio-methanation. These technologies provide 20 to 30% energy conversion efficiency for electricity. Moreover, emerging technologies such as microbial fuel cells (MFCs) and microbial electrolysis cells (MECs) shows potential in converting organic waste into electricity or biofuel. These waste conversion technologies not only address the problems associated with MSW management but also contribute to sustainable development by providing renewable energy sources and reducing greenhouse gas emissions by 1 tonne of CO_2 equivalent per tonne of waste produced. However, successful implementation of these technologies requires careful evaluation of technological, economic, and environmental factors. Hence, this chapter explores the WtE technologies, their economic and environmental impacts.

Keywords

Energy · Circular economy · Municipal solid waste · Microbial fuel cells · Waste-to-energy

1 Introduction

Several unseen environmental issues have a major influence on the natural world that must be understood. These environmental challenges prevalent on the planet lead to global warming and climate change (Zandalinas et al. 2021). The management of municipal solid waste (MSW) is a major problem that developing nations are facing due to increasing urbanization and population increase. Improper management of waste leads to different types of environmental pollution, biodiversity loss and possess a risk to human health. Along with serious impact on human health, the increasing waste stream also puts an economic burden on the environment. There is a need to find out eco-friendly alternatives for waste management that not only reduces the waste but also produces energy in the present times (Mujtaba et al. 2024).

Solid waste management (SWM) is a significant problem that the government and policymakers are facing worldwide, impacting countries development. Based on the literature survey, it has been found that the current global population produces approximately 2.01 billion tonnes of waste per year. Furthermore, this number is expected to rise by 16% in high-income countries and 40% in low to middle-income countries by 2050. Around one-third of this

S. Kalia (✉) · N. Chandra
Department of Biotechnology, IMS Engineering College, Ghaziabad, India
e-mail: Shweta.kalia@imsec.ac.in

N. Chandra
e-mail: Neha.Chandra@imsec.ac.in

A. Mandpe et al. (eds.), *Sustainable Waste Management*, Advances in Science, Technology & Innovation,
https://doi.org/10.1007/978-3-031-85685-3_17

waste is not being adequately managed in a way that is safe for the environment which possess a significant risks to our economy, environment and health. Developed countries are working in finding a cost effective and eco-friendly ways to reduce MSW. However, developing countries are still facing issues with inefficient waste management practices. These factors contribute to the inefficiency of the segregation, collection, storage, treatment, and disposal processes. Factors that contribute to the issue include inadequate infrastructure, a lack of legislative backing for recycling and limited public awareness (Agaton et al. 2020).

Although much is known about the response of plants towards individual stressors, less is known about their response to concurrent multifactorial stressors. Recent studies show that a rise in a variety of co-occurring stress factors significantly reduces the richness of vital microbiomes and severely impairs plant development and survival. This troubling trend emphasizes how urgently society must take steps to lower pollution, stop global warming, and improve agricultural resistance to combinations of complex stresses. The intricate interactions between many elements call for thorough knowledge and preventative actions to protect our future ecosystem. Addressing municipal solid waste management is essential in reducing greenhouse gas emissions and combating global warming. Utilizing WtE technologies can effectively reduce the amount of waste dumped in landfills, leading to more sustainable waste management practices (Mujtaba et al. 2024). WtE technologies not only reduce the amount of waste but produce valuable energy sources such as electricity and heat that reduce the environmental problems. Also, the plants generate job for operation and maintenance. The reduction of waste and production of renewable energy proposes a route to circular economy.

With the exponential growth of population, urbanization, and economic activities, the management of municipal solid waste (MSW) has become a pressing challenge globally, particularly in developing nations. This chapter addresses the dual crises of environmental degradation and energy scarcity by exploring the potential of waste-to-energy (WtE) technologies as a transformative solution. The improper handling of MSW not only exacerbates environmental pollution, biodiversity loss, and health risks but also imposes significant socio-economic burdens.

The chapter contributes to the discourse by presenting WtE technologies as a path towards sustainable development. It examines how these technologies can mitigate the environmental and economic issues associated with traditional waste disposal practices by converting waste into valuable energy resources such as electricity and heat. Through this process, WtE promotes circular economy principles, reduces greenhouse gas emissions and supports job creation in operational and maintenance sectors.

The chapter is structured to provide a comprehensive analysis of WtE technologies, detailing their conversion pathways, statistical significance, and implications for environmental and economic performance. Beginning with the composition and characterization of global waste streams, it discusses thermal, biological, chemical, and electrochemical conversion methods, followed by an assessment of their techno-economic feasibility and environmental impacts. The goal is to offer a holistic framework for policymakers, researchers, and practitioners to advance sustainable MSW management.

2 Composition, Characterization of Wastes Generated

Based on World Bank's Report 2.0, 2.1 billion tonnes of municipal waste are generated annually around the globe out of which 33% is not managed properly. Based on the trajectory, the global waste is projected to increase up to 3.88 billion/tonnes by 2050 (Ahmed et al. 2024). Due to increasing industrialization, economic development and population growth, solid waste generation is increasing constantly. There is an urgent need to treat the waste generated around the globe. For the effective management of the generated waste, it is essential to know its composition and characteristics (Yadav and Samadder 2017). The type of WtE technology depends upon the composition of wastes. The waste with high moisture content is well suited for anaerobic digestion as it produces biogas or pyrolysis for fuel production whereas low moisture content waste idea for incineration to produce heat electricity. The municipal waste composition varies with respect to the topographical area, population in the area, people's standard of living (Beyene et al. 2018). MSW generated from developing nation contains higher organic content which is biodegradable. The region of North Africa contains < 30% of organic waste whereas dry waste is > 55%. Regions of Europe and Central Asia, East Asia, and Pacific Latin America and Caribbean produce around 1/3 of dry waste (Sharma and Jain 2020). Factors such as particle size, density, moisture, calorific value are important for development of efficient waste-to-energy technology (Aleluia and Ferrão 2016). The waste generated from developed nations contain less moisture compared to the waste generated from developing nations such as China, India (Mohee and Mudhoo 2012). However, the calorific value of the waste generated from developed countries is higher (2000–4000 kcal/kg) than developing countries due to the presence of papers, other dry organic waste (Patumsawad and Cliffe 2002). Low and middle-income countries primarily produce MSW consisting of a higher proportion of organic and biodegradable materials whereas in developed countries organic content is low. Higher density waste indicates environmentally friendly materials and decomposes faster compared to lower density waste, which contains more combustible materials such as paper, cardboard, and plastics.

3 Waste-to-Energy Conversion

The main goal of waste utilization is energy recovery and residual disposal according to standard disposable limits. The calorific value of the waste is one of the important factor responsible for energy determination. There is a necessity of a reliable calorific value for efficient function and maintenance of waste-to-energy technology (Shi et al. 2016). There is a need of better waste conversion technology. There are various available technologies such as thermal conversion (pyrolysis, gasification), biological conversion (fermentation, anaerobic digestion), chemical conversion, landfilling used for energy production from waste stream (Kalyani and Pandey 2014). Characteristics of waste such as higher moisture content make it appropriate for the processes like anaerobic digestion, landfill gas recovery, and bio-digester gas production (Zuberi and Ali 2015). For instance, significant amounts of food waste, kitchen scraps, and leftovers from households, eateries, workplace areas, and markets fall into the organic fraction of municipal solid waste (OFMSW) due to their high moisture content and biodegradability (Beylot et al. 2013). Incineration involves simply burning of waste at high temperature o produce heat whereas in gasification and pyrolysis involves an advance thermal process to produce fuels in oxygen deficit condition. Anaerobic digestion involves microbes that degrade organic waste and produces biogas. The flow chart of different municipal waste conversion technologies is shown in Fig. 1.

4 Thermal Conversion Technologies

Thermal conversion process involves the thermal treatment of organic matter present in the MSW to produce heat energy. The large molecules of MSW are break down into smaller molecules and further oxygen is combined with produced carbon and hydrogen molecules to produce large amount of energy (Qazi et al. 2018). The technology is particularly useful for managing dry waste with low moisture content and a substantial portion of non-biodegradable organic matter. In order to transform waste into heat, electricity and valuable products high temperature are required (Shahnazari et al. 2020). Different type of technologies based on thermal conversion are incineration, pyrolysis, and gasification.

4.1 Incineration

Incineration is commonly used technology for waste-to-energy conversion. United States Environmental Protection Agency (US EPA) declared MSW incineration as a clean source of energy due to reduced amount of emission from the waste after combustion (Leme et al. 2014). In this techniques, organic matter present in the waste is completely combusted at high temperature and convert it into heat energy in the presence of oxygen (Beyene et al. 2018). The incineration takes place at 850–1100 °C in the incinerator plant. MSW having low moisture content can directly fed into the incinerator; however, MSW with higher moisture content needs mechanical pre-treatment to improve the combustible quality of the waste (Lombardi et al. 2015). As high moisture content in waste reduces its calorific value due to latent heat of vaporization. Hence, pre-treatment removes the moisture and reduces the ash content of the waste. Tan et al. (2014) reported that incinerators are suitable for combustion of non-biodegradable waste having less moisture content.

MSW comprises of various materials that have different size, physical, and chemical characteristics which affects the combustion time. Due to the presence of large particles in MSW, complete combustion takes longer time period compare to other solid fuels (Ruth 1998). Based on the composition and degree of certain materials recovery, incinerators reduce around 70% of total waste biomass and 90% of total volume (Anderson et al. 2016).

The MSW incineration facility produces heat and hot flue gas, containing carbon dioxide (CO_2), nitrogen (N_2), oxygen (O_2), water vapour (H_2O), among others. Additionally, ash is a significant by-product, constituting approximately ~ 25% of the total input material (Atwadkar et al. 2014). During incineration, the generated heat and hot flue gases are utilized to produce high-pressure steam, which is subsequently used in rotating the steam turbine to generate electricity (Tan et al. 2014). Following this, the flue gases undergo to mitigate toxicity and control environmental pollution (Jurczyk et al. 2016). The generated ash can be utilized as a raw material in the cement industry or disposed of in landfills (Qazi et al. 2018).

Case study of Campinas, Brazil reported of population over 1 million reported generation of around 1034 tonne/day of municipal waste. According to the reports, the waste collected from 39% residencies generates sufficient energy to fulfil the electricity needs. It can generate $5.7*10^6$ US dollars per month revenue (elective collection index 20%) from generated electricity and produce 114.092 kt CO_2/year emissions (Lino and Ismail 2017).

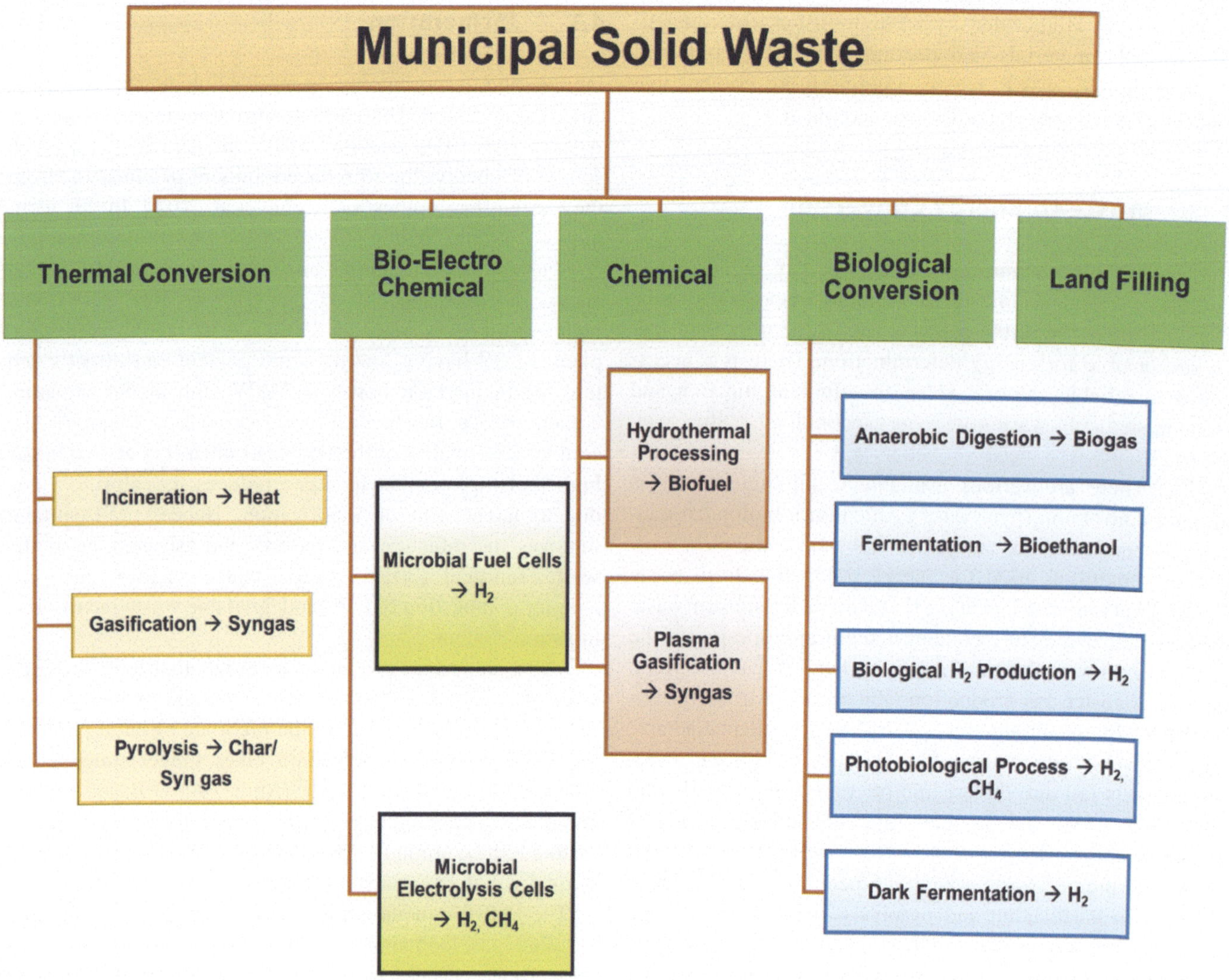

Fig. 1 Different waste-to-energy conversion technologies for municipal solid waste

4.2 Pyrolysis

Pyrolysis is a process that occurs in the absence of oxygen at a temperature in a range of 300–1300 °C. Depending upon temperature, heating rate, particle size, and residence time, pyrolysis is of three types. First is conventional pyrolysis occurs temperature 500–950 °C, residence time 450–550 s, particle size 5–50 mm, and heating rate 0.11 °C/s. Second is fast pyrolysis occurs at 850–1250 °C, residence time 0.5–10 s, particle size < 1 mm and heating rate 10–200 °C/s. Third is flash pyrolysis 1050–1300 °C, residence time < 0.5 s, particle size < 0.2 mm, and heating rate 200–1000 °C/s (Farooq et al. 2021). It converts MSW's organic content into gaseous product, pyrolytic liquid, and char. The yield and composition of the product depends on several factors such as the feedstock, temperature range, heating rate, and the type of reactor used (Lombardi et al. 2015). At a temperature ranging from 500 to 550 °C, the primary products of pyrolysis are oil, wax, and tar, while at > 700 °C, pyrolysis gases become the predominant products.

Pyrolysis reaction expressed in equation:

$$\text{CxHyOz} + \text{Q} \rightarrow \text{Char} + \text{Liquid} + \text{Gas} + H_2O$$

where Q is the heat provided for the reaction.

The feedstock will undergo decomposition.

$$Q = W * 2260\ \text{kJ/kg}$$

where W is the moisture content of feedstock.

A pyrolysis plant set up in Edmonton, Canada in June 4, 2014, handle 100,000 tonne of waste annually and convert it into energy. The facility is a part of MSW to biofuels initiative by Edomonton city and Alberta Innovates-Energy and Environment Solutions initiative partnership. The facility produces biomethanol and in 2015 a module was added to convert biomethanol into ethanol. The process has low GHG emissions as compared to the incineration. However, the initial cost of the process and maintenance was too high (Shareefdeen et al. 2015).

Energy conversion of MSW through pyrolysis process has several challenges, including air pollution from substances like HCl, H_2S, NH_3, SOx, NOx, as well as odour and exhaustive emissions. To address these issues, strategies to control emission are consistently integrated into pyrolysis operations. Additionally, efforts should be made to enhance the quality of products, making pyrolysis an eco-friendly technique (Gedam and Regupathi 2012).

4.3 Gasification

Gasification (modified version of pyrolysis) process converts MSW into hydrocarbon gases such as hydrogen (H_2), carbon monoxide (CO), carbon dioxide (CO_2), methane (CH_4), and hydrocarbon oils, char in the presence of steam. The process occurs at around 800–1200 °C in controlled O_2/steam concentration. Various factors influence its efficiency, including temperature, feedstock particle size, moisture content, and the type of gasification agent used (Beyene et al. 2018). This technology transforms plastics and combustible organic components of MSW into a clean and highly valuable syngas, consisting primarily of CO and H_2. In addition to CO and H_2, trace amounts of CO_2, water vapour, N_2, and CH_4 are also present (Qazi et al. 2018). The purified syngas generated through the process can be utilized to rotate gas turbine to generate combined heat and power (CHP). Furthermore, it serves as a secondary raw material in fertilizer and chemical industries (Moya et al. 2017; Qazi et al. 2018).

A Teesside Waste Gasification plant set up in South Wales, and United Kingdom convert around 350,000 tonnes of waste produced annually into energy. It handles variety of waste such as plastics with less harmful emissions. However, due to the high initial and operational cost, there is a need to have advance technology for efficient and stable operations (Sanna et al. 2012).

5 Biological Conversion Technologies

5.1 Anaerobic Digestion

The different types of biomass includes plants, animals, humans, and their by-products. Furthermore, biomass wastes can also originate from food processing, agro-industrial, and industrial activities. Based on the characteristics of these wastes, the methods such as combustion, gasification, and anaerobic digestion were determined to convert waste into energy or fuel, (Wilkie 2014). Anaerobic digestion is a process where microorganisms break down organic matter in the absence of oxygen. This technology is can be used to treat sewage sludge and other aqueous organic wastes. The major product released is biogas, primarily containing methane, which can be used to generate electricity or heat. Also, the digestate can be utilized as a fertilizer. This offers a sustainable loop of energy and resource recovery from wastewater (Manyi-Loh et al. 2013).

Nitrogen and phosphorus are essential nutrients in anaerobic digestion that directly impact microbial growth and activity. The optimum concertation of these nutrients are required optimal digestion performance and stable biogas production.

Nitrogen, the main element of amino acids, proteins, and nucleic acids, is commonly derived from proteins and urea in feedstocks such as food waste, animal manure, and sewage sludge. Phosphorus is essential for nucleic acids synthesis, ATP, and phospholipids, primarily comes from organic materials like food waste and animal manure, as well as phosphate salts (Rajlakshmi et al. 2023). In anaerobic digestion, hydrolytic microorganisms depend on nitrogen and phosphorus to secrete extracellular enzymes that break down complex organic matter such as carbohydrates, proteins, and fats into simpler compounds. Further, acidogenic bacteria convert the simple organic compounds into volatile fatty acids, alcohols. The VFAs produced undergo acetogenesis with the help of microorganisms that convert them into acetic acid. At last, Methanogens utilize acetic acid, hydrogen, and carbon dioxide to produce methane but are sensitive to nutrient imbalances, particularly high ammonia levels (Manyi-Loh et al. 2013). This process is widely used of energy production and waste management.

5.2 Fermentation

Fermentation involves the conversion of organic matter into energy with the help of microorganisms in anaerobic conditions. In the presence of water-rich substrates like diluted organic waste or wastewater, fermentation processes can produce alcohols (ethanol), which can be used as fuel. In WtE, the main challenge is to efficiently extract energy from highly diluted solutions. For this, it is essential to develop highly efficient microbial strains or reactor designs that can handle large volumes of water with low organic content (Chng et al. 2019). In this process, the waste material undergoes pre-treatment such as physical, chemical of enzymatic to break the complex organic matter into simpler substances. After that, the microorganisms metabolize the simple sugars and convert them into alcohols. At the end of the fermentation process, the alcohol are separated and purified. The remaining spent waste can be used as a fertilizer.

The co-fermentation involves synthesis of bioproducts by using multiple substrates simultaneously. In wastewater treatment plants (WWTPs), waste-activated sludge (WAS) is used as the primary substrate along with food waste as a co-substrate. The interactions between the microbes work together to enhance the metabolic pathways, resulting in increased synthesis of valuable bioproducts. Co-fermentation depends on various key factors:

Optimizing pH: A slightly acidic pH (5.5–6.5) is favourable in order to maximize SCCA yields by finding the right balance between acidogenic and methanogenic activities.

Regulating Temperature: The ideal temperature for microbial growth and activity are within the range of 30–40 °C. However, temperatures above 45 °C can increase the rate of hydrolysis and fermentation.

Substrate Mixing Ratio: Nutrients and carbon ratio maintained by balancing WAS with food waste or other co-substrates promotes microbial synergy.

Lignocellulosic Biomass: Depending upon the biomolecular structure of the feedstock different conversion technologies is applied. Physicochemical pre-treatment is required to disrupt the lignin structure in lignocellulosic biomass followed by enzymatic saccharification to releases fermentable sugars. Through co-fermentation, these sugars can be converted into SCCAs or biofuels.

Food Waste: Food waste are rich in simple sugars and proteins. The addition of WAS during co-fermentation results in a nutrient-rich environment, which leads to increased microbial productivity (Awogbemi and Kallon 2022; Chandra and Mallick 2022).

Waste-Activated Sludge (***WAS***): It includes microbial biomass and complex organics thermal, chemical, or enzymatic pre-treatment can be applied to enhance its digestion and fermentation efficiency.

Industrial Wastewater: Due to varied concentration of industrial wastewater, so it is essential to pre-treat it and monitor the parameters such as pH, COD, and nutrient balance. This is necessary to prevent toxicity and ensure optimal takes place (Zabed et al. 2017).

5.3 Dark Fermentation

Dark fermentation focuses on producing hydrogen gas from organic substrates in anaerobic conditions in the absence of light. This process converts organic content present in wastewater directly into hydrogen. The efficiency of hydrogen production in dark fermentation can significantly enhance the energy recovery process from water-based waste streams (Ahmad et al. 2024).

Biochemical processes such as hydrolysis, acidogenesis, acetogenesis, and methanogenesis occur during conversion of an organic substrate. Organic molecules are hydrolyzed by hydrolytic bacteria into simpler compounds. Acidogenic bacteria further ferment these monomers to produce a combination of hydrogen (H_2) and low molecular weight volatile fatty acids (VFAs). H_2 generation can also be aided by the reversible conversion of CO_2 and H_2 into acetate by acetogens and homoacetogens. Ultimately, during the methanogenesis process, acetoclastic methanogens transform these organic acids into CO2 and methane. Acetogens typically thrive symbiotically with hydrogenotrophic methanogens, maintaining a low H_2 partial pressure to facilitate favourable conditions for interspecies H_2 transfer during acidogenesis. Consequently, methanogenic activity must be inhibited in order to prioritize H_2 as a significant metabolic by-product.

Biohydrogen production from MSW is significant for continuously enhancing H2 generation. The yield of biohydrogen depends on various factors, including reactor conditions, MSW components, and substrate types. It produces hydrogen from a variety of organic wastes, mainly through the breakdown of monosaccharides and glycerol. Carbohydrate-rich substrates tend to exhibit higher H_2 yields compared to other organic fractions, reflecting their suitability for dark fermentative H_2 production (Lin et al. 2012). Among the organic fractions of MSW, carbohydrates emerge as the primary substrate for dark fermentative H_2 production, outperforming proteins, lipids, and lignocellulosic materials in terms of H_2 production potential (Cheng et al. 2015). According to the literature, the composition of the organic fraction of MSW significantly influences H_2 potential, with carbohydrate-rich fractions showing high H_2 potentials while protein and lipid-rich fractions demonstrate lower potentials; while methane production is comparatively less affected (Alibardi and Cossu 2015). However, problems persist in dark fermentation for H_2 production from waste, primarily from low generation rates and yields.

5.4 Biological Hydrogen Production

Hydrogen act both as an energy carrier and a component in vital goods throughout the shift towards climate neutrality. Nevertheless, the most efficient and sustainable technology for producing hydrogen is not yet determined owing to difficulties in achieving high energy conversion efficiency, managing economic costs, and addressing the additional costs of reducing emissions (Li et al. 2024). Waste-derived fuels are renewable that also mitigate the environmental impacts associated with the treatment process. MSW has organic constituents such as include paper, cards, horticulture, and food waste that makes it a potential bioenergy production source (Meng et al. 2021). This method includes other biological pathway for hydrogen production, i.e. photo fermentation. In WtE systems, biological processes are a clean source of hydrogen production from water-based organic materials. The integration of this technology with wastewater treatment or organic waste management systems can lead to sustainable hydrogen fuel generation.

5.5 Photobiological Process

Photobiological processes involve photosynthetic microorganisms, such as blue-green algae and green algae, that produce hydrogen directly from sunlight and water. This process can be integrated into WtE systems where these microorganisms are cultivated in photobioreactors using wastewater. Not only this process will treat wastewater, but also produces a valuable fuel. These microorganisms use solar energy through photosynthesis and utilize specialized enzymes like hydrogenase and nitrogenase to generate hydrogen gas. The basic reaction for hydrogen production via direct biophotolysis is:

$$2H_2O + \text{photon} \rightarrow 2H_2O_2$$

Biodegradable municipal solid waste (BMSW) is a rich source of nutrient used for photosynthetic microorganisms cultivation. The organic compounds provide essential nutrients, promoting microbial growth and enhancing hydrogen production. This sustainable approach not only reduces waste but also contributes to clean biofuel production by converting waste streams into hydrogen (Li et al. 2024).

5.6 Ethanol Production

In the context of WtE, ethanol production is effective while using water-based biomass such as algal biomass or diluted organic waste. Fermenting of waste not only reduces the waste but also produces value added product such as renewable fuel. Advances in biotechnology that increase yields and fermentation efficiency in high-water-content systems are critical for making this process economical. Ethanol has historically been made from biomass sources including energy crops, sugarcane, and maize. But worries about land usage, food shortages, and the growing need for biofuels in transportation have spurred research into other feedstocks. With regard to waste management, CO_2 reduction, water quality and quantity control, land use, and biodiversity, using biodegradable municipal solid waste (BMSW) in place of traditional biomass for ethanol production has major environmental benefits. By turning trash into lucrative biofuel, BMSW not only helps keep garbage out of landfills but also lowers greenhouse gas emissions. This strategy also reduces the conflict between the production of food and fuel, enabling agricultural area to concentrate on food crops and improving biodiversity. Moving to BMSW as the main feedstock may therefore help to solve important sustainability concerns and advance the circular economy in the manufacture of biofuel (Lil and Khraisheh 2007).

6 Chemical Conversion Technologies

6.1 Hydrothermal Processing

In hydrothermal process (HTP), the solid waste is heated at high temperature range from 120 to 550 °C and pressure range from 20 to 150 bar in the presence of oxygen or inert nitrogen. Higher pressure is applied to keep the water in the liquid phase and high temperature convert the complex organic compound into simpler ones. Due to the use of water in the process, removal of moisture before processing is not required hence reducing the energy usage for the process compared to other processes such as pyrolysis (Zaman et al. 2017).

HTP is classified into 4 types: thermal hydrolysis (TH), wet oxidation (WO), hydrothermal carbonization (HC), and hydrothermal liquefaction (HTL). TH operates at moderate temperatures (approximately 120–200 °C) and pressures (around 20–150 bar) in a nitrogen-rich environment. It is a pre-treatment step that aims to dissolve raw materials like carbon and further prepare them for subsequent processes such as anaerobic digestion and bio-refining. Through TH, valuable chemicals can be produced, alongside the creation of natural fertilizers (Bonmatí et al. 2001). WO is an oxidative reaction occurred at elevated temperatures (typically around 150–300 °C) and pressures (ranging from 20 to 150 bar) in the presence of oxygen. It hydrolyses the raw materials into valuable chemicals such as organic acids, recovery of nitrogen and phosphate, and the reduction of solid content in raw materials (Hii et al. 2014; Munir et al.

2017). HTC is a non-oxidative hydrothermal process that function at a temperature ranging from 150 to 350 °C and pressures from 13 to 70 bar. It converts the biomass into a solid product which is abundant in carbon, known as bio-char. Additionally, hydrothermal carbonization can yield nanostructured carbon and organic acids (Funke and Ziegler 2010). HTL is also a non-oxidative hydrothermal process that function at temperature range from 250 to 550 °C and pressures from 50 to 250 bar. This method facilitates the conversion of biomass into a variety of products, including crude oil, sugars, amino acids, and fatty acids (Gollakota et al. 2018).

HTP converts solid waste into various products such as bio-char, bio-oil, and biogas depending on temperature, pressure, reaction time, and catalyst used in the process. HTP involves less energy, time consumption, and efficient as it eliminates the cost of waste drying (Kumar et al. 2018). The fuels produced by HTP have a density similar to that of fossil fuels except bioethanol. The fuel produced is almost near the purity level.

6.2 Plasma Gasification

Plasma gasification technique pyrolysis of the solid waste into syngas using AC and DC torches as a heat source (Mukherjee et al. 2020). The technology can use various wastes, such as MSW and hazardous waste (Sikarwar et al. 2016). Waste management, the plasma gasifier, gas cleaning, and conversion unit operations are some of the sequential steps involved in the plasma gasification of MSW. Following waste handling, the feed or MSW undergoes exposure to plasma heat in the presence of an oxidant. This process converts the waste into syngas, primarily composed of carbon monoxide and hydrogen, along with inert slag (Sanlisoy and Carpinlioglu 2017). In solid waste, fluidized bed plasma gasification is showing promise as a way to increase the efficacy and efficiency of gasification procedures (Du et al. 2018).

Plasma gasification reactions occur when the oxygen to carbon molecules ratio < 1. In oxygen-deprived conditions, some of the MSW undergoes partial oxidation, releasing heat that initiates most of the mentioned endothermic reactions. High-temperature plasma gasification promotes these endothermic reactions further (Arena 2012). Plasma gasification transforms nitrogen and sulphur in the feedstock into nitrogen and hydrogen sulphide, preventing the formation of nitrogen oxides (NOx) and sulphur dioxide (SO_2). Additionally, inorganic acid halides are produced from halogen feedstock.

Plasma activation yields higher amounts of syngas compared to conventional gasifiers. Operating at exceedingly high temperatures, typically exceeding 5000 °C, it effectively eliminates inorganic waste components by transforming them into inert vitrified slag, significantly reducing the leaching of toxic elements (Rajasekhar et al. 2015). Compared to incinerator ash in landfills, plasma gasification demonstrates significantly reduced toxicity in its slag leachate (Mountouris et al. 2003).

7 Bio-Electrochemical Process

7.1 Microbial Fuel Cells

Microbial fuel cells (MFCs) are bio-electrochemical devices designed to harness the metabolic capabilities of electrogenic microorganisms in order to generate electricity from various organic substrates. It encourages biohydrogen production using bacteria in aerobic and anaerobic conditions. Typically, two types of MFCs are utilized, featuring either one or two chambers. In the case of a two-chamber MFC, the anodic chamber operates under anaerobic conditions, where microorganisms degrade the organic substrate to generate electrons and protons (H^+). These hydrogen ions traverse a proton exchange membrane (PEM) to reach the cathodic chamber. Within the cathodic chamber, a solution containing electron acceptors such as oxygen is present, and they combine to form water (Khan et al. 2017). These acceptors react with the electrons from the anodic chamber via a copper wire and the H^+ ions from the anodic chamber via the PEM to complete the half-reaction (Beyene et al. 2018). Microbial fuel cells use a variety of organic matter such as domestic waste, animal waste, and sewage sludge as feedstock (Beyene et al. 2018).

7.2 Microbial Electrolysis Cell

The microbial electrolytic cell (MEC) is a potentially disruptive green technology that can both meet energy demands and combat global warming. By using electrochemically active bacteria, MEC efficiently converts MSW into hydrogen (H_2) and various valuable chemicals like methane, acetate, hydrogen peroxide, ethanol, and formic acid. To initiate the reaction at the electrodes, power source is required as since no oxidation is allowed at the cathode. Electrons produced are transferred to the cathode to reduce the protons and produce hydrogen. Compared to MFC, dark and photo fermentation, MEC has higher hydrogen recovery and wide substrate diversity (Rahimnejad et al. 2015). It has the capacity to convert a range of soluble organic substances into hydrogen or methane while simultaneously treating wastewater effectively (Beyene et al. 2018). According to Lu and Ren, (2016), MEC shows huge potential for waste biorefineries. MECs convert biodegradable waste into useful energy carriers and bioproducts, possibly making the system carbon–neutral and energy-positive. Integration with the fermentation process enhances the yield and rate of MEC. Additionally, advancements in materials, reactor designs, and costs contribute to reduced expenses and improved system efficiency. Experimental findings deepen our understanding of the synergistic and competitive relationships among diverse microorganism groups. Therefore, techniques for optimizing and using systems maximize syntrophic interactions or reduce energy and product losses can be developed (Beneyene et al. 2018).

8 Landfilling Process

The landfilling process basically involves the disposal of waste generated on land with minimum space and decreased to a lower permissible limit by biogas recovery and leachate management. This is primarily practiced in developing nations. However, it poses a severe environmental threat such as land pollution, water pollution, and air pollution (Chen et al. 2019). (Kumar and Chakrabarti 2010) reported that wastes are disposed of in the low-lying areas of developing countries, causing land degradation, environmental impact, health impact, and groundwater contamination. Nevertheless, developed nations have begun discouraging the practice of landfilling waste by implementing strict regulations and promoting waste reduction and recycling initiatives.

A landfill undergoes two distinct phases: the operational stage, during which municipal solid waste (MSW) is deposited, and biogas generation begins, and the closed stage when the volume of waste reaches its peak and biogas production diminishes. When municipal solid waste (MSW) is disposed off in a landfill, its organic content undergoes biochemical degradation through anaerobic digestion, which unfolds in three successive stages, resulting in the generation of methane in the form of biogas. Initially, bacteria convert carbohydrates, fats, proteins, and other insoluble fibrous materials into soluble substances during liquefaction. Then specific bacteria transform these soluble organic materials into volatile acids. Finally, in the biogas production stage, methanogens convert these VFAs into biogas. Beyene et al. (2018) reported that Methane (50–75%), ammonia (0.1–1%), nitrogen (3.9–4.1%), oxygen (0.9–1.1%), non-methane volatile organic compounds (0.01–0.60%), carbon dioxide (25–50%), and water vapour (6–75%) are the biogas components. Majorly, methane and carbon dioxide produced from landfill sites are called land filled gases (LFG), which are effected by temperature, pH, moisture content, age, availability of organic components that decompose, and composition of waste (Fazeli et al. 2016). LFG is used as a energy sources in internal combustion engines, gas turbines, and steam boilers to facilitate production of electricity or heat (Scarlat et al. 2015). However, the breakdown of solid waste on landfilling resulted in the production of methane gas, which is a potent greenhouse gas (GHG) than carbon dioxide. According to literature, different technologies used for solid waste conversion are shown in (Table 1).

9 Environment and Techno-Economic Assessment

The functioning of a process is possible with the involvement of a particular device in the system. Life cycle assessment of the systems such as pyrolysis, gasification, incineration, landfill used for waste conversion has been extensively used in past decades for studying its environmental impacts. For life cycle assessment, all the input raw materials, energy, and output, such as emissions and products formed, are identified and quantified. The effectiveness and reliability of solid waste LCA modelling depend on comprehensive and accurate data (Diaz and Warith 2006; Feo and Malvano 2009). The choice and completeness of data sources significantly impact the outcomes of waste-to-energy process analyses. For analysing the environmental impact of WtE technologies, various LCA software such as SimaPro (Turconi et al. 2011), GaBi (Margallo et al. 2014), Umberto (Kulczycka et al. 2015), and EASEWASTE (Kirkeby et al. 2007) are have been reported in literature. It consists of Ecoinvent database comprising more than 2500 processes (Frischknecht and Rebitzer 2005). Also, Open LCA (an open-source LCA software) is also extensively used for life cycle assessments (Mukherjee et al. 2020). The diverse range of outcomes observed in existing LCA studies of waste-to-energy technologies underscores the importance of gaining a thorough understanding of the particular systems analysed in these studies, along with the LCA modelling principles employed in each case.

The techno-economic assessment (TEA) of a process depends on the function, maintenance, and cost involved in it. Advance Simulator for Process Engineering (Aspen Plus), which is based on mass and energy balance, is used of estimation of technology cost. It consists of extensive database having diverse stream properties essential for modelling material flows within a gasification plant. Users

Table 1 Shows different technologies used for solid waste conversion

Waste	Technology	Parameters	Energy efficiency	References
Food waste	Anaerobic digestion	Moisture content of 70–90%, volatile solids to total solids ratio (VS/TS) of 85–95%, and carbon to nitrogen ratio (C/N) of 9–36.4	45 GWh	Nguyen et al. (2014)
Municipal solid waste	Microbial fuel cells	Feb batch mode, Temperature- 35 °C, current density 355.4 mA/m^2	Power density production 47.6 mW/m^2 at 100 Ω external resistor	Karluvali et al. (2015)
Solid waste activated sludge	Microbial electrolysis	–	Maximum hydrogen yields reached to 4.66 ± 1.90 mg-H2/g VSS	Sun et al. (2014)
Municipal solid waste	Hydrothermal processing	Saturated steam at about 2 MPa for about 60 min holding time	Solid fuel has better uniformity, higher density, and better drying performance	(Yoshikawa and Prawisudha 2014)
Municipal solid waste	Pyrolysis and gasification	Reaction temperature- and moisture content-0.4 temperature- 650 °C	Energy efficiency- 68.5%	Dong et al. (2016)
Municipal solid waste and slaughter-house waste	Dark fermentation	Temperature- 35 °C, HRT- 3 abd 5 d	Hydrogen yield- 52.5–71.3 N L kg^{-1} VSrem	Gómez et al. (2006)

can include this database with proprietary data for more accuracy of the process. For more precise calculations, FORTRAN is devised. In literature, various TEA models based on biomass WtE processes were developed using Aspen Plus (Fivga and Dimitriou 2018; Gao et al. 2018).

10 Discussion

The transition from traditional waste management practices to advanced waste-to-energy (WtE) technologies represents a pivotal shift in addressing global sustainability challenges. This chapter highlights the critical role of WtE in not only managing municipal solid waste (MSW) but also contributing to energy generation and reducing environmental pollution. The findings emphasize the importance of aligning technology selection with waste characteristics, environmental regulations, and economic feasibility to maximize the benefits of waste valorization.

One of the key takeaways is the significant variability in MSW composition across regions, influenced by economic development, lifestyle, and population density. For instance, the higher organic content in developing nations provides opportunities for biological processes like anaerobic digestion, while the low moisture and high calorific value of waste in developed countries favour thermal processes like incineration and gasification. Understanding these regional differences is crucial for tailoring WtE technologies to specific contexts, ensuring optimal performance and environmental benefits.

While thermal conversion technologies such as incineration and pyrolysis are widely implemented due to their energy efficiency and scalability, they face challenges related to emissions control and high capital costs. Emerging biological technologies, such as microbial fuel cells and microbial electrolysis cells, offer promise for cleaner energy production but require further research to enhance efficiency and reduce costs. Similarly, chemical methods like hydrothermal processing show potential for handling wet waste streams but need optimization for widespread application.

The integration of advanced technologies such as plasma gasification and photo-biological systems highlights the future direction of WtE. These innovations can improve resource recovery rates and reduce environmental impacts but require significant investments in research and infrastructure. Furthermore, the application of life cycle assessment (LCA) and techno-economic analysis (TEA) as decision-making tools can help identify the most sustainable and cost-effective approaches for different waste types and scenarios.

11 Conclusion

There is a critical challenge associated with the management and conversion of municipal solid waste (MSW) globally. Around 2.1 billion tonnes of MSW is generated annually which is estimated to increase to 3.88 billion tonnes by 2050. For effective treatment of the waste, it is essential to estimate its composition and characterization. Developed nations generated waste is of high calorific value whereas developing nation generate waste has high organic content. For conversion of these wastes to energy, several technologies such thermal conversion (incineration, pyrolysis, and gasification), biological conversion (anaerobic digestion and fermentation), chemical conversion (hydrothermal processing and plasma gasification), and bio-electrochemical processes (microbial fuel cells and microbial electrolysis cells) have been reported. Each technology has distinct advantages and challenges, from energy recovery to environmental impact. For evaluation of environmental impacts and economical viability of the conversion technologies, life cycle assessment (LCA) and techno-economic assessment (TEA) are crucial methodologies that helps in making decision towards sustainable solutions. For sustainable management of MSW, a comprehensive approach of integrating various technologies specific to generated waste compositions and prioritizing resource recovery, energy efficiency, and environmental protection is need of an hour.

Acknowledgements The authors are grateful to the Director of the institute for his kind support during manuscript preparation.

References

Agaton CB, Guno CS, Villanueva RO, Villanueva RO (2020) Economic analysis of waste-to-energy investment in the Philippines: a real options approach. Appl Energy 275:115265. https://doi.org/10.1016/j.apenergy.2020.115265

Ahmed MM, Hossan MN, Masud MH (2024) Prospect of waste-to-energy technologies in selected regions of lower and lower-middle-income countries of the world. J Clean Prod 450:42006

Ahmad A, Rambabu K, Hasan SW, Show PL, Banat F (2024) Biohydrogen production through dark fermentation: Recent trends and advances in transition to a circular bioeconomy. Int J Hydrogen Energy 52:335–357. https://doi.org/10.1016/j.ijhydene.2023.05.161

Aleluia J, Ferrão P (2016) Characterization of urban waste management practices in developing Asian countries: a new analytical framework based on waste characteristics and urban dimension. Waste Manag. https://doi.org/10.1016/j.wasman.2016.05.008

Alibardi L, Cossu R (2015) Composition variability of the organic fraction of municipal solid waste and effects on hydrogen and methane production potentials. Waste Manag. https://doi.org/10.1016/j.wasman.2014.11.019

Anderson E, Addy M, Ma H, Chen P, Ruan R (2016) Economic screening of renewable energy technologies: incineration,

anaerobic digestion, and biodiesel as applied to waste water scum. Bioresour Technol. https://doi.org/10.1016/j.biortech.2016.09.076
Arena U (2012) Process and technological aspects of municipal solid waste gasification. A Review Waste Manag. https://doi.org/10.1016/j.wasman.2011.09.025
Atwadkar R, Jadhav L, Wagh M, Shinde N (2014) A multi criteria ranking of different technologies for the waste to energy of municipal solid waste in the city of Kolhapur. Int J Emerg Technol Adv Eng
Awogbemi O, Kallon DVV (2022) Biofuel generation from potato peel waste : current state and prospects
Beyene HD, Werkneh AA, Ambaye TG (2018) Current updates on waste to energy (WtE) technologies: a review. Renew Energy Focus. https://doi.org/10.1016/j.ref.2017.11.001
Beylot A, Villeneuve J, Bellenfant G (2013) Life Cycle Assessment of landfill biogas management: sensitivity to diffuse and combustion air emissions. Waste Manag. https://doi.org/10.1016/j.wasman.2012.08.017
Bonmatí A, Flotats X, Mateu L, Campos E (2001) Study of thermal hydrolysis as a pretreatment to mesophilic anaerobic digestion of pig slurry. Water Sci Technol. https://doi.org/10.2166/wst.2001.0193
Chandra N, Mallick N (2022) A synergistic "waste-to-wealth" approach towards a cyanobacterial biorefinery via valorizing potato peels for the cultivation of marine Synechococcus elongatus. Biomass Convers Biorefinery. https://doi.org/10.1007/s13399-022-03281-8
Chen G, Sun Y, Xu Z, Shan X, Chen Z (2019) Assessment of shallow groundwater contamination resulting from a municipal solid waste landfill-a case study in Lianyungang. China Water (Switzerland). https://doi.org/10.3390/w11122496
Cheng J, Ding L, Xia A, Lin R, Li Y, Zhou J, Cen K (2015) Hydrogen production using amino acids obtained by protein degradation in waste biomass by combined dark-and photo-fermentation. Bioresour Technol. https://doi.org/10.1016/j.biortech.2014.11.109
Chng LM, Teo KSK, Juinn D, Chan C, Lee KT, Toh PY, Rahman A (2019) Fermentation of microalgae biomass through mild acid pretreatment for bioethanol. Production 2:15–22
Diaz R, Warith M (2006) Life-cycle assessment of municipal solid wastes: development of the WASTED model. Waste Manag. https://doi.org/10.1016/j.wasman.2005.05.007
Dong J, Chi Y, Tang Y, Ni M, Nzihou A, Weiss-Hortala E, Huang Q (2016) Effect of operating parameters and moisture content on municipal solid waste pyrolysis and gasification. Energy Fuels. https://doi.org/10.1021/acs.energyfuels.6b00042
Du C, Qiu R, Ruan J (2018) Scientific and industrial application of plasma fluidized bed. Adv Top Sci Technol China. https://doi.org/10.1007/978-981-10-5819-6_8
Farooq A, Haputta P, Silalertruksa T, Gheewala SH (2021) A framework for the selection of suitable waste to energy technologies for a sustainable municipal solid waste management system. Front Sustain. https://doi.org/10.3389/frsus.2021.681690
Fazeli A, Bakhtvar F, Jahanshaloo L, Che Sidik NA, Bayat AE (2016) Malaysia's stand on municipal solid waste conversion to energy: a review. Renew Sustain Energy Rev 58:1007–1016. https://doi.org/10.1016/j.rser.2015.12.270
Feo GD, Malvano C (2009) The use of LCA in selecting the best MSW management system. Waste Manag. https://doi.org/10.1016/j.wasman.2008.12.021
Fivga A, Dimitriou I (2018) Pyrolysis of plastic waste for production of heavy fuel substitute: a techno-economic assessment. Energy. https://doi.org/10.1016/j.energy.2018.02.094
Frischknecht R, Rebitzer G (2005) The ecoinvent database system: a comprehensive web-based LCA database. J Clean Prod. https://doi.org/10.1016/j.jclepro.2005.05.002
Funke A, Ziegler F (2010) Hydrothermal carbonization of biomass: a summary and discussion of chemical mechanisms for process engineering. Biofuels, Bioprod Biorefining. https://doi.org/10.1002/bbb.198
Gao W, Wang S, Li DX, Liu JB, Farahini MR, Huo Y, Imran M, Doranehgard M (2018) Techno-economic evaluation of biomass-to-synthesis gas (BtS) based on gasification. Energy Sources, Part B Econ Plan Policy. https://doi.org/10.1080/15567249.2016.1241839
Gedam VV, Regupathi I (2012) Pyrolysis of municipal solid waste for syngas production by microwave irradiation. Nat Resour Res. https://doi.org/10.1007/s11053-011-9161-1
Gollakota ARK, Kishore N, Gu S (2018) A review on hydrothermal liquefaction of biomass. Renew Sustain Energy Rev. https://doi.org/10.1016/j.rser.2017.05.178
Gómez X, Morán A, Cuetos MJ, Sánchez ME (2006) The production of hydrogen by dark fermentation of municipal solid wastes and slaughterhouse waste: a two-phase process. J Power Sources. https://doi.org/10.1016/j.jpowsour.2006.01.006
Hii K, Baroutian S, Parthasarathy R, Gapes DJ, Eshtiaghi N (2014) A review of wet air oxidation and Thermal hydrolysis technologies in sludge treatment. Bioresour Technol. https://doi.org/10.1016/j.biortech.2013.12.066
Jurczyk M, Mikus M, Dziedzic K (2016) Flue gas cleaning in municipal waste- to-energy plants—part I. Infrastruct Ecol Rural Areas. https://doi.org/10.14597/infraeco.2016.4.1.086
Kalyani KA, Pandey KK (2014) Waste to energy status in India: a short review. Renew Sustain Energy Rev. https://doi.org/10.1016/j.rser.2013.11.020
Karluvali A, Köroğlu EO, Manav N, Çetinkaya AY, Özkaya B (2015) Electricity generation from organic fraction of municipal solid wastes in tubular microbial fuel cell. Sep Purif Technol. https://doi.org/10.1016/j.seppur.2015.10.042
Khan MZ, Nizami AS, Rehan M, Ouda OKM, Sultana S, Ismail IM, Shahzad K (2017) Microbial electrolysis cells for hydrogen production and urban wastewater treatment: a case study of Saudi Arabia. Appl Energy 185:410-420
Kirkeby JT, Birgisdottir H, Bhander GS, Hauschild M, Christensen TH (2007) Modelling of environmental impacts of solid waste landfilling within the life-cycle analysis program EASEWASTE. Waste Manag. https://doi.org/10.1016/j.wasman.2006.06.017
Kulczycka J, Lelek Ł, Lewandowska A, Zarebska J (2015) Life cycle assessment of municipal solid waste management—comparison of results using different LCA models. Polish J Environ Stud. https://doi.org/10.15244/pjoes/26960
Kumar M, Olajire Oyedun A, Kumar A (2018) A review on the current status of various hydrothermal technologies on biomass feedstock. Renew Sustain Energy Rev. https://doi.org/10.1016/j.rser.2017.05.270
Kumar S, Chakrabarti T (2010) Effective municipal solid waste management in India. Waste Management. https://doi.org/10.5772/8459
Leme MMV, Rocha MH, Lora EES, Venturini OJ, Lopes BM, Ferreira CH (2014) Techno-economic analysis and environmental impact assessment of energy recovery from Municipal Solid Waste (MSW) in Brazil. Resour Conserv Recycl. https://doi.org/10.1016/j.resconrec.2014.03.003
Li Y, Lin R, O'Shea R, Thaore V, Wall D, Murphy JD (2024) A perspective on three sustainable hydrogen production technologies with a focus on technology readiness level, cost of production and life cycle environmental impacts. Heliyon 10:e26637. https://doi.org/10.1016/j.heliyon.2024.e26637
Lil A, Khraisheh M (2007) Municipal solid waste used as bioethanol sources and its related environmental impacts. Assoc Environ Heal Sci 23rd Annu Int Conf Soils, Sediments Water 13:167–171
Lin CY, Miyake J, Susilaningsih D (2012) The 2011 Asian bio-hydrogen and biorefinery symposium (2011ABBS). Int J Hydrogen Energy. https://doi.org/10.1016/j.ijhydene.2012.09.018

Lino FAM, Ismail KAR (2017) Incineration and recycling for MSW treatment: case study of Campinas, Brazil. Sustain. Cities Soc. https://doi.org/10.1016/j.scs.2017.09.028

Lombardi L, Carnevale E, Corti A (2015) A review of technologies and performances of thermal treatment systems for energy recovery from waste. Waste Manag. https://doi.org/10.1016/j.wasman.2014.11.010

Lu L, Ren ZJ (2016) Microbial electrolysis cells for waste biorefinery: a state of the art review. Bioresour Technol. https://doi.org/10.1016/j.biortech.2016.03.034

Manyi-Loh CE, Mamphweli SN, Meyer EL, Okoh AI, Makaka G, Simon M (2013) Microbial anaerobic digestion (bio-digesters) as an approach to the decontamination of animal wastes in pollution control and the generation of renewable energy. Int J Environ Res Public Health 10:4390–4417. https://doi.org/10.3390/ijerph10094390

Margallo M, Aldaco R, Irabien A, Carrillo V, Fischer M, Bala A, Fullana P (2014) Life cycle assessment modelling of waste-to-energy incineration in Spain and Portugal. Waste Manag Res. https://doi.org/10.1177/0734242X14536459

Meng F, Dornau A, Mcqueen Mason SJ, Thomas GH, Conradie A, McKechnie J (2021) Bioethanol from autoclaved municipal solid waste: Assessment of environmental and financial viability under policy contexts. Appl Energy 298:117118. https://doi.org/10.1016/j.apenergy.2021.117118

Mohee R, Mudhoo A (2012) Energy from biomass in Mauritius: overview of research and applications. Green Energy Technol. https://doi.org/10.1007/978-1-4471-2306-4_12

Mountouris A, Voutsas E, Magoulas K, Vardavoulias M, Tassios D (2003) Plasma waste treatment: process design and energy optimization. In: 8th International conference on environmental science and technology

Moya D, Aldás C, López G, Kaparaju P (2017) Municipal solid waste as a valuable renewable energy resource: a worldwide opportunity of energy recovery by using Waste-To-Energy Technologies. Energy Procedia. https://doi.org/10.1016/j.egypro.2017.09.618

Mujtaba MA, Munir A, Imran S, Nasir MK, Muhayyuddin MG, Javed A, Mehmood A, Habila MA, Fayaz H, Qazi A (2024) Evaluating sustainable municipal solid waste management scenarios: a multicriteria decision making approach. Heliyon 10:e25788. https://doi.org/10.1016/j.heliyon.2024.e25788

Mukherjee C, Denney J, Mbonimpa EG, Slagley J, Bhowmik R (2020) A review on municipal solid waste-to-energy trends in the USA. Renew Sustain Energy Rev. https://doi.org/10.1016/j.rser.2019.109512

Munir MT, Li B, Boiarkina I, Baroutian S, Yu W, Young BR (2017) Phosphate recovery from hydrothermally treated sewage sludge using struvite precipitation. Bioresour Technol. https://doi.org/10.1016/j.biortech.2017.04.129

Nguyen HH, Heaven S, Banks C (2014) Energy potential from the anaerobic digestion of food waste in municipal solid waste stream of urban areas in Vietnam. Int J Energy Environ Eng. https://doi.org/10.1007/s40095-014-0133-1

Patumsawad S, Cliffe KR (2002) Experimental study on fluidised bed combustion of high moisture municipal solid waste. Energy Convers Manag. https://doi.org/10.1016/S0196-8904(01)00179-0

Qazi WA, Abushammala MFM, Azam MH, Younes MK (2018) Waste-to-energy technologies: a literature review. J Solid Waste Technol Manag. https://doi.org/10.5276/JSWTM.2018.387

Rahimnejad M, Adhami A, Darvari S, Zirepour A, Oh S (2015) Microbial fuel cell as new technology for bioelectricity generation: a review. Alexandria Eng J

Rajasekhar M, Rao NV, Rao GC, Priyadarshini G, Kumar NJ (2015) Energy generation from municipal solid waste by innovative technologies—plasma gasification. Procedia Mater Sci. https://doi.org/10.1016/j.mspro.2015.06.094

Jadhav RDA, Dutta S, Sherpa KC, Jayaswal K, Saravanabhupathy S, Mohanty KT, Banerjee R, Kumar J, Rajak RC (2023) Chapter 7–Co-digestion processes of waste: Status and perspective. In: Hussain CM, Kushwaha A, Bharagava RN, Goswami L (Eds.), Bio-based materials and waste for energy generation and resource management, advanced zero waste tools. Elsevier, pp 207–241. https://doi.org/10.1016/B978-0-323-91149-8.00010-7

Ruth LA (1998) Energy from municipal solid waste: a comparison with coal combustion technology. Prog Energy Combust Sci. https://doi.org/10.1016/S0360-1285(98)00011-2

Sanlisoy A, Carpinlioglu MO (2017) A review on plasma gasification for solid waste disposal. Int J Hydrogen Energy. https://doi.org/10.1016/j.ijhydene.2016.06.008

Sanna A, Dri M, Hall MR, Maroto-Valer M (2012) Waste materials for carbon capture and storage by mineralisation (CCSM)–a UK perspective. Appl Energy. https://doi.org/10.1016/j.apenergy.2012.06.049

Scarlat N, Motola V, Dallemand JF, Monforti-Ferrario F, Mofor L (2015) Evaluation of energy potential of municipal solid waste from African urban areas. Renew Sustain Energy Rev. https://doi.org/10.1016/j.rser.2015.05.067

Shahnazari A, Rafiee M, Rohani A, Bhushan Nagar B, Ebrahiminik MA, Aghkhani MH (2020) Identification of effective factors to select energy recovery technologies from municipal solid waste using multi-criteria decision making (MCDM): a review of thermochemical technologies. Sustain Energy Technol Assessments. https://doi.org/10.1016/j.seta.2020.100737

Shareefdeen Z, Elkamel A, Tse S (2015) Review of current technologies used in municipal solid waste-to-energy facilities in Canada. Clean Technol Environ Policy. https://doi.org/10.1007/s10098-015-0904-2

Sharma KD, Jain S (2020) Municipal solid waste generation, composition, and management: the global scenario. Soc Responsib. J. https://doi.org/10.1108/SRJ-06-2019-0210

Shi H, Mahinpey N, Aqsha A, Silbermann R (2016) Characterization, thermochemical conversion studies, and heating value modeling of municipal solid waste. Waste Manag. https://doi.org/10.1016/j.wasman.2015.09.036

Sikarwar VS, Zhao M, Clough P, Yao J, Zhong X, Memon MZ, Shah N, Anthony EJ, Fennell PS (2016) An overview of advances in biomass gasification. Energy Environ Sci. https://doi.org/10.1039/c6ee00935b

Sun R, Xing D, Jia J, Liu Q, Zhou A, Bai S, Ren N (2014) Optimization of high-solid waste activated sludge concentration for hydrogen production in microbial electrolysis cells and microbial community diversity analysis. Int J Hydrogen Energy. https://doi.org/10.1016/j.ijhydene.2014.09.163

Tan ST, Hashim H, Lim JS, Ho WS, Lee CT, Yan J (2014) Energy and emissions benefits of renewable energy derived from municipal solid waste: analysis of a low carbon scenario in Malaysia. Appl Energy. https://doi.org/10.1016/j.apenergy.2014.06.003

Turconi R, Butera S, Boldrin A, Grosso M, Rigamonti L, Astrup T (2011) Life cycle assessment of waste incineration in Denmark and Italy using two LCA models. Waste Manag Res. https://doi.org/10.1177/0734242X11417489

Wilkie AC (2014) Biomethane from biomass, biowaste, and biofuels. Bioenergy, pp 195–205. https://doi.org/10.1128/9781555815547.ch16

Yadav P, Samadder SR (2017) A global prospective of income distribution and its effect on life cycle assessment of municipal solid waste management: a review. Environ Sci Pollut Res. https://doi.org/10.1007/s11356-017-8441-7

Yoshikawa K, Prawisudha P (2014) Hydrothermal treatment of municipal solid waste for producing solid fuel. https://doi.org/10.1007/978-3-642-54458-3_14

Zabed H, Sahu JN, Suely A, Boyce AN, Faruq G (2017) Bioethanol production from renewable sources: current perspectives and technological progress. Renew Sustain Energy Rev 71:475–501. https://doi.org/10.1016/j.rser.2016.12.076

Zaman CZ, Pal K, Yehye WA, Sagadevan S, Shah ST, Adebisi GA, Marliana E, Rafique RF, Johan RB (2017) Pyrolysis: a sustainable way to generate energy from waste. Pyrolysis. https://doi.org/10.5772/intechopen.69036

Zandalinas SI, Fritschi FB, Mittler R (2021) Global warming, climate change, and environmental pollution: recipe for a multifactorial stress combination disaster. Trends Plant Sci 26:588–599. https://doi.org/10.1016/j.tplants.2021.02.011

Zuberi MJS, Ali SF (2015) Greenhouse effect reduction by recovering energy from waste landfills in Pakistan. Renew Sustain Energy Rev. https://doi.org/10.1016/j.rser.2014.12.028

Socio-environmental Impacts of Sustainable Waste Management

Anjali Singh, Rishabh Joshi, Anju Patel, Sandhya Maurya, and Shaib Abdulazeez Shehu

Abstract

Millions of metric tons of waste are produced because of urbanization. Sustainable waste management is a product of circular economy concept that is beneficial to the society, environment, and economy. It solves the waste disposal issues with improved waste management practices and reduced human footprint on environment leading to clean air, water, and soil. It involves the monitoring, collection, transportation, valorization, and disposal of waste, such that environment and human health of future generations are not compromised. It protects the environment from the impacts of resource extraction, promote economical and environment friendly processes and thereby improving air and water quality along with reduction in greenhouse gas emissions, embarking on renewable energy sources like biogas, solve energy crisis to support huge population loads through the concept of waste-to-energy and cradle-to-cradle to render the environment waste free. The 4R system, reduce, reuse, recycle, and recover, helps in minimizing waste and environmental costs. In return, it boosts the economy and generate employment opportunities for the society, increasing per capita income, improve the overall human life by preventing health issues, increasing good and healthy lifestyle, ensuring surplus availability of resources to all the people, reducing food wastes, ensuring good food and water quality available to all, and providing a blissful environment to live in. Hence, an effective and integrated sustainable waste management system needs to be implemented and strictly enforced through environmental regulations. It requires active involvement of local government, Non-Government Organizations, residents, private sector, donor agencies, garbage collectors, and scavengers.

Keywords

Sustainable waste management · Social and environmental impacts · Waste to energy · Smart waste management

1 Introduction

Sustainable waste management aims to increase the useful life of resources by reducing the products from getting wasted and then burned or dumped in landfills. It can be used to reduce the detrimental effects of careless resource consumption on society, the environment, and the economy. A linear economy generates waste even before the products get manufactured.

Sustainable waste management is significant for maintaining circular economy. It is a systemic approach in comparison to a take-make-waste paradigm that contributes in economic development by keeping the expansion and consumption of limited resources at bay. It also helps in catering to the more important concerns of society characterized by linear economy apart from answering to the various challenges of waste management.

According to Environmental Protection Agency (EPA) reports, the largest portion of Municipal Solid Waste

A. Singh · R. Joshi (✉) · A. Patel · S. Maurya
Faculty of Civil Engineering, Shri Ramswaroop Memorial University, Barabanki, Uttar Pradesh, India
e-mail: rishabhjoshi.ce@srmu.ac.in

A. Singh
e-mail: anjalisingh.ce@srmu.ac.in

S. A. Shehu
Department of Civil and Mining Engineering, Confluence University of Science and Technology, Osara, Nigeria

School of Materials and Mineral Resources Engineering, Universiti Sains Malaysia, Nibong Tebal, Penang, Malaysia

A. Mandpe et al. (eds.), *Sustainable Waste Management*, Advances in Science, Technology & Innovation,
https://doi.org/10.1007/978-3-031-85685-3_18

(MSW) in United States is occupied by paper and paperboard items. However, paper is biodegradable and less harmful in comparison to plastics but they can cause environmental stress due to wasteful usage like its production requires rigorous deforestation and use of noteworthy amount of energy and water. However, recycling the paper using efficient recycling programs can help save 100 million tons of wood annually.

EPA suggests that 21.59% of MSW comprises food waste that has notable negative effects on the environment, society, and the economy. 40% of the food supply in the United States gets converted to waste. The irony is that while 10% of US households experienced food insecurity in 2019 but a large portion of edible food gets wasted (Bruno 2023). These waste materials are landfilled where they decompose to produce carbon dioxide and methane, which contribute the most to greenhouse gases and subsequently global warming. Sustainable waste management practices like composting can help in curbing such wasteful practices. Food production is hugely responsible for greenhouse gas emissions; freshwater withdrawals cause 25% of world emissions. This largely contributes to inefficient solid waste management. If the overproduction of waste can be managed at the source, then the collection systems can handle the sheer volume of waste created on any given day.

The third largest component of MSW is plastic that is currently the major havoc on the environment. The single-use plastic items, when dumped on the land and in the seas, jeopardize the linear economy because of their nondegradable property. Around 2.5 trillion dollars of resources are expected to be lost in extracting and creating plastics from petrochemicals. Single-use plastics should be reduced and recycled to sustainably manage the plastics.

The garbage is produced at an alarming rate annually on a worldwide scale. According to an estimate, 2.02 billion metric tons of MSW produced globally in 2016 is expected to increase to 3.4 billion metric tons by 2050 (Bruno 2023). Based on the collected data, a strong correlation shows that the top contributors are typically from important nations that have a significant economic impact on the world market. However, when looking at trash generation globally by region, the Middle East and North Africa generate the least waste around 6%, while East Asia and the Pacific generate the highest around 23% (Bruno 2023). If the existing regime is maintained, the projected trend will only become worse; therefore, a 70% increase in the world's annual garbage generation is anticipated. Ensuring the prosperity of the next generation necessitates the international community working together to address this issue (McVeigh et al. 2019).

The chapter is based on the concept about sustainable waste management and the social and environmental impacts of the same. However, to understand sustainability in waste management, it is necessary to discuss the trends of waste generation. Then this chapter discusses the concept of sustainable management of waste and social life cycle assessment, waste management hierarchy, and waste minimization strategies. The chapter further delves into details about various sustainable waste management techniques such as composting, recycling, landfilling, and incineration and the technological innovations in sustainable waste management, like waste-to-energy technologies and smart waste management system. Finally, this chapter focuses on the various socioeconomic factors influencing waste management and its impact on environmental health and society. The major objective of this chapter is to spread awareness about the various sustainable waste management techniques, its implementation, and its impact on the ecosystem with special reference to socioeconomic factors. This chapter also aims to spread awareness about the importance of sustainable waste management in current era.

2 Waste Generation Trends

2.1 Waste Generation in World

According to a World Bank analysis, there would be 3.4 billion tons of trash generated per year in 217 countries by 2050. Landfills continue to be the most popular and cost-effective method of disposing of MSW worldwide, although they also have the worst environmental effects (Michaels and Krishnan 2018). According to World Bank Energy Council report, the Municipal Solid Waste (MSW) generation is predicted to increase to 6 million tons of rubbish per day by 2025. The rising prices of fossil fuels and oil does not affect the ever-growing energy demand. The price is increasing because of their less availability due to depletion. Fossil fuel-based power plants have been developed due to increased energy needs, particularly electricity production. These thermal plants cause climate change and global warming since they release huge amounts of greenhouse gases in the atmosphere, especially carbon dioxide (Lipu et al. 2013).

The ineffective waste management practices widely across the world have been linked to numerous known health problems. Among the public health issues are the respiratory problems, bacterial infections, hemoglobin deficiency, allergies, nose, throat, sinuses, chronic lung disease, and reduced resistance to illnesses all contribute to the general public's poor health.

2.2 Waste Generation in India

According to a report published in 2017, India generated over 53 million metric tons of trash annually, or 1,45,626 tons each day. The nation is anticipated to have generated 71.15 million metric tons (MT) by 2021. By 2041, the estimated amount will be 160.96 million MT, by 2031 it will be 107.01 million MT, and by 2036 it will be 31.24 million MT and likewise, it is anticipated that 27 million MT of trash will be produced globally in 2050. According to research forecasts, India's total garbage creation is predicted to reach 0.7 kg per person by 2025 in comparison to 1999, there will likely be an increase that is at least four or six times greater. These results demonstrate the impact of growing populations and garbage production.

Public health has suffered because of the unfavorable pollution that has been brought about by this increase in trash output. India's traditional waste disposal methods, such as chaotic landfilling, mass burning, and careless dumping, have negatively affected human health and the ecosystem. However, the nation has acknowledged the regrettable outcomes and detrimental repercussions of such practices and has stepped up to support economical and environmentally appropriate alternatives for the discard of waste.

3 Sustainable Waste Management

3.1 Circular Economy (CE) in Sustainable Waste Management

As urban infrastructure development progresses worldwide, sustainability has become a buzzword and a pressing issue (Anastasiades et al. 2020). As a result, ecology is affected by rising pollution levels (Jhatial et al. 2020). Project finances and the environment are improved considerably due to sustainability. Therefore, sustainable development is defined as a guarantee that a project will meet the current generation's demands without jeopardizing future generations' needs (Anastasiades et al. 2020). The planet, people, and profit are the three pillars supporting sustainability ideals. It should be ensured that profits are maximized while development meets people's demands by utilizing limited resources, maintaining ecology, and maintaining good environmental conditions. Sustainable development goals are socially, environmentally, and economically feasible and manageable.

The concept of circular economy has developed from understanding the necessity to increase public awareness of the various threats that the exorbitant use of resources and subsequent waste production imposes on the environment. CE focuses on an innovative approach to address the frivolous exhaustion of natural resources. The CE concept was originally started with the induction of 3Rs, i.e., reduce, reuse, and recycle which developed into 4R structure emphasizing on the raw material reduction, waste reuse, recycling, and recovery (Anastasiades et al. 2020). In the CE, resources are refurbished and recycled to be utilized in other processes as opposed to the linear economy.

Therefore, reducing, reuse, recycle along with avoiding the surplus raw material purchase and changing the design specifications are the means to achieve circular economy. The recycling and reduce of waste can be effective in arresting the present global environmental deterioration.

3.2 Social Life Cycle Assessment (S-LCA) for Sustainable Waste Management

S-LCA helps evaluate the social impacts of the LCA, i.e., life cycle assessment of any product. The life cycle of a product includes the extraction of raw materials, its processing, market distribution, usage by the people, re-use, maintenance, recycling, and then final disposal on losing its productivity completely. S-LCA is one of the proxy indicators of sustainable development. It ensures that the people residing near the waste disposal sites and the workers involved in the waste handling are not negatively impacted because of the improper waste management practices. Since many marginalized, poor, lower caste, socially vulnerable society is at stake, their livelihood, health, standards of living, and recognition in society, all depends upon the waste management services (Mattos and Calmon 2023). Hence, a social assessment of the;ife cycle of a product is necessary to develop the policies to address people's issues and the ways to include the public in this process to ensure sustainable development.

4 Waste Management

4.1 Waste Management Hierarchy

Waste needs to be reduced as much as possible since it causes pollution in the air and water (Branz 2014). Hence, utilizing wastes to create energy is among the numerous approaches to enhance waste management. The availability of so many different factors need to be considered for a comprehensive and strategic comparison to select the most effective thermal MSW treatment technology among the many available options appears to be challenging regularly. An organization's life cycle evaluation or the region's development strategy are two fundamental efficiency trends in studying Municipal Solid Waste energy use provisions. For both industrialized and developing nations, this technology offers a practical and environmentally responsible way to generate renewable energy while also offering a viable answer to MSW management issues (Vukovic and Makogon 2022).

Gasification, incineration, and pyrolysis are the available methods that assist in waste management and burning-based techniques that are widely used but have raised serious problems in political, social, and environmental research, including refuse-derived fuel (RFD), anaerobic digestion, composting, bio-methanation, and landfill gas recovery.

Proper disposal techniques must preserve the harmony of ecology, technology, and economics. To achieve balance, continual research and development are needed to use cutting-edge methods, apparatus, and tools for the waste-to-energy (WTE) conversion (Vukovic and Makogon 2022). The percentage contribution of the global MSW treatment methods in waste treatment has been shown in Fig. 1.

The waste hierarchy is one of the fundamental ideas of a circular economy. It ranks different approaches to waste reduction according to importance, prioritizing the most workable and efficient strategies in cutting waste. This widely used hierarchy system starts at the most desirable waste-reducing approach, i.e., minimizing the waste at the source. This suggests cutting down waste from packing materials or finding better uses. The industrial units manufacturing these goods or other infrastructure are responsible for handling this issue. The second step in the hierarchy is reusing materials after their productive life is over. This requires a variety of contributions from organizations for manufacturing materials with a long lifespan or designing for deconstruction on a construction site (Pal and Bhatia 2022).

Recycling or composting is the next course of action. Material rendered useless are recycled by giving the resources a new appealing form for use in other applications. Recycling of organic matter through its breakdown into nutrient-rich soil with the help of microorganisms is

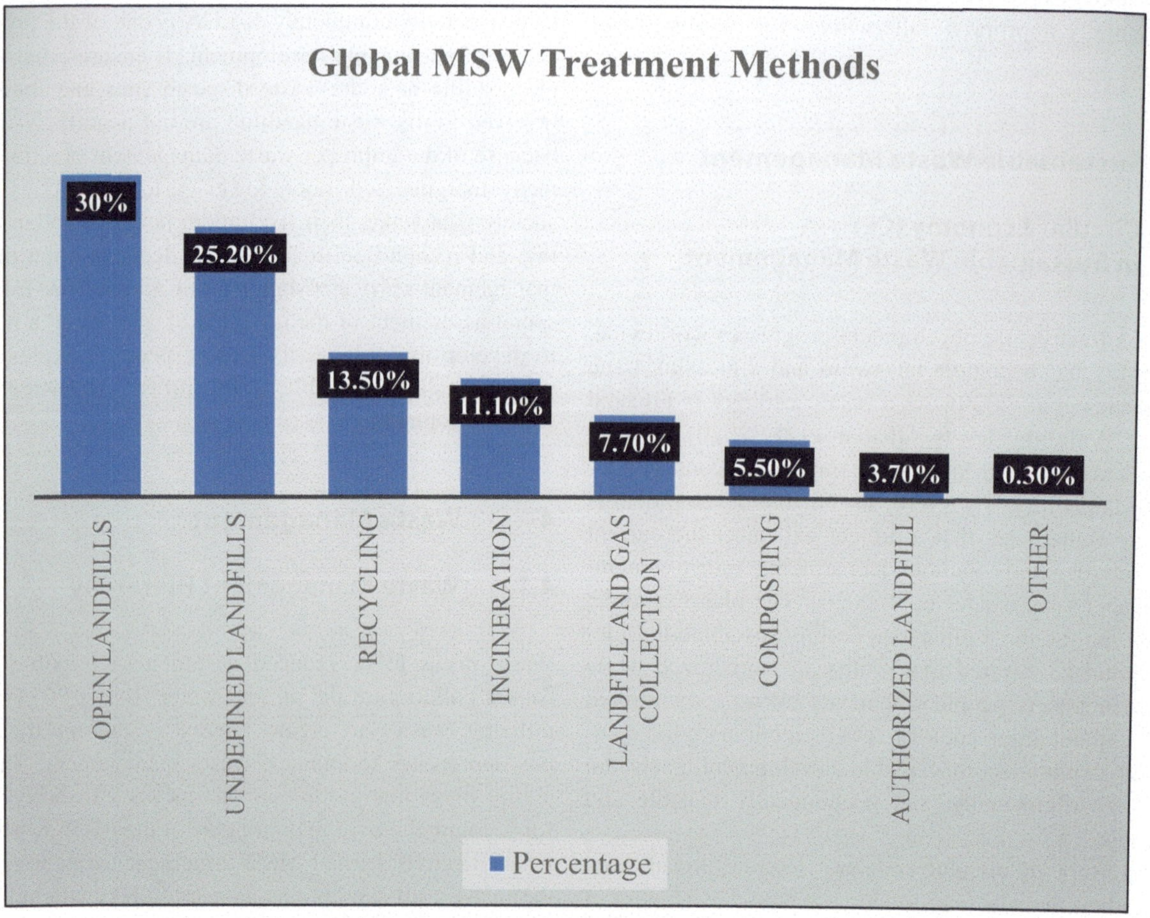

Fig. 1 Percentage contribution of different treatment techniques used for MSW waste at the global scale (Vukovic and Makogon 2022)

known as composting. Then recovery of useful trash is done like recovering precious metals from gadgets or burning municipal solid trash to generate energy. The last and least desirable option in this hierarchy is disposal. The rubbish that is no longer useful is disposed off (Gillespie 2015).

4.2 Waste Minimization Strategies

The four Rs of waste management—Reduce, Recycle, Recover, and Reuse—are used by modern trash management systems. This requires minimizing the waste generation, garbage recycling, recovering and reusing the materials that still has productive value. Waste minimization involves preventing garbage from being produced on building sites. On the other hand, recycling and reusing are linked to the effective and sustainable use of wastes in a workable way to incorporate sustainability in building projects. Recovering the materials from garbage that can be used otherwise instead of being destroyed. These methods greatly reduce the quantity of waste to be dumped in landfills, which helps to preserve the environment by lowering the quantity of physical and nonphysical resources needed for waste disposal while also lowering the problems related to the planet's ecology. Some ecology and construction professionals advocate prioritizing the minimization of waste generation on sites since it is better than developing unapproachable broad frameworks for waste treatment. Although reusing and recycling techniques ensure that the raw waste materials are utilized effectively before dumping them in trash, their scope does not cover preventing creation of waste in the first place. Still, they are significant contributors in cutting down the handling and disposal of huge quantities of waste. Although these are not ideal methods to decode the waste problem, organizations can take steps to incorporate waste minimization and recycling on site (Kazerooni Sadi et al. 2012).

5 Sustainable Waste Management Techniques

The following schematic diagram (Fig. 2) shows the various waste management techniques that utilize sustainability measures.

5.1 Composting

Composting is a biological waste management technique that utilizes naturally existing microorganisms, such as fungi and bacteria, to convert organic wastes into simpler forms. However, it reduces organic wastes to at least half but produces more carbon dioxide than other disposal techniques (Ayilara et al. 2020). It refers to the biological degradation of organic matter in a controlled environment to turn it into a stable humus. It can also be considered a form of recycling. It is an aerobic process in which the microorganisms need enough oxygen to convert the organic waste into humus, heat, and gas. However, this process can be both aerobic and anaerobic. It produces manure, a valuable soil conditioner for gardening and agriculture because of its nitrogen and phosphate content. The input for the composting process comes from kitchen waste, grocery stores, trimmed grass, and bushes. This can be easily carried out by common people in their households to recycle organic waste, which nourishes the garden and increases their yield when used by farmers in agricultural fields. Since this process has zero or very little side effects on the environment and produces great output, it is an important sustainable waste management process (Ayilara et al. 2020). This compost-enriched soil can help prevent pests and improve crop health. This makes it a very economical process, reduces the use of pesticides, and saves the environment. All the disease-causing organisms are removed and converted into nutrient-rich products. This technique is now extended to utilizing municipal solid waste and sewage sludge. It helps in restoring contaminated soils, controlling Odors, degrading volatile organic compounds, and managing storm water. This compost can improve the greenery in the city, and transportation costs are also reduced with a considerable reduction in the usage of pesticides and chemical fertilizers.

The solid waste needed for composting is first received, sorted, and separated; plastics, rubber, and leather are removed; the organic waste is shredded, and nutrients and moisture are added. The decomposition of these wastes is mostly carried out by aerobic micro-organisms. The windrow composting technique takes 6 to 12 months, depending on the environmental conditions and organic matter composition. The process results in methane production that can be collected in an engineered and controlled way and can further be utilized as energy input.

Composting is environmentally more friendly than anaerobic digestion and combustion (Seo et al. 2004). In some of the research, the environmental impact of composting was determined to be 2.4 times higher than incineration's (Seo et al. 2004). Composting creates several health problems, such as asthma, alveolitis, allergies, clogged noses, dry coughs, and sore throats. Since most waste materials are not biodegradable, composting has not been considered a principle in many prior waste management hierarchies.

Fig. 2 Waste management rechniques

5.2 Recycling

Recycling produces useful products after the collection, separation, and processing of the materials. It helps in reprocessing waste materials, products, or substances into reusable materials, products, or substances for the original purpose or alternative uses. The collection and separation of wasted materials and processing them to produce products find market value again. This process turns waste into valuable materials again (Kumar et al. 2024). The raw materials are separated and sent for further processing into new materials. It requires resources, energy, and funds to turn waste into usable materials. It prevents the emission of harmful greenhouse gases and pollutants into the environment, supplies raw materials for industries by enhancing their life cycle and usability, reduces landfilling and combustion of waste, resources are protected for future generations, reduces the extraction of raw materials, green technologies are promoted, helps generate jobs, etc.

5.3 Landfilling

Worldwide, most of the municipal solid waste is dumped outdoors or in filthy landfills. For instance, even during COVID-19 pandemic situation when most of the cities or states worldwide were showing positive results on the environment yet there is a rise in the quantity of waste getting landfilled because of decreased recycling rates in Brazil. Landfilling damages natural wildlife habitats and decimates them while affecting the public health in South America, Malaysia, and Johor. Urban and Nakada (Urban and Nakada 2021) indicate that improper handling of 35% of medical waste led to endangering public health and perhaps supporting the spread of COVID-19 according to research conducted in 30 cities in Brazil. Landfilling and open dumping can also lead to methane emission (CH_4), a significant greenhouse gas. Landfills and wastewater treatment plants are major contributors of about 17% of the world's methane emissions.

On a worldwide level, the annual methane emissions from landfills amount to around 29 metric tons, which is approximately 8% of the estimated global emissions. 1.3 metric tons of these emissions are released from Africa's landfills (Scarlat et al. 2015). Most of the gas production from landfills gradually increases due to the building up of MSW. The release of methane and ammonia gas causes health risks such as respiratory disorders. Methane also poses a risk of explosion and fire due to its high flammability.

Open disposal sites containing organic waste act as breeding grounds for flies, rodents, and mosquitoes—disease spreading vectors such as dengue, zika virus, and malaria. Furthermore, waterborne diseases such as diarrhoea, leptospirosis, intestinal worms, and hepatitis are spread through the leaching of these wastes to groundwaters and runoff into streams (Silva and Morais 2021).

The physical appearance and odors from landfill sites affect the lives of the people in the vicinity of living areas devaluing their properties, endangering their health and livelihoods (Ramachandra et al. 2018). Furthermore, ammonia emissions from these landfilling sites harm the plant leaves and the inherent composition of certain species. Additionally, the soil condition is also threatened and it is a potential source of dust in the nearby living areas.

Hot and humid rainy season affect air and water severely due to runoff, disease-carrying leachates, and the release of bad odors. The skin, nose, eyes, and respiratory systems are significantly affected due to interaction with high levels of CO_2, sulfur oxides and methane from landfill sites. Ammonia emissions can cause blindness also. In China, landfill methane recovery is less than 20%, but in Western countries, it can reach 60%.

Serious health risks are associated with landfill leachate that contaminates water sources used for drinking and other domestic purposes. For instance, a study in Pudong, China revealed that 160–180 m^3 of leachate was getting released form landfill sites in 2006 (Hong et al. 2006). Conversely, a well-protected design of the waste disposal facility can help avoid clogging of drainage systems, safeguard public health, prevent leachate migration, protection of significant environmental assets like air, surface and groundwater, wildlife, and humans. Furthermore, seasonal changes like the heating effect in summers can accelerate the rate of breakdown of biodegradable organic material by microorganisms and release offensive odors.

Disposal of medical and industrial waste along with municipal waste increases the risk of getting connected illness such as tetanus, chemical and radioactive threats, hepatitis B and C, and HIV infections. Furthermore, careless solid waste disposal can result in psychological issues, allergies, genetic disorders, gastrointestinal, respiratory, and dermatological diseases, as well as cholera, chest pains, skin, nose and eye irritation, and diarrhoea (Sankoh et al. 2013).

5.4 Incineration

In this technique, combustible waste materials are burned during the oxidation process, which yields heat, oxygen, and carbon dioxide. Due to airborne emissions of dangerous particles like volatile organic carbons, furans, carbon monoxide, dioxins, sulfur dioxide, hydrogen fluoride, and heavy metals, as well as the fact that it is the only method of disposing of hazardous waste that is acceptable for a variety of hazardous wastes like toxic, infectious, highly flammable, and volatile substances, this process is met with strong public opposition.

Open burning of MSW contributes significantly to air pollution and respiratory illnesses such as inflammation and infections of the nose, throat, and chest, as well as breathing problems, anemia, lowered immunity, asthma, and allergies. For instance, open burning produces over 22,000 tons of pollution a year in Mumbai (Kumar et al. 2017). In Thailand, Mongkolchaiarunya (Mongkolnchaiarunya 2005) documented smells coming from burning garbage and air pollution. Furthermore, burning plastic garbage releases dioxins and hydrochloric acid in harmful amounts that can lead to cancer, allergies, and hemoglobin shortage (Hong et al. 2006). Even for those who live distant from dumpsites, smoke from open burning and dumpsites contributes significantly to air pollution.

6 Technological Innovations in Sustainable Waste Management

Environmental pollution is one adverse consequence of economic progress that might counteract its beneficial effects on people. There are three categories of environmental contaminants, dependent upon their origin, i.e., soil, water, and air. Another factor contributing to pollution in the environment is improper waste management (Sasana et al. 2022). Poor management of solid waste causes serious environmental problems, slowed economic growth, and hampered health and overall lives of people. Hence, Municipal Solid Waste management is an essential part of human life and thus, it is imperative to create a well-thought-out infrastructure around MSW. As the urbanization, standards of living, incomes, and consumption rate rise, the amount of waste generation also rises along with them. The problem is getting worse and calls for a quicker response. The global sustainable development goals (SDGs) "Affordable and clean energy" (SDG 7) and "Sustainable cities and communities" (SDG 11) are supported by waste-to-energy programs and recycling measures Ternald (2020). The production of energy from municipal solid waste (MSW) provides communities with other sources of energy, particularly in limited natural energy supply areas. In metropolitan areas, sustainable development is promoted through the rise of waste-to-energy generation networks. Urban areas are encouraged to expand sustainably by the growth of waste-to-energy generation networks. The COVID-19 pandemic has made things more difficult for towns, which now need to preserve their MSW in addition to developing new waste treatment systems (Vukovic and Makogon 2022). Figure 3 shows the various technological innovations for managing waste sustainably.

6.1 Waste-To-Energy (WTE) Technologies

The waste-to-energy industry makes a substantial contribution to achieving the goals of waste management, along with the prevention and recycling strategies. These conversion technologies include landfills, thermal conversion methods, and biochemical conversion methods (Tozlu et al. 2016). These systems ultimately recover heat, solids, liquids, electricity, and fuel gasses. Combinations of two or more techniques may be used in real-world situations. WTE is expected to have strong growth and favorable policy backing from the government as traditional Power plants using fossil fuels face growing uncertainties in transition. The government is proactively supporting waste-to-energy initiatives by providing funding and incentives. The Government of India's Ministry of New and Renewable Energy (MNRE) offers projects that not only helps to reduce pollution but also produces electricity from garbage (Mani and Singh 2016). WTE is expected to witness vigorous expansion and government support for policy making in the future as there is a rising transition risk in the conventional fossil fuel power projects. It turns the nonrecyclable garbage into useful goods like transportation fuels, heat, and power as a part of waste management hierarchy. India is expected to meet its climate commitments easily. The Ministry additionally encourages the establishment of programs for the energy generation from garbage using all available technological means. Proper disposal techniques must preserve the harmony of ecology, technology, and economics. To achieve balance, continual research and development is needed to use cutting-edge methods, apparatus, and tools for the WTE.

These days, biomedical wastes are a potential disaster to public health due to their toxic environmental effects. Many technologies, such as autoclaving and incineration, are used to manage biomedical wastes. Waste is still regarded as refuse that has lost all its value, but with the aid of WTE technology, it was transformed into energy-producing heat source and is now used as a feedstock for similar methods. Thus, by turning garbage into valuable power, fuels, and fertilizers. WTE facilities use waste as feedstock not only

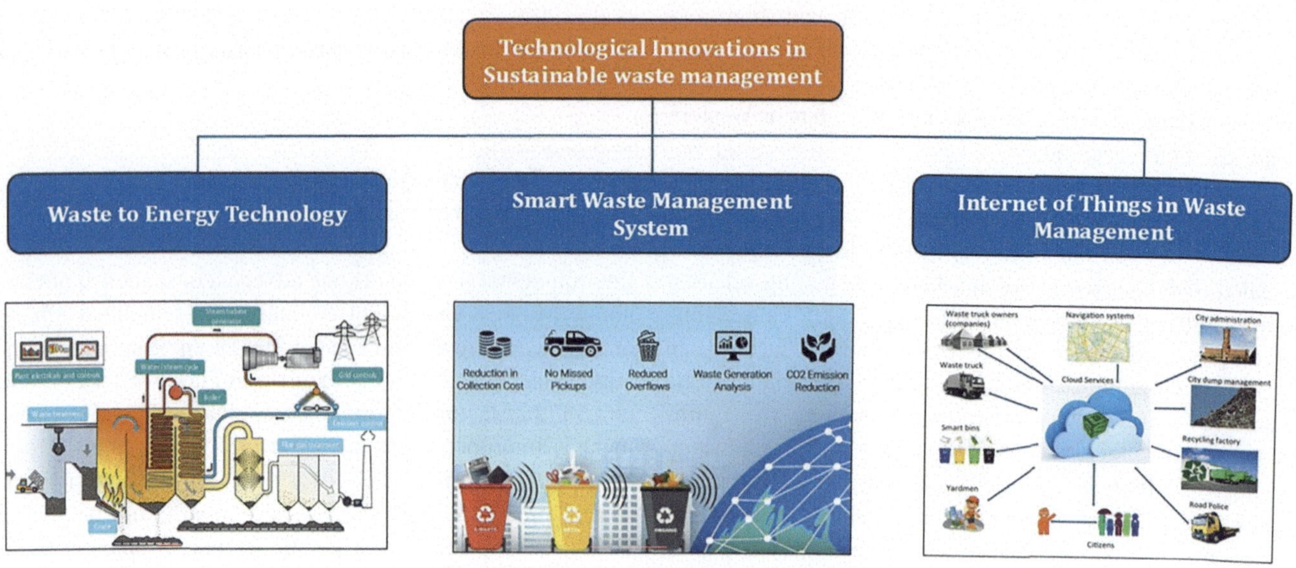

Fig. 3 Technological innovations in sustainable waste management

to reduce the space occupied by landfill and the waste disposal expenses and management, but also to make the waste useful. Energy is produced during the first stages of waste treatment, either as heat or electricity. Utilizing a variety of methods, including thermal, biochemical, thermochemical, and electrochemical conversion, WTE recovers this energy. These methods depend on various factors, including the quantity and quality of waste, the desired energy source, customer demands, national environmental ministry regulations, standards, and the financial status of the government or private businesses handling the WTE (Singh et al. 2011).

Two categories of technological approaches are thermochemical and biochemical.

6.1.1 Thermochemical Technologies

Thermochemical techniques comprise of incineration, pyrolysis, and thermal gasification. Anaerobic digestion, fermentation, gas-captured landfills, and microbial fuel cells are examples of biochemical processes. The esterification is the chemical process. These methods break the bonds between molecules in the waste materials and cause them to convert into smaller ones by applying sufficient heat energy to the materials inside a closed vessel. Then recombination is done which releases more energy than converting the molecular structure of waste by combining oxygen with freed carbon and hydrogen atoms from the previous molecular configuration (Qazi et al. 2018).

(a) **Incineration**

As it has already been discussed in previous section, incineration is a sustainable waste management technology, it is also an established method for energy recovery by burning the organic materials completely to lower the MSW volume. Furthermore, potential toxicity and infectious qualities of hazardous and medical waste is reduced (National Research Council 2000). The incineration processes like mass burning and combustion generates heat and power both. Thermal gasification process produces hydrogen, methane, and syngas. Pyrolysis produces gases, syngas, char, and aerosols. A significant amount of oxygen is needed for waste incineration to ensure full oxidation. In an incinerator, the temperature of combustion is typically around 850 °C, with CO_2 and H_2O being the primary byproducts. It is necessary to separate non-combustible materials from MSW prior to the incineration process because noncombustible materials, such as glass and metal etc. that are typically included in MSW cause solid slag (incinerator bottom ash) to form after incineration. As a result, requires greater energy input, such as an electromagnetic belt.

The four primary components of the incinerator process are the input system, exhaust combustion chamber, exhaust gas system, and residue disposal system. But a primary combustion chamber along with continuous feeder mechanism and incinerators with heat resistant lined moving grates is used in the updated system. However, for incineration to be a financially viable option 50,000 metric tons of combustible waste must be annually generated for MSW treatment through self-sustaining incinerator with supplemental fuel. Two principal byproducts of incineration of MSW are heat and ash. In steam turbines, the heat generated is often utilized to create steam for generating power (Atwadkar et al. 2014). The ashes that are produced, however, are either fly ash or bottom ash. The produced ash is regarded as a noteworthy by-product because it makes about 20–25% of the feedstock's overall weight. One-third to one-half of the total weight of ashes is made up of bottom ash, or slag. The majority of its composition consists of partially burnt and noncombustible trash, such as metals and glass, with fly ash making up only 10 to 25% of the overall weight of ash. The heavy metals and dioxins adhere to fly ash; hence, it is more poisonous than bottom ash.

As a result, cutting-edge technologies are necessary to regulate the fly ash emissions which raises the cost of construction and maintenance. Incineration induced bottom and fly ash can be used to make both landfills and secondary raw materials. Policies pertaining to environmental protection typically aim to minimize disposals in landfills in developed nations with strict environmental regulations and limited land. The anaerobic digestion process can be combined with incinerator such that the waste produced from anaerobic digestion can be processed in incinerators. The efficiency of the system is believed to rise from 32.6 to 63% using this combination. However, it would be more expensive considering the capital expenditures but with a shorter payback period. Merits of this technology are (Ouda and Raza 2014):

(i) Technology is using practically all waste kinds.
(ii) It works well in both urban and rural settings.
(iii) It has low capital requirement comparatively and also does not require much skilled labor as compared to other WTE technologies.
(iv) Excellent power generating efficiency and reduced emission rates because of Air pollution control systems in the incinerator that helps in adhering to current emission limitations.
(v) It is one of the advanced technologies when compared to others because incinerator plants may produce energy that can sometimes be used in place of power plants.
(vi) The incineration process generates bottom ash that can be utilized as an infilling material or inexpensive aggregate for building projects including roads, bridges, and motorways.

(vii) It can reduce waste volume by 80–95% and weight by about 70–80%, which greatly reduces the land area needed for trash disposal.

The demerits of incineration are as follows (Margallo et al. 2015):

(i) Release of some harmful gases like dioxins and furans in spite of the developed and modified model of the incinerators. This degree, nevertheless, cannot entirely be avoided.
(ii) The necessary equipment for pollution control and combustion changes to lower emissions raises construction and operational expenses.
(iii) Incineration also produces heavy metals which can release residues into the flue gas if not treated correctly and are extremely harmful. Therefore, to avert contamination of surface and ground water, such heavy metals and wastes must be properly disposed of at landfill sites that are managed efficiently.
(iv) Since incinerators need highly qualified personnel, it is frequently doubtful if waste in underdeveloped nations is suitable for internal combustion engines.
(v) Variability in trash over time could lead to issues with the incineration process' operation.

(b) **Gasification**

Gasification is a waste-to-energy technology that converts biomass into different types of energy using thermochemical process. It creates pure, functional syngas from solid and liquid components (Narnaware et al. 2017). This conversion involves the chemical reactions of carbon-based materials (feedstock) with a small amount of oxygen. This combination results into breaking of the organic components into simpler molecules forming syngas. The entire process produces gaseous products with a high yield of up to 85% at a temperature of about 800 °C, which are utilized directly. Gasification efficiency and yield are affected due to parameters such as reaction temperature, heating rate, moisture, feedstock particle, gasification agent, and technique. Product's heating value, gas yield, and chemical composition are majorly affected by temperature and gasification agent. A mixture of hydrogen and carbon monoxide, along with lesser contents of water, carbon dioxide, methane, and nitrogen combine to form syngas. This is used to generate electricity, fertilizers, vehicle fuel, and other valuable products. Gasifier feedstock comprises various waste materials such as tires, automotive shredded residue (ASR), black liquor, MSW, coal, biomass, refuse derived fuel (RDF), sludge, polyvinyl chloride (PVC), and refinery leftovers. But before the feedstock enters the gasifier, it must undergo pre-processing to eliminate any inorganic items that cannot be gasified, like metal, glass, impurities, and inert (Narnaware et al. 2017). Waste must be shredded into very fine particles that are consistent with uniform particle size. The food waste should be either removed or feedstock should be dried to perpetuate the moisture content less than 15% (Warren 2013).

The merits of the gasification are:

(i) The byproducts of incineration process like dioxins and furans are prevented from being created in the low oxygen atmosphere provided by gasification process.
(ii) It is a low-energy process that uses a small percentage of the stoichiometric oxygen.
(iii) An affordable and minor gas cleaning equipment is needed since the process gas volume is modest.
(iv) The pollutants present in the exhaust have a higher pressure due to small volume of gas making it easier to capture particles and fully absorb them.
(v) Gasification produces 8–12% ash, whereas incineration produces about 15–20% ash but still gasification reduces the waste volume by 90%.

The demerits of gasification are.

(i) It can result in issues with operations and the environment by releasing harmful substances into the generated syngas, such as tar, alkaline, heavy metals, and halogens.
(ii) Alkalis can destroy gas turbines during combustion and cause agglomerated fluidized beds used in a few gasification systems.
(iii) High molecular weight organic gases and tar create slagging or refractory surfaces causing harm to ceramic filters, metals, boilers, sulphur removal systems, and reforming catalysts.
(iv) Halogens because of their corrosive nature, can produce acid rain when they are released into the environment.
(v) Built-up of heavy metals released into the environment.

(c) **Pyrolysis**

Syngas, a mix of carbon monoxide, carbon dioxide, methane, and hydrogen along with oxygen less thermally decomposed products like char and oxygenated oils is produced through pyrolysis of carbonaceous materials (Náthia-Neves et al. 2018). It is possible to alter the composition of products since the heating rate, temperature, size of waste particles, and vapor resident time all affect the end products' composition. Higher temperatures during the pyrolysis process led to the development of syngas, whereas relatively lower temperatures produce liquid products like pyrolytic oils. Three varieties of pyrolysis are slow, flash, and rapid. These types of pyrolysis depend on the feedstock residence time, temperature, and heating rate.

In conventional pyrolysis, low rate of heating (0.1–2 °C/s) of the feedstock, 450–550 s of solid residence times, and approximately 550 °C of low temperatures are maintained. Tar and char are produced as the primary products. The main products of fast pyrolysis are Tar and bio-oils which occurs at moderate temperatures (586–977 °C), residence times of 0.5–10 s, and heating rates exceeding 2 °C/sec. In flash pyrolysis, strong heating rates of 200 to 105 °C/sec, temperatures between 777 and 1027 °C, and a very brief solid residence time are involved. The high-quality byproducts like ethylene-rich gases that are being created can be used to make gasoline and alcohol. However, because of the nature of its products, fast pyrolysis is widely used technology (Qazi et al. 2018).

Syngas produced from pyrolysis of waste can be burned to provide thermal energy, valuable materials, and steam to create power. The main feedstock for pyrolysis is RDF, MSW, tires, biomass, coal, ASR, sludge, polyvinyl chloride, wastewater, and chloride.

A steady supply of feed particles with roughly the same moisture content, size, and composition over an extended length of time is necessary for the pyrolysis process. Preprocessing, which includes a sorting line, shredder, and dryer, is required for the feedstock before it is fed into the pyrolysis system. All inorganic items, including metal, glass, impurities, and inert materials, need to be eliminated by the sorting line. While the remaining waste components are delivered to the shredder, where they are uniformly and small shredded into particles as tiny as 50 mm in diameter to create consistent waste. The feedstock is then delivered to a dryer to reduce the moisture content to 20% (Qazi et al. 2018). This can also be accomplished by eliminating food waste.

The merits of pyrolysis are (Qazi et al. 2018)

(i) Dangerous byproducts like furans and dioxins are prevented from being created due to the oxygen devoid.
(ii) It requires low working temperatures and no air with relative insensitivity to input materials resulting in less flue gas.
(iii) The syngas and pyrolytic oils generated from the MSW pyrolysis is utilized as fuel boiler directly, or as heat and power facilities to produce electricity and for higher-quality applications such as chemicals, post-refining adhesives, and motor fuels.
(iv) Volume reduction of waste: For example, pyrolysis treatment of MSW can reduce its volume by 70–90%.
(v) Better metal recovery is possible with lower pyrolysis process temperatures before the residuals reach the chamber with high temperatures.
(vi) Except plasma arc gasification, pyrolysis has the lowest pollution output and the highest environmental value.

The demerits of pyrolysis are (Qazi et al. 2018):

(i) Air pollutants, toxic ash, and hazardous residues are released during the pyrolysis process, which harms the ecosystem by producing pollution, depleting natural resources, and accelerating climate change.
(ii) High investment and operating expenses.
(iii) Use of syngas, causes tarring issues in high operational and capital pyrolysis.
(iv) Tar deposition can lead to operational difficulties, bottlenecks, inefficiencies, and failure of plants in pilot and commercial scale operations.
(v) Since pyrolysis is highly susceptible to air, a reaction chamber that is adequately sealed is necessary for a safe pyrolysis procedure.
(vi) Accidental air intrusions carry process disruptions and the risk of explosive responses.
(vii) Pyrolysis produces low energy and there are very few commercial pyrolysis facilities worldwide that take municipal solid waste (MSW); this is because the technology is not yet developed.
(viii) This technology is appropriate for developed countries due to the need for highly skilled individuals.

6.1.2 Biochemical Technologies

In this technology, the microorganisms are used to create a growth conducive environment, and further the trash can be broken down using biochemical technologies. Usually, the waste's organic component is broken down by this procedure and generating carbon-rich byproducts which gets changed into gaseous products like water molecules and other gases like CH_4. Biochemical reactions are typically affected by factors such as pH, oxygen availability, moisture content, and C/N ratio. This technology includes process like anaerobic digestion and fermentation.

(a) **Anaerobic digestion**

The process of converting biodegradable materials in an oxygen-free environment using bacteria is known as anaerobic digestion. This technique is carried out in specialized reactors with precisely controlled humidity, pH, and temperature levels. Under these circumstances, microorganisms can proliferate and accelerate the breakdown process to generate biogas, bioliquid, and fiber digestate.

This is an efficient and sustainable way for transforming waste (organic) to energy (Biogas). The organic wastes obtained from households, industries, agricultural activities, and animal manure are utilized. This technology offers good waste management, optimized energy consumption by communities, better ecology in rural regions, and reduced health issues caused by reduced pathogens through better sanitation and waste removal. Biogas comprises of carbon dioxide, methane, and some other trace gases. However, this composition varies depending on pH, organic loading, temperature, hydraulic retention time, digested substrate type, and digester design. The nutritional substrate composition activates the micro-organisms. Hence, the efficiency of the process depends upon the type of substrate (Náthia-Neves et al. 2018).

(b) **Fermentation**

The fermentation process entails the breakdown of biodegradable materials by microbes at a temperature of 35 to 40 degrees Celsius and oxygen less atmosphere by several biological mechanisms. Various metabolic process converts sugars into alcohols, acids, or gasses using yeast and bacteria. Along with this, lactic acids in oxygen-starved muscle cells go through this process. Two methods for the fermentation process to occur are batch and continuous. In the continuous reactor, continuous feedstock loading is done, and the products achieved through fermentation are discharged in a steady stream to maintain the reaction volume. A balance is maintained between the feed and outflow during the continuous fermentation process. In this operation, less vessel filling and cleaning are needed (Kasiński et al. 2024).

6.2 Smart Waste Management System

This is a system based on Internet of Things (IoT) in which a dustbin is deployed and smartly used for effective disposal. In this system, the Sensors are used to rectify the filling amount of garbage in the bins. If it gets full, an automated message is sent to the system's server via the Arduino SIM module, which uses the Arduino board's application and when the server receives the message, it gets transmitted to the worker in charge; if the worker is available, he would accept the assignment and travel to the needed location. In this way, smart dustbins information may be accessed anytime and anywhere which helps in making an informed decision. The implementation of this proposed system results in cost savings, resource optimization, and effective use of smart dustbins. This approach consequentially reduces the city traffic. Several other smart waste management systems are based on the on an infrared sensor, a CPU, and a Wi-Fi module have been proposed. The primary issues with the current solid waste collection and handling system like this procedure is a complex process, several factors need to be taken control off as every unit is linked to one another, and the execution process is higher as compared to other technologies (Kadus et al. 2020).

Intelligent waste management system is made possible by Internet of Things. Many organizations are directing their efforts in adopting IoT to reduce the impacts on the environment and society. The best way to implement this technology is to use sensors on garbage cans to maintain real time monitoring of waste and monitor the amount of waste produced by each individual and then altering the price of waste management appropriately. Chronological input-based data using real-time information about waste gathering can be used to do likelihood prediction using IoT based machine learning algorithms. Radio-frequency identification (RFID) tags and RFID readers can be used as an additional data transmission alternative and as a data storage medium. The scanners today have been programmed effectively to identify the type of waste based on irregularities, dimensions, shapes, and contamination level (Umer et al. 2022).

7 Socioeconomic Factors Influencing Waste Management

Municipal solid waste (MSW) has been diverted from landfills recently. These days, the prime goal of solid waste management (SWM) is to increase the quality and recovery performance of the materials gathered through recycling. However, it seems to be a convoluted process incorporating several environmental and socioeconomic factors. Waste Management is a complex issue that requires a solid understanding of social, economic, environmental, cultural, and technical issues (Wang and Darendeliler 2013) One of the key measures to ensure sustainable development is the conversion of linear economy to circular economy (Velis et al. 2023). Poor countries that are environmentally developing still face the havoc of mismanaged Municipal Solid Waste. The socioeconomic condition of any country decides the performance of sustainable waste management and the development of the circular economy. Solid waste management is perceived as an indicator of urban governance (Velis et al. 2023).

The significance of social indicators that involve members of society is highlighted for the assessment of social sustainability in technology, and a method for presenting the findings in conjunction with ecological and economic indicators is proposed. A focus on sustainable SWM is made (Wong et al. 2016). In particular, it has long been known that socioeconomic status affects MSW amount and composition when fabricating an efficient SWM plan for a city. To determine solid waste formation mechanism and predict future dynamics in the sector, a thorough analysis of the factors influencing recycling rate and solid waste production is thought to be essential (Grazhdani 2016). The association between various socioeconomic characteristics and the quantity and content of generated garbage has been assessed using statistical approaches. The significance of accurate and pertinent data has been emphasized in several studies. Socioeconomic factors such as family size, income, and education level should be considered when evaluating solid waste production (Khan et al. 2016). Furthermore, the amount and makeup of garbage can be influenced by various factors, such as living habits, social attitudes, seasonal changes, religious and cultural beliefs, and even peculiarities unique to a particular area.

The Waste management system faces significant obstacles due to effective citizens' engagement in recycling programs, which seem essential for meeting household recycling goals. However, finding acceptable land locations for recycling containers is consistently hampered as people do not want them established in their back yard (Byrne and O'Regan 2014). Even though people generally approve of SWM initiatives, they are against SWM facilities being built near their homes. It is important to address the factors influencing attitudes toward participation. Answers to questions on what kinds of items homeowners are willing to recycle, how many recycling bins they need at their premises, and whether to impose financial incentives can provide valuable insights. Understanding the attitudes and perspectives of various social, economic, and cultural groups toward recycling can be aided by studying the procedures, practices, and regulations surrounding recycling at the local level and the challenges experienced during the recycling process. Community-based involvement in education and awareness-raising campaigns is necessary to encourage people to be a part of home recycling programs (Kotzé 2015) actively. Despite efforts to implement integrated SWM in elementary schools, there appears to be a lack of effective planning that documents the performance and development of recycling programs. As a result, it is essential to assess current programs and determine the stages needed to create a successful program. Creating a recycling framework for the execution in an elementary school can yield useful methods and suggestions (Ward et al. 2014).

Social life cycle assessment can help assess all the waste management process to help identify the social factors affecting the process and also its impact on the process, so that decision makers can make an informed decision (Mattos and Calmon 2023). Similarly economic analysis of the process helps in assessing the economic factors and impacts on the waste management and hence help in bringing out proper policies for efficient waste management system.

7.1 Social Factors Affecting Waste Management

Long-term sustainability of any waste management project requires an understanding of the social dimensions involved in the procedure (Kotzé 2015). Hence, social assessment of such projects is done to identify the social risks and impacts and propose interventions and designs to maximize the benefits and remove any harmful effects. Several important social factors influence waste management listed as follows (Kotzé 2015):

(a) **Health and Safety of workers**—Open dumping of waste, contaminated sites, and poor hygiene measures for waste management workers lead to exposure to toxic materials and vermins, affecting their health. Hence, proper social assessments must be done to understand these threats faced by informal and formal waste management workers, proper health and safety regulations for these people, and the availability of insurance and safety for coping with such diseases.

(b) **Urban Health and Public Participation**—It is necessary to understand that waste collection and disposal can influence the health of the urban environment. The selection of disposal sites or landfill sites is a big social issue because no one wants to live near these sites. Proper public consultation should be done to ensure that the local communities disseminate the information about the sites, comparative analysis of alternate disposal sites, the reason for the selection of a particular site transparently, grievance redressal mechanism is not just a formality but practicality, and the adverse environmental impacts of the disposal sites. Public participation should be mandatory to maintain transparency at the local, community, and federal levels. The people should be able to address their issues and be compensated adequately by providing gardens, schools, and health centers in addition to such waste disposal sites.

(c) **Lack of willingness to pay**—The access to poor people or marginalized communities is mainly because of the incurring cost of waste management services. Data should be gathered to analyze the ability of people to pay for these services, their willingness to pay, and their preferences.

(d) **Gender issues**—Any waste management project has the potential to affect men and women differently. It is necessary to address the issues of differentiation imposed on women in different waste management activities, on waste collection sites, the difference in pay, workload, employment opportunities, safety, livelihood, health, and knowledge.

(e) **Opportunities and Services for poor people**—Waste management is just a dream for people living in slums near the cities. The people of such areas face differentiation at the hands of waste management organizations. Many such groups are intentionally excluded from waste management services because they do not have a political voice and organizational structure and mostly because of the discrimination of other people in the society with higher stake holding. The opportunities for employment in this sector is highly disproportionate since it excludes the vulnerable population from any access to these opportunities, The workers of waste management system need thorough attention for their upliftment. The poor people, low caste, minorities, or immigrants living near the waste disposal sites often face critical health hazards increasing the risk of mortality, disability, and morbidity.

(f) **Marginalized waste pickers**—The recyclable waste at disposal sites, or roads or other places in informally collected by waste pickers. These workers are marginalized despite their valuable services in maintaining this informal economy. These people come from poor families, socially backward people, children, migrants, and women workers. It is necessary to address this group of people to develop social regard for them, improve their working conditions, reduce their vulnerability to serious health hazards, provide access to all the basic facilities and social recognition, arrange alternative work opportunities, increase their access in voicing their issues and concerns in the country.

(g) **Recycling behavior of the people**—An expanded theory of planned behavior can be used to examine the primary variables influencing people's intentions and self-reported recycling of printed paper and packaging waste. Psychology uses this idea to develop a connection between beliefs, attitudes, and behavior. The results can be implemented to creating more effective public awareness campaigns and recycling programs. Based on the self-determination theory, a correlation has been discerned between the function of social influence and worldview on the self-reported and observed recycling behavior. To increase people's support for sustainable waste management and recycling, Thomas and Sharp (Wong et al. 2016) investigated the impacts of policies, habits, norms, and the recycling facilities availability on people's recycling behavior. Methods for altering behavior to decrease the quantity of non-targeted content have been contrasted. For instance, a three-way comparison has been made between: door stepping, incentives, and individualized feedback delivery. It was discovered that the customized feedback was significantly the most economical option, and it has also shown a high degree of efficacy in lowering contamination (Thomas and Sharp 2013).

7.2 Economic Factors Affecting Waste Management

The lack of political will to raise funding, solve problems, and develop proper waste management rules and regulations is one of the factors causing unsustainable waste management. Developing countries have no directed roles in developing circular economy which has a crucial role to play in waste management. The successful linear economy can collapse in the twenty-first century. Linear economy believes in production, supply, and disposal, whereas circular economy believes in reusing the waste again as raw material for the production process to reduce waste. This helps in solving environmental problems. To achieve sustainable waste management, all countries must adopt the practical approach of circular economy (Wikurendra et al. 2022). Economic analysis helps estimate the costs incurred by pollution caused by waste, determine the benefits of waste management on the environment, and develop cost-effective resolutions for effective waste management (Begum and Ehsan 2020). The following factors affect the waste management process:

(a) **Socioeconomic factors and level of education**—Various socioeconomic factors such as education level, per capita income, policy making, and population density affect waste management practices. Public awareness and education through fundraising can help make people understand the harmful effects of improper waste disposal and the advantages of sustainable waste management. The per capita income of the society decides their willingness to finance for efficient waste management services. Population density decides the design of the waste management (Jagun et al. 2022).

(b) **Job generation**—Sustainable waste management can support creation of formal and informal jobs in this industry. Local economies can benefit greatly by enhancing public trade and boosting tourism due to proper waste management. A skilled workforce is necessary for efficient waste management. New industries and businesses can prosper due to the expansion of recycling of waste. Domestic industries and trade benefit the most due to the recycling business (Allen 2023).

(c) **Market availability and Financial Incentives**—Efficient waste management is hampered due to a lack of markets for recycled products and financial help for stakeholders (Islam et al. 2024). People do not prefer to buy recycled products; hence, the markets do not find a place for such products. The lack of political will to raise financial incentives to develop sustainable waste management services has led to improper waste management.

(d) **Waste management cost**—In 2020, the global direct waste management cost was measured to be USD 252 billion. However, it raises to USD 361 billion on considering the indirect costs incurred in pollution, health, and change in climate due to poor waste management (2024). It is estimated to double by the year 2050. But incorporation of circular economy can help make a net gain of USD 108.5 Billion per year (2024).

(e) **Conversion of linear to circular economy**—circular economy promotes a closed-loop system for utilizing the productive resources till the exhaustion of their useful life. The linear economy transition into circular by prioritizing reusing and recycling materials. It promotes sustainability, a resilient and competitive economy, and innovation, reduces waste generation, and conserves resources (Allen 2023).

(f) **Life Cycle Cost Analysis**—Recently, life cycle cost (LCC) analysis concepts and models have been presented to evaluate sustainability in SWM technologies. LCC can offer comprehensive cost items for every important technology used in contemporary SWM systems (Byrne and O'Regan 2014). It was also shown how applicable models utilizing statistical methods and the multivariate econometric approach are for managing and forecasting MSW. Additionally, waste management pinch analysis was suggested to investigate how recycling and cost-cutting goals might lead to better waste management planning and a deeper understanding of SWM strategy (Martinez-Sanchez et al. 2015). In addition, it is anticipated that the potential introduction of further solid waste resource recovery, such as nutrients from food waste, will increase cost recovery and strengthen the financial sustainability of MSW management (Tan et al. 2015). These models attest that public policies, initiatives, socioeconomic traits, and climate variables all play a complex role in MSW formation.

(g) **Stakeholders' initiatives and policy making**—Decisions on the sustainability of waste management can be influenced by looking at how home trash disposal relates to environmental issues, public knowledge, and the degree of satisfaction with local capacity (Timlett and Williams 2008). Techno-economic analysis can recommend a more gradual switch from an integrated to a classic SWM system. This is especially the case in emerging nations. Convenience, costs, and communication are reportedly important characteristics of an intervention encouraging household collaboration in trash management for policymakers (Al-Khatib et al. 2015). Moreover, it is frequently necessary to promote the involvement of many societal stakeholders in decision-making processes. These stakeholders include government, local governments, businesses, specialists, and the general public (Briguglio 2016).

(h) **Revenue generation and Cost Saving**—Waste Management costs can be reduced using efficient waste management practices. This can help businesses and households and help the government save money at the country and state levels. Recycling, refusing to create waste, and reusing waste can save costs associated with waste collection, transportation of waste, and disposal. Sustainable methods to handle waste are more cost-effective than traditional methods, and they can generate revenue through recycled product markets. This can also aid in generating revenue through taxes and fees imposed by the Governments. Optimization of waste management systems helps reduce financial burdens, utilize resources more efficiently, and use the revenue for the social and economic benefit of the society.

Economically, various models and statistical methods can help save costs through better waste planning and management strategies. Social surveys are important for evaluating society's attitude, perception, readiness, awareness, and behavior to achieve social acceptance of sustainable waste management. Participants should be selected from various stakeholders, using appropriate sampling techniques and data gathering through structured questionnaires, interviews, or focus groups.

8 Socio-environmental Impact of Sustainable Waste Management Practices

A vision of a better environment and society can be achieved through sustainable waste management. This approach entails waste generation reduction, material reuse and recycling, and efficient disposal methods for accomplishing sustainable development goals. This can promote better air, soil, and water quality, natural resource conservation, and protected ecosystems. The key to better implementation is the incorporation of the circular economy. This reduces resource extraction and minimizes the landfilling or incineration of waste. Environmental consciousness and the necessity for efficient resources play an important role in making the basis for adopting sustainable waste management practices (Greenify hub team. n.d.).

Society can experience overall well-being, save money, improve community participation, and reduce waste footprint. Adoption of these sustainable practices by society, commercial sectors, and industries can positively impact the environment, create a healthy environment, provide economic benefits, and a sustainable future for everyone.

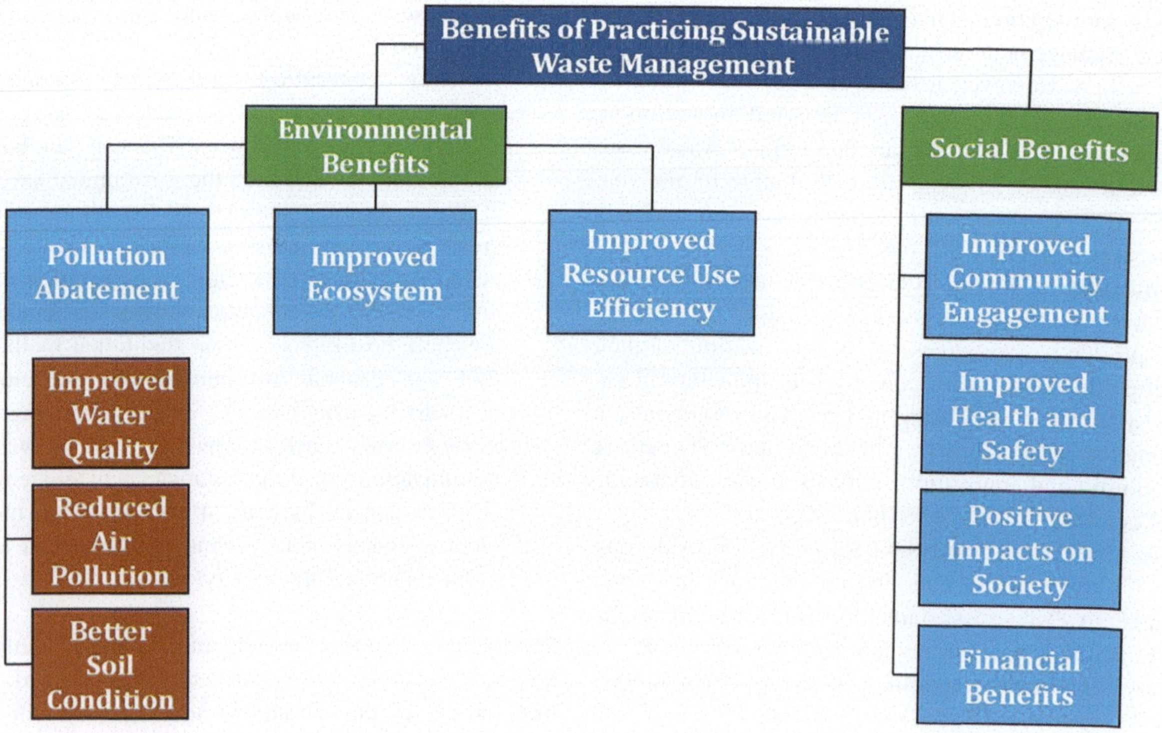

Fig. 4 Benefits of practicing sustainable waste management

Hence, Sustainable waste management can be a game-changer for the society and environment. The benefits are shown in Fig. 4.

8.1 Impact on Environmental Health

8.1.1 Improved Quality of Soil–water–air Nexus

(i) **Improved Water Quality**—Water contamination can be decreased using sustainable waste management techniques. Appropriate waste management and disposal shields water bodies from pollutants and toxins. This guarantees the maintenance of aquatic ecosystems, essential for biodiversity and ecological balance, and helps to maintain the water source quality. Water is considered an essential resource for any development. Water pollution and scarcity are side effects of population growth, industrialization, and urbanization. This has been aggravated due to human-induced climate change. Sustainable waste management techniques include the management of not only solid waste but also liquid and gaseous waste. The new innovative green technologies for wastewater treatment are also revolutionary for improving water quality. Conversion of waste water to reusable water has had a significant impact in betterment of the quality of water and also quantity of water (Silva 2023). Traditionally, wastewater was disposed off in nearby water bodies without treatment which led to various water-borne diseases that killed significant communities. However, with the development of modern wastewater management techniques, this field has been revolutionized. Still, 80% of global sewage enters the environment without treatment, especially in less developed countries where raw waste is dumped into waterbodies. Marine ecosystems get badly affected due to such unsustainable wastewater management practices. Hence, last decade has seen the development of various sustainability methods. The greenhouse gas (GHG) emissions from wastewater management facility are eliminated using solar panels. The waste from these units can be turned to energy-hydrogen, graphite which in turn can be used to run the same units, hence reducing runoff, surface pollution, and GHG emissions. The municipal solid waste can also be used as a source of energy which can be used to run power public transportation, power systems for small buildings, etc. The reuse of treated water from the waste water plants can help minimize water scarcity hence sustain the ever-growing population (Singh et al. 2018).

The attempts to utilize wastewater and optimize its usage is very critical to achieve and is one of the sustainable development goals. Clean water, globally enhanced safe recycled and reuse of water and reduction in disposed waste water is crucial to protect environment, public health, improve quality of life for all (Silva 2023). All these measures will always lead to abundant clean water resources for all, enhanced green areas, improved hydrological cycle, lesser dependency on groundwater to meet the needs of population, surplus water for self-cleaning mechanism of the river systems, lesser drought, and a positive impact on the overall climate of the earth.

(ii) **Reduced Air Pollution**—Waste disposal significantly contributes to air pollution, affecting human health and nature. Open dumping and burning create mismanagement of waste. Sustainable waste management practices are necessary to ensure clean air for future generations. Mismanaged waste releases harmful substances into the air, affecting human health and the environment. Every year, respiratory and cardiovascular disorders account for the deaths of over 6.7 million individuals due to air pollution, which also has a major negative influence on mental health. Since the primary sources of air pollution are the combustion of fossil fuels for transportation, industry, agriculture, and cooking stoves, which produce waste products that enter the atmosphere. Still, they also meet out many of our fundamental requirements.

The high levels of air pollution demonstrate the unsustainable lifestyle of today's generation. This is just one more challenge that must be overcome to meet the goal of SDGs (i.e., sustainable development goals) of creating a more equitable and liveable society. Air pollution and climate change are intimately related because they are caused mainly by human sources. Ambitious climate mitigation initiatives are not without expense, but they often result in cheaper healthcare expenditures and better air quality. It is believed that the social costs of air pollution avoided by lowered exposure levels due to climate mitigation measures alone exceed the upfront costs of these policies. The climate system is also physically impacted by air pollution; particles in the atmosphere impact clouds, precipitation, and surface temperatures. Thus, climate change can potentially worsen air pollution even in the areas where the air quality is improving (Gómez-Sanabria et al. 2022).

Exposure inequality happens both internationally and within individual nations. Historical and systemic forms of discrimination frequently result in increased exposure and, thus, increased health burdens for marginalized communities worldwide. Green infrastructure and sustainable waste management have the potential to reduce air pollution. However, unequal access can limit improvements for marginalized communities.

Hence, it is necessary to manage the waste using composting, recycling, and safe waste treatment. Development of efficient waste management plans, preventing them from open dumping and burn sites, improvement in waste handling techniques by local waste collection teams, and normalizing the use of waste as resources through recycling, reuse, and reduction are the various initiatives taken by United Nations Environment Programme (UNEP) International Environment Technology Centre to protect air. Many children are hired to collect, sort, and dispose of waste in unsafe conditions. They are not provided with protective gear and are unaware of the risks they might face in this task. UNEP proposes safer and environmentally friendly methods to help them and their families benefit by reducing health and well-being risks. Sustainable waste management not only involves the role of individuals by practicing the 3Rs, but corporations can help contribute to this cause by creating products from recycled materials, producing less waste, and using less packaging or environment-friendly packaging materials. Governments also play a significant role by establishing strong regulations to control open dumping and burning of waste.

(iii) **Better Soil Conditions**—Soil plays an important role in provision of essential services for maintaining environment quality, carbon sequestration, climate regulation, biodiversity maintenance, plant growth, and habitats of animals. However, soil pollution, acidification, salinization, and erosion has led to soil degradation throughout the world. Sustainable soil management practices can help in managing the climate since soil is a large terrestrial carbon pool that regulates carbon cycle and also affect the level of greenhouse gases in the environment. Thus, sustainable measures to manage the soil and the waste dumped on them is important along with strong inputs from end-users and practitioners who need to be informed and made knowledgeable. This social learning process can be enforced by scientists and policy makers.

Most of the waste generated from municipality, hospitals, wastewater treatments, industries, etc., get disposed off on land either in landfills, or for composting, or dumped insensitively. These practices cause climate change, health risk to humans, affect agricultural productivity, water source contamination, loss of biodiversity. Soil organic carbon plays a significant role in maintaining the earth's carbon cycle hence affecting the climate. In turn, soil properties and vegetation are affected like global warming accelerates soil erosion because it affects the micro-organisms, animals, and plant species dependent on them. Different soil types are differently sensitive to temperature changes, hence affecting the wildlife and ecosystem. The increased input of soil organic carbon (SOC) through biosolids and composts leads to increased greenhouse gas emissions. This level can be increased due to many conservative farming practices. Soil carbon sequestration can increase due to dumping of crop residue on farmlands (Das et al. 2021).

The unsustainable practices in waste management lead to contamination of soil, pollutants affecting the food chain and hence health problems such as cancer, neurological disorders, and organ damage can directly impact the living beings, the diverse range of microorganisms can be killed disrupting the ecosystem balance- nutrient cycle, decomposition, soil structure formation, reduced soil fertility, increased susceptibility to erosion, stunted crop growth, increased susceptibility of plants to diseases and pests, the pollutants leaching from ground enter the groundwaters and surface waters polluting them. The disastrous impacts of soil pollution can be arrested by using sustainable waste management measures. Proper disposal of waste material, recycling of plastics, electronics, and hazardous wastes. Reusing the materials to decrease the rate of generation of waste and hence the disposal on ground.

Landfills need to be properly designed to prevent leaching of pollutants from reaching the ground waters and surface water. Remediation and restoration techniques can be utilized to clean up already contaminated soil to improve its health and fertility. Bioremediation utilizes microorganisms to convert pollutants into less harmful substances. Phytoremediation uses plants to absorb and remove contaminants from soil. Nanotechnology is a revolutionary remediation tool for pollutant removal for sustainable waste treatment. Circular economic approach is important for handling waste, reducing them, and improving resource use efficiency (Das et al. 2021).

8.1.2 Improved Ecosystem

Ecosystems can be severely impacted by improper waste management, upsetting the delicate balance of nature. Reducing trash generation, encouraging recycling, and composting, and implementing appropriate techniques for waste disposal are sustainable waste management practices that can help reduce these effects. By doing this, we not only preserve biodiversity but also ensure the protection of natural habitats, and the general well-being of our ecosystems. Businesses and communities can improve environmental health and build a sustainable future for future generations by embracing sustainable waste management techniques (Abubakar et al. 2022). It is a crucial step in protecting the environment and building a more sustainable, healthy living space for all. Climate change can be influenced by greenhouse gas emissions, such as methane generated during the breakdown of organic waste.

Businesses and communities can improve environmental health and build a sustainable future for future generations by embracing sustainable waste management techniques (Greenify hub team. n.d.). It is a crucial step in protecting the environment and building a more sustainable, healthy living space for all. Climate change can be influenced by greenhouse gas emissions, such as methane generated during the breakdown of organic waste. A number of mitigation strategies must be put into place in order to lessen potential adverse effects on the environment. Effective planning is a crucial first step, as is choosing suitable sites for waste processing and landfills while avoiding delicate ecological areas. To avoid contaminating the air and water, strict supervision is necessary during the waste processing and disposal process.

Improved waste management innovation and technology can also help lessen environmental effects; for instance, by cutting emissions or streamlining recycling procedures. Appropriate guidelines must be integrated for sustainable waste management (Perkumienė et al. 2023). Thorough planning needs to identify safe locations for final disposal and take environmental factors into account. It is necessary to pursue more advanced technology, such as that which lowers toxic emissions while processing. Raising public awareness and educating people about the value of appropriate waste management is another essential step in minimizing adverse effects. Innovative waste management techniques and technology have a variety of complex environmental effects. When evaluating an impact, a number of factors are taken into account, such as greenhouse gas emissions, water pollution, and possible habitat destruction. Nevertheless, employing cutting-edge tactics and technology has a lot of potential environmental advantages, such as a decrease in plastic waste, a decrease in emissions, and the creation of new resources from trash. Mitigation strategies, including thorough preparation, vigilant observation, and technology advancement, will lessen potential adverse effects. An essential first step toward a greener and more sustainable future in addressing the environmental issues of today is creative and sustainable waste management.

8.1.3 Improved Resource Use Efficiency

Resource efficiency can be improved by the implementation of reuse, recycling, reducing, and composting of waste under the purview of Sustainable waste management. Reuse and Recycling prevents the material from being thrown to waste if it still has some productive value and it also reduces the extraction and manufacturing of new resources. By reducing waste generation, resource utilization can be maximized, and resources can be significantly conserved. Organic waste is composted to reduce the need of synthetic fertilizers since they develop nutrient-rich soil. These methods support a circular economy that minimizes environmental effect, maximizes resource utilization, and fosters long-term sustainability. Resource efficiency can be ensured by incorporating these components: Material efficiency, energy efficiency, water efficiency, supply chain made green, promotion of industry code of practice.

Material efficiency ensures utilizing the materials up to its scrap value since material is considered as an important resource that should not be thrown to waste until its productive value is completely exhausted. This is made sure by collection, remanufacturing, reuse, processing of the waste material, and promotion of resource recovery firms. Energy efficiency can be ensured by practicing the usage of renewable energy, including solar, biomass, hydrogen fuel, and wind (Mehmood 2022). Private and government sectors should be encouraged to invest in technologies that are energy efficient. Water efficiency can be ensured by implementing regulations, policies, and technological solutions to control the release of wastewater into water bodies. Rainwater harvesting should be promoted along with the recycling of treated water. Integrated water resource management should be promoted along with ensuring efficiency in agriculture, domestic and industrial supplies. It should be ensured that all the industries and their associated suppliers and vendors follow environmental standards and adopt environmental practices.

Conventional practices of resource use have led to environmental degradation, high pollution levels and depletion of natural resources. In order to practice sustainable development, circular economy, efficient resource use, and achieving low carbon economy is the crucial key. A framework for the technological and structural changes needs to be developed. Resource use efficiency can be improved by increasing the rate of reuse and recovery. Life cycle assessment tool can be used for decision making to enhance resource use efficiency in any product development activity. Mass balance concept is a basis for design of sustainable future by understanding the past, it helps in detection of actions that need to be taken to manage waste in the global flow and energy conversion.

8.2 Impact on Society

Developing an understanding of the community perspective on waste management, its services, regulatory measures, and the obstacles in achieving sustainable waste management is crucial for taking effective measures for society to handle and manage waste. Socio-Ecological Model (SEM) can be used to investigate the components that affect Sustainable WASTE MANAGEMENT practices like intra personal and inter-personal system, community and institutional system, and government level or macro-system (Sewak et al. 2021).

The factors affecting sustainable waste management at microlevel are negligence, lack of personal responsibility, awareness, and civic pride. Meso- and exo-systems include structural factors like recycling infrastructure and inadequate waste collection services. Population should be made aware and educated is the best solution to encourage pro-environmental behaviors. However, this development in education is a tough task that needs some type of educational tools and frequent inputs and educational workshops. The inclusion of sustainable measures in waste management affects the society on a large scale. The effects are listed below.

8.2.1 Improved Health and Safety

The implementation of reuse, recycling, and resource efficient technologies are helpful to develop a circular economy which minimizes disposal and efficient handling of waste with energy recovery. Adopting sustainable waste management techniques improves occupational safety and public health in addition to being good for the environment. Through the reduction of trash disposal's adverse effects, communities and companies may establish a safer and healthier environment for all. The unsustainable waste management techniques such as open dumping, open burning, unsustainable landfilling methods, all lead to major health hazards. Open dumping of Municipal Solid Waste gets leached out within the soil and reaches groundwater. This polluted groundwater is a direct and indirect source of infectious diseases (Mahajan 2023).

Openly dumped wastes releases greenhouse gases including methane causing gradual air pollution and hence contributing to global warming. The animals, insects, rodents that feed on such openly dumped waste increase the risk of transmission of fatal diseases through bite or contact. Most severe health risks are posed by biomedical wastes dumped openly on the ground or in water bodies. These are very infectious and highly risky wastes. The rag pickers are exposed to cuts, injury, and infection through these mismanaged and unsafe handling of wastes. Burning of these wastes causes air pollution and subsequently severe cancer and respiratory diseases. Openly dumped wastes on soil are a breeding ground for rodents and insects which cause malaria, dengue, and yellow fever, etc. Openly burning such toxic and hazardous waste can lead to disastrous fires, release of toxic gases in the air which if inhaled by living beings can be fatal. Landfilling is a good solution for managing waste for the wastes that cannot otherwise be reused, recycled, or reduced or incinerated for energy generation. However, landfills have a potential to impact the groundwaters through leachate and biogas. Therefore, sanitary landfilling must be practiced which are engineered in such a way that they have a provision of bottom lining, leachate collection and removal systems, and final covers in which municipal waste is deposited in layers through controlled mechanisms (World health organization 2023).

Burning induces carbon, carbon monoxide and dioxides adding to greenhouse gases and air pollution subsequently changing the environment and affecting human health. These wastes when dumped into water bodies disturb the marine ecosystem, affecting their life and habitat both. These marine animals are in turn consumed by humans and animals that lead to transfer of diseases via food chain. In today's era, a significant rise can be observed in the industrial waste, e-waste, and construction and demolition waste, etc. and if they are not handled properly, they further impose serious health issues on humans. E-waste comprises of toxic materials such as lead, cadmium, mercury, lithium, and barium which negatively affects the human health causing failure of brain, hearth, nervous system, reproductive system, skeletal system, and kidneys.

The productivity of good soil is reduced causing its collapse and landslides due to construction and demolition waste. They also are a storehouse of carbon and release carbon dioxide which is a major air pollutant responsible for causing global warming. These wastes comprise of asbestos, chlorine, plastics, lead, mercury which are toxic for respiratory, nervous system, and skin. Heavy metals affect soil, groundwater, and surface water through surface runoff of such wastes harming the human health.

Landfills in the developing countries are unsanitary and have unsorted, uncontrolled open dumping, improper separation of waste, poor leachate collection system or no collection system at all and unsanitary disposal systems. Such practices contribute to polluted groundwaters, hence adverse health impacts on consumers causing nausea, asthma, diarrhea, cardiovascular and renal diseases, throat infections, neurological diseases, etc. (World health organization 2023).

Open composting release bio-aerosols which cause asthma, chronic pulmonary diseases, bronchitis, and carcinogenic diseases. These impacts mostly affect the people working in the composted areas. Material recycling also if not handled properly by the workers or if they are not provided with protection to handle electrical and electronic waste or for recovering valuable metals. Some of the methods pose great risks to pickers causing toxic effects on human health.

Therefore, seeing the innumerable reasons and effects of mishandled wastes and due to unsustainable waste management methods, sustainable waste management is a necessity and crucial need of the hour (Alam and Ahmade 2013). The positive impacts of good practices for waste management on human health are:

(a) **Safer Workplaces**: Workplaces can be made safer and more hospitable to employee well-being by implementing sustainable waste management techniques. The risks of mishaps such as trips, falls, and slides are decreased when waste is properly separated, handled, and stored. A safer workplace is also a result of putting safety procedures into place and offering sufficient training on waste management techniques.
(b) **Reduced Health Risks**: Sustainable waste management practices like proper treatment and disposal of waste help prevent the emission of hazardous substances into air, soil, and water, hence reducing the risks associated with waste-related health issues. The risk of waterborne diseases, health complications, and respiratory problems is also lowered.
(c) **Healthy Communities**: Communities' general health and well-being are enhanced by sustainable waste management. The quality of the soil, water, and air is improved by minimizing the release of hazardous compounds and lowering pollution. This improves the citizens' quality of life and reduces the incidence of ailments, all of which are related to their health.
(d) **Establishing a Culture of Safety**: Businesses and communities may foster a culture of safety by implementing sustainable waste management techniques. Organizations can foster a sense of responsibility among residents and staff and increase knowledge of the importance of appropriate management of waste and its influence on individual well-being by giving priority to health and safety considerations.
(e) **Preventing Contamination**: The major aim of sustainable waste management techniques is to avoid environmental contamination, which could otherwise be harmful to human health. By using appropriate waste disposal techniques, communities can be guaranteed clean, safe drinking water by averting the leaching of dangerous substances into groundwater sources.

Recall that implementation of sustainable waste management techniques is essential for maintaining safer workplaces and protecting public health in addition to being a responsible environmental measure. Industries and communities can help ensure a safer and healthier future for all by adopting waste management techniques and making thoughtful decisions.

8.2.2 Improved Community Engagement

A key impact of sustainable waste management programs is community engagement. Participating in these initiatives with community members allows us to promote civic pride, social solidarity, and a sense of ownership (Greenify hub team. n.d.). Sustainable waste management involves the following positive impacts on community engagement:

(a) **Community Awareness and Education**—Educating and increasing awareness of the value of sustainable waste management can be accomplished by involving the community. Community members can comprehend the impact of their actions through the conduction of seminars, events, educational initiatives, and learn practical waste reduction measures.
(b) **Long-term Effect**—By fostering sustainable behaviors and habits, community involvement has a long-term effect. Active community involvement in waste management increases the likelihood that these behaviors will be sustained over time, protecting the environment, and promoting health and cleanliness for coming generations.
(c) **Social Cohesiveness**—By bringing people together, sustainable waste management programs may foster cooperation and social contact. Clean-up drives and recycling initiatives, for example, provide a platform for community members to interact and work together toward a common objective while also fostering a sense of pride in the town.
(d) **Community collaboration-based solutions**—Getting involved in the community promotes cooperation between local companies, government, and citizens. Communities can create waste management plans collaboratively that are specifically suited to their available resources and requirements.
(e) **Shared Responsibility**—Active community involvement in waste management initiatives, such as composting or recycling, fosters a sense of shared accountability for environmental stewardship. This feeling of collective ownership strengthens and strengthens the community.

8.2.3 Financial Benefits

Businesses might benefit monetarily from implementing sustainable waste management strategies (Greenify hub team. n.d.). Here are some important financial advantages to think about:

(a) **Income development**—Recycling programs are one way that sustainable waste management techniques offer chances for income development. Businesses can make money by selling recyclable resources to recycling companies by correctly sorting and processing these commodities.
(b) **Lower Waste Disposal Expenses**—Businesses can minimize trash generation and hence lower waste disposal expenses by using sustainable waste management strategies. Recycling and composting are two efficient waste reduction techniques that can assist keep materials out of landfills, where disposal costs might be high.
(c) **Increased Operational Efficiency**—Putting sustainable waste management ideas into reality frequently entails streamlining operations, cutting costs, and improving waste management procedures. Increased productivity can result in financial savings on labor, energy, and waste transportation, among other things.
(d) **Customer retention and Brand Reputation**—Adopting environmentally friendly waste management techniques can improve a business's standing and draw in eco-aware customers. Businesses that put sustainability first are more likely to obtain a competitive edge and win over loyal customers in the face of the growing demand for eco-friendly solutions.
(e) **Tax Benefits**—Organizations actively using sustainable waste management strategies can receive tax incentives from several state and municipal governments. These incentives may come in the form of grants, tax credits, or deductions, which offer extra financial advantages.

Businesses may make well-informed decisions to adopt environmentally friendly waste management techniques that not only improve the environment but also generate profits by taking these financial advantages into account.

8.2.4 Positive Effects

Ecological waste management techniques have a lot of positive social effects and help make communities safer and healthier places to live. Communities and businesses may benefit society by putting green waste management techniques into practice. The following are some major societal benefits:

(a) **Community Involvement**—Including the community in trash management programs promotes social cohesion and a feeling of civic pride. It motivates locals to take an active role in establishing a clean and sustainable environment.
(b) **Better Quality of Life**—Pollution is decreased and water and air quality are improved through sustainable waste management. As a result, inhabitants have better health outcomes and a higher standard of living.
(c) **Safer Environments**—Hazardous materials are controlled with the aid of appropriate waste management techniques, enhancing public safety and lowering health hazards. This improves people's and communities' general quality of life (Kneeland and Knutson 2012).
(d) **Education and Awareness**—Putting into practice sustainable waste management techniques offers chances to teach the public about the value of disposing of garbage responsibly and the advantages of recycling and composting.

(e) In general, the execution of Sustainable WASTE MANAGEMENT procedures has the capacity to generate favorable social effects through the amelioration of environmental conditions, community engagement, and individual health and safety.

9 Conclusion

This chapter suggests that an integrated sustainable waste management plan should be prepared and integrated emphasizing on improvement in the municipality operation to manage all the steps to achieve sustainability, i.e., segregation, transportation, sorting, treatment, and disposal. This initiative can be made successful only when all the stakeholders are involved pro-actively while recognizing the technical, legal, economic, social, and environmental aspects suitable for any local setting (Teshome et al. 2022). The life cycle assessment is a proven method to prepare waste management plan. One of the important aims of these waste management plans should be health of society and overall preservation of the environment.

Use of environment friendly technologies such as sustainable landfilling, incineration, reduce, recycle, and reuse along with smart management of resources using Internet of Things is the need of the hour. Incineration is particularly useful for nonrecyclable waste. These methods have been found to be environment friendly since they result into 70.82% reduction in green house gas (GHG) emissions. The best results for reduction in GHGs are achieved by 50% incineration of the waste, 30% landfilling, and 20% recycling (Xin et al. 2020). Landfilling should also be upgraded with improved techniques to prevent leakage and leaching of waste along with collection of biogases. Composting should be used for organic waste. An integrated treatment approach to handle solid waste through mechanical and biological means like anaerobic treatment of waste water, waste-to-energy conversion, use of smart management schemes is significant to protect our environment and social beings.

These new measures not only provide means of livelihood for society but also help in curbing the harmful health impacts of unsustainable practices and protecting environment for the future generations. The lack of proper management plans for waste and its inefficient management systems are the major barriers in the effective waste management systems. These include collection, segregation, recycling of waste, insufficient funds, lack of skilled workers and professionals, weak policies, and lack of intent to enforce strict rules and regulations. Their efficiency also depends on the rate of waste generation, population, financial status, commercialization, urbanization, culture, and region in which waste is generated (Ziraba et al. 2016).

Environmental regulations and making civilians responsible for making sustainable waste management a success along with mitigating the potential health impacts of the same can be achieved by strict enforcement by local government bodies. It must be ensured that any breach of rules and regulations is severely punished by law. Most importantly the policies should focus on waste reuse, reduce, and recycle to realize sustainable development and a circular economy (Abubakar et al. 2022).

The first step on the part of society is to desist from unsustainable practices and follow waste management religiously. However, to develop such positive attitudes public awareness campaigns, provision of waste segregation bins and waste separation at source must be encouraged. Waste reduction, sorting, separation at source, material reuse, and then recycle can be promoted by community-based organizations.

These measures are impactful and help in improved environmental health like minimal impacts on ecosystem, better air quality, improved soil ecosystem, and reduced water pollution. These practices offer several financial benefits also like job creation, improving the reputation of companies practicing sustainable waste management, attracting environmentally conscious people to products made in sustainable and environment friendly enterprises (Greenify hub team. n.d.).

The most important impact of using these sustainable measures is developing a shared responsibility among all citizens toward saving environment, creating awareness and education, bringing collaborations among local people, industries, and government. It also promotes cooperation among all sections of society creating platforms for people to connect and work toward a common target of saving environment. The resource use efficiency is greatly enhanced. There are minimized health risks, safer workplaces, healthy communities, developing a culture of safety hence ensuring improved quality of life for all, safer surroundings, revenue generation, increased operational efficiency, and reduced waste disposal cost. These measures are sure to create a lasting impact on society such that people become habitual to acting sustainably and impart these values to upcoming generations as well making sure the sustainable development of the society. By adopting sustainable waste management measures, we as a society can make conscious choices and adopting sustainable practices to fabricate a waste free world, good quality life, and better future for future generations.

References

Abubakar IR, Maniruzzaman KM, Dano UL, AlShihri FS, AlShammari MS, Ahmed SMS, Al-Gehlani WAG, Alrawaf TI (2022) Environmental sustainability impacts of solid waste management practices in the global south. Int J Environ Res Public Health 19:12717. https://doi.org/10.3390/ijerph191912717

Alam P, Ahmade K (2013) Impact of solid waste on health and the environment. Int J Sustain Develop Green Eco 2:165–168

Al-Khatib IA, Kontogianni S, Nabaa HA, Alshami N, Al-Sari MI (2015) Public perception of hazardousness caused by current trends of municipal solid waste management. Waste Manage 36:323–330. https://doi.org/10.1016/j.wasman.2014.10.026

Allen C (2023) Advances in recycling and waste management. Econ Environ Benefits Efficient Waste Manag. https://doi.org/10.37421/2475-7675.2023.8.297

Anastasiades K, Blom J, Buyle M, Audenaert A (2020) Translating the circular economy to bridge construction: lessons learnt from a critical literature review. Renew Sustain Energy Rev 117:109522. https://doi.org/10.1016/j.rser.2019.109522

Atwadkar RR, Jadhav LD, Wagh MM, Shinde NN (2014) A multi criteria ranking of different technologies for the waste to energy of municipal solid waste in the city of Kolhapur. Int J Emerg Technol Adv Eng 4:937–942

Ayilara M, Olanrewaju O, Babalola O, Odeyemi O (2020) Waste management through composting: challenges and potentials. Sustainability 12(11):4456. https://doi.org/10.3390/su12114456

Begum RA, Ehsan S (2020) Economics of waste minimization, recycling, and disposal. In: Oxford research encyclopedia of environmental science. Oxford University Press. https://doi.org/10.1093/acrefore/9780199389414.013.424

Beyond an age of waste Turning rubbish into a resource (2024). https://wedocs.unep.org/20.500.11822/44939

BRANZ (2014) Waste reduction—demolition. BRANZ Limited: Porirua, New Zealand

Briguglio M (2016) Household cooperation in waste management: initial conditions and intervention. J. Economic Surveys 30:497–525. https://doi.org/10.1111/joes.12156

Bruno A (2023) Global waste generation–statistics and facts. Available online: https://www.statista.com/topics/4983/waste-generation-worldwide/#topicOverview

Byrne S, O'Regan B (2014) Attitudes and actions towards recycling behaviours in the Limerick. Ireland Region Resour Conserv Recy 87:89–96. https://doi.org/10.1016/j.resconrec.2014.03.001

Das S, Pradhan AK, Beura K, Das R, Singh R (2021) Improvement of soil quality by solid waste recycling: a global perspective. In: Soil science: fundamentals to recent advances, pp 637–667. Springer Singapore. https://doi.org/10.1007/978-981-16-0917-6_32

Gillespie A (2015) Waste policy: international regulation. Comparative and Contextual Perspectives. Edward Elgar Publishing, Cheltenham, UK

Gómez-Sanabria A, Kiesewetter G, Klimont Z, Schoepp W, Haberl H (2022) Potential for future reductions of global GHG and air pollutants from circular waste management systems. Nat Commun 13(1):106. https://doi.org/10.1038/s41467-021-27624-7

Grazhdani D (2016) Assessing the variables affecting on the rate of solid waste generation and recycling: An empirical analysis in Prespa Park. Waste Manage 48:3–13. https://doi.org/10.1016/j.wasman.2015.09.028

Greenify hub team: How sustainable waste management can benefit your business and the environment. https://greenifyhub.com/waste-management/sustainable-waste-management/

Hong RJ, Wang GF, Guo RZ, Cheng X, Liu Q, Zhang PJ, Qian GR (2006) Life cycle assessment of BMT-based integrated municipal solid waste management: case study in Pudong. China Resour Conserv Recycl 49:129–146. https://doi.org/10.1016/j.resconrec.2006.03.007

Islam N, Sandanayake M, Muthukumaran S, Navaratna D (2024) Review on sustainable construction and demolition waste management—challenges and research prospects. Sustainability 16:3289. https://doi.org/10.3390/su16083289

Jagun ZT, Daud D, Ajayi OM, Samsudin S, Jubril AJ, Rahman MSA (2022) Waste management practices in developing countries: a socio-economic perspective. Environ Sci Pollut Res 30:116644–116655. https://doi.org/10.1007/s11356-022-21990-5

Jhatial AA, Goh WI, Mohamad N, Rind TA, Sandhu AR (2020) Development of thermal insulating lightweight foamed concrete reinforced with polypropylene fibres. Arab J Sci Eng 45:1–10. https://doi.org/10.1007/s13369-020-04382-0

Kadus T, Nirmal P, Kulkarni K (2020) Smart waste management using IOT. Int J Eng Res Technol 9:738–741. https://doi.org/10.17577/IJERTV9IS040490

Kasiński S, Szuszkiewicz J, Rudnicki M (2024) Regional strategies for implementing methane fermentation technology in waste management: environmental, technological, and social perspectives. Sustainability 16(20):9034. https://doi.org/10.3390/su16209034

Kazerooni Sadi MA, Abdullah A, Navazandeh Sajoudi M, Kamal MFM, Torshizi F, Taherkhani R (2012) Reduce, reuse, recycle and recovery in sustainable construction waste management. Adv Mat Res, 446–449, 937–944. https://doi.org/10.4028/www.scientific.net/AMR.446-449.937

Khan D, Kumar A, Samadder SR (2016) Impact of socioeconomic status on municipal solid waste generation rate. Waste Manage 49:15–25. https://doi.org/10.1016/j.wasman.2016.01.019

Kneeland K, Knutson B (2012) Chapter 6. Waste management for social and environmental benefits in Ethiopian cities. Environmental Policy Update 2012: Development Strategies and Environmental Policy in East Africa

Kotzé DA (2015) Women's perceptions and attitudes about recycling: a case study in South Africa. Int J Environ Sustain 11:41–50. https://doi.org/10.18848/2325-1077/CGP/v11i02/55123

Kumar S, Smith SR, Fowler G, Velis C, Kumar SJ, Arya S, Cheeseman C (2017) Challenges and opportunities associated with waste management in India. R Soc Open Sci 4:160764. https://doi.org/10.1098/rsos.160764

Kumar NM, Hait S, Priya A, Bohra V (2024) Resources conservation, recycling and waste management, Nallapaneni MK, Hait S, Priya A, Bohra V (Eds.). MDPI. https://doi.org/10.3390/books978-3-7258-0903-5

Lipu MSM, Uddin MS, Miah MAR (2013) A feasibility study of solar-wind-diesel hybrid system in rural and remote areas of Bangladesh. Int J Renew Energy Res 3:892–900

Mahajan R (2023) Environment and health impact of solid waste management in developing countries: a review. Current World Environment 18:18–29. https://doi.org/10.12944/CWE.18.1.3

Mani S, Singh S (2016) Sustainable municipal solid waste management in India: a policy agenda. Procedia Environ Sci 35:150–157. https://doi.org/10.1016/j.proenv.2016.07.064

Margallo M, Taddei MBM, Hernandez-Pellon A, Aldaco R, Irabien A (2015) Environmental sustainability assessment of the management of municipal solid waste incineration residues: a review of the current situation. Clean Techn Environ Policy 17. https://doi.org/10.1007/s10098-015-0961-6

Martinez-Sanchez V, Kromann MA, Astrup TF (2015) Life cycle costing of waste management systems: overview, calculation principles and case studies. Waste Manage 36:343–355. https://doi.org/10.1016/j.wasman.2014.10.033

Mattos F, Calmon JL (2023) Social life cycle assessment in municipal solid waste management systems with contribution of waste pickers: literature review and proposals for new studies. Sustainability 15(2):1717. https://doi.org/10.3390/su15021717

McVeigh P, Jones N, Ponniah S (2019) Auckland future skills. Martin, Jenkins & Associates Limited: Wellington, New Zealand

Mehmood U (2022) Environmental sustainability through renewable energy and banking sector development: policy implications for N-11 countries. Environ Sci Pollut Res 30:22296–22304. https://doi.org/10.1007/s11356-022-23738-7

Michaels T, Krishnan K (2018) Directory of waste-to-energy facilities. Energy Recovery Council: New York, NY, USA

Mongkolnchaiarunya J (2005) Promoting a community-based solid-waste management initiative in local government: Yala municipality, Thailand. Habitat Int 29:27–40. https://doi.org/10.1016/S0197-3975(03)00060-2

Narnaware SL, Srivastava NSL, Vahora S (2017) Gasification: an alternative solution for energy recovery and utilization of vegetable market waste. Waste Manage Res 35:276–284. https://doi.org/10.1177/0734242X16679257

Náthia-Neves G, Berni M, Dragone G, Mussatto SI, Forster-Carneiro T (2018) Anaerobic digestion process: technological aspects and recent developments. Int J Environ Sci Technol 15(9):2033–2046. https://doi.org/10.1007/s13762-018-1682-2

Ouda O, Raza SA (2014) Waste-to-energy: solution for municipal solid waste challenges- global perspective. In: 2014 International symposium on technology management and emerging technologies (ISTMET 2014), Bandung, Indonesia. https://doi.org/10.13140/2.1.5027.9680

Pal MS, Bhatia M (2022) Current status, topographical constraints, and implementation strategy of municipal solid waste in India: a review. Arab J Geosci 15(12):1176. https://doi.org/10.1007/s12517-022-10414-w

Perkumienė D, Atalay A, Safaa L, Grigienė J (2023) Sustainable waste management for clean and safe environments in the recreation and tourism sector: a case study of Lithuania. Turkey Morocco Recycl 8:56. https://doi.org/10.3390/recycling8040056

Qazi WA, Abushammala M, Azam MH, Younes MK (2018) Waste-to-energy technologies: a literature review. J Solid Waste Technol Manag 44:387–409. https://doi.org/10.5276/JSWTM.2018.387

Ramachandra TV, Bharath HA, Kulkarni G, Han SS (2018) Municipal solid waste: generation, composition and GHG emissions in Bangalore. India Renew Sustain Energy Rev 82:1122–1136. https://doi.org/10.1016/j.rser.2017.09.085

Sankoh FP, Yan X, Tran Q (2013) Environmental and health impact of solid waste disposal in developing cities: a case study of granville brook dumpsite, Freetown. Sierra Leone J Environ Prot 4:665–670. https://doi.org/10.4236/jep.2013.47076

Sasana H, Prasetyanto PK, Priyono N, Novitaningtyas I (2022) The importance of sustainable waste management due to socio-economic changes. Int J Energy Econ Policy 12:170–174 (2022). https://doi.org/10.32479/ijeep.12665

Scarlat N, Motola V, Dallemand JF, Monforti-Ferrario F, Mofor L (2015) Evaluation of energy potential of municipal solid waste from African urban areas. Renew Sustain Energy Rev 50:1269–1286. https://doi.org/10.1016/j.rser.2015.05.067

Seo S, Aramaki T, Hwang Y, Hanaki K (2004) Environmental impact of solid waste treatment methods in Korea. J Environ Eng 130:81–89. https://doi.org/10.1061/(ASCE)0733-9372(2004)130:1(81)

Sewak A, Deshpande S, Rundle-Thiele S, Zhao F, Anibaldi R (2021) Community perspectives and engagement in sustainable solid waste management (SWM) in Fiji: a socioecological thematic analysis. J Environ Manage 298:113455. https://doi.org/10.1016/j.jenvman.2021.113455

Silva JA (2023) Wastewater treatment and reuse for sustainable water resources management: a systematic literature review. Sustainability 15(14):10940. https://doi.org/10.3390/su151410940

Silva WDO, Morais DC (2021) Transitioning to a circular economy in developing countries: a collaborative approach for sharing responsibilities in solid waste management of a Brazilian craft brewery. J Clean Prod 319:128703. https://doi.org/10.1016/j.jclepro.2021.128703

Singh RP, Tyagi VV, Allen T, Ibrahim MH, Kothari R (2011) An overview for exploring the possibilities of energy generation from municipal solid waste (MSW) in Indian scenario. Renew Sustain Energy Rev 15:4797–4808. https://doi.org/10.1016/j.rser.2011.07.071

Singh J, Saxena R, Bharti V, Singh A (2018) The importance of waste management to environmental sanitation: a review. Adv Biores 9:202–207. https://doi.org/10.15515/abr.0976-4585.9.2.202207

Tan ST, Ho WS, Hashim H, Lim JS, Lee CT (2015) Waste management pinch analysis (WAMPA) with economic assessment. Chem Eng Trans 45:145–150. https://doi.org/10.3303/CET1545025

Ternald D (2020) International environmental technology centre. In: Annual report 2020. United Nations Environment Programme, Osaka, Japan

Teshome ZT, Ayele ZT, Abib MI (2022) Assessment of solid waste management practices in Kebridehar city Somali regional state. Ethiopia Heliyon 8:e10451. https://doi.org/10.1016/j.heliyon.2022.e10451

Thomas C, Sharp V (2013) Understanding the normalisation of recycling behaviour and its implications for other pro-environmental behaviours: a review of social norms and recycling. Resour Conserv Recy 79:11–20. https://doi.org/10.1016/j.resconrec.2013.04.010

Timlett RE, Williams ID (2008) Public participation and recycling performance in England: a comparison of tools for behaviour change. Resour Conserv Recy 52:622–634. https://doi.org/10.1016/j.resconrec.2007.08.003

Tozlu A, Ozahi E, Abusoglu A (2016) Waste to energy technologies for municipal solid waste management in Gaziantep. Renew Sustain Energy Rev 54:809–815. https://doi.org/10.1016/j.rser.2015.10.097

Umer H, Ahmed A, Ali F, Ali SS, Khan MA (2022) A systematic literature review on smart waste management using machine learning. In: 2022 Mohammad Ali Jinnah University international conference on computing (MAJICC), Karachi, Pakistan, pp 1–9. https://doi.org/10.1109/MAJICC56935.2022.9994104

Urban RC, Nakada LYK (2021) COVID-19 pandemic: solid waste and environmental impacts in Brazil. Sci Total Environ 755:142471. https://doi.org/10.1016/j.scitotenv.2020.142471

Velis CA, Wilson DC, Gavish Y, Grimes SM, Whiteman A (2023) Socio-economic development drives solid waste management performance in cities: a global analysis using machine learning. Sci Total Environ 872:161913. https://doi.org/10.1016/j.scitotenv.2023.161913

Vukovic N, Makogon E (2022) Waste-to-energy generation: complex efficiency analysis of modern technologies. Sustainability 14:13814. https://doi.org/10.3390/su142113814

Wang C, Darendeliler Z (2013) Conducting social assessments in urban solid waste management projects. www.worldbank.org/socialdevelopment

Ward MN, Wells B, Diyamandoglu V (2014) Development of a framework to implement a recycling program in an elementary school. Resour Conserv Recy 86:138–146. https://doi.org/10.1016/j.resconrec.2014.02.013

Warren K (2013) Waste to energy background paper. Report for Zero Waste SA, Ricardo-AEA Ltd

Wikurendra EA, Ferto I, Nagy I, Nurika G (2022) Strengths, weaknesses, opportunities, and threats of waste management with circular economy principles in developing countries: a systematic review. Environ Qual Manage 32:87–94. https://doi.org/10.1002/tqem.21846

Wong JWC, Surampalli RY, Zhang TC, Tyagi RD, Selvam A (2016) Sustainable solid waste management. American Society of

Civil Engineers (ASCE), Reston, United States. https://doi.org/10.1061/9780784414101

World health organization: Assessing the health impacts of waste management in the context of the circular economy. Copenhagen: World Health Organization, Regional Office for Europe (2023). https://iris.who.int/handle/10665/366667

Xin C, Zhang T, Tsai S-B, Zhai Y-M, Wang J (2020) An empirical study on greenhouse gas emission calculations under different municipal solid waste management strategies. Appl Sci 10:1673. https://doi.org/10.3390/app10051673

Ziraba AK, Haregu TN, Mberu B (2016) A review and framework for understanding the potential impact of poor solid waste management on health in developing countries. Arch Public Health 74:55. https://doi.org/10.1186/s13690-016-0166-4